Molecular Basis of Nerve Activity

Molecular Basis of Nerve Activity

Proceedings of the International Symposium
in Memory of David Nachmansohn (1899–1983)
Berlin (West), Germany, October 11–13, 1984

Editors
J.-P. Changeux · F. Hucho
A. Maelicke · E. Neumann

Walter de Gruyter · Berlin · New York 1985

Editors

Jean-Pierre Changeux, Professor, Dr.
Institut Pasteur
28, Rue du Dr. Roux
F-75724 Paris Cedex 15

Ferdinand Hucho, Professor, Dr.
Freie Universität Berlin
Fachbereich Chemie
Institut für Biochemie
Thielallee 63
D-1000 Berlin 33

Alfred Maelicke, Professor, Dr.
Max-Planck-Institut für Ernährungsphysiologie
Rheinlanddamm 201
D-4600 Dortmund

Eberhard Neumann, Professor, Dr.
Universität Bielefeld
Fakultät für Chemie
Physikalische und Biophysikalische Chemie
Postfach 86 40
D-4800 Bielefeld 1

Library of Congress Cataloging in Publication Data

International Symposium in Memory of David Nachmansohn,
1899 - 1983 (1984 : Berlin, Germany)
Molecular basis of nerve activity.
Bibliography: p.
Includes indexes.
1. Neurochemistry--Congresses. 2. Acetylcholine--Receptors--Congresses. 3. Nachmansohn, David, 1899 - --Congresses.
I. Changeux, Jean-Pierre. II. Nachmansohn, David, 1899 -
III. Title. QP356.3.I547 1984 599'.0188 85-10301
ISBN 0-89925-043-2 (U.S.)

CIP-Kurztitelaufnahme der Deutschen Bibliothek

Molecular basis of nerve activity : proceedings of the Internat. Symposium in Memory of David Nachmansohn (1899 - 1983), Berlin, Fed. Republic of Germany, October 11 - 13, 1984 / ed. J.-P. Changeux . . . [Sponsored by: Max-Planck-Ges. zur Förderung d. Wiss. . . .]. - Berlin ; New York : de Gruyter, 1985.
ISBN 3-11-010345-1 (Berlin)
ISBN 0-89925-043-2 (New York)

NE: Changeux, Jean-Pierre [Hrsg.]; International Symposium in Memory of David Nachmansohn (1899 - 1983) <1984, Berlin, West>; Nachmansohn, David:

ISBN 3 11 010345 1 Walter de Gruyter · Berlin · New York
ISBN 0-89925-043-2 Walter de Gruyter, Inc., New York

 Printing: Gerike GmbH, Berlin. - Binding: Dieter Mikolai, Berlin. Printed in Germany.

Preface

Last year, the many good friends and colleagues of David Nachmansohn jointly suggested to several prominent European scientific societies he had been affiliated with that they organize a symposium in his memory. The Weizmann Institute of Science in Israel, the Société Française de Chimie Biologique, the Medical Faculty of the Free University of Berlin and the Max-Planck-Gesellschaft z.F.d.W., München, immediately agreed. This led to the invitation to prominent scientists in the field to come to the Dahlem district of Berlin where, during the prewar decade, the historical research center of the Kaiser-Wilhelm-Gesellschaft, the forerunner of the Max-Planck-Gesellschaft,had been located.

Indeed, it was in the immediate neighborhood, at the famous department of Otto Meyerhof at the Kaiser-Wilhelm-Institut für Biologie, that David Nachmansohn, as a young postdoctoral student, was introduced, during the twenties, by his great teacher into the fundamental problems of the new biological sciences. His colleagues were Hermann Blaschko, Severo Ochoa, Fritz Lipmann, Ken Iwasaki and Paul Rothschild. Directors and senior scientific members of the other departments of the institutes were Prof. Correns, Prof. Goldschmidt, Prof. Hartmann, Prof. Warburg and Prof. Mangold. Just around the corner the famous Haber Colloquia were held. This was the scientific forum for the most brilliant intellectual endeavours of the time. The colloquia were attended by Einstein, von Laue, Hahn, Nernst, Michaelis, Meyerhof, Warburg, Rona and Schrödinger, to name only a few members of the scientific community of Berlin who came to them. David Nachmansohn reports: "The Haber Colloquia had an entirely different and extraordinary character ... Everything from the helium atom to the flea could be the topic ... The Colloquia provided an ideal opportunity for Haber to realize his aim of breaking down the barriers between physics, chemistry, physical chemistry and the biological sciences".

When I had the opportunity of visiting David Nachmansohn in 1953 in New York together with Annemarie Weber, while we were discussing the problems on acetylcholine and cholinesterase that he was working on he changed the subject, going over to his personal experiences in the Kaiser-Wilhelm-Institutes. Over the years all his colleagues came to learn just how deeply committed he was to the spirit of the scientific community in the old capital of Prussia.

Indeed, during that time in Berlin, David Nachmansohn experienced the first scientific revolution of this century. Berlin's Kaiser-Wilhelm-Gesellschaft and its relatively young University was the center of a scientific excellency. That had been the fruit of a unique cultural evolution: begun more than a century earlier, it had been nourished by an equally unique German-Jewish community which led and tremendously enhanced the development of modern science and culture. And here David Nachmansohn writes "but whatever the basis of the great achievement of German-Jewish scientific collaboration has been, both Germans and Jews have every reason to look back with great pride on this glorious chapter of their history".

It was a great tragedy that this brilliant academy of scientific research was destroyed by the brutal force of the Nazi movement, leaving in its wake a disaster-struck world of horror and lies. As did many other scientists, David Nachmansohn left Germany with his family. He became a pioneer himself, of molecular neurobiology, and one of the most prominent citizens of the scientific community. Today we inherit not only his scientific work, but also his penetrating and brilliant history: "The German-Jewish Pioneers in Science".

This chronicle of the scientific culture prevailing then is of the highest rank, and is moving to read. It serves as a primary source in illuminating the history of Berlin and of a brutalising time in the development of our country; in its broader context it contributes to an understanding of the historical movement of science. This history shows that the science of yesterday issues in the civilization and culture of both the present and the future.

There is a community and continuity among scientists and their thinking that can and sometimes must traverse barriers, whether they be geographical, ideological or political or invariably greater than all three combined. The Max-Planck-Gesellschaft and the German scientific community are indebted to David Nachmansohn for crossing and thus obviating these barriers and for helping to regenerate the spirit of science in modern Germany.

David Nachmansohn's life and scientific achievements were addressed on several occasions during this symposium: Prof. H. Herken in his opening address to the scientific session provided us with an in-depth account of Nachmansohn's relationship to Berlin, and Prof. E. Katzir presented us in his afterdinner speech with a lively picture of Nachmansohn's relationship to Israel and the Weizmann Institute.

A special memorial session was held in Harnack-Haus, the former meeting place of the Kaiser-Wilhelm-Gesellschaft. Relatives, friends, scientific and political representatives, and the participants of the symposium, came together to honor David Nachmansohn and to share recollections. During this session, Prof. B. Hess, Vicepresident of the Max-Planck-Gesellschaft, Prof. W.A. Kewenig, Senator für Wissenschaft und Forschung, Berlin, Prof. H. Kewitz, Freie Universität Berlin, Prof. E. Katzir, Weizmann Institute of Science, Rehovot, and Prof. H. Blaschko, University, Department of Pharmacology, Oxford, addressed the community of David's friends and Prof. Jean-Pierre Changeux, Institut Pasteur, Paris, delivered a memorial lecture which appears as the introductory chapter of this volume. The volume contains the written contributions of the participants in the Symposium and gives evidence of the progress in neurobiological and neurobiochemical sciences to date, in which David Nachmansohn pioneered with his lifetime's work.

March 1985 Benno Hess

Symposium Guests of Honour:

Dr. Edith Nachmansohn
Lady Margaret Krebs
E. Katzir
S. Ochoa
H. Blaschko
J. Cohn
R. Wurmser

Conference Scientific Committee:

J.-P. Changeux (Chairman)
M. Eigen
E. Heilbronn
B. Hess
U. Littauer
E. Schoffeniels

Organizing Committee:

H. Herken
F. Hucho
H. Kewitz
A. Maelicke
E. Neumann

Sponsored by:

Max-Planck-Gesellschaft zur
Förderung der Wissenschaften

Deutsche Forschungsgemeinschaft

Société Française de Chimie Biologique

Der Senator für Wissenschaft und
Forschung der Stadt Berlin

Freie Universität Berlin

Symposium activities:

(1) Opening address by H. Herken, Berlin

(2) Lecture and poster sessions

(3) Evening lecture by M. Eigen, Göttingen: "Physical Principles in Molecular Biology"

(4) Academic memorial session for David Nachmansohn (March 17, 1899 - November 2, 1983), Harnack-Haus, Berlin-Dahlem; greeting addresses:
B. Hess, Vice President of the Max-Planck-Society
W.A. Kewenig, Senator of Science and Research, Berlin
H. Kewitz, Free University of Berlin
E. Katzir, Weizmann Institute of Science, Israel
H. Blaschko, University Department of Pharmacology, Oxford, U.K.

(5) David Nachmansohn memorial lecture by J.-P. Changeux, Paris

(6) Symposium concluding remarks by E.A. Barnard, London.

David Nachmansohn in Leningrad, 1965

1. David Nachmansohn: 18 years old
2. In Professor Rona's laboratory, Charité Hospital, Berlin, around 1923. David Nachmansohn is the first one on the left in the standing row.
3. At the Kaiser-Wilhelm-Institut, Dahlem, 1929. David Nachmansohn is the second from the left, between Fritz Lipmann and Severo Ochoa.

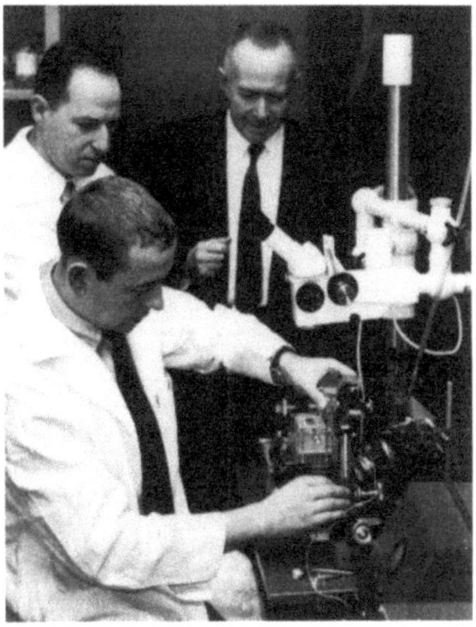

4. David Nachmansohn and Wilhelm Feldberg assaying acetylcholine liberated during the discharge of Torpedo electric organ, July 1939

5. J.F. Fulton and D. Nachmansohn in the fifties.

6. David Nachmansohn with Irwin Wilson and Ernest Schoffeniels (sitting) in the fifties.

On David Nachmansohn's Relationship to Berlin

from the Opening Address by H. Herken, Berlin

"... Anyone who is familiar with David Nachmansohn's career knows how much this city meant to him. After receiving his honorary doctorate from the Medical Faculty of the Free University Berlin, and being elected honorary member of the Berliner Medizinische Gesellschaft, he wrote me: 'Berlin hat mich entscheidend geformt, in meinem Denken und meiner Weltanschauung, in Kunst und Wissenschaft. Auch wenn die Verzweiflung über die törichte und kurzsichtige Politik europäischer Regierungen es einer Verbrecherbande möglich machte, die Macht an sich zu reißen, was ja nicht nur für Juden, sondern für ganz Deutschland und Westeuropa ein furchtbares Unglück bedeutete, kann und will ich nicht vergessen, was Berlin, was die deutsche Kultur mir bedeutet hat und noch bedeutet'. 'Es ist immer meine Überzeugung gewesen, daß es eine der vornehmsten und befriedigendsten Aufgaben der Wissenschaftlicher ist, Brücken zwischen den Völkern zu bauen. Sie sind dafür besonders geeignet, dank des internationalen Charakters der Wissenschaft. Sie können helfen, die furchtbaren Wunden zu heilen im Interesse der kommenden Generationen, die durch Katastrophen hervorgerufen sind, wie solche durch die Nazis'.

... Whoever has heard David Nachmansohn talking enthusiastically about the happy time he spent working in Berlin can judge what it meant for him to have to leave Germany in order to escape the approaching Nazi terror. He was given a hospitable reception in France. The cultivated academic atmosphere at the Faculty of Sciences of the Sorbonne and the support he received from his colleagues, René Wurmser, Henri Langier, and Jean Perin, helped him to get over the hard time he went through after his expulsion from Germany ...

... In 1972, in the introduction to his autobiographical essay 'Biochemistry as Part of My Life', David Nachmansohn wrote: 'Character, emotions, literary and artistic experience, philosophy, and political involvements form an integral part of a personality. Since scientists are human, all these factors determine their reactions, their way of thinking, and must be essential elements in the formation of scientific ideas and views, motives and attitudes. Knowledge alone of a special scientific field, however, solid and profound, provides only the tools. What is achieved with these tools depends to a very large extent on the complex factors of a personality.' This was the life he exemplified to us. In the occupation with science he saw the expression of supreme human dignity. Even the dreadful injustice which he experienced did not deprive him of the firm believe in imperishable ideas and possessions befitting this dignity, namely freedom, truth, justice and decency. He was a great personality. An exceedingly successful, scientific career came to an end on November 2, 1983 but David Nachmansohn's work and ideas live on amongst us and their influence will continue into the future".

CONTENTS

II. THE ACETYLCHOLINE RECEPTOR: STRUCTURAL ASPECTS

III. THE ACETYLCHOLINE RECEPTOR: FUNCTIONAL ASPECTS

IV. IMMUNOLOGY OF THE ACETYLCHOLINE RECEPTOR: MYASTHENIA GRAVIS AND SYNAPTIC PLASTICITY

DAVID NACHMANSOHN

(1899-1983):

a pioneer of neurochemistry

J.-P. Changeux
Institut Pasteur, Laboratoire de Neurobiologie Moléculaire
25, rue du docteur Roux, 75015 Paris

Almost a year ago, on the second of November 1983, David Nachmansohn died of pneumonia, at the Jewish Home and Hospital for the Aged in New York. With him disappeared one of the founders of neurochemistry, a foremost pioneer in the molecular biology of nerve activity, the master of several generations of scientists who, inspired by his ideas, have contributed to the development of a new field of research, of which the ambition is no less than to understand brain functions at the molecular level.

David Nachmansohn had been a professor at Columbia University, College of Physicians and Surgeons since 1942. He was a member of the National Academy of Sciences and several other academies including the Deutsche Forschung Leopoldina. He received honorary degrees from the Free University of Liège. He was an Honorary Fellow of the Weizmann Institute

Molecular Basis of Nerve Activity

of Sciences, Rehovot, Israel, and a member of the Board of Governors of the same institution. He received several medals, the Pasteur medal, Paris, the Neuberg medal, New York, the gold medal, Madrid and the Albrecht-von-Graefe medaille, Berlin.

To retrace the scientific work and life of a man of such creativity and exceptional dimensions would have been a challenge, if David Nachmansohn himself, had not left several essential testimonies. In 1954, he gave a, still, much quoted, Harvey lecture ; in 1959, he wrote a fundamental book "Chemical and Molecular basis of nerve activity"; in 1972 he presented a "Prefatory chapter" of Annual Reviews of Biochemistry entitled "Biochemistry as part of my life" ; then, an essay about German-Jewish Pioneers in Science (1900-1930) : an historical outlook on "German Jews in Physics, Chemistry and Biochemistry". One year before his death he finished an autobiography which will be published soon by Springer Verlag. Reading this ultimate work has been an inspiring and poignant experience and I will often refer to it in my presentation. I wish to thank Edith Nachmansohn and Professor Severo Ochoa for communicating this precious manuscript before publication.

David Nachmansohn was born in 1899 in Jekaterinoslav, today called Dnjepropetrovsk, Russia. On his father's side, the grand parents were orthodox Jews, whereas on his mother's side, the family was westernized and liberal. The family included several lawyers or physicians, but his father was a merchant. They spoke several languages. David learned German as his mother tongue. Before he began school, his parents in fear of the pogroms, had moved to Germany and settled in Berlin. Anxious to give him the best education, they sent him to the humanistic Gymnasium. There, the emphasis was not on science but on Latin and Greek and later on philosophy. Interestingly, as in the case of Jacques Monod, a teacher of Greek had a great impact on his developing personality, stimulating him to read Homer, Sophocles and, of course, Plato. At 16 he begins Spinoza's ethics and at 17 Faust. According to his "Prefatory Chapter", reading Goethe was a stirring experience. Faust's message that "creative work is more valuable and counts more than anything in life" remained a guideline for his life.

In 1918, David graduates from the Gymnasium and enters the University of Berlin where he attends lectures on history and philosophy. His parents, however, worried about the bleak outlook for a Jew in a purely academic profession. Antisemitism was becoming strong and frightening. They urged him to study medecine which would make him economically more independant. The transition was difficult. With reluctance he registers in the Medical school, while at the same time he continues to attend lectures on Plato or Kant. However, his readings on the history of Science, on the life and work of Claude Bernard, Pasteur, Helmholtz, Ehrlich ... stimulate his imagination. Biomedical research seemed to offer opportunities for creative work and also answers to many of his philosophical questions. Biochemistry was the most promising field. He thus joins the laboratory of Peter Rona professor of biochemistry in the Charité, the Medical School of Berlin University. Rona was an inspired and inspiring teacher. In his laboratory David is exposed to the most recent advances in physical chemistry and biochemistry and becomes acquainted, for the first time, to the works of Warburg, Michaelis, Meyerhof, Neuberg who were the leading biochemists of their generation.

To complete his medical studies, a four month of internship in a clinic was required. David chooses the clinic of Professor Goldscheider. Interestingly, the topic selected by Goldscheider for his dissertation concerned the physiology of the nervous system : the eventual existence of a brain center controlling sleep. The dissertation David wrote was purely theoretical but he already freely criticized the authorities in the field, demonstrating, according to his Biography, the "inadequacy of the evidence presented and the many fallacies of the interpretation". The reasonings and discussion impressed Goldscheider and the faculty members and he received the doctor degree eximia cum laude, a mark rarely given, even for an experimental work.

At the Goldscheider clinic, David meets Hans Kebs and this encounter was the beginning of a collaboration which turned to a life-long friendship. Back in Rona's laboratory, they begin together experiments in the Ehrlich tradition : the idea is to compare the ability of dyes to stain cells, actually Paramecia, with the capacity of absorption by aluminium silicates. A correlation did exist and this was the opportunity for David to write his first experimental paper : "Vitalfaerbung und adsorpsion". At about the same time, in 1925, Hans Weber joins Rona's laboratory and David begins to work with him on the hydration of proteins. Their results indicated that the same water space was occupied by the ionized and electroneutral forms of proteins whereas the viscosity was dependent on the degree of ionization. David thus becomes a trained physical chemist.

After three years spent with Rona in the Charité Hospital, David has thus experienced both, theory about nervous processes, with his dissertation and, laboratory work on protein chemistry and physical-chemistry. He thus acquired there the two main facets of his scientific personality. A third, his interest for thermodynamics and energetics, developed in another laboratory in Berlin. Although the Rona environment was outstanding (Hans Krebs, Ernst Chain, Rudolf Schoenheimer, Hans Weber, Fritz Lipmann were trained in his laboratory), David felt the desire to leave. Otto Meyerhof's work fascinated him by his systematic attempt to explain cellular functions in terms of physics and chemistry, following a long tradition in German science which florished with Helmholtz, Du Bois-Reymond, Liebig, Büchner, Ostwald and many others. This physicalist point of view appealed to him so much that one day, following the advice of Hans Weber, he goes, uninvited, to see Otto Meyerhof at the Kaiser Wilhelm Institute. According to David's Biography, the encounter was rather cool. Meyerhof asked him about his training. Five years of Medecine, three years with Rona. "Usually I don't accept beginners" said Meyerhof. But after one hour of discussion, Meyerhof, recognized David's knowledge of physical chemistry and thermodynamics, he finally accepted him into his laboratory. This was a decisive point in David intellectual and scientific life.

The years in Dahlem with Otto Meyerhof.

The Meyerhof group was small though very active : five or six investigators, among them, Hermann Blaschko, Fritz Lipmann, Severo Ochoa . Meyerhof himself spent most of his time working at the bench. But he also shared David's interest in philosophy. He was an active participant of the neo-Kantian group of Fries. Discussions were numerous on scientific matters but also on literary publications, theater, opera, political events. In the several Kaiser Wilhem-Institutes in Berlin, coexisted physical chemists, chemists, and biologists. Nearby were the laboratories of Warburg and of Neuberg. The director of the Institute of Physical-chemistry, Fritz Haber, created strong scientific animation through the organization of seminars.The famous Haber Colloquia took place every second Monday afternoon, first in Haber's Institute then, after March 1929, in a new guest house, a building originally offered as some kind of birthday present to President Adolf Von Harnack, and thus called Harnack House. Haber was a strong personality. He had developed a method for producing ammonia (and thus the nitrate fertilizers) from atmospheric nitrogen and hydrogen. This was of great importance for agriculture and industry and resulted in his receiving the Nobel Prize. His aim was to bring together physicists, chemists, biologists, "from the helium atom to the flea", to create contacts, abolish barriers between disciplines and stimulate research in frontiers areas. The discussions were very lively, with constant controversies. As David states in his Biography "There, I really learned how dangerous rigidity is and how essential it is for a scientist to be flexible and to have an open mind, to adjust his ideas when new facts emerged".

In such an exciting scientific environment, David's experimental work developed very actively. Otto Meyerhof and his associates had just discovered the difference between two types of phosphate derivatives depending upon the amount of energy released upon hydrolysis : some were classified as energy rich such as creatine phosphate (and of course ATP), other as energy poor. Breakdown of creatine phosphate during muscle activity had just been demonstrated. Meyerhof thus suggested to David to further explore this question and compare quantitatively phosphocreatine breakdown and lactic acid formation as a function of muscle tension. David found that the amount of phosphocreatine split decreases with the number of contractions and the length of the tetanus in contrast to lactic acid formation. Temperature also diminished phosphocreatine breakdown but had little effect on lactic acid formation. Then, he began to suspect that phosphocreatine breakdown was related to a characteristic parameter of contraction : its speed. Exploring the possibilities of zoology, David found that the muscles from turtle, toad and frog, work at increasing speed and have indeed increasing content in phosphocreatine. Subsequent work by Lohman on the intervention of ATP confirmed David's views. In intact muscle, ATP hydrolysis preceeds phosphocreatine breakdown. Phosphocreatine is rapidly used to restore ATP from AMP during contraction. Then, the greater the speed of contraction, the greater has to be the concentration of phosphocreatine to regenerate ATP. David was right.

Meanwhile, David became acquainted with the enzyme work going on in Meyerhof laboratory. The isolation of the enzymes involved in the glycolytic pathway had for him - he says in his Biography - a significance "comparable to Buchner's isolation of zymase for the elucidation of alcoholic fermentation". With an already solid experience both in biological energetics and in enzymology David was ready to embark in what was to become his lifework : the biochemistry of electrogenesis. The transition did not occur in Meyerhof laboratory but in Paris.

The year is now 1929. The political, economic and social situation in Germany is deteriorating rapidly. Anti-semitic movements and forces are becoming more and more powerful. David is engaged to a young medical student Edith Berger who becomes his wife in July 1929. Meyerhof was planning to move to a new building in Heidelberg and offered to David a position of auxiliary assistant. A decision had to be made. Thinking that, as a physician, he would have more security for his family everywhere, David chooses to become a clinician, first in Frankfurt, then in Hamburg and Berlin. Hitler was ready to take over the government. Ruth was born. The family Nachmansohn could no longer stay safely in Germany. They decide to emigrate. In January 1933 David makes a trip to Palestine, meets Weizman, who urges him to wait for a few years before moving there and encourages him to continue his research where the conditions appear favorable. David had offers from England, Holland and France. Only in France would he be able to work as an independent investigator. He decided in favour of Paris.

Paris, Arcachon : The early work on cholinergic transmission.

In the "Laboratoire de Physiologie générale" headed by Louis Lapicque at the Sorbonne, David now had three good sized laboratories as well as a skillful technician, a young French student, Annette Marnay. The scientific contacts in Paris, to some extent, resembled those he formerly had in Dahlem. At the Institut de Biologie Physico chimique, David meets Jean Perrin, René Wurmser, Louis Rapkine and colleagues from the Institut Pasteur : André Lwoff, Daniel Bovet. For two years, he continues to work on intermediary carbohydrate metabolism of muscle. As Paris is only a few hours by train from London, he is able to visit former friends from Dahlem in Cambridge or Oxford : Hans Krebs, Herman Blaschko, Ernst Chain and attends the meetings of the Physiological Society. There he becomes aware of the work done by Sir Henry Dale and his associates. Their hypothesis, based on the Elliot scheme, is that acetylcholine serves as a chemical transmitter at the neuromuscular junction. Electrical impulses propagate in the motor nerve as in the muscle, while, at the level of the neuro muscular junction, acetylcholine is released, crosses the non-conducting gap and depolarizes the muscle membrane. The idea, based mainly on pharmacological experiments, was challenged by physiologists including Sherrington or Eccles. Nobody yet had approached the problem from a biochemical point of view. David, with his strong background in physical chemist and enzymology becomes naturally interested in the proteins associated with acetylcholine function. His main concern is to find biochemical evidence in favor of Dale's hypothesis of chemical transmission. In less than three years David will perform a most brilliant series of experiments which have retained to this day their entire validity. They can be subdivided into three groups.

First, if the assumption that acetylcholine plays the role of transmitter at the synapses is correct, the acetylcholine liberated should be removed in a few milliseconds : the duration of the refractory period of voluntary muscle. To answer this question, David, with the help of Annette Marnay, divides frog sartorius muscle into two regions : one which contains the endplates, the other, aneural, which does not receive any nerve endings. The result is unambiguous. The concentration of cholinesterase is several thousand times higher at the endplates region than in the surrounding, aneural, part of the muscle fibers. It is sufficiently high to account for the disappearance of acetylcholine during transmission. They conclude that "One of the main obstacles opposed to chemical transmission, thus, falls".

The result is subsequently confirmed by Couteaux and Nachmansohn in 1940 who carefully cut frozen guinea pig gastrocnemius into 11 slices of similar thickness and weight. They further showed that 14 days after denervation, a high concentration of esterase (60-70 %), persists in the former endplate region and remains there for months. This finding, extensively confirmed since then, in particular by the quantitative studies of Eric Barnard and collaborators, indicate that "the enzyme, at the endplate is localized outside the nerve endings" in a region that we know to day is the basal lamina which covers the folded postsynaptic membrane.

In the second series of experiments David, still with the collaboration of Annette Marnay, makes use of the zoological diversity of the animal world, as he first did in his studies on creatine phosphate. The esterase is present in brain of rats and rabbits, and in the sympathetic superior cervical ganglion of the cat. It also occurs in embryonic tissue. In chick embryos, the content of esterase increases strikingly between the 12th and 14th day of incubation and this increase coincides with the period when movements of the muscles start. In the course of these studies they casually assayed the esterase content of muscle from the lizard and also of tissue from a rare creature which was going to become famous : the electric fish. Then comes the surprise. David tells us about it : "When writing my paper on the cholinesterase concentration in muscle and at the neuromuscular junction in the spring of 1937, I read a review paper of Lindhard about muscle ... There he described the electric organs of electric fish and referred to them as to a kind of an accumulation of motor endplates ... At that time in 1937, there was a scientific exhibition where I saw several specimen of Torpedo marmorata at the International World Fair in Paris. Dr. Catherine Veil, from the laboratory at the Sorbonne where I worked, was in charge of this exhibition and I asked her permission to use a specimen for my experiments.Annette Marnay determined the enzyme concentrations. The result was simply stunning : 1 kg of electric tissue (fresh weight) hydrolyzed 3-4 kg of acetylcholine per hour, although the tissue is 92 % water and only 3 % protein"... "I felt immediately that this discovery was of paramount importance and one of the most fascinating observations I had ever made". In student lectures and private conversations David often liked to quote, with his customary sense of humour, the Danish physiologist, August Krogh : "Nature has created quite a few animals with the special purpose to help biologists to solve their problems". Indeed, the electric organ from fish has since then been used by generations of biochemists and molecular biologists in their studies of the proteins involved in acetylcholine metabolism and action. I counted that, in this meeting, more than 50 % of the speakers use fish electric organ in their laboratory.

Of course, David made immediate profit of this observation. In collaboration with Edgar Lederer, in the winter of 1937-1938, he achieves the extraction of cholinesterase from electric tissue and begins its purification. Short after the publication of the high content in esterase of the electric organ, David meets Alfred Fessard, a renowned electrophysiologist working at the Collège de France. Fessard became quite excited about the discovery and David and him plan together to submit the Elliott or Dale model of chemical transmission to a decisive test. Is the nerve supplying the electric organ acting by the liberation of acetylcholine ? The planned experiments involve the close-range arterial injection of acetylcholine. Wilhelm Feldberg, a pharmacologist from Dale's group, knew the technique and is invited to join the team. The three of them meet, in June 1939, in Arcachon, a small city near Bordeaux. In two weeks the experiments are completed. They appear as a short summary in the Proceedings of the Physiological Society. The data are clear.

"Cholinesterase content : 170-250 mg of acetylcholine hydrolysed in 1 hour by 100 mg of ground up tissue...

"Acetylcholine content : 60-100 μg of acetylcholine per gram of fresh tissue containing about 92 % water...

"Liberation of acetylcholine : Perfusate collected during nervous stimulation caused strong contractions equivalent to those produced by solutions of acetylcholine of 1 : 15 millions to 1 : 40 millions...

"Effect of eserine on the discharge : The addition of eserine to the perfusate fluid lengthened the single discharge...

"Effect of acetylcholine : close range arterial injections of acetylcholine into the perfused organ connected with an amplifier caused potential changes in the same direction as those of the discharge"...

This was the first unambiguous and almost complete demonstration of the electrogenic action of acetylcholine and also, to my opinion, one of the best evidence ever given for the role of acetylcholine as neurotransmitter in the electric organ.

The Arcachon experiment is a moment of grace. Noises of soldiers boots become increasingly louder. Nazi and fascist parties had taken power in several places in Europe and made anti-semitism legal. To stay in Europe under such circumstances was taking an enormous risk. Even though David was appointed "Maître de Recherche", he and his family realized that they had to leave France. "The decision to leave Paris was not an easy one- David tells us in his Biography- My wife and I were greatly attached to the city ; we had several good friends ; Meyerhof had just settled in Paris ; I was fourty years old and deeply rooted in European civilization". The family nevertheless abandons Europe for the United States. Fulton had offered a position to David at Yale University Medical School. At the end of the summer 1939 David joins his department.

The first preoccupation of David in Yale is to persevere on the purification of acetylcholinesterase. He had brought with him about fifteen kilograms of electric organ from Arcachon. The purification progressed satisfactorily and, in the early 1940, he was able to obtain nearly 80 % pure enzyme.

1942 : the discovery of choline acetyltransferase.

Meanwhile, David, loyal to the Meyerhof tradition, worries about the energetics of the system he is working with. Where does the energy of fish electric discharge and, in general, of electrical activity of excitable, nerve, muscle, cells come from ? His axiom is clear : "There is no manifestation of life conceivable without molecular change, i.e, without chemical reactions". Nerve excitability and bioelectricity in general could not be exceptions. It was for sure an axiom that specific chemical reactions must be responsible for this change in ion permeability assumed to take place in the excitable membrane. Axioms are fundamental principles which cannot be rigidly proved or disproved but are nevertheless real since there exists an overwhelming amount of experimental evidence to support them. Axioms are a necessity in all fields of science. One may consider them as an act of faith".

On the basis of this axiom, he naturally was interested in the origin of the energy involved in bioelectrogenesis. Among the reactions which contribute to the production of energy, which one provides the energy ?. David naturally returns to his first love : phosphocreatine. He used now Electrophorus electricus that the naturalist Christopher Coates, the curator of New-York aquarium provided for him from South America. The first surprise, pleasant of course, is that the concentration of phosphocreatine in the electric organ is higher than in muscle.

Moreover, the concentration of ATP is about the same, in spite of the extremely low protein content of the electric organ. Then, handling the eels under conditions where they produce 3000 to 8000 discharges, Nachmansohn, Cox, Coates and Machado, find that a strong breakdown of phosphocreatine coincides roughly with electrical activity. By analogy with what happens in muscle, they assume that phosphocreatine is used for the resynthesis of ATP split during activity. A large part of the energy resulting from ATP hydrolysis is indeed accounted for the restoration of the ionic concentration gradient via Na^+ K^+ ATP ase. Is that all ? Could another part of it serve for the biosynthesis of acetylcholine ?

The answer to this question will be at the origin of a major discovery of David Nachmansohn. But, meanwhile, David had left Yale for a position of Research Associate at Columbia University College of Physicians and Surgeons. The first experiment he performs after the settling of his new laboratory at the third floor, in September 1942, indeed concerns the use of ATP energy to acetylate choline. ATP was not, at that time, commercially available ; he synthesizes it. The first experiment done with extracts of frog brain or eel electric tissue looks deceptive. Why? Is it possible that the ATPase abundant in these tissues hydrolyzes ATP before energy transfer could happen. Severo Ochoa had shown in 1941 that sodium fluoride inhibits ATPase activity. Sodium fluoride is added to the extracts. Surprise ! In its presence an important production of acetylcholine takes place : 100 to 150 μg per gram of tissue are synthesized per hour. The finding is of importance. It is the first enzymatic synthesis of a neurotransmitter achieved *in vitro* . In his Prefatory Chapter David mentions that the paper was submitted to Science, Journal of Biological Chemistry, Proceedings of the Society for Experimental and Biological medecine and successively turned down by the referees. As Anatole France wrote in l'Ile des Pingouins "Si vous avez une vue nouvelle, une idée originale... vous surprendrez le lecteur. Et le lecteur n'aime pas être surpris. Il ne cherche jamais dans une histoire que les sottises qu'il sait déjà".("If you have a new point of view, an original idea ... you will surprise the reader. And the reader does not like to be surprised. In a story, he only looks for the silly things he already knows".) The paper signed by Nachmansohn and Machado was finally accepted by Fulton for publication in the Journal of Neurophysiology.

The finding was not immediately accepted. For instance, David mentions that Fritz Lipmann, an other former Dahlem student who at that time was working on acetylphosphate, could not believe it. To make up his mind, at the end of 1944, Fritz Lipmann came to David's laboratory with a box containing several phosphate derivatives. All the compounds were tested. Only ATP was indeed effective. Nevertheless, the mechanism of acetylation remained obscure. Nachmansohn and Machado soon find that on dialysis the enzyme activity rapidly disappears but is recovered when a co-enzyme preparation from liver,heart or brain is added back to the system. Lipmann subsequently shows that the co-enzyme of the choline acetylation reaction is the same in all acylations. It can be identified as co-enzyme A. The complete scheme is finally elucidated by Paul Berg in 1956 who finds that the donor of the acetyl group of co-enzyme A is acetyladenylate formed by the reaction of ATP with acetate. Several enzymatic steps are thus involved. The last enzyme, cholineacetyltransferase has, since then, been purified and extensively studied by several groups which include that of Henry Mautner and Hans Thoenen.

The catalytic site of acetylcholinesterase and the discovery of 2-PAM.

In parallel, David continues to work on the biochemistry and physical chemistry of his favorite enzyme : acetylcholinesterase. One should not forget that at the time choline acetylase is discovered, the second World War devastates Europe. According to his Biography, one day of 1943 three Army officers enter his office. They ask him to work on a "top secret" military project. The nazi army had developed an extremely toxic "nerve gas" and it is of primary strategic importance to find an antidote. The compound is diisopropyl fluorosphosphate, DFP, a derivative of organophosphate insecticides. British scientists had shown that DFP inhibits cholinesterases. David, struck by the possibility offered by DFP to investigate the properties and function of the enzyme and its role in bioelectrogenesis, responds positively to the Army officers. This is the beginning of a long and very productive series of experiments to which contributed first a young couple of Swedish postdoctoral fellows, the Augustinsson's, then in 1948, the Israeli biochemists Shlomo Hestrin, and Felix Bergmann and particularly the physical organic chemist Irwing Wilson, with whom, from 1949, a very creative collaboration developed for more than 15 years. In retrospect, the primary impact of this applied work on DFP appears as a remarkable progress in the basic understanding of acetylcholine active site.

The first experiment that David did was to confirm the inhibition of several types of cholinesterase by DFP and to show the irreversible character of the inhibition upon dilution or dialysis. Careful examination by the Augustinssons' of the substrate saturation curves, then, led to the clear-cut distinction between the acetylcholinesterase from excitable tissues and the cholinesterase from serum. Among other differences, the former gave characteristic bell-shape curves typical of substrate inhibition at high concentration which are lacking in the "non specific esterases". With the purified acetylcholinesterase from electric tissue, the effects of the irreversible inhibitors DFP and tetraethyl pyrophosphate is then compared with those of reversible competitive inhibitors, such as neostigmine or eserine. Incubation of the enzyme with neostigmine fully protects the enzyme against DFP irreversible inhibition, suggesting that the two types of inhibitors, react with the same groups in the enzyme active site.

The well known contribution of Hestrin in David's laboratory is the development of a colorimetric assay for acetylcholine which bears his name and universally replaced the tedious measurements of frog rectus contractions !. The method is based on the formation of acethydroxamic acid in the presence of hydroxylamine at alkaline pH finally leading to a brown-purple compound upon addition of Ferric chloride. This gave Hestrin the opportunity to study the interaction of hydroxylamine on acetylcholinesterase, and to show that purified acetylcholinesterase does catalyse the condensation of acetate and hydroxylamine into acethydroxamic acid. DFP and prostigmine inhibit the reaction. The possibility was even evoked of the formation of an acyl enzyme which might also be a step in the hydrolytic mechanism of acetylcholinesterase. Hydroxylamine would then react with the active acetyl group then formed. This idea later became decisive in the development of an antidote against DFP poisoning.

As a consequence, David becomes increasingly interested in physical organic chemistry. In his biography, he mentions the "many private lessons" he received in Woods Hole from Leonor Michaelis, another refugee from Rona's group. He becomes acquainted with nuclophilic and electrophilic atoms, electron pushing and pulling etc ... Meanwhile, in the laboratory, Felix Bergman and Irvin Wilson develop intense and very creative work. I will mention only a few of the results they collect in these highly productive years. First of all, in 1950, the effects of pH on the properties of the esterase are investigated in detail. Prostigmine is a quaternary ammonium ion positively charged at all pH ; eserine, on the other hand, is a tertiary amine which is neutral at pH 10 and charged at pH 6. The inhibition of the enzyme by prostigmine does not change with pH while that by eserine is strongly pH dependent, suggesting that the molecule has to be ionized to inhibit the enzyme. Thus, a complementary negatively charged group, or anionic subsite, exists within the active site of the enzyme close to the group which splits - or forms - the esters referred to as esteratic subsite. The pH dependance of the inhibition by the organo phosphate, tetraethyl pyrophosphate, is found similar to that of acetylcholine.The inhibitor thus behaves as a substrate: it reacts with a nucleophilic atom in the esteratic subsite and forms a strong convalent bond, thus phosphorylating the enzyme. Two years later, in 1953, Schaffer and collaborators identify serine as the phosphorylated amino acid. The nucleophilic atom in the esteratic subsite is thus the oxygen of serine. In the normal hydrolytic process a C = 0 bond is formed and hydrolyzed in less than a millisecond while, with organophosphate, DFP in particular, the P = 0 bond formed is stable. A detailed picture of acetylcholinesterase active site progressively emerges from the extensive studies which finally gave to David worldwide recognition.

The basic knowledge on the enzyme became sufficient to return to the initial project suggested by the three army officers in 1943 : to find an antidote against organophosphate poisoning. Hestrin's observation

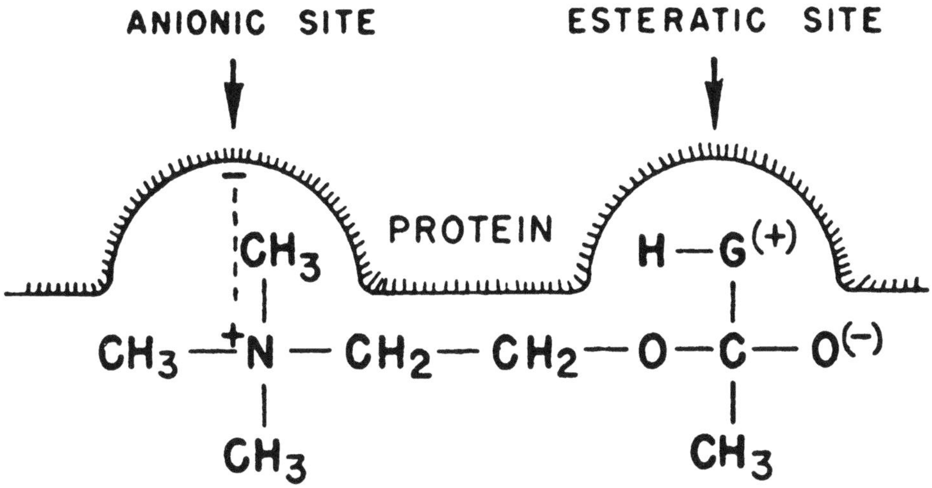

that hydroxamic acid forms in the presence of hydroxylamine and that hydroxylamine attacks the carbon of the carbonyl group lead to an other decisive experiment done in David Nachmansohn laboratory : to test the action of hydroxylamine on the activity of the phosphorylated enzyme. The effect is dramatic. In the hands of Irving Wilson close to 50 % recovery is obtained after one hour incubation, over 80 % in 5 hours. Reactivation of the phosphorylated enzyme is indeed possible. A new phase of the research was opened.

The next step is to find a compound which would approach the catalytic site by its interaction with the anionic subsite and exert a nucleophilic attack on the phosphorus atom at the nearby esteratic subsite. The organic chemists Estelle Meischlich and Sara Ginsburg, also trained by Rona in Berlin, join David's laboratory. A variety of compounds are synthesized. Among them, Wilson and Ginsburg synthesize pyridine-2-aldoxime methiodide (2-PAM). _In vitro_ at 10^{-5} M, 2-PAM completely restored in one minute the activity of the DFP inhibited acetylcholinesterase. As David said : "This was a superb achievement".

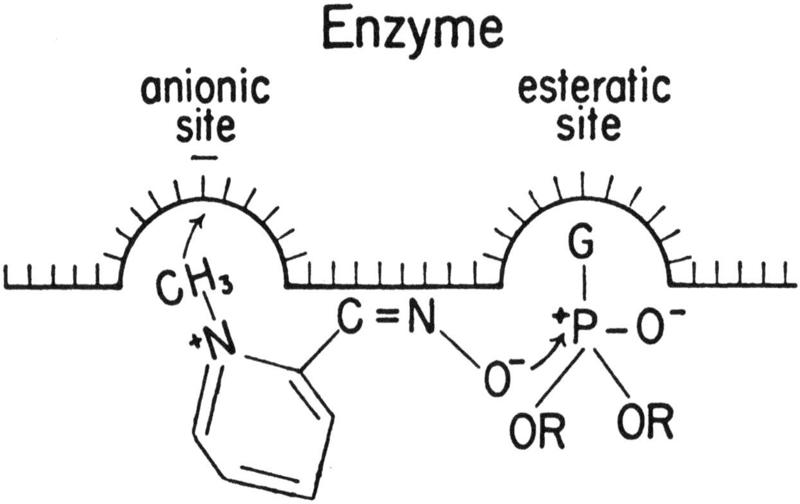

The demonstration that 2-PAM would work in vivo, on the whole animal as an antidote against DFP poisoning, however, remained to be done. Helmut Kevitz, a young pharmacologist of the Free University of Berlin, did it with David. Le me quote David's biography :"His first experiments proved to be a spectacular success. He injected 20 mice with an absolutely sure lethal dose of DFP. Ten of the mice received injections of 2-PAM ; these ten survived without any symptoms ; the other 10 were dead within 20-30 min. As we said, we did not need any statistics". In combination with atropine PAM proved to be a powerful antidote against DFP as well as against many insecticides poisoning. First tried in Japan in 1958 on humans, victims of insecticides intoxication, 2-PAM has, since then, saved many lives.

For David, the discovery of PAM had essential theoretical implications, as if not more important to him, than its practical utilisation. DFP was thought by most pharmacologists to exert a "general toxic effect". It was no longer the case. DFP using Claude Bernard words acted as a "chemical lancet" on a highly "localized" site : the active site of acetylcholinesterase. By blocking the enzyme, DFP interfered with neural processes, thus fulfilling David's axiom that nerve excitability and bioelectricity in general cannot be conceived without chemical reactions.

David Nachmansohn theoretical views.

This conclusion can be taken as a brilliant illustration of one of the most salient and, to me, attractive feature of David's scientific personality : his concern and involvment in theoretical work. In physiology and biochemistry empiricism and phenomenology have often been the rule, the basic paradigms underlying the experiments remaining in general implicit without explicit formulation. Impregnated throughout his life by the intellectual atmosphere of Dahlem, David often quoted Einstein's remark to Heisenberg "It is the theory which decides what we observe". This reminds me of the many lively discussions we had together in his laboratory at Columbia or at Woods Hole during the Summer 1967 and I can attest to the reality of the statement he wrote in his Biography "I never objected when my associate did not share my views. I tried to correct mistakes, but academic freedom includes freedom to have his own views. As Haber used to say, students are not receptacles to be filled, but fires to be kindled"...

To my opinion, David's theoretical views on bioelectrogenesis have two basic aspects. First, the already mentioned axiom that "electrical activity, as all cellular function, is affected by chemical reactions", and also that enzymes and proteins in general, play a crucial role. The data, I briefly presented up to now illustrate the validity of these fundamental axioms.

A second aspect of David's theoretical views concerns the detailed mechanisms of the propagation of nerve impulses along the axon. His ideas on this question evolved significantly throughout his career. At the time where, in France, he discovered the very high content of the endplate in acetylcholinesterase and when he carried the experiments in Arcachon with Feldberg and Fessard, David shared Dales's view on chemical transmission at the synapse. He explicitly stated that the experiments were performed "to satisfy the requirements of the theory that acetylcholine may be involved in the transmission of nerve impulses from neurone to neurone or from neurone to striated muscle fibre".

A significant evolution of David's ideas and thoughts follows his emigration to the United States. In two papers published in 1941 with Bettina Meyerhof or with Coates and Cox, he writes already. "Recent investigations suggest that the original theory must be altered to account for the acetylcholine metabolism which closely parallels the electrical changes occuring everywhere or at near the neuronal surface". According to him, "only a quantitative difference exists between the concentration of acetylcholinesterase in nerve fibers and that at synapses". Enzyme measurements indeed reveal significant amounts of acetylcholinesterase in axons located, in the case of squid giant axon, in the sheat not in the axoplasm. According to him, "acetylcholine is an essential link in the generation of the electrical changes recorded during activity". Acetylcholine would thus play the role of some kind of general co-factor a "specific operative substance", as he said, of any manifestation of bio-electricity. With the discovery of cholineacetyl transferase, the scheme becomes even more complete. In 1955, David includes within the scheme a new hypothetical molecule : the "receptor" for acetylcholine which is "essential for the change of ionic permeability".

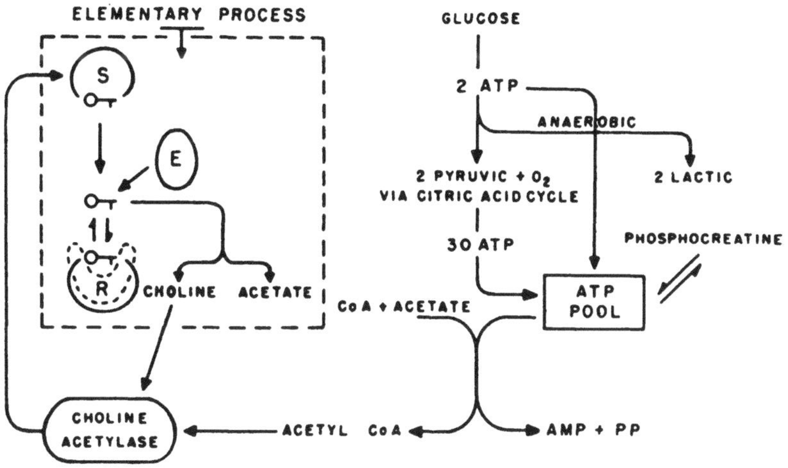

The sequence of reactions, or as he writes, "energy transformations", which account for bioelectrogenesis is integrated into the metabolic pathways. In my opinion the whole picture shows striking similarities with the Krebs or Lynen cycles. According to David, acetylcholine would play an universal role in bioelectrogenesis comparable to that of ATP or acetyl CoA in cellular metabolism except that the acetylcholine cycle would occur <u>within</u> the excitable membrane rather than in the cytoplasm.

The evolution of David's ideas in the forties thus appear to be influenced by both what I may call the Dahlem paradigms and, of course, by his own experimental work on acetylcholinesterase. Were there other influences ?. In the paper he signs in 1943 with Fulton, his host in the USA, it appears that a polemic existed at that time among electrophysiologists about chemical transmission. Fulton, Lorente de Nó and, also, Eccles, still rejected Dale's views. "The new concept, they write, is easily compatible with the conclusions of the American and Australian electrophysiologists cited above". Did the sociological environment of American physiology, replacing that of France and England, have significant impact on David intellectual evolution ?.

Rapidly, however, after Sir John Eccles spectacular conversion, most physiologists, even the American, adopted Dale's views. In my opinion, the nature of the debate shifted. The analysis of electric signal propagation and transmission by electrophysiological methods made considerable progress, thanks to the work of Sir Bernard Katz, Stephen Kuffler, Stephen Thesleff and others. On the other hand, David's neurochemical work had just began and the intrinsic difficulties and problems of biochemistry imposed a slower rythm. The techniques and even the language became on both sides so specialized that communications problems arose. Is that a sufficient reason to explain the radicalisation of David views leading him to abandon the role of acetylcholine as a transmitter, at the synapse while preserving it for conduction ?

Whatever the reasons, David's attitude is clear. For him, the gap between electrophysiology and biochemistry has to be filled and the answer to the questions raised by the theory will come from investigations at the biochemical and molecular level. During the Summer months at Woods Hole, he thus establishes collaborations with electrophysiologists including Ted Bullock and many others. Attempts are made to show effects of inhibitors of acetylcholinesterase such as DFP on conduction of nerve impulses in axons. With squid giant axon or bullfrog sciatic nerve, DFP does block conduction. However, the concentrations required seem extremely high : 5 to 20 millimolar ! David's explanation for such high concentrations was that "structural barriers" may interfere with the action of the compounds used. Is this the final answer ? In any case, to overcome these difficulties one needed a preparation which would satisfy the requirements of both the electrophysiologists and of the biochemists.

The isolated electroplax and the acetylcholine receptor.

Ernest Schoffeniels provided it during the years he stayed in David laboratory at Columbia University from 1957 to 1959. With a remarkable skill, he was able to dissect the single cellular unit, or electroplax, of *Electrophorus* electric organ, to insert it in an adequate perfusion chamber and to record electrical potentials under well-defined conditions. Providing access to physiological and pharmacological experiments on material exceptionally favorable for biochemical studies, the isolated electroplax opened a new field of research.

Henry Higman, Eva Bartels, Tom Podleski, Arthur Karlin and many others, including myself, greatly benefited from this unique preparation to assay in a quantitative manner a large variety of pharmacological agents, which were subsequently used for biochemical studies. Depolarizing compounds or agonists were distinguished from competitive antagonists. Local anesthetics, which typically block axonal conduction, were also shown by Eva Bartels and Tom Podleski to affect the acetylcholine response in a non competitive manner. Arthur Karlin demonstrated the sensitivity of the acetylcholine response to thiol group reagents. Finally, the snake venom α-toxins initially characterized by Chen Yuan Lee on vertebrate muscle, were shown on the electroplax set up now installed in Paris to block the acetylcholine response, thereby, leading to the isolation and purification of the acetylcholine receptor from the electric organ.

During the last ten years of his life David's interest on the acetylcholine receptor never declined ; on the contrary, he had the satisfaction to see two of his major theoretical predictions about this molecule experimentally fulfilled. As early as 1951, struck by the ability of the receptor "to recognize and react in a specific way with such a biologically potent compound as acetylcholine" and by analogy with acetylcholinesterase, he postulates that the acetylcholine receptor is a protein but distinct from the enzyme. The recent deciphering of the complete sequence of the four subunits of the acetylcholine receptor achieved by several groups, in particular that of Numa in Kyoto, brilliantly illustrates the validity of David's view. In his Harvey Lecture, in 1953, he further suggests, "to account for the generation of the bioelectric potential".. "that the effect of acetylcholine is to produce a change in the configuration of the protein". Thanks to the work done in several laboratories, in particular with rapid mixing techniques, this prediction has received today solid experimental support.

Finally, one essential question raised by David still deserves, to my opinion, an experimental test. Are there cholinergic proteins involved in axonal conduction ?. The effects of DFP and curare on conduction cannot be neglected even if they occur in a rather high range of concentration. Are they due to the <u>same</u> protein as those acting at the synapse or to <u>homologous</u> ones evolutively related to the cholinergic proteins ?. For instance, would the tetrodotoxin-sensitive $^{+}$ Na channel from axons or electric organ possess, even locally, stretches of sequences common with, or at least highly similar to, those present in the acetylcholine receptor or acetylcholinesterase? Would such homologies account for instance for the effect of curare and DFP on axonal conduction or of local anesthetics on the acetylcholine receptor ? The subdivision of the chromosomic gene of the receptor α subunit into 9 exons encoding different domains of the transmembrane polypeptide chain make such evolution plausible. Indeed some of these exons look specific for the acetylcholine recognition site, others for the traversy of the membrane and possibly for the ion channel. Would the cholinergic system be the most primitive one, from which evolved subsequently the molecular systems involved in axonal conduction ? The molecular genetics of the synaptic and axonal proteins will soon bring an answer to these crucial questions, raised by David Nachmansohn.

David Nachmansohn personality.

I would not like to close this lecture on David Nachmansohn's scientific work without evoking his fascinating and attractive personality, a personality that many of us here today - colleagues, pupils or simple friends - had the opportunity to appreciate. His prefatory chapter starts with the following sentences : "Characters, emotions, literary and artistic experience, philosophy and political involvements form an integral part of a personality. Since scientists are human, all these factors determine their reactions, their way of thinking, and must be essential elements in the formation of scientific ideas and views, motives and attitudes". David knew philosophy. In the world of biochemistry and physiology, classically dominated by a close-to-fact empiricism, David Nachmansohn seemed exceptional because of his taste for ideas, his expertise in utilizing them. David also loved literature and art. His many visits to Paris were opportunities to see the most recent exhibitions, to learn about the latest novels. His way to honour guests in New York was to take them to the Metropolitan Opera. Many of us, here, experienced these unique evenings.

David was, what one may call, a "man of culture". But his cultural interests were not restricted to a unique tradition. In his Biography he writes "The period which most decisively influenced my own scientific formation were the years of Dahlem... because I was brought up in German culture... Nevertheless, I feel fortunate to have experienced French and English civilizations and to be fluent in three languages. I also am fond of my Jewish heritage and became a devoted

Zionist at an early age". Very early, indeed, David was aware of his family origins. After reading Blumenfeld and Buber he spent considerable time and energy in Zionists movements, visiting Chaïm Weizman himself in Israel before the war and often coming repeatedly to Israel "helping" - using his words - "to lay the foundations of a Jewish home in Palestine".

This constructive attitude towards Israel was accompanied by an almost as strong attitude towards Europe and in particular Germany. Despite the tragic memories of the Nazi period, very soon he came back to visit Berlin and Heidelberg. The terms with which he describes this first visit after the war are poignant and one of his most happy achievements was to recreate collaborations between Israeli and German scientists. For him, I quote, "One of the noblest functions of Science is to build bridges between nations, to avoid future disasters". "Man of culture", David had been, one of the most efficient I know, "man of reconciliation".

David was also "un homme d'amitié". "Frienships, he often said, were an essential part of my life just as science and art". In a world where science is viewed more as competition than as cooperation, this attitude of David strikes by his "grandeur". Often he mentioned, in his conversations, the old Dahlem friends : Hans Krebs, Severo Ochoa, Ernst Chain, and more recent ones, Ephraim Katchalsky, Feodor Lynen, Manfred Eigen, and many others. Most of his former pupils and collaborators shared with him a similarly strong friendship. This rare sense of frienship of course accompanied an equally strong attachment to family life. And my last words will be for Edith, to welcome her in Dahlem and to assure her of our most affectionate feelings.

Scientist of exceptional stature, man of culture and reconciliation, beloved friend and teacher, David Nachmansohn will remain an outstanding example for all of us. The French philosopher, Michel Foucault, once said that "il faut faire de sa vie une oeuvre d'art". (One should make his own life a work of art). David did it in the most exemplary manner.

I. The Motor Endplate and Presynaptic Mechanisms

REMODELLING OF NEUROMUSCULAR JUNCTIONS DURING REPAIR OF MUSCLE FIBRES

R. Couteaux, J.-C. Mira

Laboratoire de Cytologie,Université Pierre-et-Marie-Curie,
7 quai Saint-Bernard, 75005 Paris,France ; Service de Technologie appliquée à la Microscopie électronique,C.N.R.S.

Introduction

At the neuromuscular junction,the nerve fibre and the muscle fibre are so intimately connected and subjected to such complex interactions that it is sometimes very difficult to know whether any structural or chemical change at this junction is of nervous or muscular origin.

In some experiments the primary cause of the junctional changes has been clearly defined. This is the case of the nerve terminal degeneration following the cutting of the motor nerve. Symmetrically it has been also possible by using appropriate physical or chemical agents to bring about degeneration of muscle fibres and of the postsynaptic component of the neuromuscular junctions without destroying the motor nerve fibres and their endings.

The first experiments of selective degeneration of muscle fibres by means of chemical agents were carried out with local anaesthetics, in particular with bupivacaine (1) and methyl-bupivacaine (2,3,4). Afterwards myotoxins of snake venoms were also used for the same purpose,mainly cardiotoxin (5,6), notexin (7,8) and taipoxin (9).

These various myotoxic drugs as well as mechanical injuries (10,11,12) were able to cause necrosis of muscle fibres while leaving the motor nerve structurally and functionnally intact as demonstrated after a bupivacaine or methyl-bupivacaine treatment (3,13). The necrosis was soon followed by regeneration of muscle fibres. The surviving nerve terminals established synaptic connections with the new muscle fibres. The structural characteristics of the new neuromuscular junctions seemed to depend on the nature of the chemical agent. After injection of cardiotoxin in mouse hindleg

Molecular Basis of Nerve Activity

muscles,the motor endplates of the regenerated muscle fibres appeared morphologically different from the normal endplates (6). The structural changes were not observed at the endplates of regenerated muscle fibres of lumbrical muscles after injection of methyl-bupivacaine in the mouse foot muscles (4).

In the experiments of muscle fibre degeneration the highly differentiated subsynaptic part of the sarcoplasm,where the cholinergic receptors are concentrated, disappeared completely. To study the effects of muscular lesions on the neuromuscular junctions,including the subsynaptic part of the sarcoplasm, we resorted to local injuries to mouse muscles and in these experiments we made a point of preserving both the nerve terminals and the muscle fibre portions on which these terminals were located. It was thus possible to follow the changes induced at endplates by the lesions of the muscle fibres.

Material and methods

Experiments were mainly performed on mouse external intercostal muscles, whose superficial fibres were frozen or cut near one of their rib attachments without injuring the motor innervation. Freezing was done with a cryode which ended in silver spherule and was supplied with liquid nitrogen.

For the silver staining of the motor endplates (Bielschowsky-Gros method) and the cytochemical demonstration of cholinesterase activities and nicotinic cholinergic receptors at the light microscope level,the muscles were fixed with buffered solutions of formaldehyde (paraformaldehyde or methanol-free formalin) ; for the ultrastructural study the primary fixa-

---

Figs. 1 and 2.- Mouse motor endplates. External intercostal muscles. t, branch or bouton of a motor nerve terminal ; m,muscle fibre. Electron micrographs of endplates at 3 days after a localized freezing of muscle fibres which caused a very marked regression of the muscle fibres' organization,notably at subsynaptic areas. The nerve terminal branches are almost normal in appearance (Mx10,720). Fig.1: a branch of a nerve terminal is seen in longitudinal section.Fig.2: two branches of a nerve terminal are seen in cross-section.

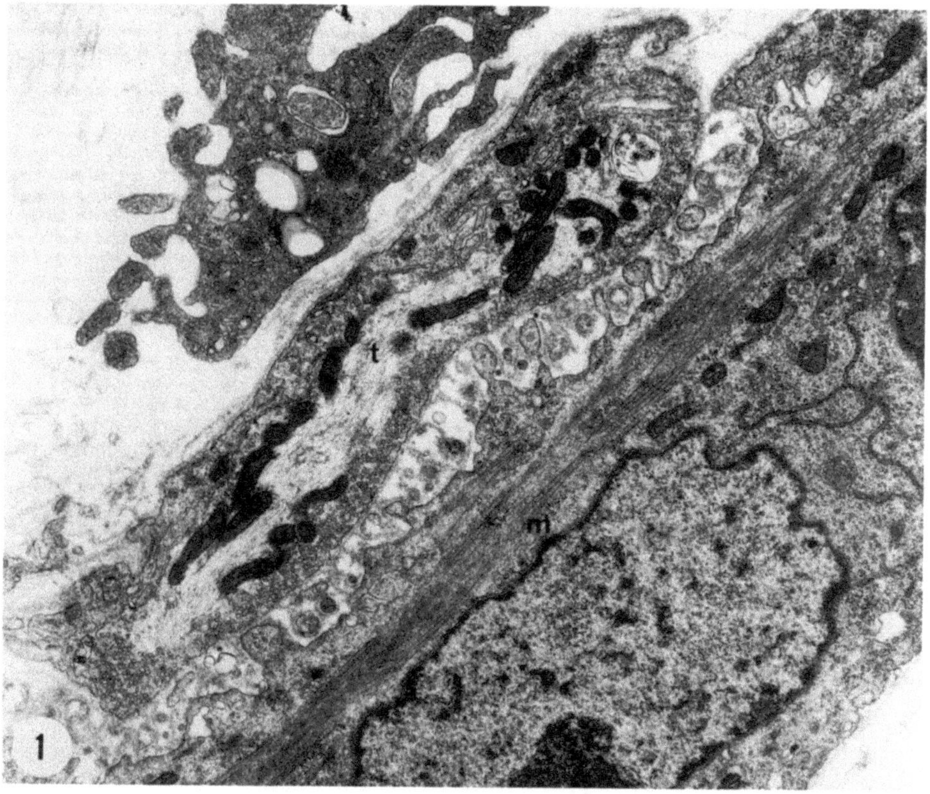
t
m
1

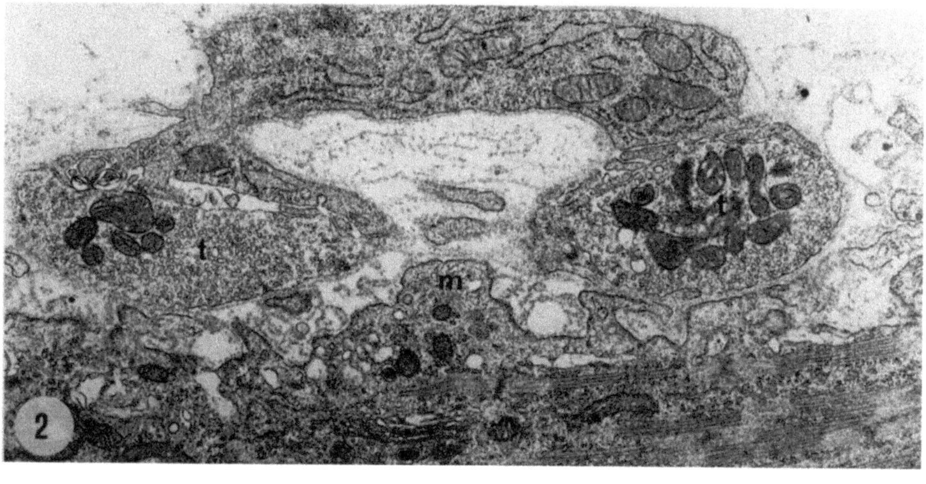
t
m
t
2

tion was carried out with a buffered solution of glutaraldehyde.

The cholinesterase activities were localized with a modification of the Koelle's thiocholine method (14) . The distribution of cholinergic receptors was studied by means of the fluorescence microscopy using alpha-bungarotoxin labeled by tetra-methyl-rhodamine isothiocyanate (15),prepared by Dr. M. Vigny (Laboratoire de Neurobiologie de l'Ecole Normale Supérieure).

Results

Within two or three days of freezing or cutting the muscle fibres underwent very different degrees of structural regression (Figs.1-5),affecting in particular the myofibrils and T-system. When the regression of the fibres was very pronounced these fibres resembled the myotubes formed during the development and regeneration of the muscle. Nevertheless,they were easily distinguishable from myotubes by their basal lamina folds,clearly visible on cross-sections,and by the nature of their connections with the motor nerve terminals.

During the regression period,the neuromuscular junctions also underwent marked changes,principally affecting their postsynaptic portion. Electron microscopy showed that their subsynaptic folds gradually became scarce.In the case of severe regression,they disappeared completely,often due to the thinning out of the interfolds and the merging of secondary synaptic clefts. This led to the formation of very wide clefts in which the still visible basal lamina proved to be the most stable remnant of the subsynaptic folds.

This phase of involution of muscle fibres and neuromuscular junctions, which in many respects resembled the muscle dedifferentiation that Lentz

---

Figs. 3 to 5.- Electron micrographs of endplates at 6 days after a localized freezing of muscle fibres. The importance of the changes observed in injured muscle fibres is extremely different from one fibre to another. Fig.3: in this fibre the dissociation of the neuromuscular junction is almost complete (Mx8,640). Fig.4: the number of subsynaptic folds probably has been reduced (Mx16,000). Fig.5: the junction looks almost normal (M x10,720).

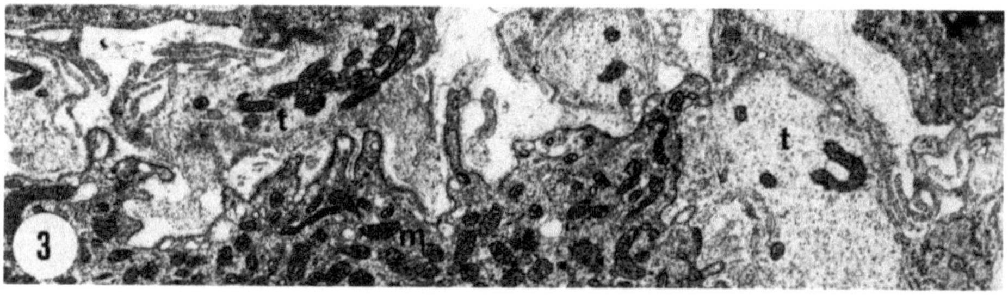
3
t
t
m

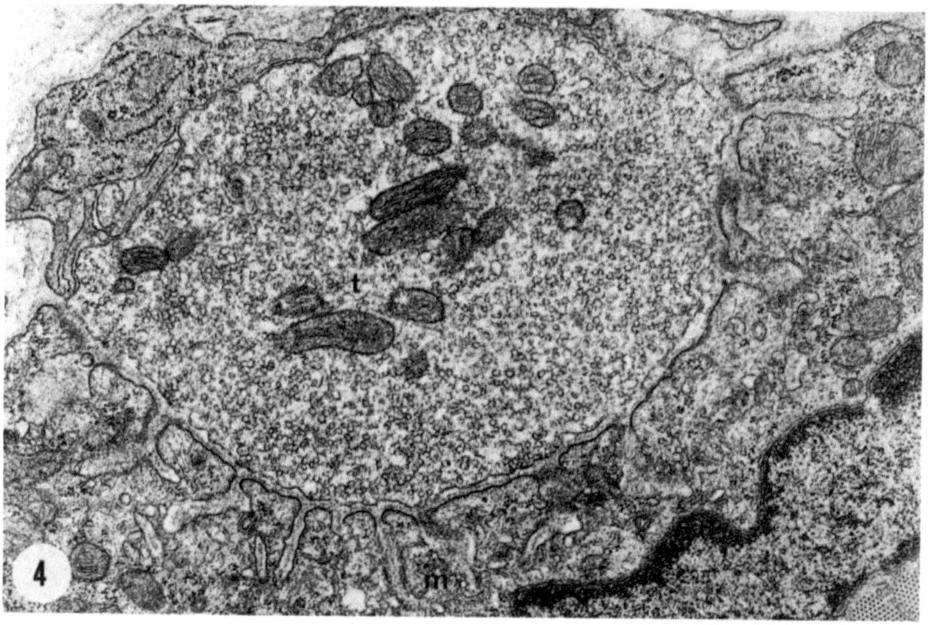
4
t
m

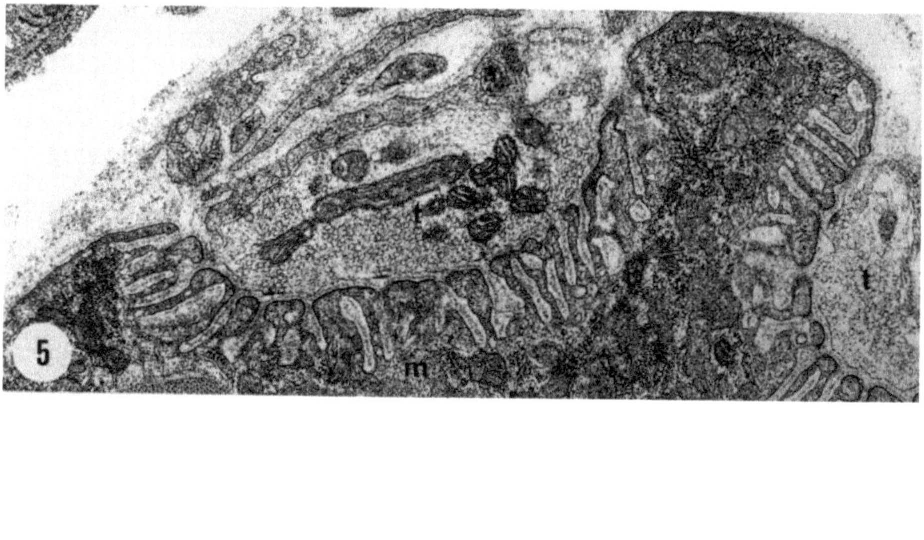
5
t
t
m

observed after cutting a newt limb (16),was followed by a phase of recovery characterized by intensive formation of new myofibrils. The muscle fibres began to acquire the features of growing fibres with considerable development of the granular endoplasmic reticulum and an increase in the number of Golgi cisternae stacks . Towards the end of the first week,varicosities and very short branches ending in boutons started to appear along the original branches of the nerve terminal. Silver staining of the nerve terminals,cytochemical demonstration of cholinesterase activities and cholinergic receptors,as well as electron microscopy, showed that this sprouting culminated in new synaptic connections (Figs. 6-13). The subsynaptic apparatus was formed by a cluster of cupules instead of the synaptic gutters generally seen in mouse intercostal muscles (17). The structure of the nerve terminal is reminiscent of the "terminaison en grappe" first described in reptiles (18) which is also seen in normal muscles of other vertebrates - and very frequently - in human muscles. In the case of repaired muscle fibres,this aspect of the nerve terminal did not seem to be a transitional stage since it had not changed five months after the operation. These endplates also exhibited ultrastructural peculiarities;while nearly all the normal endplates of the intercostal muscles had numerous deep subsynaptic folds,many of the transformed endplates only displayed a few shallow folds.

All of the results of the experiments reported here were similar,whether they were obtained by freezing or cutting.

Experiments carried out on the m.rectus abdominis and m.gastrocnemius brought about the same changes at neuromuscular junctions.

Figs. 6 to 13.-Structural and cytochemical characteristics of transformed endplates at one or several months after a localized freezing of muscle fibres. Figs.6 and 7: silver impregnation of a normal endplate,with a ramified nerve ending (Fig.6) and of a transformed endplate with synaptic boutons (Fig. 7)at 30 days after freezing (Mx640). Figs. 8 and 9 : distribution of cholinergic receptors at a normal endplate (Fig.8) and a transformed endplate (Fig.9) 154 days after freezing (Mx640). Fig. 10 and 11 : distribution of cholinesterase activities at a normal endplate (Fig. 10) and at a transformed endplate (Fig. 11) 95 days after freezing (Mx640). Fig. 12 and 13: electron micrographs of transformed endplates at 109 days after freezing ; below the terminal boutons the subsynaptic folds are very little developed,which is normally very rare in these muscles (Mx10,720).

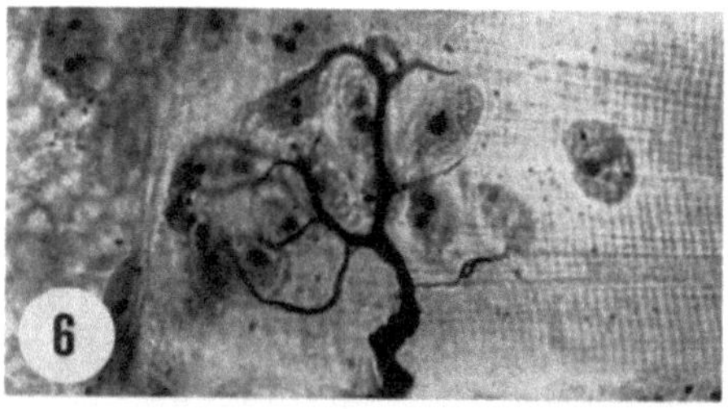
6

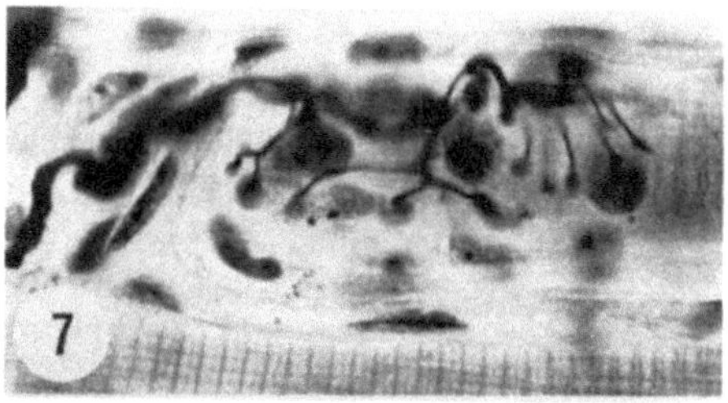
7

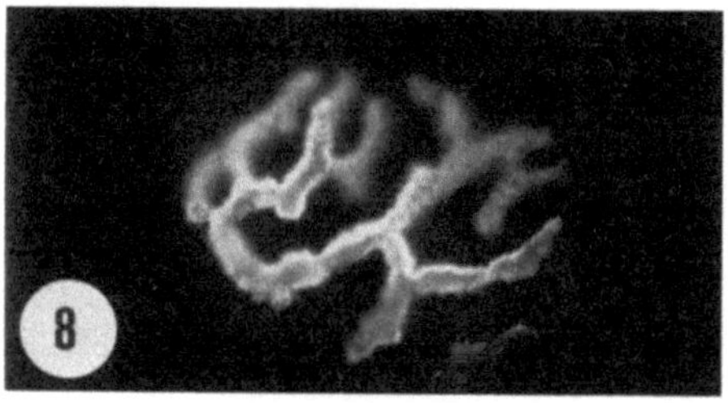
8

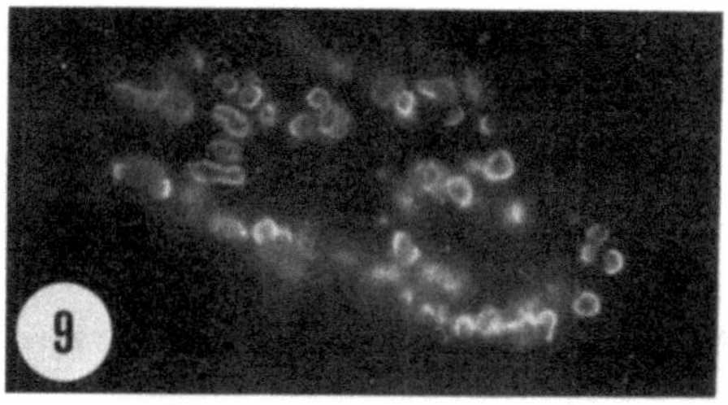
9

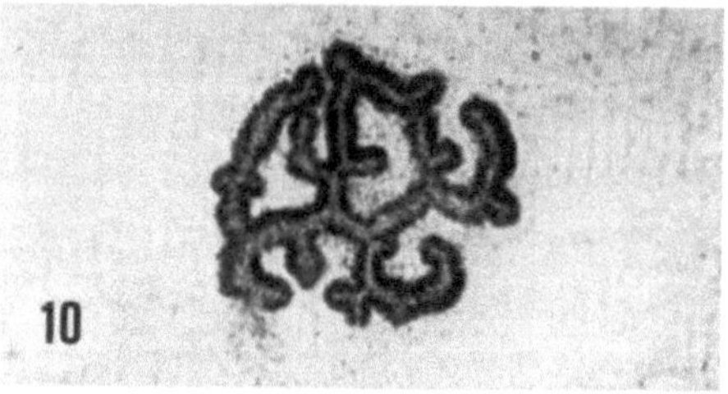
10

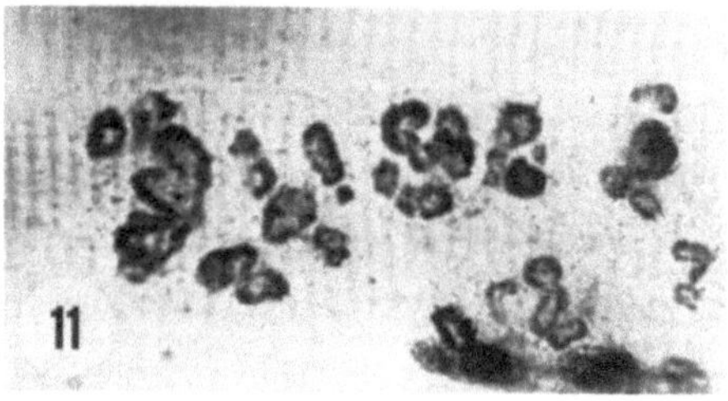
11

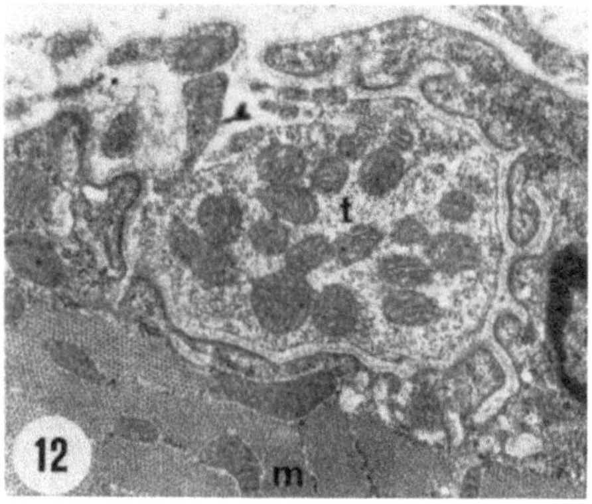
t
m
12

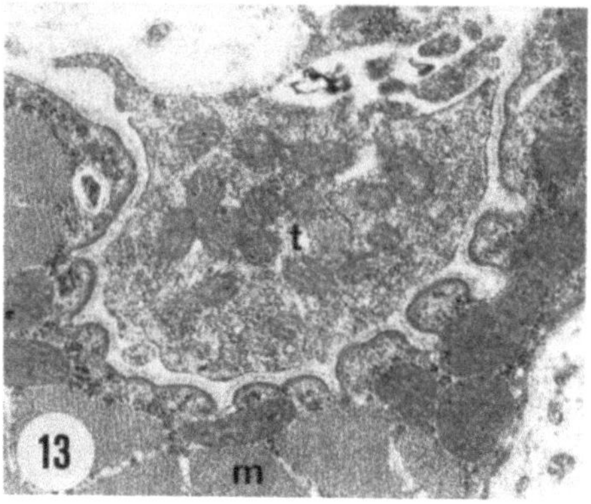
t
m
13

Discussion

The transformed endplates described here are remarkably similar in many ways to the ectopic motor endplates induced in normally nerve-free portions of rat muscles by the proximal stump of the cut motor nerve (19,20). To avoid motor reinnervation at old denervated endplates,the innervated portion of the muscle was removed by cutting in such a way that the retained portion was damaged just before reinnervation. It is therefore possible that the peculiarities of these ectopic endplates were at least partly caused by lesions to the muscle fibres.

Abnormal endplates of the same type were also observed during the regeneration of mouse muscles subjected to treatments that caused necrosis of muscle fibres without affecting their innervation(6,12). Regarding these experiments,as well as the above-mentioned case of ectopic innervation, the question arises whether the muscle fibres which acquired the abnormal endplates stemmed from the survival and reorganization of injured fibres as in our experiments, or from the fusion of undifferentiated cells as in normal myogenesis , or from both.

Experiments carried out on mouse intercostal muscles with the cardiotoxin of the venom of *Naja massombica massombica** incline us to think that the last of these hypotheses is the most probable .

What is the mechanism of the terminal sprouting observed during repair of muscle fibres ?

We have seen that local injuries were sometimes followed by a considerable regression of the injured muscle fibres and at the same time by a dedifferentiation of the subsynaptic areas equivalent to a complete or almost complete dissociation of the myoneural synapse.The terminal sprouting that followed the regression of synaptic neuromuscular connections is to some extent comparable to the sprouting observed in partially denervated or paralysed muscles(21,22,23). As in these experiments,the effect induced by muscular lesions on nerve terminals can also probably be explained by the muscle fibres' own action.

*This cardiotoxin was prepared in purified form by Dr.Pierre Bougis (Groupe de recherche sur les toxines animales,ERA 617 , CNRS ,Marseille)

It seems possible that the release by the muscle fibres of a growth factor intervenes in the mechanism of the nervous sprouting leading to a readjustment of the endplates. This assumption fits in the framework of explanations proposed by many authors to account for the reinnervation of denervated fibres by sprouts of the nerve terminals of neighbouring active muscle fibres as well as for the innervation of myotubes during the normal myogenesis and the muscle regeneration.

The possible roles of both the terminal Schwann cells and remnants of the original endplates must be taken into account. Especially important in this respect is the synaptic basal lamina,whose inductive properties may trigger the mechanisms ensuring the differentiation of regenerating axons into nerve terminals (11) and the differentiation of subsynaptic areas in regenerating muscles (24) .

Summary

Local anaesthetics,cardiotoxin and mechanical injuries may cause necrosis of muscle fibres while leaving the motor nerve fibres and their terminals intact .With local injuries to mouse muscles carried out by freezing or cutting we made a point of preserving both the nerve terminals and the muscle fibre portions on which these terminals were located. It was thus possible to follow the changes induced at endplates by these lesions.Within two or three days of the freezing or cutting,the muscle fibres underwent very different degrees of regression of the contractile material and T-system. The neuromuscular junctions also underwent changes,principally affecting their postsynaptic portion,in particular the folds of the subneural apparatus. After dedifferentiation of subsynaptic areas,we observed sprouting of the nerve terminal on muscle fibres which survived the amputation of one end and formed actively new myofibrils. This sprouting restored synaptic connections at the original sites,but with new structural features and correlative changes in the distribution of cholinergic receptors and cholinesterases. It is probable that after a phase of involution followed by a phase of recovery,the injured muscle fibres triggered off the nerve terminal sprouting which led to the remodelling of the endplates.

References

1. Benoit,P.W. and P.Belt.1970. Destruction and regeneration of skeletal muscle after treatment with a local anaesthetic,bupivacaine (Marcaine). J.Anatomy (London)107 ,547-556.

2. Libelius,R.,B.Sonesson,B.A.Stamenovic and S.Thesleff.1970.Denervation-like changes in skeletal muscle after treatment with a local anaesthetic (Marcaine).J.Anatomy (London),106,297-309.

3. Jirmanova, I.and S.Thesleff.1972.Ultrastructural study of experimental muscle degeneration and regeneration in the adult rat.Z.Zellforsch. mikroskop.Anatomie 131, 77-97.

4. Jirmanova,I.1975.Ultrastructure of motor end-plates during pharmacologically-induced degeneration and subsequent regeneration of skeletal muscle J.Neurocytology 4,141-155.

5. Duchen,L.W.,B.J. Excell,R.Patel and B.Smith.1973. Light and electron microscopic changes in mouse muscle fibres and motor end-plates caused by the depolarizing fraction (cardiotoxin) of the venom of *Dendroaspis jamesoni*.J.Physiol.(London)234 ,1-2P.

6. Duchen,L.W.,B.J.Excell,R.Patel and B.Smith.1974.Changes in motor end-plates resulting from muscle fibre necrosis and regeneration.A light and electron microscopic study of the effects of the depolarizing fraction (cardiotoxin) of *Dendroaspis jamesoni* venom.J.neurol.Sciences 21,391-417.

7. Harris,J.B.,M.A.Johnson,E.Karlsson.1975.Pathological responses of rat skeletal muscle to a single subcutaneous injection of a toxin isolated from the venom of the Australian tiger snake *Notechis scutatus scutatus*.Clin. Exp.Pharmacol.Physiol.2,383-404.

8. Harris,J.B.and M.A. Johnson.1978.Further observations on the pathological responses of rat skeletal muscle to toxins from the venom of the Australian tiger snake,*Notechis scutatus scutatus*.Clin.Exp.Pharmacol.Physiol. 5,587-600.

9. Harris,J.B. and C.A.Maltin.1982. Myotoxic activity of the crude venom and the principle neurotoxin,taipoxin,of the Australian taipan,*Oxyuranus scutellatus*.Br.J.Pharmacol. 76,61-75.

10.Marshall,L.M.,J.R.Sanes and U.J.McMahan.1977.Reinnervation of original synaptic sites on muscle fibre basement membrane after disruption of the muscle cells.Proc.Natl.Acad.Sci.USA 74,3073-3077.

11.Sanes,J.R.,L.M.Marshall and U.J.McMahan.1978.Reinnervation of muscle fiber basal lamina after removal of myofibers.Differentiation of regenerating axons at original sites.J.Cell Biol. 78,176-198.

12.Summers,P.J. and C.R.Ashmore.1983.Regeneration and reinnervation of the dystrophic mouse soleus muscle.A light- and electron-microscopic study. Acta Neuropathol.59,207-215.

13.Sokoll,M.D.,B.Sonesson and S.Thesleff.1968.Denervation changes produced in an innervated skeletal muscle by long-continued treatment with a local anaesthetic.Europ.J.Pharmacol.4 ,179-187.

14.Couteaux, R.and J.Taxi.1952Recherches histochimiques sur la distribution des activités cholinestérasiques au niveau de la synapse myoneurale.Arch. Anat.Micr.41 ,352-392.

15.Anderson,M.J. and M.W.Cohen.1974. Fluorescent staining of ACh-receptors in vertebrate skeletal muscle.J.Physiol.(London)237 , 385-400.

16.Lentz,T.L. 1970.Development of the neuromuscular junction.II.Cytological and cytochemical studies on the neuromuscular junction of dedifferentiating muscle in the regenerating limb of the newt Triturus.J.Cell Biol.47 ,423-436.

17.Couteaux, R.1946.Sur les gouttières synaptiques du muscle strié.C.R. Soc.Biol. 140, 270-273.

18.Tschiriew,S.1879.Sur les terminaisons nerveuses dans les muscles striés. Arch.Physiol.norm.pathol. ,Sér.2 ,6, 89-115.

19.Koenig, J.1963.Innervation motrice expérimentale d'une portion de muscle strié normalement dépourvue de plaques motrices chez le Rat.C.R.Acad.Sci. (Paris)256 ,2918-2920.

20.Koenig,J.In: Couteaux,R.1963.The differentiation of synaptic areas.Proc. R.Soc.London,B158, 457-480.

21.Brown,M.C. and R.Ironton.1978. Sprouting and regression of neuromuscular synapses in partially denervated mouse muscles.J.Physiol. (London) 278,325-348.

22.Betz,W.J.,J.H. Caldwell and R.R.Ribchester.1980.Sprouting of active nerve terminals in partially inactive muscles of the rat.J.Physiol. (London) 303,281-299.

23.Slack,J.R. and S.Pockett.1981.Terminal sprouting of motoneurones is a local response to a local stimulus. Brain Res. 217, 368-374.

24.McMahan,U.J. R.M. Marshall and C.R.Slater.1983. Acetylcholine receptor accumulation at original sites on regenerating frog skeletal muscle fibres induced by basal lamina.J.Physiol. (London) 342,56-57P.

INTERMITTENT, CALCIUM INDEPENDENT RELEASE OF ACETYLCHOLINE FROM MOTOR NERVE TERMINALS

S. Thesleff

Department of Pharmacology, University of Lund, S-223 62 Lund, Sweden

Introduction

In his studies of neuronal function and neuromuscular transmission Professor David Nachmansohn assigned to acetylcholine a much broader physiological role than merely to serve as a chemical transmitter for the nerve impulse across the synaptic cleft. In accordance with that view I should like in this presentation to outline the results of some recent studies which indicate that acetylcholine, in addition to serving neuromuscular transmission, is secreted by a separate mechanism to act as a chemical signal with a possible neurotrophic function.

Ample documentation exists regarding the role and function of acetylcholine in neuromuscular transmission (1). Suffice it to say that calcium ions are necessary for nerve impulse-transmitter release to occur. Here, I will describe another type of acetylcholine secretion which is unaffected by extra- or intracellular calcium concentrations and which therefore is uninfluenced by nerve impulses. It has the characteristics of a spontaneous intermittent release of acetylcholine which becomes particularly prominent whenever normal neuromuscular transmission is blocked or impaired. This type of secretion is selectively stimulated by the drug 4-aminoquinoline.

The Calcium Independent Acetylcholine Secretion

Characterization

The calcium independent type of acetylcholine secretion is characterized by intermittent release of variable, but generally quite large, amounts of acetylcholine (2,3,4). In the post-synaptic muscle cell this is mani-

Molecular Basis of Nerve Activity

fested as slow-rising, large-amplitude spontaneous potentials as illustrated in Fig. 1B. Typically, such potentials have a time-to-peak exceeding more than twice that of calcium dependent miniature end-plate potentials or nerve impulse evoked end-plate potentials. Their amplitude is highly variable, with many more smaller and larger events compared to the uniform--sized miniature end-plate potentials representing calcium dependent transmitter release, compare Fig. 1A and B. In many instances the potential amplitudes are as large as 10-12 mV and their times to peak as long as 10-15 ms.

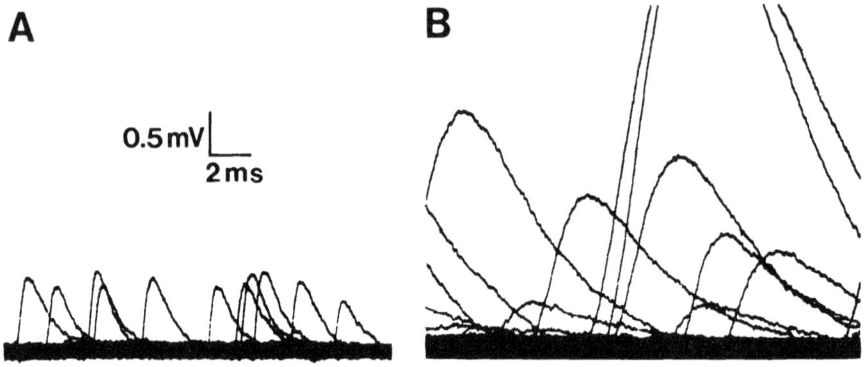

Fig. 1. Examples of spontaneous miniature end-plate potentials caused by calcium dependent transmitter release (A) and calcium independent transmitter release (B). Record A is from a normal extensor digitorum longus muscle of an adult rat and record B from a similar muscle in which the calcium activated transmitter release was blocked 13 days previously by the use of botulinum toxin. Superimposed oscilloscope sweeps. 30°C.

The potentials are caused by acetylcholine since they are abolished by curare, and their amplitude and duration are enhanced by cholinesterase inhibitors. The release originates from the nerve since it is abolished by denervation and does not reappear until reinnervation (2).

Much evidence indicates that acetylcholine is stored in specialized membraneous structures, known as synaptic vesicles, inside the nerve terminal. Elevation of the intracellular calcium concentration provokes the exocytotic release of transmitter at the so-called active zones in the nerve terminal membrane which are in close proximity to the post-synaptic receptor area (5). This type of calcium dependent vesicular release of acetylcholine accounts for the fast-rising, uniformed sized miniature end-plate potentials and for the end-plate potential in response to nerve stimuli (1).

A slow-rising acetylcholine potential would result if the distance between transmitter release and receptor was longer than normal or if the exocytotic process was prolonged (6). On that basis we have suggested that the calcium independent type of acetylcholine release might originate from areas outside of the active zones, i.e. at some distance from the post-synaptic receptor, and that in addition the release process is prolonged possibly as a result of successive exocytosis of synaptic vesicles in a cluster or in a row (2). It should, however, be mentioned that in ultrastructural studies of the nerve terminal no clustering of vesicles nor large-sized vesicles or cisternae have been detected to account for such a release (7). Gated release of acetylcholine from a cytoplasmic pool of free transmitter would be another, hitherto unexplored, possibility to account for this type of transmitter secretion. In connection with this it is interesting that the temperature dependence of the two release processes are quite different. Calcium dependent acetylcholine release is much less enhanced by temperature, with a Q_{10} of about 3, than the calcium independent type, which has a Q_{10} of about 12 (8).

The calcium independence of the described type of acetylcholine secretion is indicated by the failure of nerve impulses to release such quanta of acetylcholine. Furthermore, procedures which alter extra- or intracellular calcium concentrations in the nerve terminal fail to affect the frequency of this release while the same procedures markedly alter the frequency of the miniature end-plate potentials which constitute the basis for calcium activated release of acetylcholine. Thus, high extracellular concentrations of potassium (20 mM) or calcium (8 mM) or the presence of ethanol (0.5 M), ouabain (0.2 mM) or hypertonic solutions (2 times), which all greatly enhance calcium dependent miniature end-plate potential frequency fail to affect the frequency of the slow-rising, large amplitude potentials (8).

The frequency of the intermittent, calcium independent secretion of acetylcholine varies greatly under different conditions (3). When at maximum, however, it approaches 1 Hz and seems quantitatively at least as large as the spontaneous calcium dependent quantal release of acetylcholine.

Occurrence

Calcium independent transmitter release is present at a low frequency at normal adult neuromuscular junctions. Already in 1957, Liley (9) observed in rat muscle unusually large spontaneous end-plate potentials which he called giant miniature end-plate potentials. They varied in frequency between fibres and muscles and constituted on the average a few percent (0-20%) of the total number of spontaneous potentials recorded at a junction (3,10).

The drug 4-aminoquinoline stimulates selectively this type of transmitter release (11). When 4-aminoquinoline is added to a mammalian neuromuscular preparation, in a concentration of 100-200 µM, it induces within minutes the appearance of spontaneous potentials with prolonged times-to-peak and larger than normal amplitude and all the other characteristics of a calcium independent secretion of acetylcholine. This effect occurs without changes in the number of calcium dependent miniature end-plate potentials or alterations in nerve impulse evoked transmitter release. In the presence of 4-aminoquinoline up to 50% of all spontaneous potentials may be of the calcium independent type. The mechanism by which 4-aminoquinoline exerts this effect is unknown.

Calcium independent transmitter release is stimulated whenever the calcium dependent type of release is blocked or impaired. The application of botulinal neurotoxins to skeletal muscle causes complete block of calcium activated transmitter release resulting in total paralysis of the muscle lasting, with type A toxin, several weeks. During that period the calcium independent type of transmitter release gradually increases and becomes the only type of spontaneous, intermittent release of acetylcholine present at the end-plate (3,10). Many of these potentials are sufficiently large to trigger an action potential in the muscle cell. When the effects of the toxin subside calcium activated acetylcholine release returns and the independent type of release is again reduced to a low level (3).

A somewhat similar picture is seen at early stages of regenerated neuromuscular junctions (10,12). At such junctions calcium evoked transmitter release is impaired, nerve stimuli releasing subnormal amounts of acetylcholine. An examination of spontaneous end-plate potentials reveals the presence of slow-rising potentials of the calcium independent type which in some fibres constitute up to 60% of all the potentials present. Similar potentials also characterize skeletal muscle of chickens curarized during early development (13) and have been reported to be present at neuromuscular junctions of dystrophic mice (14). It is also of interest that growth cones of embryonic cholinergic neurons in culture, in the absence of target tissue, spontaneously and intermittently release large amounts of acetylcholine which can be detected by sensitive probes at some distance from the nerve (15,16)

Physiological role

Since the described secretion of acetylcholine is independent of calcium fluxes it is also uninfluenced by nerve impulses and therefore does not participate in synaptic impulse transmission nor in the immediate control of muscle activity. The nerve, however, has other effects on muscle, not primarily associated with impulse transmission. For example, it influences the chemical and electrical sensitivity of the muscle membrane, the composition of contractile proteins and the rate of protein degradation. In addition, the nerve is able to locate and to preferentially innervate its appropriate muscle and to induce in the muscle membrane the morphological and biochemical specialization of a synapse. Such influences are by a common term called neurotrophic and are considered to be exerted by a combination of the chemical influences originating from the nerve and the activity pattern imposed on the muscle cell by the innervation (17,18).

Acetylcholine is believed to be one neurotrophic factor (19,20). Therefore, it is tempting to assume that a spontaneous intermittent release of relatively large amounts of this substance could constitute a chemical signal which during development or reinnervation helped the nerve to locate responsive targets. Once a synapse was formed this release could be suppressed in favor of calcium dependent transmitter release and would only be reactivated when synaptic transmission for some reason was suppressed or impaired. It is also possible that the release might be a consequence of the incorporation into the nerve terminal of synaptic vesicular membranes during neurite extension or sprouting, but this idea has so far not received experimental support since procedures which prevent sprouting and nerve growth have failed to affect the frequency of such potentials (3,8).

Summary

The acetylcholine release responsible for impulse transmission across the synaptic cleft is a calcium requiring process. However, there also exists a calcium independent type of acetylcholine secretion. This neurosecretion is recorded as spontaneous, slow-rising, frequently large amplitude potentials at the post-synaptic end-plate. The occurrence of such potentials is uninfluenced by nerve impulses and by procedures affecting the intra- or extracellular calcium concentration of the nerve terminal. At normal adult neuromuscular junctions the frequency of this type of secretion is low with variability between fibres. However, when neuromuscular transmission is blocked or impaired calcium independent acetylcholine release is greatly accelerated and becomes quantitatively large. The drug 4-aminoquinoline selectively stimulates the calcium independent secretion of acetylcholine. It is proposed that the spontaneous, intermittent and calcium independent type of acetylcholine secretion is a primitive embryonic type of release with a neurotrophic function.

References

1. Katz, B. 1969. The Release of Neural Transmitter Substances. Liverpool University Press.

2. Thesleff, S. and J. Molgó. 1983. Commentary. A new type of transmitter release at the neuromuscular junction. Neuroscience 9. 1-8.

3. Kim, Y.I., T. Lømo, M.T. Lupa and S. Thesleff. 1984. Miniature end--plate potentials in rat skeletal muscle poisoned with botulinum toxin. J. Physiol. In press.

4. Thesleff, S. and M.T. Lupa. 1985. Calcium independent quantal transmitter release at the neuromuscular junction. In: Calcium, Neuronal Function and Transmitter Release (R. Rahamimoff, ed.). Martinus Nijhoff Publ., Boston.

5. Heuser, J.E., T.S. Reese, M.J. Dennis, Y. Jan, L. Jan and L. Evans. 1979. Synaptic vesicle exocytosis captured by quick freezing and correlated with quantal transmitter release. J. Cell Biol. 81. 275-300.

6. Gage, P.W. 1976. Generation of end-plate potentials. Physiol. Revs. 56. 177-247.

7. Pécot-Dechavassine, M. and J. Molgó. 1982. Attempt to detect a morphological correlate for the "giant" miniature end-plate potentials induced by 4-aminoquinoline. Biol. Cell. 46. 93-96.

8. Thesleff, S., J. Molgó and H. Lundh. 1983. Botulinum toxin and 4-aminoquinoline induce a similar abnormal type of spontaneous transmitter release at the rat neuromuscular junction. Brain Res. 264. 89-97.

9. Liley, A.W. 1957. Spontaneous release of transmitter substance in multiquantal units. J. Physiol. 136. 595-605.

10. Colméus, C., S. Gomez, J. Molgó and S. Thesleff. 1982. Discrepancies between spontaneous and evoked synaptic potentials at normal, regenerating and botulinum toxin poisoned mammalian neuromuscular junctions. Proc. R. Soc. Lond. B. 215. 63-74.

11. Molgó, J. and S. Thesleff. 1982. 4-aminoquinoline-induced "giant" miniature end-plate potentials at mammalian neuromuscular junctions. Proc. R. Soc. Lond. B. 214. 229-247.

12. Bennett, M.R., E.M. McLachlan and R.S. Taylor. 1973. The formation of synapses in reinnervated mammalian striated muscle. J. Physiol. 233. 481-500.

13. Ding, R., J.K.S. Jansen, N.G. Laing and H. Tønnesen. 1983. The innervation of skeletal muscles in chickens curarized during early development. J. Neurocytol. 12. 887-919.

14. Carbonetto, S. 1977. Neuromuscular transmission in dystrophic mice. J. Neurophysiol. 40. 836-843.

15. Hume, R.I., L.W. Robe and G.D Fischbach. 1983. Acetylcholine release from growth cones detected with patches of acetylcholine receptor--rich membranes. Nature. 305. 632-634.

16. Young, S.H. and M. Poo. 1983. Spontaneous release of transmitter from growth cones of embryonic neurones. Nature. 305. 634-637.

17. Gutmann, E. 1976. Neurotrophic relations. Ann. Rev. Physiol. 38. 177--216.

18. McArdle, J.J. 1983. Molecular aspects of the trophic influence of nerve on muscle. Progr. Neurobiol. 21. 135-198.

19. Thesleff, S. 1960. Effects of motor innervation on the chemical sensitivity of skeletal muscle. Physiol. Revs. 40. 734-752.

20. Drachman, D.B., E.F. Stanley, A. Pestronk, J.W. Griffin and D.L. Price. 1982. Neurotrophic regulation of two properties of skeletal muscle by impulse-dependent and spontaneous acetylcholine transmission. J. Neurosci. 2. 232-243.

PROSTAGLANDINS MEDIATE THE MUSCARINIC INHIBITION OF ACETYLCHOLINE RELEASE FROM TORPEDO NERVE TERMINALS

I. Pinchasi, M. Burstein, D.M. Michaelson

Department of Biochemistry, the George S. Wise Faculty of Life Sciences, Tel Aviv University, Ramat Aviv 69978, Israel.

Introduction

Torpedo nerve terminals are endowed with presynaptic muscarinic acetylcholine receptors (mAChR) which regulate acetylcholine (ACh) release by a negative feedback mechanism (1-4). The available data suggests that activation of these receptors inhibits release by interfering with the stimulus-secretion coupling at a stage distal to Ca^{2+} entry (1-3). It is therefore likely that the muscarinic effects on the nerve terminal are mediated by a second messenger.

Binding studies utilizing muscarinic agonists and antagonists, have shown that the *Torpedo* presynaptic mAChR is probably of the M_2 type (4) and that although the receptor is coupled to a GTP-regulatory protein (4-6), the presynaptic adenylate cyclase activity is not affected by muscarinic ligands (7). This finding, together with the observation that membrane permeable analogs of cAMP do not mimick the muscarinic effects on ACh release imply that the latter are not mediated by cAMP.

In many mammalian systems, activation of mAChR results in increased turnover of phosphatidylinositol (PI) (for review see (8) and (9)), and in enhanced synthesis of prostaglandins (PG's), particularly of the E series (10-16). Since PG's are oxydation products of arachidonate, which is abundant in PI, it was suggested that one of the functions of the "PI effect" is to trigger the synthesis of these PG's which, in turn, mediate various physiological effects, depending on the system investigated (for review see (16)). This hypothesis and the report by Bleasdale et al. (17) on a "PI effect" in *Torpedo*, prompted us to investigate the possibility that PG's are involved in the muscarinic regulation of ACh release in the *Torpedo*.

Molecular Basis of Nerve Activity

Results and Discussion

1) Characterization of the muscarinic inhibition of ACh release from Torpedo nerve terminals

Intact prisms of the Torpedo electric organ (tissue slices) release ACh in a Ca^{2+}-dependent manner ($EC_{50} \simeq 1$ mM) when depolarized by K^+. Addition of the muscarinic agonist oxotremorine to tissue slices results in inhibition of ACh release (Fig. 1). This oxotremorine-induced inhibition is transient,

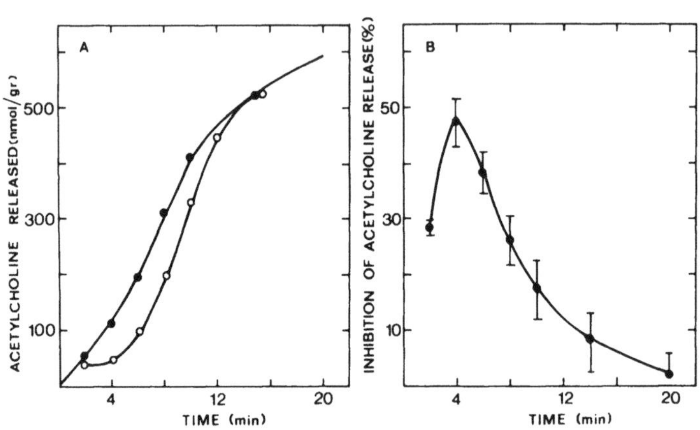

Figure 1: The effect of oxotremorine on ACh release from electric organ slices.

A. Ca^{2+}-dependent ACh release from K^+-depolarized electric organ slices in the absence (•) and presence(o) of oxotremorine (10 µM).

B. Percent inhibition of ACh release by oxotremorine. Results presented are the mean ± S.D. of 5 experiments, one of which is shown in A.

Prisms of the electric organ (0.25 gr) were excised and washed at 25° for 30-60 min in modified Torpedo buffer (modified TB), which contained (in mM): NaCl, 250; KCl, 4.8; $MgCl_2$, 2.4; D-glucose, 10; sucrose, 200; Na-phosphate buffer, 1.2; pH = 7.2. The tissue slices were transferred to fresh modified TB which contained 5 mM $CaCl_2$. Following a 30 min period of incubation, the slices were stimulated by transferring them into a K^+-modified TB, which contained 5 mM Ca^{2+}. The composition of the K^+-modified TB was identical to that of modified TB, except that it contained 125 mM KCl and 125 mM NaCl. Medium samples were withdrawn at the designated intervals and their choline content, which was produced by hyrolysis of the released ACh, was determined by the chemiluminescence method of Israel and Lesbats (18).

Oxotremorine was added 15 min prior to stimulation of release and was also present throughout the release experiment.

namely, oxotremorine slows down the rate of release without affecting the size of the releasable pool. Accordingly, the muscarinic effect on ACh release can be characterized by two parameters: maximal inhibition and time to peak inhibition.

The inhibitory effect of oxotremorine on ACh release from tissue slices is totally reversed by the muscarinic antagonist atropine (Fig. 2). By contrast, α-bungarotoxin and curare, at concentrations which completely block the postsynaptic nicotinic ACh receptor, do not affect the oxotremorine-induced inhibition of release (Fig. 2). These results show that the effect of oxotremorine is mediated by a mAChR and suggest that this receptor is presynaptic. Further proof for the presynaptic location of mAChR in the *Torpedo*

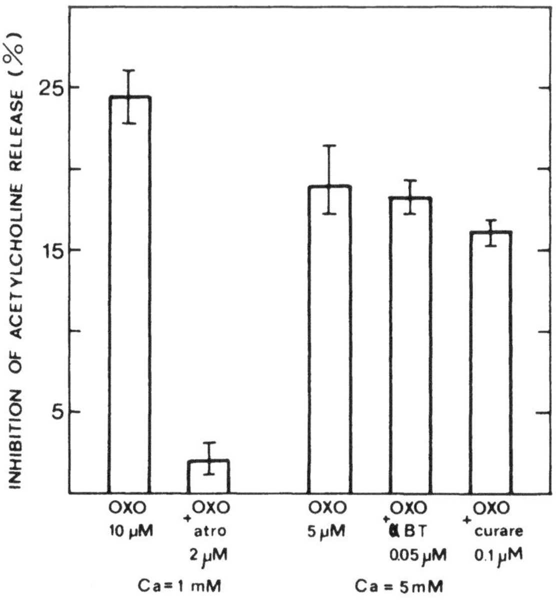

Figure 2: The effects of atropine, α-bungarotoxin (α-BT) and curare on the inhibition of ACh release by oxotremorine.

ACh release from K^+ (125 mM)-depolarized electric organ slices was measured in the presence of the designated concentrations of Ca^{2+}, as in Fig. 1. Results presented are maximal inhibition induced by the specified concentration of oxotremorine, which was added 15 min prior to stimulation. Atropine, α-BT and curare were added 30 min prior to stimulation. Results are the mean ± S.D. of 3 experiments.

electric organ comes from previous binding studies, in which mAChR binding sites and isolated nerve terminals (synaptosomes) copurified upon subcellular fractionation of the electric organ (4, 19). Furthermore, oxotremorine also inhibits ACh release from isolated synaptosomes (1, 2) and is as efficient in the latter as it is in tissue slices. This is demonstrated in Fig. 3, in which the dose/reponse curves for oxotremorine in the two preparations are superimposable ($EC_{50} \sim 2.5\ \mu M$). The time to peak inhibition is also concentration-dependent, e.g. at 1 μM oxotremorine it is 8 ± 0.7 min, whereas at 10 μM oxotremorine it is 3 ± 0.8 min (n=3). The inhibitory effect of oxotremorine on the synaptosomal release of ACh is also transient (e.g. see Fig. 6 in reference (2)). Similar transient inhibitions were

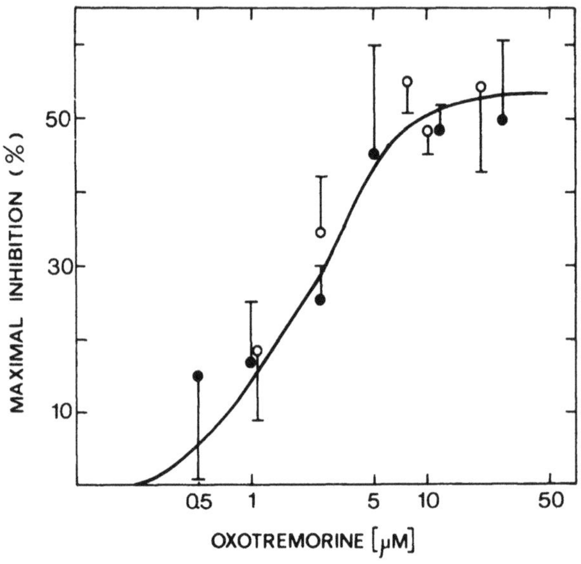

Figure 3: The concentration-dependence of the inhibition of ACh release by oxotremorine.

ACh release from electric organ slices (- ● -) and from purified synaptosomes (- o -) was measured as in Fig. 1 (tissue slices) and as described by Michaelson et al. (2) (synaptosomes). Release was induced by 125 mM K^+ in the presence of 5 mM Ca^{2+} (tissue slices) or 2 mM Ca^{2+} (synaptosomes). Oxotremorine was added 15 min (tissue slices) and 2 min (synaptosomes) prior to stimulation. Results presented are the maximal inhibition induced by the specified concentrations of oxotremorine and are the mean ± S.D. of 4 experiments.

previously reported for the muscarinic effects on ACh release from cortical and hippocampal slices (20, 21) and from the guinea pig myenteric plexus (22). They therefore seem to be a general characteristic of the regulation of ACh release by muscarinic receptors. It should be noted that in the Torpedo, the sensitivity of ACh release to muscarinic regulation varies circa - annually (23). Therefore all the results presented in this communication were obtained from experiments which were performed during the "peak" season (December - March).

In principle, the transient profile of the muscarinic inhibition of ACh release can result from time-dependent changes either in the release processes or in the transducing mechanism which mediates the effects thereon of receptor activation. One explanation which is consistent with the first possibility is that oxotremorine can inhibit the release of only a given amount of ACh. If this was true, one would expect a smaller inhibition (expressed in percent) when the extents of release are larger, e.g. when the concentration of Ca^{2+} is higher. However, as can be seen in Fig. 4, the opposite is true: The inhibitory effect of oxotremorine increases as the concentration of Ca^{2+}, and consequently ACh release, increase and the time to peak inhibition decreases accordingly. In fact, the Ca^{2+}-dependency of the oxotremorine-induced effect is very similar to that of ACh release itself ($EC_{50} \simeq$ 1 mM). The possibility that the transient profile of the muscarinic effect is due to receptor desensitization occuring either prior to, or concomitantly with, the induction of ACh release, was investigated by examining the effect of the time of preincubation with oxotremorine on the pattern of inhibition of release. As can be seen in Fig. 5, exposing the tissue to oxotremorine for 60 min under non-releasing conditions (e.g. prior to depolarization) results in a very effective inhibition of ACh release, which was even higher than that obtained after preincubation of 15 min with the agonist. This suggests that the mAChR is not desensitized under non-depolarizing conditions. However, when oxotremorine was added to the slices concomitantly with the high K^+ buffer, no inhibition occurred. The latter observation suggests either that the receptor is desensitized rapidly upon depolarization in the presence of Ca^{2+} or that there is a critical initial phase in the ACh release process which is sensitive to

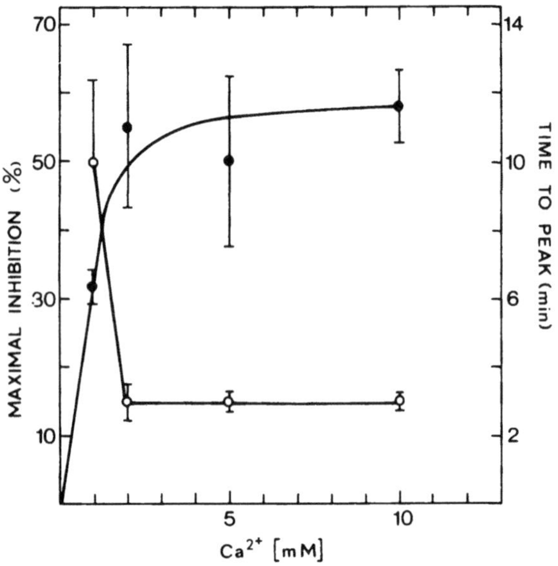

Figure 4: The Ca^{2+}-dependence of the oxotremorine-induced inhibition of ACh release

ACh release from electric organ slices was measured at the designated Ca^{2+} concentrations in the presence of oxotremorine (10 µM) as in Fig. 1. Data presented are the maximal inhibition (- ● -) and the time to maximal inhibition (- o -) of release induced by oxotremorine. Results are the mean ± S.D. of 3 experiments.

muscarinic regulation. In order to distinguish between these possibilities, the post-binding events by which receptor activation affects release need to be studied. To this end we now turned to investigate the role of PG's, which we have previously suggested as putative second messengers (24), in the regulation of ACh release.

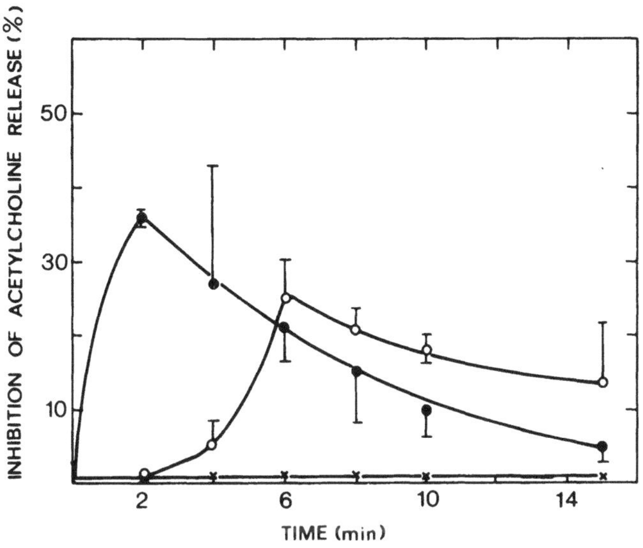

Figure 5: The effect of the time of exposure to oxotremorine on the inhibition of ACh release.

ACh release from electric organ slices was induced by K^+ (125 mM) and Ca^{2+} (5 mM) as in Fig. 1. Oxotremorine (2.5 μM) was added either 60 min (- ● -) or 15 min (- o -) prior to, or concomitantly with (- x -) stimulation. Results are the mean ± S.D. of 3 experiments.

2) The role of PG's in the muscarinic regulation of ACh release

The following criteria need to be met for a PG to be a mAChR second messenger:

a) Blockage of PG's synthesis should abolish the ability of muscarinic agonists to inhibit ACh release.

b) Exogenous PG's should mimick the effect of receptor activation on ACh release.

c) Activation of the mAChR should increase the synthesis of endogenous PG's.

The synthesis of PG's can be blocked by the addition of indomethacin and aspirin which inhibit cyclooxygenase activity. Incubation of tissue slices with indomethacin for up to 2 hrs, has no effect on the Ca^{2+}-dependent K^+-mediated ACh release (Fig. 6). However, indomethacin effectively reverses the inhibition of ACh release elicited by oxotremorine (Fig. 6). Similar results were obtained with purified synaptosomes (24).

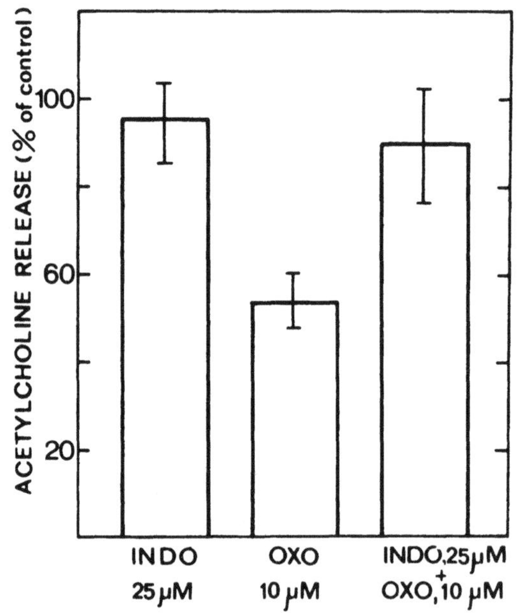

Figure 6: The effect of indomethacin on the oxotremorine-induced inhibition of ACh release.

ACh release from electric organ slices was induced by K^+ (125 mM) and Ca^{2+} (5 mM) as in Fig. 1. Oxotremorine and indomethacin were added at the designated concentrations, 15 min and 60 min, respectively, prior to stimulation. Control ACh release (100%) = 350 nmol/gr·10 min. Results are the mean ± S.D. of 5 experiments.

Addition of PGE_2 to tissue slices results in inhibition of K^+ + Ca^{2+}-induced ACh release (Fig. 7). The extent of inhibition is maximal at 2 min after depolarization and declines rapidly. This transient inhibition pattern is similar to that observed following activation of the mAChR by oxotremorine (Fig. 1). The inhibitory effect of PGE_2 is concentration-dependent: Maximal inhibition is ∿ 70% and $EC_{50} \simeq 0.05$ µM PGE_2 (Fig. 8). Similar results were obtained when the effect of exogenous PGE_2 was tested on

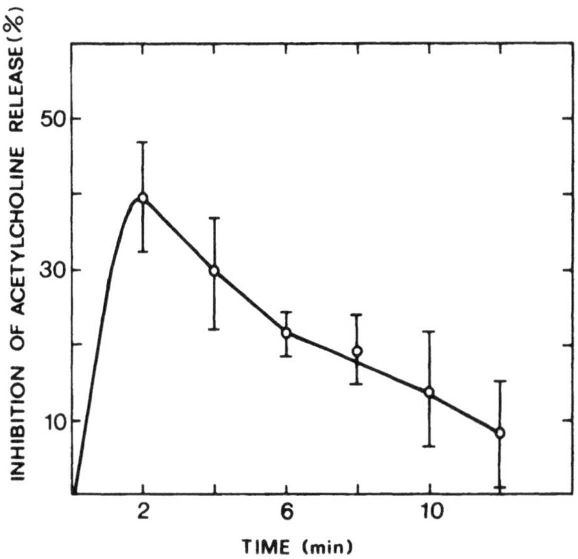

Figure 7: Inhibition of ACh release by exogenous PGE_2.

ACh release from tissue slices was induced by K^+ (125 mM) and Ca^{2+} (5 mM) as in Fig. 1. PGE_2 (0.15 μM) was added 15 min prior to stimulation. Results are the mean ± S.D. of 3 experiments.

purified synaptosomes (24), thus demonstrating that the effects of PGE_2 are due to direct interactions with the nerve terminal.

The specificity of the effect of PGE_2 was examined by testing the effects on release of other exogenous PG's: PGE_1, $PGF_{2}\alpha$, PGA_2 and PGD_2. None of these arachidonate metabolites (tested at 0.15 μM), were found to affect ACh release from tissue slices.

The results obtained this far satisfy the first and second criteria outlined above, and suggest that a cyclooxygenase derivative - similar or identical to PGE_2 - mediates the effect of mAChR activation on release. Moreover, the similar kinetic profiles of the effects of exogenous PGE_2 and of oxotremorine on release suggest that the transient nature of the muscarinic inhibition stems, at least in part, from the existence of a critical initial phase in the release process which is suceptible to inhibition.

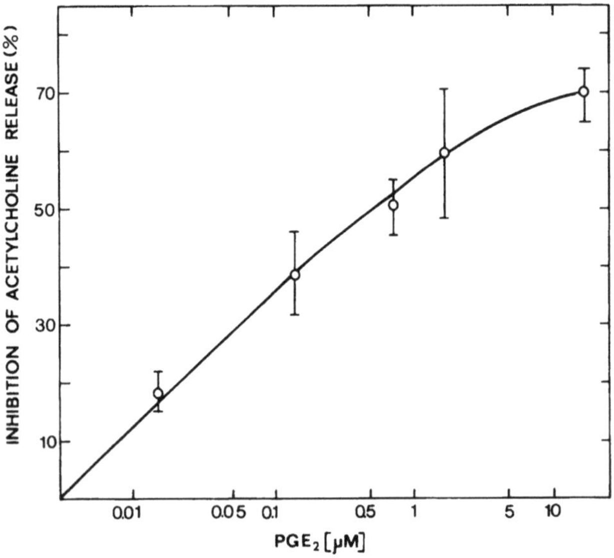

Figure 8: The concentration-dependence of the inhibition of ACh release by exogenous PGE_2.

ACh release from tissue slices was induced by K^+ (125 mM) and Ca^{2+} (5 mM) as in Fig. 1. PGE_2 was added at the designated concentrations 15 min prior to stimulation. Results presented are the maximal inhibition of ACh release induced by the specified PGE_2 concentrations and are the mean ± S.D. of 3 experiments.

For a PG to be the muscarinic second messenger, its intraterminal level should be modulated by muscarinic ligands. In view of the above findings, we first examined the possibility that PGE_2 is the muscarinic second messenger. To this end we studied the release of PGE_2 from tissue slices and synaptosomes and the effects thereon of muscarinic ligands. This was pursued by means of a highly specific $[^{125}I]$-PGE_2 radioimmunoassay (RIA).

When freshly-excised tissue slices are incubated in a physiological buffer (modified TB), they release PGE_2 into the medium. This release is increased 2-fold in the presence of extracellular Ca^{2+} and is abolished by indomethacin - thus demonstrating that PGE_2 is synthesized de-novo by the slices (Fig. 9). Prolonged preincubations of the slices in modified TB ("aging") result in a time-dependent decrease in the rate of synthesis of PGE_2, down

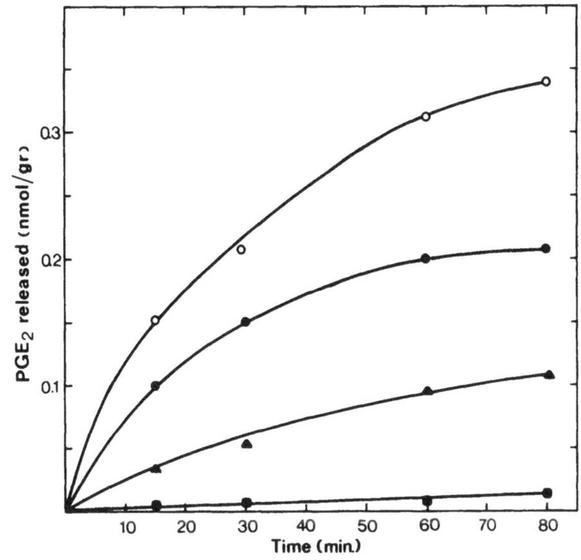

Figure 9: PGE_2 release from tissue slices.
Tissue slices (0.25 gr) were washed at 25° for 60 min in modified TB after which they were transferred to either fresh modified TB (- ● -), modified TB which contained 1 mM EGTA (- ▲ -), or modified TB which contained 1 mM free Ca^{2+} (- o -). Samples were then withdrawn from the medium at the designated intervals and their PGE_2 content was determined by a highly specific $[^{125}I]$-PGE_2 RIA (New England Nuclear). (- ■ -) corresponds to tissue slices in modified TB which were preincubated for 60 min with indomethacin (25 μM) prior to the initiation of experiment.

to 50 percent of control after a 3 hr preincubation and to 5 percent of control following an over-night preincubation.

Isolated synaptosomes also contain PGE_2 (8.12 ± 1.5 pmole/gr organ; 29 ± 5.4 pmole/mg protein, n=8). However, in contrast to tissue slices, the synaptosomal PGE_2 content is not affected by indomethacin (25 μM) or Ca^{2+} (1 mM) thus demonstrating that no detectable synthesis of PGE_2 occurs under these conditions.

Unexpectedly, when oxotremorine is added to freshly-excised tissue slices, it decreases, and not increases. the synthesis of PGE_2 (Fig. 10). This effect of oxotremorine is reversed by atropine (Fig. 10) and its concentration-dependence ($EC_{50} \simeq 2.5$ μM) is similar to that of the oxotremorine-

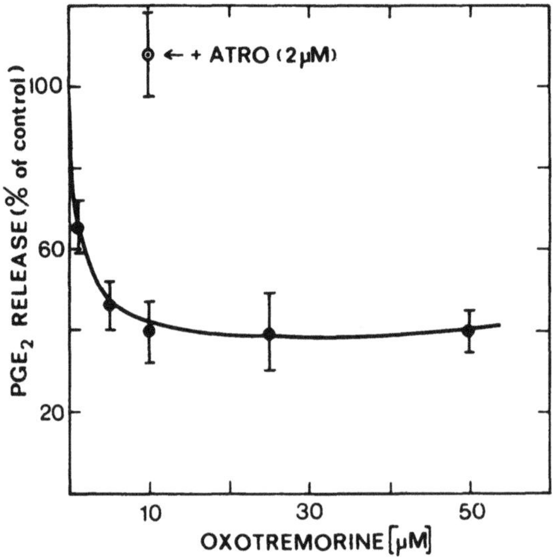

Figure 10: The effect of oxotremorine on the release of PGE_2 from tissue slices.

Tissue slices were washed, incubated for 30 min in modified TB + Ca^{2+} (5 mM) and then transferred to K^+ (125 mM) modified TB + Ca^{2+} (5 mM). The amount of PGE_2 released into the medium was determined 10 min after K^+-depolarization in the absence and presence of the designated concentrations of oxotremorine (●), which was added 15 min prior to depolarization. Atropine (◉) was added where specified, 30 min prior to depolarization. Control PGE_2 release (100%) = 0.18 ± 0.05 nmol/gr·10 min.

induced inhibition of ACh release (Fig. 3). Furthermore, it requires external Ca^{2+}, with a concentration-dependence (EC_{50} ≈ 1 mM Ca^{2+}) similar to that of the oxotremorine-induced inhibition of ACh release (Fig. 4).

When we repeated these experiments with the purified synaptosomes, oxotremorine (0.1 - 50 μM) was found to have no effect on their PGE_2 level (Fig. 11). These results, obtained with both tissue slices and with purified synaptosomes, exclude the possibility that PGE_2 itself is the mAChR second messenger. It thus seems that the muscarinic regulation of ACh release is mediated by a PG which may be similar - but not identical - to PGE_2. This possibility was probed by utilizing a broad-specificity [^{3}H]-PGE RIA, which enables measurement of the total synaptosomal PGE's (see legend to Fig. 11).

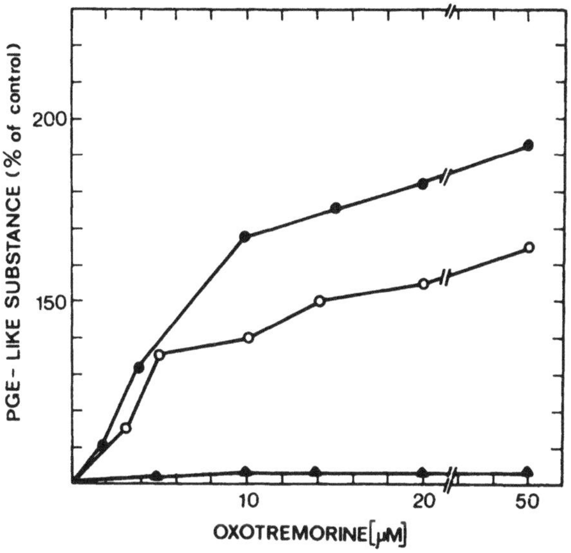

Figure 11: The effect of oxotremorine on the level of synaptosomal PGE.

The level of synaptosomal PGE and the effects thereon of oxotremorine were determined by rat stomach bioassay (25) (- ● -); [^{3}H]-PGE RIA (- o -) and [^{125}I]-PGE_2 RIA (- ▲ -). Purified synaptosomes (fraction a_2, as in (26)) were preincubated for 10 min at 25° in modified TB after which they were diluted into K^+ (125 mM) modified TB which contained 2 mM Ca^{2+}. Oxotremorine was added at the designated concentrations 2 min prior to stimulation and the level of synaptosomal PGE was determined 1 min following stimulation. In the experiments with the rat stomach bioassay, the reaction was stopped by acidification to pH 3.5 and the PG's were extracted and assayed as described elsewhere (24). In the experiments with the [^{3}H]-PGE RIA (Clinical Assays), the PGE's were converted to PGB's by boiling at pH 11.6 and assayed without extraction. The antiserum employed, which is directed against PGB_1, is not specific, e.g. it has a 25% cross reactivity with PGB_2. In the experiments with the highly specific [^{125}I]-PGE_2 RIA the samples were vigorously mixed with methanol and assayed without extraction. 100% corresponds (in pmol PGE/mg protein) to: 36 ± 8, 33.8 ± 6.9 and 29 ± 5.4 for the bioassay, [^{3}H]-PGE RIA and [^{125}I]-PGE_2 RIA, respectively.

As can be seen in Fig. 11, this assay enables the detection of an oxotremorine-induced increase in the level of synaptosomal PGE (EC_{50} ≃ 2.5 μM oxotremorine). This increase is reversed by atropine (1 μM), abolished by indomethacin (25 μM) and inhibited (∿ 70 percent) by mepacrine (100 μM) (not shown). Furthermore, the oxotremorine-induced increase in the level of PGE is transient: It reaches a maximum within about 30 sec following stimulation and declines to control levels within 2-3 min (Fig. 12). Thus, oxotremorine induces de novo synthesis of a synaptosomal PGE-like substance with kinetics and with a potency very similar to those of its effect on ACh release. Similar results were obtained when the level of the synaptosomal PGE-like substance was monitored by a PGE-bioassay (Fig. 11; see also (24)). Interestingly, the basal levels of synaptosomal PGE's determined by the [^{3}H]-PGE RIA and by the PGE-bioassay (33.8 ± 6.9 and 36 ± 8.6 pmol/mg protein, respectively), are very similar to that of PGE_2, as determined by the highly specific [^{125}I]-PGE_2 RIA (29 ± 5.4 pmol/mg protein). Hence it seems that in the absence of oxotremorine, the synaptosomes contain only PGE_2, which is detected by all 3 assays, while the PGE-like substance, detected by the former 2 assays, is synthesized only in the presence of oxotremorine.

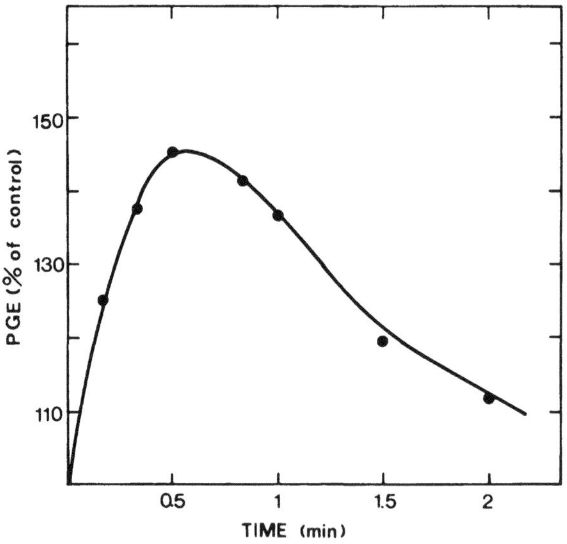

Figure 12: The kinetic profile of the effects of oxotremorine on the level of synaptosomal PGE.

Synaptosomes were incubated in the presence and absence of oxotremorine and their PGE content was assyed by the [^{3}H]-PGE RIA at the designated times following K^+ (125 mM) depolarization in the presence of Ca^{2+} (2 mM), as in Fig. 11. Oxotremorine (10 µM) was added 2 min prior to stimulation. 100% corresponds to 33.8 ± 6.9 pmol/mg protein.

In summary, the present findings suggest that a cyclooxygenase product, similar but not identical to PGE_2, is formed transiently following mAChR activation. This PGE-like substance mediates the muscarinic regulation of ACh release and its physiological action can be specifically mimicked by exogenous PGE_2. Thus, the PGE-like substance fulfills the 3 criteria presented above for the mAChR second messenger. We now address the question of how muscarinic activation is coupled to the synthesis of this substance.

Theoretically, activation of the mAChR can affect either the oxydation of the fatty acid precursor to the PGE-like substance, or its liberation from phospholipids. The first possibility was tested by comparing the pattern of oxydation of $[^{14}C]$-arachidonate in the absence and presence of oxotremorine. Tissue slices oxydize $[^{14}C]$-arachidonate to yield $[^{14}$-$C]$-PGE_2 as the major product (22 ± 4 percent of the initial radioactivity) and very low levels of $[^{14}C]$-$PGF_{2\alpha}$ and $[^{14}C]$-PGD_2 (3.4 ± 1.5 percent and 4.5 ± 2 percent of initial radioactivity, respectively). By comparison, when synaptosomes, prepared by the usual procedure (26), are incubated with $[^{14}C]$-arachidonate, they yield low levels of $[^{14}C]$-PGE_2 (1.2 ± 0.2 percent) and of another product, designated peak II (1.8 ± 0.5 percent) which is indomethacin-resistant (24). However, this pattern is different when the synaptosomes are prepared under conditions which minimalize cyclooxygenase self-inactivation(e.g. under N_2 and in the presence of BSA). Under these conditions, $[^{14}C]$-arachidonate is substantially and exclusively converted to $[^{14}C]$-PGE_2 (4.5 ± 0.5 percent, n=3). Oxotremorine has no effect on the pattern of oxydation of $[^{14}C]$-arachidonate by any of these preparations, under both non-stimulating and stimulating conditions (Table I).

It is not known whether arachidonate or another polyunsaturated fatty acid, with which the electric organ is endowed (28), is the precursor of the PGE-like substance. The above finding, that oxotremorine has no effect on the presynaptic cyclooxygenase activity (which was assayed with $[^{14}C]$-arachidonate as substrate) suggests that the oxotremorine-induced increase in the synthesis of the PGE-like substance is due to enhanced lipolysis and not to stimulation of cyclooxygenase. This contention is supported by the observation that mepacrine, a phospholipase A_2 inhibitor, reduces the oxotremorine-induced increase in the level of the PGE-like substance as measured

Table I: Effect of mAChR activation on the oxydation of [^{14}-C]-arachidonate

Preparation	Oxydation product	Control in modified TB (% of initial radioactivity)	Effect of oxotremorine (10 μM) (% of control)	
			Modified TB	K^{+} (125 mM) modified TB + Ca^{2+} (1 mM)
"Fresh" slices	$PGF_{2\alpha}$ -	3.4 ± 1.5	105 ± 3	124 ± 18
	PGE_2 -	22.0 ± 4	109 ± 5	122 ± 23
	PGD_2 -	4.5 ± 2	112 ± 6	128 ± 9
Synaptosomes	PGE_2	1.2 ± 0.2	N.T.	104 ± 4
	peak II	1.8 ± 0.4		102 ± 5
Synaptosomes prepared in BSA	PGE_2	4.5 ± 0.5	N.T.	110 ± 7

The synthesis of [^{14}C]-PG's by tissue slices and by purified synaptosomes was determined by incubation with [^{14}C]-arachidonate followed by extraction and separation of the products by thin layer chromatography.

Tissue slices (1-1.5 gr) or synaptosomes (3.5 gr) were preincubated for 15 min in modified TB. They were then either transferred (tissue slices) or diluted 2-fold (synaptosomes) into a fresh medium of the designated composition which contained [^{14}C]-arachidonate (0.5 - 1.x10^5 cpm/ml) and incubated for 30 min. The medium was acidified to pH 3.5, extracted with chloroform and chromatographed on Silica Gel 60 plates using the AIX solvent system of Hamberg and Samuelsson (27).

Synaptosomes were purified as previously described (26). When so indicated the tissue was homogenized under N_2 and in an homogenization buffer (0.8 M glycine, 1 mM EGTA, pH 6.8) which contained 1% fatty acid-free BSA. Oxotremorine was added 15 min (slices) or 5 min (synaptosomes) prior to the addition of [^{14}C]-arachidonate. Results are the mean ± S.D. of 3 experiments.

N.T. = not tested.

by the [^{3}H]-PGE RIA (70 percent inhibition at 100 μM mepacrine) and by the following direct measurement of the effects of muscarinic ligands on the liberation of free fatty acids.

As shown in Table II, electric organ slices release free fatty acids into the medium, the most abundant of which are C16, C18 and C22:6. This basal release is Ca^{2+}-independent. Oxotremorine induces a specific, Ca^{2+}-dependent atropine-sensitive increase in the release of arachidonate (168 ± 30 percent

Table II: The effects of muscarinic ligands on the release of free fatty acids from tissue slices

Conditions:	Modified TB			Modified TB + 1 mM Ca^{2+}		
Fatty acid	Control (nmol/ gr·hr)	+ oxo (% of control)	+ oxo, + atro, (% of control)	Control (nmol/ gr·hr)	+ oxo (% of control)	+ oxo, + atro (% of control)
16	6.4 ±0.2	100±10	N.T.	6.9 ±2.45	103±44	110±42
18	3.7 ±0.8	130±42	N.T.	5.7 ±2.0	100±23	130±70
18:1	1.7 ±1.5	95±15	N.T.	2.5 ±1.1	126±54	122±45
18:2	0.25±0.1		N.T.	0.3 ±0.15	111±15	95±10
20:4	1.2 ±0.7	102±74	N.T.	1.75±1.0	168±30*	63±65
22:6	5.8 ±2.2	92±39	N.T.	6.0 ±1.3	113±20	85±29

Tissue slices (1 gr) were washed at 25° for 30 min in modified TB after which they were transferred to a fresh medium (modified TB or modified TB + 1 mM Ca^{2+}), in the absence or presence of oxotremorine (10 μM) and atropine (2 μM), in which they were incubated for an aditional 60 min period . The slices were then transferred to a fresh medium of identical composition and the released free fatty acids were extracted 90 min later and analyzed by gas chromatography as previously described (29). An internal standard (C 15:0), which was added to the medium prior to extraction, was used to calculate the amounts of free acids released. Indomethacin (100 μM) was present throughout the incubations in order to block metabolism of polyunsaturated fatty acids. Results are the mean ± S.D. of 3 experiments.

* $p < 0.05$
N.T. = not tested.

of control) (Table II). It should be noted that due to the sensitivity of the experimental procedure, prolonged incubations (> 1.5 hrs) were required for detection of the released fatty acids. Under these conditions ("aged" slices) oxotremorine induces a parallel atropine-sensitive increase in the level of free arachidonate (Table II) and in the syntheis of endogenous PGE_2 (29). This is in contrast with the results obtained with freshly-excised tissue, in which oxotremorine inhibits PGE_2 synthesis (Fig. 10). The reason for this discrepancy is not known. Interestingly, during the "aging", the ability of oxotremorine to inhibit ACh release is lost in a time-and Ca^{2+}-dependent manner (Fig. 13), so that prolonged incubations of the slices results in the uncoupling of receptor activation and the

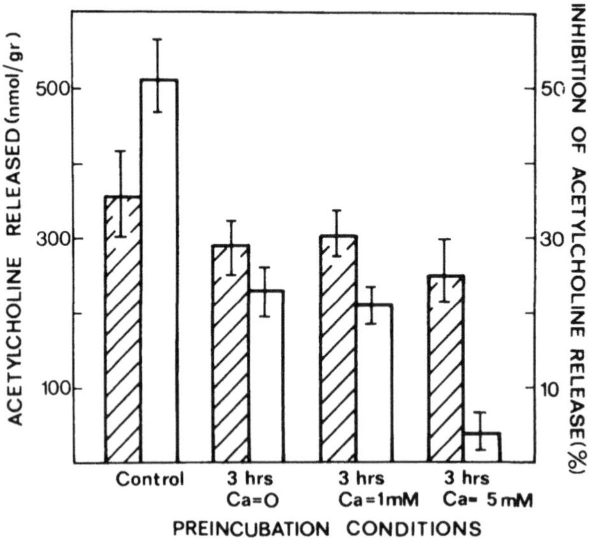

Figure 13: The effect of preincubation on ACh release from tissue slices and on the inhibition of release by oxotremorine.

Tissue slices were preincubated at 25° in the indicated buffers for the designated periods, after which they were transferred to fresh modified TB, which contained Ca^{2+} (5 mM) and incubated for an additional 30 min period. ACh release was then induced by K^+ (125 mM) and Ca^{2+} (5 mM) in the presence and absence of oxotremorine (10 μM) and measured as in Fig. 1. The amount of ACh released is presented by hatched bars, whereas the inhibition of released by oxotremorine is denoted by clear bars. Control represents the results for freshly excised slices, as in Fig. 1. Results are the mean ± S.D. of 3 experiments.

physiological effect. Thus, although the results obtained with the "aged" preparation demonstrate that receptor activation enhances the liberation of arachidonic acid, the fatty acid precursor of the PGE-like substance which is liberated following muscarinic activation in freshly-excised and coupled preparations, is probably not arachidonate. Accordingly, the reversal in polarity of the effect of oxotremorine on PGE_2 synthesis by freshly excised and "aged" tissue slices may be explained as follows: In the former, the tissue phospholipids are enriched in the precursor of the PGE-like substance which, upon liberation, competes with arachidonate on cyclooxygenase activity and thereby decreases the synthesis of PGE_2; in the uncoupled preparation the phospholipids are depleted of this fatty acid precursor due to "aging" and oxotremorine induces the liberation of arachidonate which is converted to PGE_2.

The fatty acid precursor of the PGE-like substance has not yet been identified. Nevertheless, the electric organ nerve terminals contain various polyunsaturated fatty acids (28) which can serve as precursors. Evidence for the formation and biological activity of oxydation products of $C_{20:5}$, $C_{22:4}$ and $C_{22:6}$ in other preparations was recently reported (30-32).

Future experiments with increased sensitivity, in which the muscarinic effects on lipolysis, PG synthesis and ACh release from freshly-excised preparations will be measured concomitantly, will undoubtedly resolve this problem.

Conclusions

Our results suggest that activation of the Torpedo presynaptic mAChR induces lipolysis and the subsequent oxydation of the released fatty acid to a PGE-like substance. The latter mediates the muscarinic inhibition of ACh-release by interfering with the stimulus-secretion coupling within the nerve terminals.

The transient nature of the muscarinic inhibition of ACh release stems both from the existence of an initial step in the stimulus-secretion coupling which is sensitive to oxotremorine, and from the inactivation of the receptor by rapid removal of its second messenger.

The mechanism by which receptor activation enhances lipolysis and the identity of the enzymes and substrates involved are not yet known. Binding studies have shown that the Torpedo presynaptic mAChR is coupled to a GTP-regulatory protein (4-6). Therefore it is tempting to suggest that, like in the adenylate cyclase system, this G-protein participates in the coupling between the mAChR and the lipolytic enzymes.

Acknowledgements

This work was supported in part by grants from the Dysautonomia Foundation and from the U.S.-Israel Binational Science Foundation (Grant No. 2410).

References

1. Michaelson, D.M., S. Avissar, Y. Kloog and M. Sokolovsky. 1979. Mechanism of acetylcholine release: Possible involvement of presynaptic muscarinic receptors in the regulation of acetylcholine release and protein phosphorylation. Proc. Natl. Acad. Sci. USA 76, 6336-6340.

2. Michaelson, D.M., S. Avissar, I. Ophir, I. Pinchasi, I. Angel, Y. Kloog and M. Sokolovsky. 1980. On the regulation of acetylcholine release: A study utilizing Torpedo synaptosomes and synaptic vesicles. J. Physiol. (Paris) 76, 505-514.

3. Dunant Y. and A.J. Walker. 1982. Cholinergic inhibition of acetylcholine release in the electric organ of Torpedo. Eur. J. Pharmac. 78, 201-212.

4. Dowdall, M.G., P.R. Golds and G.P. Strange. 1982. In: Presynaptic Receptors: Progress and Physiological Significance, (J. de Belleroche, ed.), Ellis Horwood, Chichester, pp. 103-113.

5. Dowdall, M.G., P.R. Golds and G.P. Strange. 1982. Properties of Torpedo electric organ muscarinic receptors. J. Physiol. (Paris) 78, 379-384.

6. Sokolovsky, M., S. Avissar, Y. Egozi, D. Gurwitz, Y. Kloog and D.M. Michaelson. 1980. In: Neurotransmitters and their Receptors, (U.Z. Littauer, Y. Dudai, I. Silman, V.I. Teichberg and Z. Vogel, eds.). John Wiley, New York. pp. 257-260.

7. Pinchasi, I. and D.M. Michaelson. 1982. Adenylate cyclase of Torpedo synaptosomes is inhibited by calcium and not affected by muscarinic ligands. J. Neurochem. 38, 1223-1229.

8. Michell, R.H.. 1982. Phosphatidylinositol breakdown in signal transduction. Neurosci. Res. Prog. Bull. 20, 338-350.

9. Berridge, M.J.. 1981. Phosphatidylinositol hydrolysis: A multifunctional transducing mechanism. Molecular and Cellular Endocrinology 24, 115-140.

10. Banschbach, M.W. and M. Hokin-Neaverson. 1980. Acetylcholine promotes the synthesis of prostaglandin E in mouse pancreas. FEBS Lett. 117, 131-133.

11. Baudin, H., N. Galand and J.M. Boeynaems. 1981. In vitro stimulation of prostaglandin synthesis in the rat pancreas by carbamylcholine, caerulein and secretin. Prostaglandins 22, 35-51.

12. Borda, E., M. del-Agostini, H. Peredo, M.F. Gimeno and A.L. Gimeno. 1983. Contractile activity of the rat vas deferens and release of prostaglandin E and F-like substances. Influence of acetylcholine and inhibitors of cyclooxygenase and phospholipase A_2. Arch. Int. Pharmacodyn. Ther. 263, 245-253.

13. Junstad, M. and A. Wennmalm.. 1974. Release of prostaglandins from the rabbit isolated heart following vagal nerve stimulation or acetylcholine infusion. Br. J. Pharmacol. 52, 375-379.

14. Marshall, P.J., J.F. Dixon and L.E. Hokin. 1980. Evidence for a role in stimulus-secretion coupling of prostaglandins derived from release of arachidonoyl residues as a result of phosphatidylinositol breakdown. Proc. Natl. Acad. Sci. USA 77, 3292-3296.

15. Trevisani, A., C. Biondi, O. Belluzi, P.G. Borasio, A. Capuzzo, M.E. Ferrati and V. Perri. 1982. Evidence for increased release of prostaglandins and the E-type in response to orthodromic stimulation in the guinea pig superior cervical ganglion. Brain Res. 236, 375-381.

16. Wolfe, L.S.. 1982. Eicosanoids: Prostaglandins, thromboxanes, leukotriens and other derivatives of Carbon-20 unsaturated fatty acids. J. Neurochem. 38, 1-14.

17. Bleasdale, J.E., J.N. Hawthorne, L. Widlund and E. Heilborn. 1976. Phospholipid turnover in Torpedo marmorata electric organ during discharge in vivo. Biochem J. 158, 557-565.

18. Israel, M. and B. Lesbats. 1981. Chemiluminescent determination of acetylcholine and continuous detection of its release from Torpedo electric organ synapses and synaptosomes. Neurochem. Inter. 3, 81-90.

19. Kloog, Y., D.M. Michaelson and M. Sokolovsky. 1978. Identification of muscarinic receptors in Torpedo electric organ. FEBS Lett. 95, 331-334.

20. Hadhazy, P. and J.C. Szerb. 1977. The effect of cholinergic drugs on $[^3H]$-acetylcholine release from slices of rat hippocampus, striatum and cortex. Brain Res. 123, 311-322.

21. Szerb, J.C., P. Hadhazy and J.D. Dudar. 1977. Release of $[^3H]$-acetylcholine from rat hippocampal slices: Effect of septal lesion and of graded concentrations of muscarinic agonists and antagonists. Brain Res. 128, 285-291.

22. Szerb, J.C.. 1980. Effect of low calcium and of oxotremorine on the kinetics of the evoked release of $[^3H]$-acetylcholine from the guinea pig myenteric plexus. Comparison with morphine. Naunyn Schmiedeberg's Arch. Pharmacol. 311, 119-127.

23. Pinchasi, I., M. Burstein and D.M. Michaelson. 1984. Circa-annual variations in the regulation of acetylcholine release by presynaptic muscarinic receptors. In preparation.

24. Pinchasi, I., B. Shanietzki, M. Schwartzman, A. Raz and D.M. Michaelson. 1982. In: Presynaptic Receptors: Progress and Physiological Significance, (J. de Belleroche, ed.), Ellis Horwood, Chichester. pp. 114-129.

25. Eckenfels, A. and J.R. Vane. 1972. Prostaglandins, oxygen tension and smooth muscle tone. Br. J. Pharmacol. 45, 451-462.

26. Michaelson. D.M. and M. Sokolovsky. 1978. Induced acetylcholine release from active purely cholinergic Torpedo synaptosomes. J. Neurochem. 30, 217-230.

27. Hamberg, M. and G. Samuelsson. 1966. Prostaglandins in human sminal plasma: Prostaglandins and related factors. J. Biol. Chem. 241, 257-263.

28. Deutsch, J.W. and R.B. Kelly. 1981. Lipids of synaptic vesicles: Relevance to the mechanism of membrane fusion. Biochemistry 20, 278-285.

29. Pinchasi, I., M. Burstein and D.M. Michaelson. 1984. Metabolism of arachidonic acid and prostaglandins in the Torpedo electric organ: Modulation by presynaptic muscarinic acetylcholine receptor. Neuroscience, in press.

30. Aveldano, M.I. and H. Sprecher. 1983. Synthesis of hydroxy fatty acids from 4,7,10,13,16,19-[1-^{14}C] docosahexaenoic acid by human platelets. J. Biol. Chem. 258, 9339-9343.

31. Sprecher, H., M. Van Rollins, F. Sun, A. Wyche and P. Needleman. 1982. Dihomo-prostaglandins and thromboxane. J. Biol. Chem. 257, 3912-3918.

32. Whitaker, M.O., A. Wyche, F. Fitzpatrick, H. Sprecher and P. Needleman. 1979. Triene prostaglandins: Prostaglandin D_3 and icosapentaenoic acid as potential antithrombotic substances. Proc. Natl. Acad. Sci. USA 76, 5919-5923.

MOLECULAR MECHANISMS UNDERLYING ACETYLCHOLINE RELEASE.

M. Israël, N. Morel, S. Birman, B. Lesbats and R. Manaranche.
Département de Neurochimie, Laboratoire de Neurobiologie Cellulaire, Centre National de la Recherche Scientifique, 91190 Gif sur Yvette, France.

Introduction

The isolation of synaptic vesicles rich in acetylcholine (ACh) (about 70 % of the total transmitter) from the electric organ of Torpedo (1) has strengthen the hypothesis that exocytosis of vesicular contents was the mechanism for the quantal release of the transmitter. But soon after it was found that the non vesicular (cytoplasmic) transmitter was released upon stimulation and also renewed by two cytoplasmic enzymes choline acetylase and a non mitochondrial acetyl CoA synthetase (3). The precursors choline and acetate in the case of neuromuscular synapses (4) (5) are recycled from the synaptic cleft where released ACh is hydrolysed by acetylcholinesterase. The vesicular ACh and the number of vesicles remained stable during physiological stimulations, and no exchange was detected with the cytoplasmic ACh for short stimulations. In addition ACh variations in the cytoplasmic pool were correlated to the kinetic of release as measured by the electroplaque discharge (see ref.6 for review) this was shown even for a few stimuli (7).

The mechanism releasing ACh from the cytoplasm in a packet form was searched at the presynaptic membrane itself. When synaptosomes were isolated from the electric organ of Torpedo (8,9) it became possible to rapidly freeze them at the peak of ACh release and study their membrane contents after cryofracture. A statistical analysis derived from about 5000

Molecular Basis of Nerve Activity

presynaptic micrographs showed that the main structural change was the occurrence of new and large intramembrane particles at the peak of ACh release and in all release conditions. This impressive change contrasted with the stability in the number of vesicles. The few endo-exocytotic pits found were not correlated to the release of ACh (10,11). A similar finding was recently reported by Y. Dunant on stimulated electroplaques (12,13).

Therefore the mechanisms of ACh release may depend of membrane proteins (perhaps those of intramembrane particles) activated by calcium. In most recent works we showed that lyophilized presynaptic membranes could be used to make proteoliposomes filled with ACh, and able to release ACh upon calcium action. A similar finding was reported by Meyer and Cooper (16). Since it is now possible to measure rapidly and continuously many ACh samples with the chemiluminescent assay (17,18), we have undertaken the purification of this membrane component by testing its ability to translocate ACh upon calcium action when inserted in liposomal membranes.

Results

Purification of the presynaptic plasma membrane. Advantage was taken from the purity of the synaptosomal fraction isolated from Torpedo electric organ (8,9) to prepare from this starting material a purified nerve terminal plasma membrane fraction (19,20). This permitted to show that most of the hydrophobic form of Torpedo electric organ acetylcholinesterase was bound to the presynaptic plasma membrane (21,22) and to demonstrate the presence of a specific protein binding the presynaptic neurotoxin of the annelid *Glycera convoluta* (24). This neurotoxin induces a large quantal ACh release at neuromuscular junctions (23). Monoclonal antibodies to membrane antigens were prepared (25) and used as probes to follow the purification of large amounts of presynaptic membranes directly from electric organ homogenates.

Preparations of presynaptic membrane proteoliposomes. The lyophilized presynaptic membrane powder was mixed with synthetic lecithin (L-α-phosphatidylcholine dipalmitoyl) in 1-butanol (lipid/protein ratio ranging from 4 to 20). After evaporation of the organic solvent, the material was resuspended in an intracellular type solution. Acetylcholine was added after inhibition of acetylcholinesterase activity.

Figure 1. Reconstituted presynaptic membrane. The proteoliposomes obtained do not contain any internal organelles. Their convex and concave faces show numerous intramembrane particles.

The suspension was sonicated at room temperature for 15 sec. The proteoliposomes were then gel filtered in an external type solution. Experimental details are given in ref.15.
The volume occluded in the proteoliposomes represents about 0.1 % of the final fraction volume. Freeze fracture analysis (Fig.1) shows that the membrane of the proteoliposmes was rich in intramembrane particles, equally distributed between the convex and concave faces. This demonstrates that presynaptic membrane proteins are really incorporated in the reconstituted membrane. The proteoliposomes do not contain any internal organelles.

Acetylcholine release from proteoliposomes. We have shown that the proteoliposomes prepared from presynaptic membranes were able to release ACh when a calcium influx was generated. The ionophore A 23187 was first incorporated to the system, the subsequent addition of calcium triggers the efflux of ACh (Fig.2a). In the absence of ionophore, no ACh release occurred upon calcium addition (Fig.2b).
We have also checked that the ACh content of the proteoliposomes decreased after ACh release; this is performed by bursting them in the reaction mixture with a detergent (Triton-X-100) added before or after triggering the release of ACh. In general, about 25 % of the proteoliposomal ACh content is released in a few minutes after addition of ionophore and calcium.
Proteoliposomes reconstituted from other plasma membranes were not able to release ACh. In addition, when presynaptic membrane proteins were replaced by serum albumin, no significant ACh release was detected.

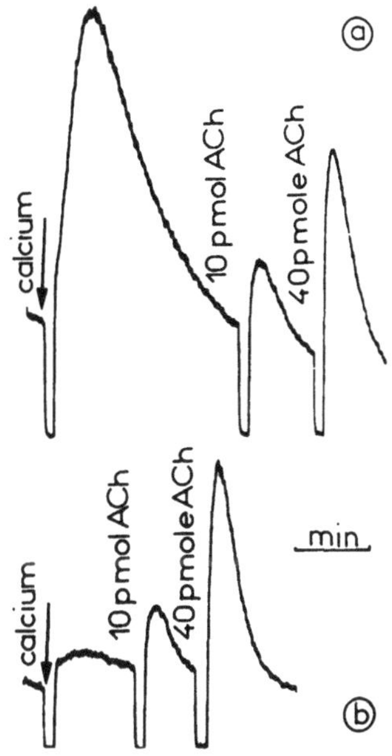

Figure 2 : Calcium influx and acetylcholine release from proteoliposomes. In a, the release of ACh is triggered by the addition of calcium to proteoliposomes which have been treated with the calcium ionophore A23187. In b, the ionophore was omitted, the calcium addition fails to induce a release of ACh comparable to the above record.

Effect of pronase treatment on acetylcholine release. When the proteoliposomes were prepared in the presence of pronase (70 % of proteins were hydrolysed during this treatment) they became unable to release acetylcholine after a calcium influx, showing that proteins are involved in this process (Fig.3).

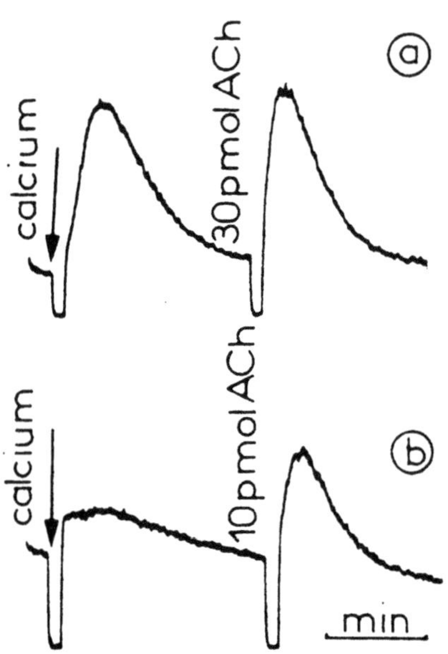

Figure 3 : Acetylcholine release from proteoliposomes prepared from presynaptic plasma membranes in the presence (b) or absence (a) of pronase. The same amount of occluded acetylcholine was present in the two conditions. The release was elicited by adding calcium after the ionophore A23187.

When pronase was added on already made proteoliposomes, the external proteolysis of liposomal proteins did not modify acetylcholine release. A similar finding was observed when intact synaptosomes were incubated with various pronase concentrations (Figure 4). This external action of pronase removed a fixed (35 %) amount of membrane proteins including most of the membrane bound acetylcholinesterase activity. In spite of this proteolysis, ACh release elicited by gramicidin was not altered after pronase treatment. This was also observed after KCl depolarization or calcium ionophore A23187 action. Therefore the proteins involved in the voltage dependent calcium entry and the ACh release have either a small ectocellular expansion or their external domain is not important for their function.

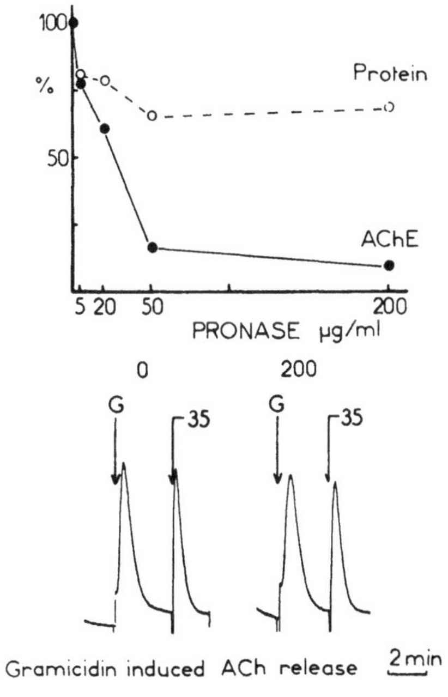

Figure 4 : Pronase treatment of intact synaptosomes. Pronase removes 35 % of the presynaptic membrane proteins (o--o and most of the membrane bound acetylcholinesterase activity (o--o). Acetylcholine release elicited by gramicidin (G) in the presence of calcium was similar in the pronase (200 µg/ml) treated sample and in the control. An ACh standard of 35 pmol. was injected to calibrate the release.

Intramembrane particle changes associated to ACh release. Since it was previously found that a category of large particles appear in the presynaptic membrane whenever ACh is released (10,11,12,13), we have tried to find out if these large particles appeared in the stimulated proteoliposomes.

The histogram of particle diameters in the convex face of controls and stimulated proteoliposomes (A23187 + Calcium) shows a significant increase after stimulation of the number of the 10 to 13 nm particles (Figure 5). The mean particle density is indicated above each histogram. The opposite concave face was not modified.

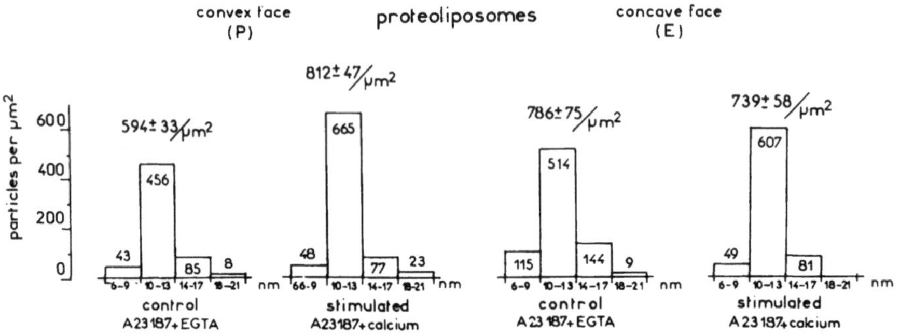

Figure 5 : Histogram of intramembrane particles diameters in controls and stimulated proteoliposomes.

Solubilization of membrane proteins which permit the calcium dependent ACh release. Presynaptic membrane proteins were solubilized by sodium cholate (1 to 3 %) and submitted to gel filtration (Figure 6).

We recovered a fraction just behind the excluded volume which, after removal of the cholate by dialysis and insertion into proteoliposomes, gave to the artificial membrane the calcium dependent ACh release capability. This fraction is taken between the two molecular weight standards thyroglobulin (670 000 dalt) and catalase (230 000 dalt.). After sodium dodecylsulfate gel electrophoresis, the protein pattern of the active fraction was much simpler than that of the whole presynaptic membrane but still too complex to allow the identification of protein chains involved.

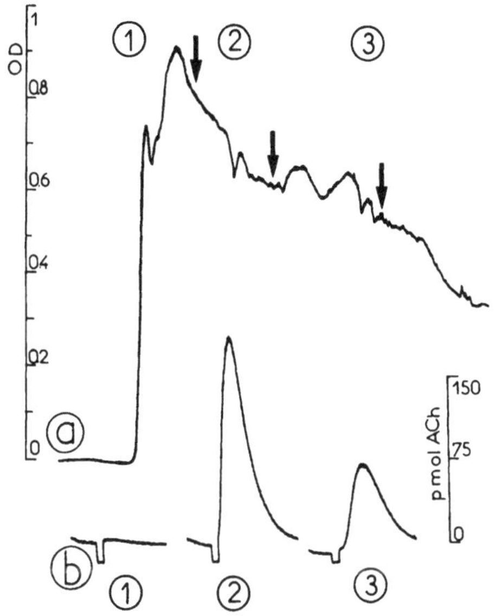

Figure 6 : Separation of a protein fraction which carries the ACh release capabilities. A cholate extract of lyophilized synaptosomal membranes was separated on ultrogel ACA 22. The record gives the optical density of the eluate at 280 nm (a). Three fractions were collected, 1,2 and 3 and incorporated to the liposomes. Fraction 2 proteoliposomes had the greatest ACh release capability as shown by the ACh release curves (b).

Discussion.

Presently most authors accept the view that there is, in the nerve terminal, a genuine cytolosic free ACh compartment, 20 to 50 % of the total ACh (6,26). Free ACh decreases and is renewed in the course of stimulation of electric organ slices (6). Starting with the whole tissue it was shown that the characteristic decay curve of the electrical discharge of a stimulated electric organ slice was associated to a characteristic variation of the free ACh compartment (6). The correlation was followed down to a few stimuli (7). The release of free ACh from isolated cholinergic synaptosomes was more recently obtained (17). Furthermore, synaptosomal sacs filled with ACh showed a calcium dependent ACh release capability (27).

Since intact tissues, synaptosomes, synaptosomal sacs were able to release their cytoplasmic or free ACh content upton calcium entry, it appeared possible to go one step ahead and try to reconstitute proteoliposomes which would have incorporated the presynaptic membrane elements that might ensure the calcium-dependent ACh permeability (14,15,16). The present report shows that it is possible to obtain, after solubilization of the presynaptic plasma membrane, a protein fraction ensuring, after insertion in an artificial membrane, the calcium dependent release of ACh.

The appearance of large intramembrane particles has been correlated with the release of ACh in intact tissue (12,13) and in synaptosomes (10,11). A similar phenomenon was also observed with proteoliposomes reconstituted from a lyophilized presynaptic membrane. Since these proteoliposomes did not contain any synaptic vesicles, the appearance of large intramembrane particles cannot result from the fusion of synaptic vesicles by exocytosis during ACh release. These particles are probably the image of a molecular event occurring in the presynaptic membrane during ACh release. We propose that, under the effect of intracellular calcium, specific proteins of the presynaptic membrane cluster to form astructure (the large intramembrane particles) which ensures the translocation of ACh. The degree of aggregation would be correlated to the quantum size leading to the observation of miniature or subminiature end-plate potentials (28).

The mechanism we propose does not agree with the view that the quantum results from the exocytosis of vesicular ACh. The endo-exocytotic cycle of synaptic vesicles has been visualized in several experimental conditions, particularly by Heuser and coworkers (29,30) at stimulated frog neuromuscular junctions.

It should however be recalled that the exocytotic pits were obtained only after 4-aminopyridine which greatly increase ACh release and that they were slightly delayed from the peak of ACh release.

Our opinion is that other roles of synaptic vesicles are revealed here. First, it was shown that synaptic vesicles are able to concentrate calcium. This calcium uptake requires the presence of ATP and occurs at calcium concentrations likely to be found in stimulated nerve terminals (31,32). Synaptic vesicles, close to the presynaptic membrane, might sequester the cytoplasmic calcium and switch off the release of ACh from the cytoplasm. The vesicular calcium would be extruded latter by exocytosis. A second role can be envisated. During intense stimulations, cytoplasmic ACh and ATP become exhausted. The low level of ATP is no longer sufficient to support neither ACh synthesis nor the synaptic vesicle calcium uptake. In addition, an acidification of the cytoplasm is likely to occur as a result of an intense glycolysis. In such conditions (increase of cytoplasmic protons and calcium) the vesicular ACh and ATP leak in the cytoplasm, restoring ACh reelase, while ATP will serve for ACh synthesis and for calcium sequestration. Synaptic vesicles might be more than a simple ACh reserve and could play a major role in the regulation of the release process by controlling the cytoplasmic calcium and ATP concentrations.

Acknowledgement

We are grateful to Dr. T. Gulik-Krzywicki and J.C. Dedieu for performing the freeze fractures and to S. Lazereg for technical assistance.

References

1. Israël, M., J. Gautron, B. Lesbats. 1968. C.R.Acad.Sci. Paris 266, 273-275.

2. Israël, M., J. Gautron, B. Lesbats. 1970. J.Neurochem. 17, 1441-1450.

3. Morot-Gaudry, Y., M.F. Diebler. 1979. Biochem.J. 180, 297-301.

4. Israël, M., S. Tucek. 1974. J.Neurochem. 22, 487-491.

5. Dreyfus, P. 1975. C.R.Acad.Sci.Paris 280, 1893-1894.

6. Israël, M., Y. Dunant, R. Manaranche. 1979. Progress in Neurobiology 13, 237-275.

7. Dunant, Y., G.J. Jones, F. Loctin. 1982. J.Physiol.Lond. 325, 441-460.

8. Israël, M., R. Manaranche, P. Mastour-Frachon, N. Morel. 1976. Biochem.J. 160, 113-115.

9. Morel, N., M. Israël, R. Manaranche, P. Mastour-Frachon. 1977. J. Cell BioL. 75, 43-55.

10. Israël, M., R. Manaranche, N. Morel, J.C. Dedieu, T. Gulik-Krzywicki, B. Lesbats. 1981. J.Ultrastructure Res. 75, 162-178.

11. Israël, M., B. Lesbats, R. Manaranche, N. Morel, T. Gulik-Krzywicki, J.C. Dedieu. 1982. J. Physiol. Paris 78, 348-356.

12. Dunant, Y., D. Muller, A. Parducz, G.J. Jones, L.M. Garcia-Segura. 1984. C.R.Acad.Sci.Paris, in the press.

13. Garcia-Segura, L.M., D. Muller, G.J. Jones, Y. Dunant. J. Cell Biol.(submitted).

14. Israël, M., B. Lesbats, R. Manaranche, N. Morel. 1983. Biochim.Biophys.Acta 728, 438-448.

15. Israël, M., B. Lesbats, N. Morel, R. Manaranche, T. Gulik-Krzywicki, J.C. Dedieu. 1984. Proc.Natl.Acad.Sci.USA 81, 277-281.

16. Meyer, E.M., J.R. Cooper. 1983. Neuroscience 3, 987-994.

17. Israël, M., B. Lesbats. 1981. J. Neurochem. 37, 1475-1483.

18. Israël, M., B. Lesbats. 1982. J. Neurochem. 39, 248-250.

19. Stadler, H., T. Tashiro. 1979. Eur.J.Biochem. 101, 171-178.

20. Morel, N., R. Manaranche, M. Israël, T. Gulik-Krzywicki. 1982. J. Cell Biol. 93, 349-356.

21. Morel N., P. Dreyfus. 1982. Neurochem.Internatioal 4, 283-288.

22. Li, Z.Y., C. Bon. 1983. J. Neurochem. 40, 338-349.

23. Manaranche, R., M. Thieffry, M. Israël. 1980. J. Cell Biol. 85, 446-458.

24. Morel, N., M. Thieffry, R. Manaranche. 1983. J. Cell Biol. 97, 1737-1744.

25. Morel, N., S. Birman, S. Rouire, J.C. Mazié, M. Israël. 1984. Neurochem. International (in press).

26. Weiler, M., I.S. Roed, V.P. Whittaker. 1982. J. Neurochem. 38, 1187-1191.

27. Israël, M., B. Lesbats, R. Manaranche. 1981. Nature 294, 474-475.

28. Kriebel, M.E., F. Llados, D.R. Matteson. 1982. J. Physiol. Lond. 322, 211-222.

29. Heuser, J.E., Reese, T.S., M.J. Dennis, Y. Jan, L. Jan, L. Evans. 1979. J. Cell Biol. 81, 275-300.

30. Miller, J.M., J.E. Heuser. 1984. J. Cell Biol. 98, 685-698.

31. Israël M., R. Manaranche, J. Marsal., F.M. Meunier, N. Morel, P. Frachon, B. Lesbats. 1980. J. Membrane Biol. 54, 115-126.

32. Michaelson D.M., I. Ophir, I. Angel. 1980. J. Neurochem. 35, 116-124.

ADENOSINE-5'-TRIPHOSPHATE AT THE CHOLINERGIC SYNAPSE: A CO-TRANSMITTER?

H. Zimmermann, E.J.M. Grondal

AK Neurochemie, Zoologisches Institut der Universität Frankfurt, Siesmayerstr. 70, D-6000 Frankfurt, Federal Republic of Germany

Introduction

ATP is stored in storage granules of a variety of non-neuronal and neuronal cell types. This includes the serotonin containing granules of blood platelets (2), peptide hormone containing granules from the posterior pituitary (22), chromaffin granules (ref. in 30), the large and small granules from adrenergic sympathetic nerves (ref. in 11, 13) as well as the cholinergic synaptic vesicles isolated from the Torpedo electric organ (3) and the electric organs of Electrophorus electricus (34) and Malapterurus electricus (28, this volume). In all these cases ATP is co-stored with physiologically active substances which are released on cell activation by exocytosis.

Also fractions of synaptic vesicles purified from brain contain a considerable amount of ATP (21). In this case, however, it is not clear, whether the nucleotide is contained in a storage particle of its own or together with a neurotransmitter substance.

Although noted a considerable time ago, the co-storage of ATP in synaptic vesicles has received rather little attention. A major reason for this certainly is the fact that no significant physiological role could be assigned to ATP and that its metabolism at the synapse was obscure. A role of ATP in the vesicular storage of any of the amines mentioned has not

Molecular Basis of Nerve Activity

yet been established. It is thus likely that the ATP exerts its physiological function after co-release together with the neurotransmitter in the synaptic cleft.

Criteria for a Co-transmitter

If ATP was a co-transmitter the following criteria should be fulfilled:

- It should be stored together with the classical neurotransmitter in the storage granule or synaptic vesicle.
- It should be released from the nerve terminal by the same criteria as the neurotransmitter. The two substances should be released in the same molar ratio as they are contained inside the vesicles.
- There should be a function in signal transmission also by the co-transmitter ATP. Possible effects would be activation or inactivation of the postsynaptic cell by a specific receptor or in general modulation of the effects of the classical neurotransmitter.
- There should be systems for terminating the action of ATP.
- In analogy to the neurotransmitter there should be systems for reuptake into the nerve terminal, possibly resynthesis and also reloading of the synaptic vesicles for renewed release.

In the following we shall briefly review the evidence for the various aspects quoted above mainly on the basis of results which have been collected at the cholinergic electromotor synapse in the Torpedo electric organ.

Nucleotide Storage in Cholinergic Synaptic Vesicles

First evidence for co-storage of acetylcholine (ACh) and ATP was obtained by co-sedimentation when cholinergic vesicles were isolated from the eclectric organ of the electric ray Torpedo marmorata (3). Molar ratios of ACh/ATP ranging between 2.5 and 7 have been reported for the various isolation procedures and species of electric rays. One synaptic vesicle may contain as many as 40 000 molecules of ATP corresponding to an intravesicular molarity of 0.1 M. The presence of ATP in the same storage granule as ACh has been verified by a variety of methods. These include co-sedimentation also in gradient media containing membrane permeable substances like glycerol and parallel elution of the two substances from columns of porous glass beads onto which fractions of synaptic vesicles had been loaded. Furthermore turnover of both vesicular ACh and ATP occurs at identical rates (ref. in 33 and below).

Analysis of vesicular nucleotides has revealed the presence not only of ATP but also of smaller amounts of ADP and AMP and particulary of GTP. The vesicular amount in GTP is 20% of that of ATP (ref. in 33). The contents in nucleotides of cholinergic synaptic vesicles are likely to reflect the specificity of the vesicular nucleotide carrier (15).

Release of ATP from Cholinergic Nerve Terminals

There have been reports for the release of ATP from tissues which carry cholinergic innervation like muscle (26), superfused cerebral tissues (33) and Torpedo electric organ (17). Furthermore release of ATP which carries the same stimulus dependencies as that of ACh has been reported for synaptosomes isolated from the Torpedo electric organ (19, 25, 27). Negative reports on the release of ATP from peripheral cholinergic sources like the perfused sympathetic ganglion (9) or synaptosomal pellets from Torpedo electric organ (18) may need reinterpretation in the light of results quoted below. There is, however, no convincing evidence to date for a release of ATP from brain synaptosomes which would follow exactly the same stimulus characteristics as that of ACh (ref. in 29).

On co-release of ACh and ATP the two substances ideally should be released in the same molar ratio as they are contained inside synaptic vesicles. There are, however, technical difficulties involved in these measurements (further details in 32). In general the measurement of molar ratios of substances released from nerve endings is difficult to determine, due to diffusion, reuptake or extracellular breakdown of substances. Furthermore those vesicles which release the substances by exocytosis may not possess them in the same molar ratio as vesicular average - the only parameter which can be determined experimentally. For Torpedo synaptosomes molar ratios of released ACh and ATP corresponding to those of synaptic vesicles have been reported by Schweitzer and Kelly (25). In initial experiments of this kind (19) molar ratios varying between 10 and 45 were reported depending on experimental conditions.

Metabolic Fate of Released ATP

It is unlikely that ATP released from the nerve terminals into the interstitial medium (synaptic cleft) can be removed directly by cellular reuptake. We infer from our results that there are nucleotidases situated at the surface of the nerve terminal which can hydrolize extracellular ATP to adenosine.

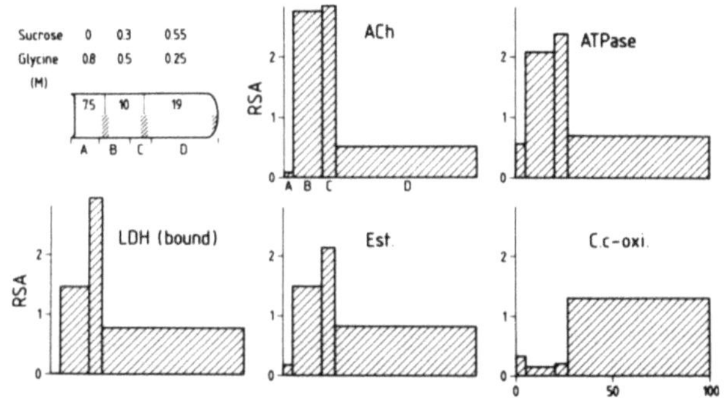

Fig. 1 Analysis of density gradient fractions derived from the Torpedo electric organ for various marker substances. LDH, lactate dehydrogenase; EST, cholinesterases; C.c-oxi, cytochrome-c-oxidase. ATPase activity was determined under isotonic conditions in the presence of ouabain (1 mM) and Ca-2+ (4 mM) and Mg-2+ (4 mM) (from 10).

Analysis of density gradient fractions derived from homogenates of the Torpedo electric organ (Fig. 1) reveals an enrichment of ATPase activity in the synaptosome fraction C which parallels that of the nerve terminal marker ACh and acetylcholinesterase. ATPase activity was measured on intact particles. In fraction C 96% of activity of lactate dehydrogenase (LDH) was in a bound form, i.e. existed in particles sealed by a plasma membrane. The activity could be released by addition of Triton X-100. The mitochondrial marker cytochrome-c-oxidase is distributed differently and mainly contained in the bottom fraction of the gradient. The interpretation that the intact synaptosomes are enriched in gradient fraction C is further substantiated by electron microscopical analysis. The intact particles contain besides synaptic vesicles numerous glycogen granules which depict their origin from the nerve terminal as well as the right (physiological) orientation of the plasma membrane.

The catalytic site of the synaptosomal ATPase appears to the orientated extracellularly. The following experiments support this view. If an unspecific protease (pronase, 1 mg/ml) is

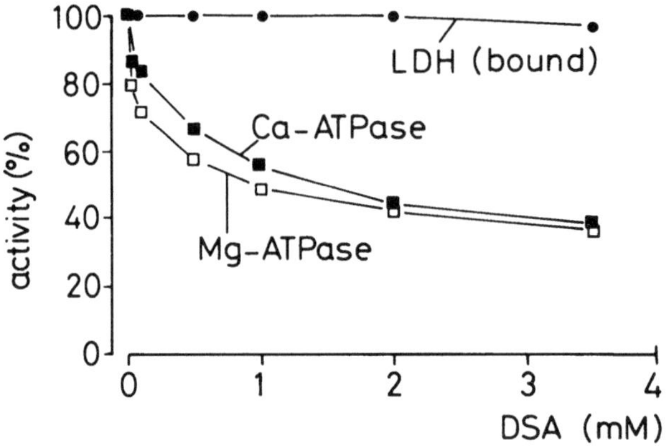

Fig. 2 Effect of incubation of intact synaptosomes (fraction C, Fig. 1) with increasing concentrations of the diazonium salt of sulfanilic acid. ATPase activity measured in the presence of Ca-2+ (4 mM) or Mg-2+ (4 mM). Activity of bound lactate dehydrogenase is inside membrane sealed particles.

added to a suspension of synaptosomes under isotopic conditions there is a time dependent loss of ATPase activity (Ca^{2+}, Mg^{2+}-activation). Whereas after 180 min of incubation ATPase activity is decreased to 19% of control, the activity of bound LDH is still unaltered (10). On incubation of intact synaptosomes with a low molecular weight substance which causes covalent inhibition of enzyme activity (diazotized sulfanilic acid, DSA) there is a concentration dependent fall in both Ca^{2+}-activated and Mg^{2+}-activated ATPase. Again activity of bound (intracellular) LDH is unaltered (Fig. 2). If synaptosomes are lyzed, however, with 1% Triton X-100, LDH-activity is reduced to the same extent (45%) as ATPase activity.

From this we conclude that the catalytic site of the ATPase faces the extracellular medium. Interestingly this parallels the findings of two studies on the localization of synaptosomal acetylcholinesterase derived from the same tissue (14, 20). Thus, the cholinergic nerve terminals in the Torpedo electric organ possess the capacity to hydrolyze via membrane bound ecto-enzymes both extracellular ACh and ATP.

In order to obtain information on the functional aspect of the ecto-ATPase activity under in situ and in vivo conditions, enzyme activity on sealed synaptosomes was further analyzed. Hydrolysis of ATP is activated by either Ca^{2+} or Mg^{2+} alone and to a similar extent (10). The apparent Km-values for ATP (between 50 amd 100 µM) suggest that intact nerve terminals in situ are capable of effectively hydrolyzing extracellular ATP. Since both Ca^{2+} and Mg^{2+} are present in the extracellular medium at millimolar concentrations the apparent Km-values for Ca^{2+} and Mg^{2+}-activation (0.61 and 0.71 mM respectively) appear to be sufficient for full enzyme activation.

Interestingly the nerve terminals not only are capable of hydrolyzing triphosphates like GTP, UTP and ITP but also of di- and mono-phosphates. The rates of hydrolysis of a variety of substrates are given in Table 2. Generally the diphosphates and monophosphates are hydrolyzed less efficiently than the triphosphates. The hydrolysis of GTP may be of physiological relevance since synaptic vesicles contain (and presumably also exocytose) GTP. GTP in turn has been shown to have an effect on agonist binding to Torpedo muscarinic receptors (4) as it has for muscarinic receptors in other tissues (5).

Table 1
Hydrolysis of Phosphates by Intact Cholinergic Synaptosomes (nmol Pi x min^{-1} x mg $protein^{-1}$)

ATP	49.1 ± 2.2	(3)	GTP	45.3 ± 1.9	(3)	IMP	14.8	(1)
ADP	34.9	(1)	GDP	31.3	(1)	pNPP	0.17	(1)
AMP	28.9 ± 7.6	(3)	GMP	12.4	(1)	PEP	0.17	(1)
G-6-P	0.26	(1)	ß-Glyc.-P	0.6 ± 0.25	(2)			

Values are means ± standard deviation or range. Number of independent experiments in parentheses. Determination was by analysis of total phosphate released.

The capacity of the nerve terminals to hydrolyze also extracellularly applied ADP and AMP is also demonstrated by tissue fractionation studies. On subcellular fractionation of electric tissue as depicted in Fig. 1 the subcellular distribution of ADPase and AMPase essentially equals that of ATPase. Thus the nerve terminals appear to possess an entire chain of enzymatic activity for hydrolysis of extracellular tri-, di-, and monophosponucleotides. The subtrate specificity for the various types of nucleotides is not very pronounced and the ecto-ATPase activity initially observed may be classified rather as an ecto-nucleotide triphosphatase activity.

Table 2 also reveals that the synaptosome preparation does not contain significant activity towards substrates for unspecific phosphatases like ß-glycerophosphate (ß-Glyc.-P); glucose-6-phosphate (G-6-P), para-nitrophenylphosphate (pNPP) or phosphoenol pyruvate (PEP).

Production and Fate of Extracellular Adenosine

When intact synaptosomes are incubated under isotonic conditions with ^{3}H-ATP there is a time dependent formation of radiolabelled derivatives in the extracellular medium. The final product of the enzyme chain which produces ADP and AMP as intermediates is adenosine (37).

We know from earlier investigations (16, 37) that synaptosomes isolated from the Torpedo electric organ can take up adenosine via a high affinity uptake mechanism. The apparent Km-value for adenosine of 2 µM equals that for uptake of the hydrolysis product of ACh, choline, which has previously been described in a variety of cholinergic tissues. The maximal uptake capacity (Vmax = 30 pmol . min^{-1} . mg $protein^{-1}$) is in the range of that reported for choline on the same tissue and by a factor of 1000 lower than the maximal velocity of hydrolysis of ATP by the ecto-enzymes present in the preparation. Nerve terminal uptake of phosphate has not yet been studied in Torpedo synaptosomes but it has been observed in synaptosomes isolated from mammalian brain (24).

Adenosine taken up from the extracellular medium is metabolized rapidly inside the nerve terminal. This can be shown either by incubation of isolated synaptosomes or by perfusion of blocks of electric tissue with ^{3}H-adenosine. The adenosine is converted to a number of nucleotides with ATP, ADP and AMP as major metabolites (37).

Extracellular Adenosine a Source for Vesicular ATP

Since synaptic vesicles have an uptake system for ATP (15) it may be expected that nerve terminal ATP which is derived from extracellular adenosine can be used for vesicle loading. This can indeed be shown by perfusion of blocks of electric tissue with ^{3}H-adenosine and subsequent subcellular isolation of synaptic vesicles (Fig. 3a). ^{3}H-ATP distributes in parallel with endogenous vesicular ATP and ACh.

On stimulation of electric tissue vesicular ATP behaves like vesicular ACh (Fig. 3b). Newly formed vesicles are denser than vesicles which have not been recently involved in exo- and endocytosis and contain the majority of both newly synthesized ACh (^{14}C-ACh) and newly synthesized ATP (^{3}H-ATP) (31, 35).

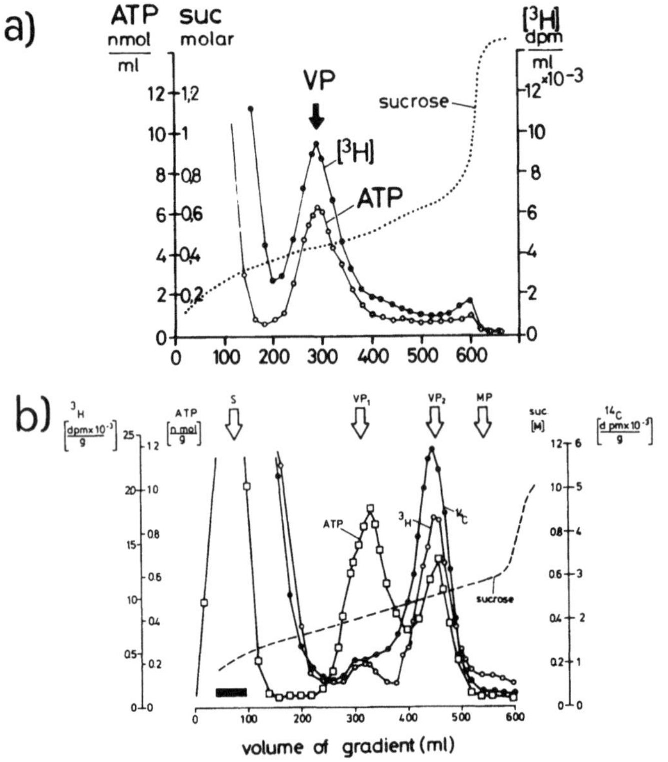

Fig. 3 Incorporation of 3-H-ATP derived from extracellular 3-H-adenosine into synaptic vesicles isolated by sucrose density gradient centrifugation. a) Co-sedimentation of 3-H-ATP and total (vesicular) ATP after previous labelling of resting tissue blocks with 3-H-adenosine. b) Co-sedimentation of newly synthesized (14-C) ACh and (3-H) ATP in recycled and denser synaptic vesicles (VP-2) after previous simultaneous labelling of tissue blocks with 14-C-choline and 3-H-adenosine and stimulation (0.1 Hz, 1080 impulses). s, soluble supernatant fractions; VP1, first vesicle peak fraction; MP, membrane peak fraction; (from 31 and 36).

Physiological Role of ATP at the Cholinergic Synapse

The observations discussed above all assign a role to ATP as a co-transmitter as far as the presynaptic aspect is concerned: There is nerve terminal synthesis, vesicular uptake and storage, most likely vesicular release, extracellular hydrolysis by ecto-enzymes, reuptake of hydrolysis products followed by nerve terminal resynthesis and vesicular reloading. These aspects are summarized in Fig. 4. But what function can be assigned to vesicular ATP? So far no role of ATP in the vesicular storage of ACh could be detected. At the presynaptic side there is, however, evidence from studies at the neuromuscular junction and also at the Torpedo electric organ that adenosine blocks the release of ACh (and possibly also of ATP?). This action is similar to that of ACh on presynaptic muscarinic receptors (ref. in 33).

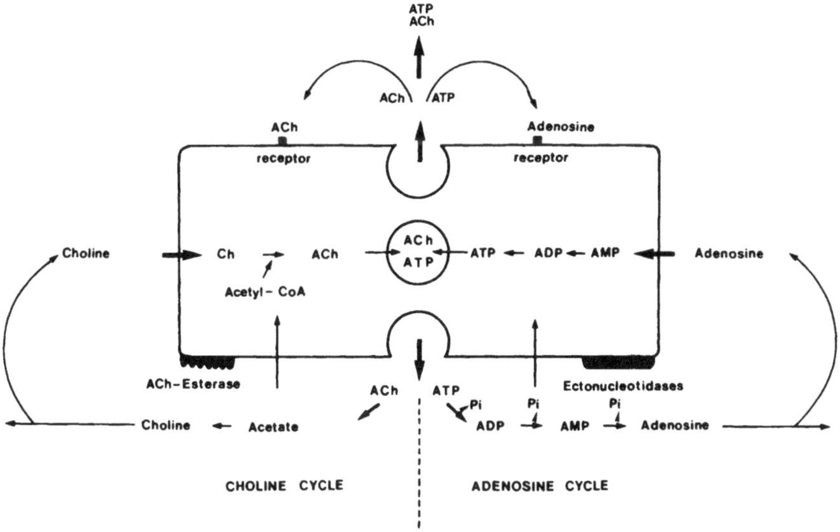

Fig. 4 Model for recycling of ACh and ATP at the cholinergic electromotor synapse: co-storage, co-release, co-hydrolysis by ectoencymes, co-uptake of choline and adenosine via respective high affinity uptake mechanisms, final resynthesis and vesicular reuptake of original compounds. Both ACh and ATP (adenosine) may act presynaptically (inhibitory) and postsynaptically (excitatory).

There is now also evidence for an action of ATP at the postsynaptic side of the neuromuscular junction. In myotubes ATP induces an increase in cation permeability similar to the action of ACh (12). Studies on skeletal muscle suggest that ATP can increase the sensitivity of the ACh-receptor (1, 6).

In addition there is first biochemical evidence for an ATP induced Rb^+-flux into muscle cells (myotubes, 8, this volume).

According to the above quoted observations it appears to be likely that ATP is a co-transmitter in the sense that it modulates the effect of the classical transmitter. There are suprisingly few studies on a possible function of ATP on cholinergic effector systems and it is likely that we shall learn considerably more about this in the future.

The release of ATP from the postsynaptic effector cell itself (17, 26) needs further characterization. It is generally observed that ATP can become released from tissues by toxic stimuli like trauma or hypoxia (7). If there should be ATP release from cholinergic effector cells under truly physiological conditions its relation to nerve terminal and vesicular release and metabolism needs further analysis.

References

1. Akasu, T., K. Hirai and K. Koketsu. 1981. Increase of acetylcholine-receptor sensitivity by adenosine triphosphate: a novel action of ATP on ACh sensitivity. Br.J. Pharmac. 74, 505-507.

2. Da Prada, M. and A. Pletscher. 1968. Isolated 5-hydroxytryptamine organelles of rabbit blood platelets: physiological properties and drug induced changes. Brit.J.Pharmacol. Chemother. 34, 591-597.

3. Dowdall, M.J., A.F. Boyne and V.P.Whittaker. 1974. Adenosine triphosphate, a constituent of cholinergic synaptic vesicles. Biochem.J. 140, 1-12.

4. Dowdall, M.J., P.G. Strange and P.R. Golds. 1983. Muscarinic acetylcholine receptors in Torpedo electric organ: effect of guanine nucleotides. J.Neurochem. 41, 556-561.

5. Ehrlet, F.J., W.R. Roeske and J.I. Yamamura. 1981. Muscarinic receptor: regulation by guanine nucleotides, ions and N-ethylmaleimide. Fed.Proc. 40, 153-159.

6. Ewald, D.A. 1976. Potentiation of postjunctional cholinergic sensitivity of rat diaphragm muscle by high-energy-phosphate adenine nucleotides. J.membr.Biol. 29, 47-65.

7. Forrester, T. and C.A. Williams. 1977. Release of adenosine triphosphate from adult heart cells in response to hypoxia. J.Physiol. Lond. 268, 371-390.

8. Häggblad, J., H. Eriksson and E. Heilbronn. 1985. In: Molecular Basis of Nerve Activity (J.-P. Changeux, F. Hucho, A. Maelicke and E. Neumann, eds.). Walter de Gruyter & Co., Berlin. New York. this volume.

9. Kato, A.C., H.S. Katz and B. Collier. 1974. Absence of adenine nucleotide release from autonomic ganglion. Nature, London, 249, 576-577.

10. Keller, F. and H. Zimmermann. 1983. Ecto-adenosine triphosphatase activity at the cholinergic nerve endings of the Torpedo electric organ. Life Sci. 33, 2635-2641.

11. Klein, R.L. 1982. In: Neurotransmitter Vesicles (R.L. Klein, H. Lagercrantz and H. Zimmermann, eds.). Academic Press, London. New York. pp.133-174.

12. Kolb, H.A. and M.J.O. Wakelam. 1983. Transmitter-like action of ATP on patched membranes of cultured myoblasts and myotubes. Nature, London, 303, 621-623.

13. Lagercrantz, H. and G. Fried. 1982. In: Neurotransmitter Vesicles (R.L. Klein, H. Lagercrantz and H. Zimmermann, eds.). Academic Press, London. New York. pp.175-188.

14. Li, Z.Y. and C. Bon. 1983. Presence of membrane bound acetylcholinesterase from a preparation of nerve endings from Torpedo marmorata electric organ. J.Neurochem. 40, 338-349.

15. Luqmani, Y.A. 1981. Nucleotide uptake by isolated cholinergic synaptic vesicles: evidence for a carrier of adenosine 5'-triphosphate. Neuroscience 6, 1011-1022.

16. Meunier, F.M. and N. Morel. 1978. Adenosine uptake by cholinergic synaptosomes from Torpedo electric organ. J..Neurochem. 31, 845-851.

17. Meunier, F.M., M. Israël and B. Lesbats. 1975. Release of ATP from stimulated nerve electroplaque junctions. Nature, London, 257, 407-408.

18. Michaelson, D.M. 1978. Is presynaptic acetylcholine release accompanied by secretion of the synaptic vesicles contents? FEBS Lett. 89, 51-53.

19. Morel, N. and F.M. Meunier. 1981. Simultaneous release of acetylcholine and ATP from stimulated cholinergic synaptsomes. J.Neurochem. 36, 1766-1773.

20. Morel, N. and P. Dreyfus. 1982 Association of acetylcholinesterase with the external surface of the pre-synaptic plasma membrane in Torpedo electric organ. Neurochem.Int. 4, 283-288.

21. Nagy, A., R.R. Baker, S.J. Morris and V.P. Whittaker. 1976 . The preparation and characterization of synaptic vesicles of high purity. Brain Res. 109, 285-309.

22. Poisner, A.M. and W. Douglas. 1968. Adenosine triphosphate and adenosine triphosphatase in hormone containing granules of posterior pituitary gland. Science 160, 203-204.

23. Pull, I. and H. McIlwain. 1972. Adenine derivatives as neurohumoral agents in the brain. The quantities liberated on excitation of superfused cerebral tissues. Biochem.J. 130, 975-981.

24. Salamin, A., J. Deshusses and R.W. Straub. 1981. Phosphate ion transport in rabbit brain synaptosomes. J.Neurochem. 37, 1419-1424.

25. Schweitzer, F.S. and R.B. Kelly. 1982. ATP release from cholinergic synapses. Soc.Neurosci.Abstr. 8, 493.

26. Silinsky, E.M. 1975. On the association between transmitter secretion and the release of adenine nucleotides from mammalian motor nerve terminals. J.Physiol.Lond. 247, 145-162.

27. Tomas, J. 1983. Calcium-induced ATP release from Torpedo cholinergic nerve endings isolated in calcium-free solutions. Revista Espagnola de Fisiologia 39, 105-106.

28. Volknandt, W. and H. Zimmermann. 1985. In: Molecular Basis of Nerve Activity (J.-P. Changeux, F. Hucho, A. Maelicke and E. Neumann, eds.). Walter de Gruyter & Co., Berlin. New York. this volume.

29. White, T., P. Potter and S. Wonnacot. 1980. Depolarization induced release of ATP from cortical synaptosomes is not associated with acetylcholine release. J.Neurochem. 34, 1109-1112.

30. Winkler, H. and S.W. Carmichael. 1982. In: The Chromaffin Granule, Vol. 1 (A.M. Poisner and J.M. Trifaró , eds.). Elsevier Biomedical Press, Amsterdam. New York. Oxford. pp.3-79.

31. Zimmermann, H. 1978. Turnover of adenine nucleotides in cholinergic synaptic vesicles of the Torpedo electric organ. Neuroscience 3, 827-836.

32. Zimmermann, H. 1979. Vesicle recycling and transmitter release. Neuroscience 4, 1773-1804.

33. Zimmermann, H. 1982. In: Co-transmission (A.C. Cuello, ed.). The MacMillan Press, London. pp.243-259.

34. Zimmermann, H. and C.R. Denston. 1976. Adenosine triphosphate in cholinergic vesicles isolated from the electric organ of Electrophorus electricus. Brain Res. 111, 365-376.

35. Zimmermann, H. and C.R. Denston. 1977. Separation of synaptic vesicles of different functional states from the cholinergic synapses of the Torpedo electric organ. Neuroscience 2, 715-730.

36. Zimmermann, H. and J.T. Bokor. 1979. Adenosine-5'-triphosphate recycles independently of acetycholine in cholinergic sanaptic vesicles. Neuroscience Lett. 13, 319-324.

37. Zimmermann, H., M.J. Dowdall and D.A. Lane. 1979. Purine salvage at the cholinergic nerve endings of the Torpedo electric organ: the central role of adenosine. Neuroscience 4, 979-993.

SOLUBILIZATION, AFFINITY LABELLING AND PURIFICATION OF THE DIHYDROPYRIDINE RECEPTOR OF THE VOLTAGE-DEPENDENT CALCIUM CHANNEL FROM RABBIT SKELETAL MUSCLE TRANSVERSE TUBULE MEMBRANES

M. BORSOTTO, J. BARHANIN, J.-P. GALIZZI, M. FOSSET, M. LAZDUNSKI

Centre de Biochimie du Centre National de la Recherche Scientifique Faculté des Sciences, Parc Valrose, 06034 NICE CEDEX.

Summary

The dihydropyridine receptor associated with the voltage-dependent calcium channel from rabbit skeletal muscle transverse tubule membranes has been solubilized by detergent extraction. A highly stable solubilized receptor preparation was obtained using 3-[(3-cholamidopropyl)dimethyl-ammonio]-1-propane sulfonate (CHAPS) as detergent with phospholipids and glycerol present as stabilizing agents. Binding of [^{3}H]dihydropyridine to the solubilized receptor was reversible and saturable. The equilibrium dissociation constant of the [^{3}H]dihydropyridine-receptor complex was very close to that determined on the membrane-bound receptor. [^{3}H]nitrendipine binding to its solubilized receptor was also inhibited by other antiarhythmic drugs such as bepridil and verapamil and enhanced by d-cis-diltiazem. These drugs conserved the same specificity and stereoselectivity as compared to the membrane-bound receptor. Since these drugs are apparent non-competitive inhibitors of [^{3}H]dihydropyridine binding it was concluded that these different sites are tightly coupled. The CHAPS solubilized extract was used as starting material for purification studies. The dihydropyridine-sensitive calcium channel has been substantially purified by a combination of filtration on Ultrogel A_2 column and lectin affinity chromatography using (+)[^{3}H]PN 200-110 as a marker. The purified material contained polypeptides of apparent molecular weights of 142 000, 32 000 and 33 000 which copurified with (+)[^{3}H]PN 200-110 binding activity. The molecular size of the native [^{3}H]nitrendipine receptor of transverse tubule membranes has been determi-

Molecular Basis of Nerve Activity

ned using the radiation inactivation technique. The [^{3}H]nitrendipine receptor was found to have a Mr of 210 000 ± 20 000. Photoaffinity labelling using (+)[^{3}H]PN 200-110 indicates the presence in the dihydropyridine receptor of a polypeptide (or polypeptides) of Mr 32 000 ± 2 000.

Introduction

Voltage-dependent calcium channels, which are known to play important roles in excitation-contraction coupling in cardiac and smooth muscles and also in a number of secretory processes, have recently become accessible to biochemical study as a result of the availability of a number of tritiated calcium channel inhibitors belonging to the dihydropyridine and the phenylalkylamine series (1, 2). Dihydropyridine derivatives and other calcium inhibitors such as verapamil, bepridil and diltiazem, are very important therapeutic agents in the treatment of cardiovascular disorders (3). Binding studies using tritiated dihydropyridine derivatives such as nifedipine, nitrendipine or PN 200-110, have identified specific high-affinity binding sites in brain, and in cardiac, smooth and skeletal muscle tissues (4, 5).

One of the difficulties associated with the biochemical characterization of macromolecules that confer electrical excitability to biological membranes, such as the voltage-sensitive calcium channel, is that they are almost always present in very low amount in the membranes. For most membrane preparations, in which the voltage-sensitive calcium channel has been characterized, the dihydropyridine receptor is only present at a density in the range of 0.1 to 1 pmol/mg protein (4). Therefore, the transverse tubule (T-tubule) membrane preparation isolated from rabbit skeletal muscle appears to be exceptional in that it contains more than 50 pmol[^{3}H]dihydropyridine binding sites/mg protein (5). Voltage-clamp analysis of frog skeletal muscle have shown that essentially all specific Ca^{2+} conductances are localized in the transverse tubular system and that the channel responsible for these conductances is inhibited by low concentrations of dihydropyridines (6). T-tubules were used as starting material for the solubilization and purification of the voltage-sensitive calcium channel.

Determination of the Molecular Size by Radiation Inactivation of the Dihydropyridine-Sensitive Calcium Channel

One approach that has been successfully applied to the determination of the molecular size of membrane-bound molecules such as voltage-sensitive Na^+ channel (7), calcium-dependent K^+ channel (8) and (Na^+,K^+)ATPase (9), is the radiation inactivation analysis. In this technique, protein samples are irradiated with high energy electron and the inactivation of the protein is assumed to be related to its size. Radiation inactivation offers two distinct advantages over conventional methods of determining molecular size is that the technique facilitates investigation of membrane bound protein in situ and is not dependent on the purity of the proteins being studied. Fig. 1A shows a simple exponential decay of $[^3H]$nitrendipine binding activity, indicative of a sole target size and in accord with the simple class of binding sites seen in equilibrium binding studies. That the observed loss of binding activity was due to a low of functional binding sites was confirmed by the observation that irradiation had no significant effect on the value of equilibrium dissociation constant K_D (Fig. 1B).

The molecular size was calculated using the empirical equation (9) $Mr = 6.4 \times 10^5/D_{37}$, where D_{37} is the radiation dose which conserved 37% of the started binding activity. D_{37}, shown in Fig. 1A, is 3 Mrads which correspond to a Mr of 210 000 ± 20 000. This smallest membrane active unit which binds dihydropyridine has the same size in skeletal muscle and in brain (10). The subunit composition cannot be determined using this technique.

Photoaffinity Labelling of the Dihydropyridine-Sensitive Calcium Channel

Dihydropyridines, including (+)PN 200-110, are light sensitive compounds. Because of this property $(+)^3H]$PN 200-110 appears as a potential photoaffinity ligand for the dihydropyridine receptor if high intensity U.V. light is used. This possibility has been tried using T-tubule membranes. After high intensity U.V. irradiation, S.D.S. polyacrylamide gel electrophoresis was performed, and the quantity of $(+)[^3H]$PN 200-110 covalently bound was determined by counting gel slices. Non-specific labelling was mesured in the presence of high concentration of unlabelled ligand.

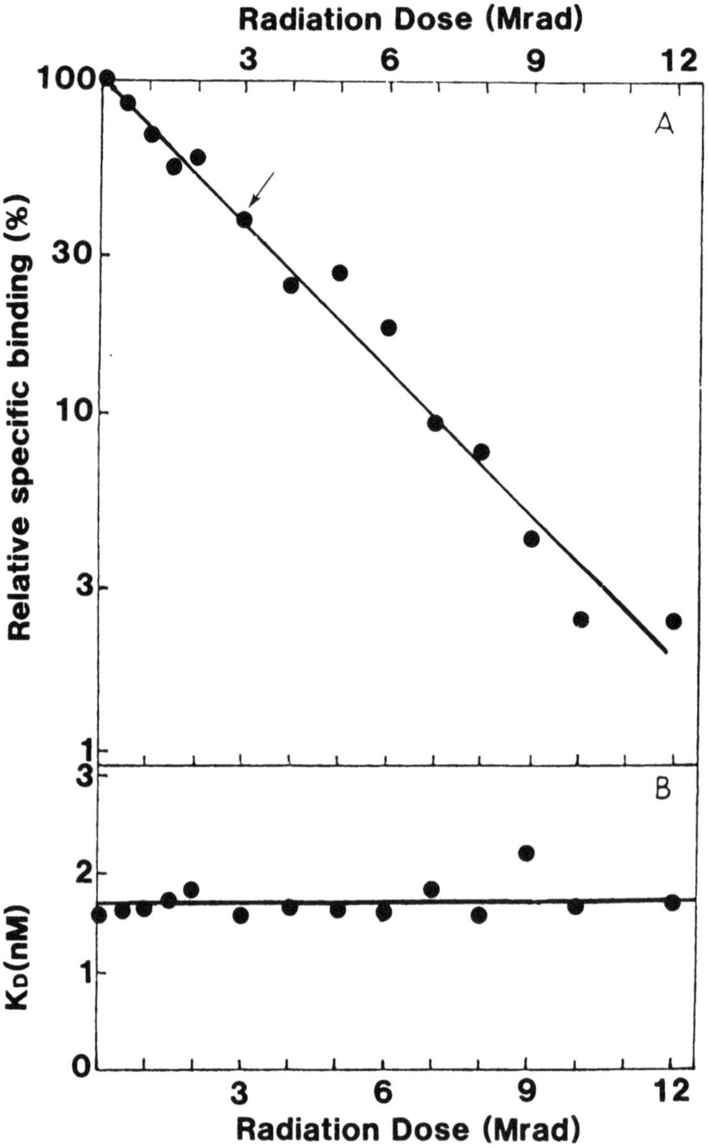

Fig. 1. A. Specific binding of [^{3}H]nitrendipine to T-tubule membranes as a function of a radiation dose. The line drawn was calculated by linear regression analysis. Arrow indicates the D_{37}. B. Shows the lack of significant effect of irradiation on the K_D value of the [^{3}H]nitrendipine-receptor complex.

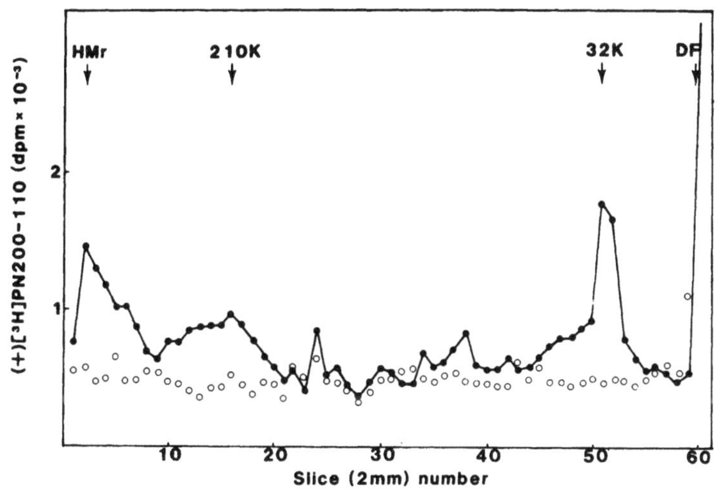

Fig. 2. Electrophoretic profile of photoaffinity labelling of T-tubule membranes with (+)[^{3}H]PN 200-110. T-tubule membranes (100 µg protein/ml) were incubated with (+)[^{3}H]PN 200-110 (10 nM) in the absence (●) and in the presence (○) of 1 µM of unlabelled nitrendipine. After high intensity U.V. irradiation, incubates were centrifuged and the pellets were solubilized in 2% SDS and 5% β-mercaptoethanol, and heated at 96°C for 10 min. Samples (100 µg protein) were subjected to electrophoresis. Gel lanes were sliced in 2 mm sections and (+)[^{3}H]PN 200-110 incorporated was determined by liquid scintillation counting following solubilization of gel slices with H_2O_2 (no staining procedure was performed before slicing). HM_r : high Mr corresponding to aggregate material which did not enter the gel. DF : dye front.

Fig. 2 shows that (+)[^{3}H]PN 200-110 is essentially incorporated into a peak which corresponds to a M_r of 32 000 ± 2 000, a minor and broader peak appears at a M_r of 210 000.

The same results were obtained using [^{3}H]diltiazem instead of (+)[^{3}H]PN 200-110 (data not shown). Our results indicate that the binding component of the voltage-sensitive calcium channel for dihydropyridines ((+)[^{3}H]PN 200-110) and for benzothiazepines ([^{3}H]diltiazem) is a protein migrating with a M_r of 32 000 in SDS gel electrophoresis.

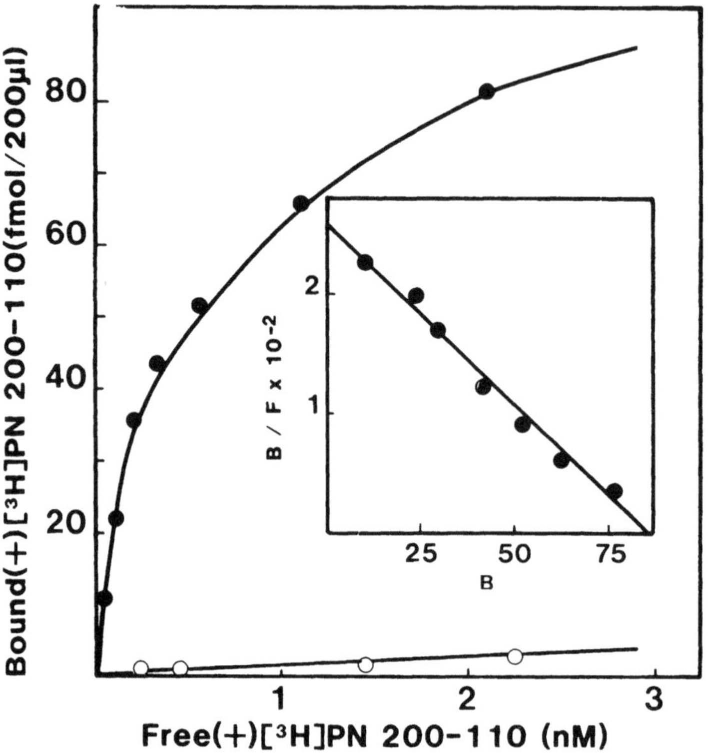

Fig. 3. Direct binding of (+)[^{3}H]PN 200-110 to solubilized T-tubule membranes. CHAPS extract (4 µg protein/ml) was incubated with increasing concentrations of (+)[^{3}H]PN 200-110 for 60 min at 4°C. (●) total binding; (○) non specific binding measured in the presence of 1 µM unlabelled nitrendipine. Inset. Scatchard plot of the specific binding.

Solubilization of Native Dihydropyridine Receptor

Among 14 detergents of varied chemical natures, only one, the 3-[(3 cholamidopropyl)-dimethyl-ammonio]-1 propane sulfonate (CHAPS) presented the required qualities for solubilization of the dihydropyridine receptor from T-tubule membranes (11). For example, Triton X-100 had by itself a high non-specific interaction with [^{3}H]dihydropyridine which interfered with the assay of the solubilized receptor. On the other hand digitonin micelles which have a relatively low non-specific interaction with the ligand give a very low recovery of the binding activity. CHAPS alone has proved to be useful in the solubilization procedure since (i) up to 25% of the initial binding activity is recovered in the extract, (ii) the non-specific component observed in binding studies is very low (Fig. 3). The more effective protein to detergent ratio was carefully investigated and found to be close to 0.25 (W/W). Adding glycerol at a final concentration of 5% during solubilization substantially increased the binding capacity of the extract (~2-fold) without increasing the protein content. Glycerol also stabilized the dihydropyridine receptor in its active form with a half-life of 34 h at 4°C (11). Half-life values were very sensitive to temperature and became lower than 10 min at 20°C. Hence, a solubilization protocol using CHAPS and glycerol was used for further investigations.

Fig. 3 shows the equilibrium binding of (+)[^{3}H]PN 200-110 to the CHAPS extract of T-tubule membranes. Non-specific binding measured in the presence of an excess of unlabelled nitrendipine, is very low. The Scatchard plot (Fig. 3, inset) is linear, indicating the presence of a single class of non-interacting solubilized binding sites. The dissociation constant (K_D) of the (+)[^{3}H]PN 200-110 receptor complex is 0.4 ± 0.2 nM, very close to that of the membrane-bound receptor (0.2 ± 0.05 nM). The maximal binding capacity (B_{max}) of 85 pmol/mg protein is also comparable to that of the starting membrane (90 pmol/mg protein).

Pharmacological Properties of the Solubilized Receptor

Different unlabelled calcium inhibitors have been tested for their ability to interfere with the [^{3}H]nitrendipine binding on the solubilized receptor. These molecules not only included dihydropyridine derivatives (nitrendipine and nifedipine) but also benzothiazepine (l-cis and d-cis diltiazem) and phenylalkylamine ((+) and (-) verapamil and bepridil) compounds which are known to bind to the calcium channel at sites allosterically related to the dihydropyridine binding sites (12).

Such competition experiments are illustrated in Fig. 4 and results are summarized in Table 1.

Evidence that the solubilization procedure used in this work is useful to solubilize the voltage-sensitive calcium channel arises from the observations that ligand specificity and stereoselectivity are very close to those known for the membrane-bound receptors.

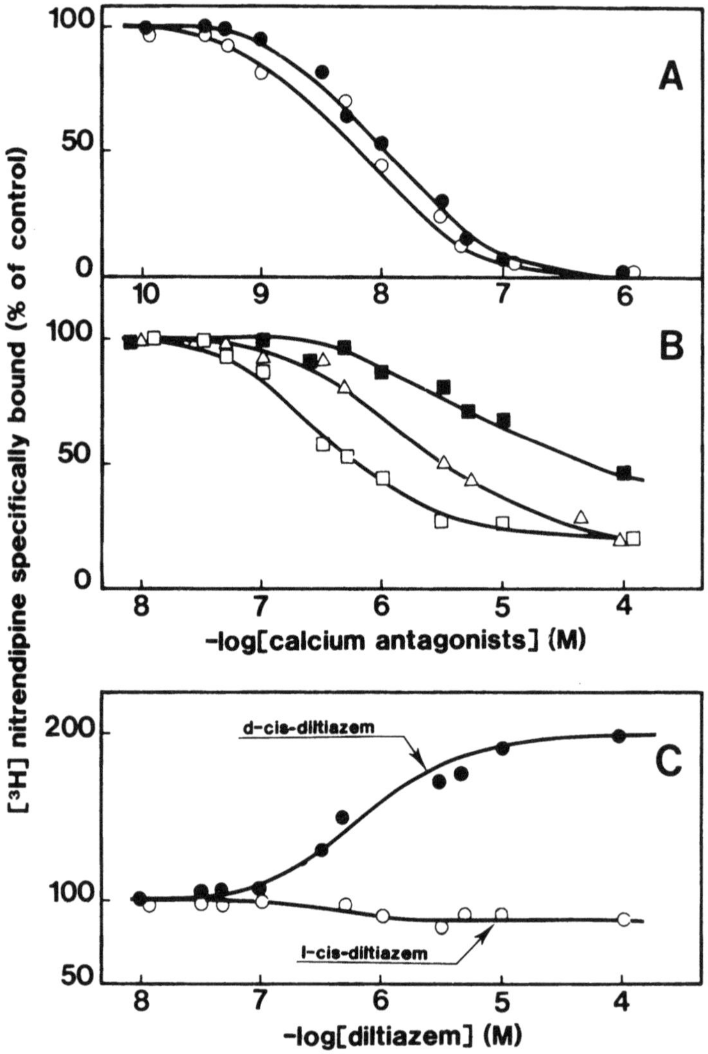

Fig. 4. Competition between [^{3}H]nitrendipine and other unlabelled drugs. [^{3}H]nitrendipine (1 nM) was incubated with solubilized T-tubule (15 μg/ml) and increasing concentrations of following unlabelled molecules. A. Nitrendipine (●) and nifedipine (○). B. Bepridil (△), (+)verapamil (□) and (-)verapamil (■). C. L-cis-diltiazem (○) and d-cis-diltiazem (●). The radioactivity bound to the solubilized receptor was measured after incubation for 60 min at 4°C.

Table 1. Comparison of binding properties of calcium inhibitors on membrane-bound and solubilized receptors.

	$K_{0.5}$ (nM)	
	membrane-bound	solubilizate
nitrendipine	1.8	7
nifedipine	4	9
PN 200-110	0.2	0.4
(+)verapamil	nd	500
(-)verapamil	nd	50 000
l-cis-diltiazem	22 000*	no effect
d-cis-diltiazem	1 600*	500
bepridil	1 000	4 000

* Values according to Glossmann and Ferry (13)

Purification of the Dihydropyridine-Sensitive Calcium Channel

The dihydropyridine receptor has been substantially purified by a combination of filtration on Ultrogel A_2 column and lectin affinity chromatography on wheat germ agglutinin (WGA)-Sepharose. After solubilization, trace amounts of (+)[^{3}H]PN 200-110 are allowed to bind to the receptor. Therefore, in the purification procedure used, a small part of the solubilized Ca^{2+} channel is specificaly labelled (less than 5% of the total receptor content) and this small proportion of [^{3}H]dihydropyridine receptor complex is used to monitor the purification steps by counting aliquots of each fraction. The total active binding site capacity can also be measured by equilibrium binding experiments to the free receptor. (+)[^{3}H]PN 200-110 has been chosen as a marker instead of [^{3}H]nitrendipine because of its high affinity (0.4 nM) for the solubilized preparation which is 300-fold lower than the binding site capacity (120 nM). In these conditions the (+)[^{3}H]PN 200-110 binds stoichiometrically to its receptor and a specific radioactivity of the total receptor population is determined which is used to quantitate the recovery of the binding activity along the purification (14).

A mixture of protease inhibitors (14) was added to all buffers to avoid proteolytic degradations. In addition, egg yolk phosphatidylcholine was found to be necessary to stabilize the active form of the receptor during purification. A detergent to phospholipid ratio of 5:1 (W/W) gave the maximal stabilization factor. Except for the washing step of the WGA-Sepharose column the buffer contained 0.1% CHAPS. The first step of purification is a gel filtration on an Ultrogel A_2 column. Fig. 5A shows that (+)[^{3}H]PN 200-110 binding sites are eluted in a single peak. Counting aliquots of each fraction or measuring the binding of (+)[^{3}H]PN 200-110 to free receptors in aliquots of each fraction resulted in the same elution profile. A typical Scatchard plot of the data from a binding experiment performed on the pooled fractions (indicated by the bar in Fig. 5) is shown in Fig. 5A inset. The K_D value of 0.5 nM is essentially unmodified as compared to the crude solubilizate. The B_{max} value of 337-414 pmol/mg protein corresponds to a 4-fold increase in specific activity. The pooled fractions from Ultrogel A_2 was adsorbed quantitatively on WGA-Sepharose. A large quantity of protein was non-specifically bound to the WGA-Sepharose and a washing step with 1% CHAPS was necessary to release it. Elution of dihydropyridine receptor was performed with 10 mM chitotriose plus 100 mM N-acetyl-D-glucosamine without reequilibration of the gel in 0.1% CHAPS (Fig. 5B).

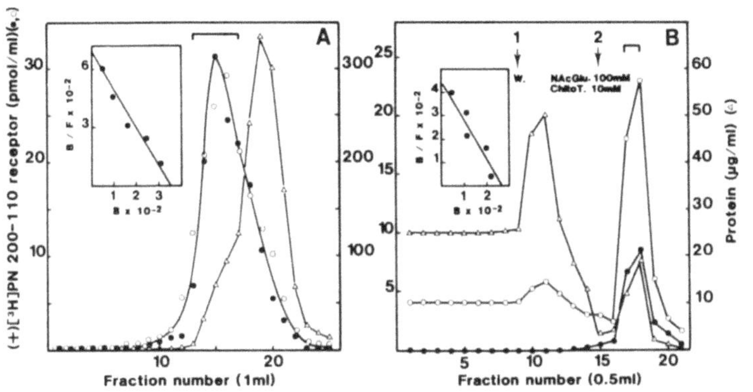

Fig. 5. A. Elution profile of solubilized Ca^{2+} antagonist receptor from Ultrogel A_2 column. (+)[^{3}H]PN 200-110 receptor concentration (pmol/ml) was determined in each fraction by binding assay (●) or counting an aliquot of the fractions (○) using the radiospecificity of the receptor calculated as indicated in the text. (△) Protein (µg/ml). Pooled fractions are indicated by the bar. Inset. Scatchard plot of the data from direct binding experiments of (+)[^{3}H]PN 200-110 to the pooled fractions (1 µg/ml). (B, pmol/mg protein; F, nM). B. Elution profile of purified Ca^{2+} antagonist receptor from WGA-Sepharose. (●) Receptor concentrations as determined by binding or (○) by counting (see Fig. 5A) washing with 1% CHAPS buffer and elution with 10 mM chitotriose plus 100 mM N-acetyl-D-glucosamine in 0.1% CHAPS buffer were commenced as indicated by arrows 1 and 2 respectively. Inset. as in 4A inset, 1 µg protein (pooled fraction)/ml.

This elution resulted in the recovery of 8% of the active receptor when measured by equilibrium binding (Fig. 5B inset) (B_{max} = 257 pmol/mg protein) and 21% (B_{max} = 837 pmol/mg protein) when the recovery of activity was measured by counting (+)[^{3}H]PN 200-110 specifically bound in each fraction.

The decrease in specific activity observed at this step is likely to be due to the washing step with 1% CHAPS. We have found that dilution of 1% CHAPS to the solubilized material results in a partial loss of the binding activity; however this washing was necessary to release non-specifically bound proteins.

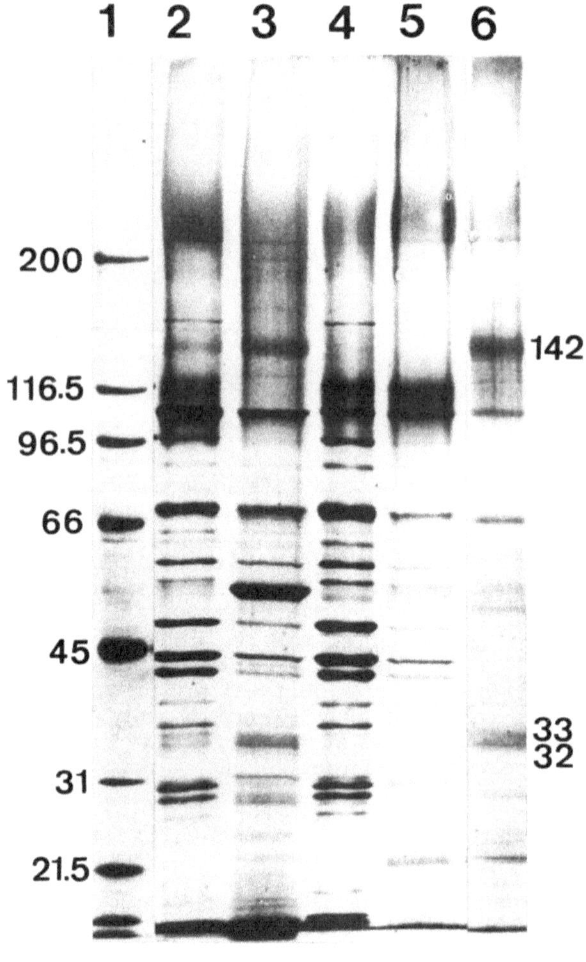

Fig. 6. SDS polyacrylamide gel electrophoresis of purified Ca^{2+} channel receptor. Electrophoresis was carried out in 4-12% polyacrylamide linear gradient gels under reducing conditions. Lanes : 1 Molecular weight standards; 2 Crude CHAPS extract (6 µg protein); 3 Pooled fraction from Ultrogel A_2 column (5 µg protein); 4 Fraction 20 from Ultrogel A_2 column (6 µg protein); 5 Break-through from WGA-Sepharose column (2 µg protein). 6 Pooled fraction from WGA-Sepharose column (2 µg protein).

Subunit Composition of the Dihydropyridine Receptor of the Voltage-Dependent Ca^{2+} Channel

After each step of purification the protein pattern of resultant material was analysed by SDS gel electrophoresis (Fig. 6). Bands of Mr 142 000 and 32-33 000 were present in active fractions (Fig. 6 lanes 2, 3 and 6) and absent in fraction where no (+)[^{3}H]PN 200-110 binding was observed (Fig. 6 lanes 4 and 5). Other components such as Mr 220 000, 105 000, 68 000 and 26 000 do clearly belong to contaminants proteins (see Fig. 6 lanes 4 and 5).

SDS gel electrophoresis analysis reveals that the probable subunit composition of the dihydropyridine receptor from T-tubule membranes consists of a protein of Mr 142 000 and a doublet of Mr 32-33 000. A 1:1:1 complex of these three polypeptides would be consistent with the Mr of 210 000 determined by radiation inactivation on the same tissue (Fig. 1 and ref. 10). This subunit composition is also in good agreement with different Mr determined by affinity labelling : 145 000 with [^{3}H]azidopine on guinea pig skeletal muscle (15), 32 000 with (+)[^{3}H]PN 200-110 and [^{3}H]diltiazem on rabbit skeletal muscle (see Fig. 2) and 33 000 with [^{3}H]nitrendipine on canine cardiac membrane (16). As for the Na^+ channel (17) reconstitution of highly purified Ca^{2+} channel into a planar bilayer will be useful to determine the minimal peptide composition responsible for physiological functions.

References

1. Belleman, P., D. Ferry, F. Lübbecke, H. Glossmann. 1981. Arzneim Forsch. 31, 2064.

2. Janis, R.A., S.C. Maurer, J.G. Sarmiento, G.T. Bolger, D.J. Triggle. 1982. Eur.J.Pharmacol. 82, 191.

3. Fleckenstein, R. 1977. Annu.Rev.Pharmacol.Toxicol. 17, 149.

4. Janis, R.A., A. Scriabine. 1983. Biochem.Pharmacol. 32, 3499.

5. Fosset, M., E. Jaimovich, E. Delpont, M. Lazdunski. 1983. J.Biol.Chem. 258, 6086.

6. Almers, W., R. Fink, P.T. Palade. 1981. J.Physiol. 312, 177.

7. Barhanin, J., A. Schmid, A. Lombet, K.P. Wheeler, M. Lazdunski. 1983. J.Biol.Chem. 258, 700.

8. Schmid-Antomarchi, H., M. Hugues, R.I. Norman, J.C. Ellory, M. Borsotto, M. Lazdunski. 1984. Eur.J.Biochem. 142, 1.

9. Richard, D.E., J.C. Ellory, I.M. Glynn. 1981. Biochem.Biophys.Acta 648, 284.

10. Norman, R.I., M. Borsotto, M. Fosset, M. Lazdunski. 1983. Biochem.Biophys. Res.Commun. 111, 878.

11. Borsotto, M., R.I. Norman, M. Fosset, M. Lazdunski. 1984. Eur.J.Biochem. 142, 449.

12. Murphy, K.M.M., R.J. Gould, B.L. Largent, S.H. Snyder. 1983. Proc.Natl.Acad. Sci.USA 80, 860.

13. Glossmann, H., D.R. Ferry. 1983. Naunyn-Schmied.Arch.Pharmacol. 323, 279.

14. Borsotto, M., J. Barhanin, R.I. Norman, M. Lazdunski. 1984. Biochem.Biophys.Res.Commun. 122, 1357.

15. Ferry, D.R., M. Rombusch, A. Goll, H. Glossmann. 1984. FEBS Lett. 169, 112.

16. Campbell, K.P., G.M. Lipshutz, G.H. Denney. 1984. J.Biol.Chem. 259, 5384.

17. Hanke, W., G. Boheim, J. Barhanin, D. Pauron, M. Lazdunski. 1984. EMBO J. 3, 509.

IS ATP A GENERAL MOLECULAR CONSTITUENT OF CHOLINERGIC SYNAPTIC VESICLES?

W. Volknandt, H. Zimmermann
AK Neurochemie, Zoologisches Institut der Universität Frankfurt, Siesmayerstr. 70, D-6000 Frankfurt, Federal Republic of Germany

Introduction

Twenty years after the first isolation of synaptic vesicles from brain nerve terminals (13) the number of tissues from which synaptic vesicles can be isolated is still limited. Furthermore the degree of purity of vesicles is quite variable between tissue sources and fractionation procedures. Amongst tissues from which isolation of vesicles has been reported are besides brain from various mammalian species, the bovine superior cervical ganglion (14), the myenteric plexus of guinea pigs (4), sympathetic axons and tissues innervated by sympathetic nerves like the spleen, vas deferens or heart (ref. in 8) as well as the optic lobes of Octopus and squid (ref. in 2) and the electric organs of the electric rays Torpedo and Narcine (ref. in 15).

The only tissue from which so far acetylcholine (ACh) containing vesicles have been isolated in pure form are the electric ray electric organs. All other tissues either yield a mixture of vesicles of different transmitter types or are used for isolation of adrenergic vesicles. Therefore all the results on the molecular characterization of cholinergic vesicles are derived solely from Elasmobranch fish and it is still an open question whether they are of validity also for higher vertebrates.

Molecular Basis of Nerve Activity

So far it has not been possible to successfully apply subcellular fractionation methods to the presynapse of the mammalian motor junction. The electric organs of the two strong electric fish Electrophorus electricus (electric eel) and Malapterurus electricus (electric catfish) do also contain cholinergic innervation but not of the same densitiy as that of electric rays. Whereas the ACh content of the Torpedo electric organ amounts to about 1000 nmoles/g of tissue that of the electric eel electric main organ was 17.5 ± 3.8 (S.E.M., n = 3) and that of the electric catfish electric organ was 1.6 ± 0.3 (S.E.M., n = 7). They provide, however, a reasonable amount of tissue for subcellular fractionation. We have therefore attempted to isolate cholinergic vesicles from the electric organs of these two bony fish which in evolutionary terms represent considerably more advanced vertebrates than the electric rays.

A typical characteristic of cholinergic vesicles from electric rays is their high content in ATP in addition to ACh (molar ratio ACh/ATP between 5 and 10 to 1). Similarly adrenergic synaptic vesicles and neurosecretory vesicles containing vasopressin contain ATP (ref. 6, 9). We report here results suggesting that ATP is also a constituent of synaptic vesicles of peripheral cholinergic motor neurones in bony fish.

Subcellular Fractionation of Electric Eel Electric Organ

A variety of procedures were attempted for fractionating the electric eel electric organ with little difference in yield. Fresh tissue was homogenized with a Waring blendor in the following media: 100 mM NaCl, 200 mM glycin, 200 mM sucrose or 250 mM sorbitol containing 10 mM Hepes, pH 7.0. The homogenate was spun at 16 000 g_{av} for 30 min. The supernatant (fraction S_{12}) was either loaded on a sucrose density gradient in a swing out rotor (Fig. 1) or in a zonal rotor (Fig. 3) or it was recentrifuged at 159 000 g_{av} for 60 min to sediment the synaptic vesicles. The pellet obtained (fraction P_3) was resuspended in the corresponding isotonic medium and further separated on a sucrose density gradient.

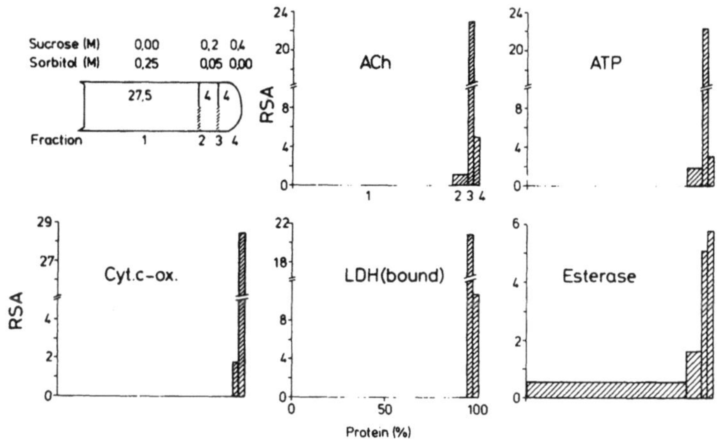

Fig. 1 Subfractionation of fraction S-12 derived from electric eel electric organ on a simple sucrose gradient (90 000 g-av x 120 min, De Duve-plot). ACh and ATP are enriched most in fraction 3. Cyt-c.-ox, cytochrome-c-oxidase; LDH, lactate dehydrogenase.

Fig. 1 shows the separation of synaptic vesicles after loading fraction S_{12} on a simple sucrose gradient. ACh and ATP show a parallel distribution on the De Duve-plot and are enriched most on top of the 0.4 M sucrose layer. In contrast activity of the mitochondrial marker cytochrome-c-oxidase, of lactate dehydrogenase (LDH) (bound, i.e. contained in membrane sealed particles) and of cholinesterases are distributed differently. When the vesicle fraction (fraction 3, Fig. 1) is further separated by chromatography (Fig. 2) on a column of Sephacryl-1000 the small amount of activity of bound LDH originally enriched in the vesicle fraction after gradient centrifugation can be separated from synaptic vesicles. Activity of bound LDH elutes with the void volume of the column as defined by

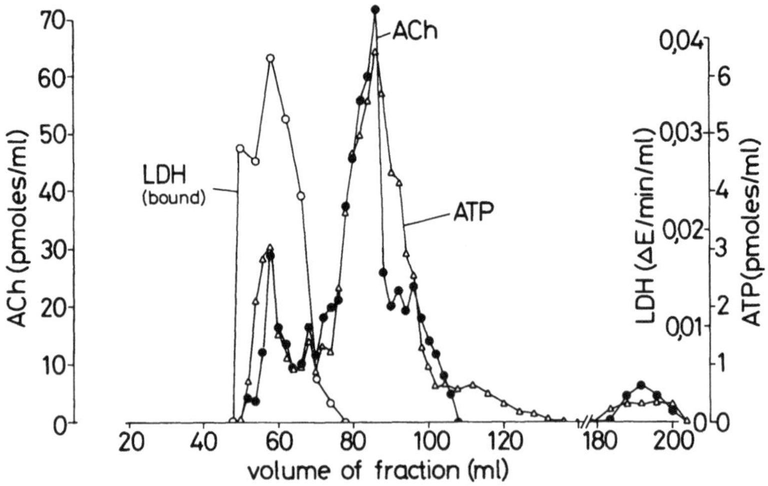

Fig. 2 Chromatography of fraction 3 (Fig. 1) on Sephacryl-1000. Note the co-elution of ACh and ATP and the separation from bound lactate dehydrogenase (LDH) which elutes with the void volume. The small peak of ACh and ATP around 190 ml corresponds to the salt peak.

elution of a bacterium of about 1 µm size. ACh and ATP are eluted in parallel on column chromatography. A similar result is obtained when synaptic vesicles are separated by zonal centrifugation (Fig. 3). When fraction S_{12} is centrifuged into a shallow sucrose gradient, ACh and ATP sediment in parallel within the synaptic vesicle region and are separated from the other marker substances. The mitochondrial marker fumarase as well as bound LDH-activity are well separated indicating that neither mitochondria nor occluded cytoplasm can be the source of the ATP detected.

The specific activity (nmoles/mg protein) of ACh in the vesicle peak fractions for the various fractionation procedures ranges between 3.5 (step gradient), 2.7 (column chromatography) and 16.6 (zonal gradient). These values are within the range of that obtained for vesicles isolated from ox superior cervical ganglion (3 to 14, ref. 14). But they are significantly

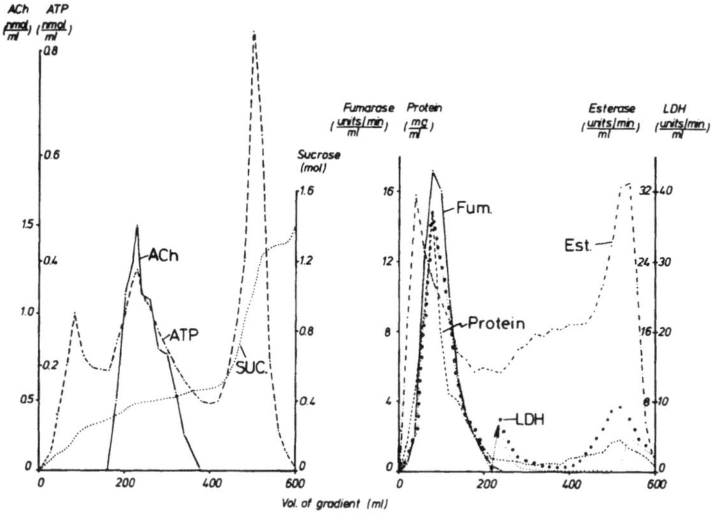

Fig. 3 Zonal separation on a shallow sucrose (suc.) density gradient of synaptic vesicles isolated from the electric eel electric organ. A concomitant peak of ACh and ATP appears at a density of 0.39 M sucrose. LDH, lactate dehydrogenase (from 16).

lower than those for electric ray synaptic vesicles (7 µmole/mg protein, ref. 12). It has to be worked out whether the difference is solely due to the contamination still present in the electric eel vesicle fraction or whether also the relative protein content of eel synaptic vesicles is higher than that of Torpedo synaptic vesicles. The molar ratios ACh/ATP of synaptic vesicle peak fractions are summarized in Table 1. For the electric eel they range between 7 and 11 and are similar to those for vesicles isolated from Malapterurus and Torpedo electric organ.

Subcellular Fractionation of the Catfish Electric Organ

The electric organ of the electric catfish was homogenized in a Waring blendor in 250 mM sorbitol (10 mM Hepes, pH 7.0).

The homogenate was spun at 14 000 g_{av} for 50 min. The supernatant fraction (S_{12}) was loaded onto a sucrose density gradient as depicted in Fig. 4 and centrifuged at 90 000 g_{av} for 120 min. Most ACh containing particles sediment on top of the 0.2 M sucrose layer but the highest relative specific activity of ACh is found in fraction 3 on top of the 0.4 M sucrose layer. ATP is enriched most in the pellet fraction

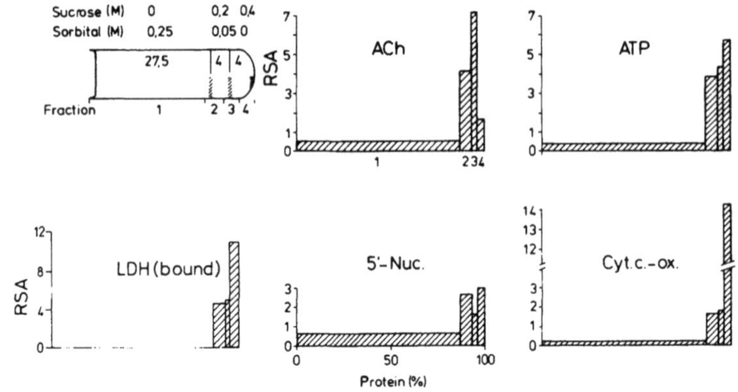

Fig. 4 Subfractionation of fraction S-12 derived from electric catfish electric organ on a simple sucrose gradient (De Duve-plot). LDH, lactate dehydrogenase; 5'-Nuc., 5'-nucleotidase; Cyt.c.-ox., cytochrome-c-oxidase.

which also contains the highest enrichment of the mitochondrial marker enzyme cytochrome-c-oxidase. 5'-nucleotidase as a membrane marker enzyme and bound LDH are clearly distributed differently from both ACh and ATP. When gradient fraction 3 (Fig. 4) is refractionated on a flotation gradient (Fig. 5) the parallel sedimentation behaviour of ACh and ATP becomes apparent and the difference in the distribution pattern of the other marker substances is clearly marked. This suggests that ACh and ATP are contained in the same neurotransmitter vesicle. The specific activities (nmoles/mg protein) of ACh range between 0.44 and 1.52 which infers that the vesicle fraction still carries considerable protein contamination. However, the analysis of the density gradients suggests that this contamination would not have an effect on the ATP content of the vesicles peak fraction. The molar ratios of ACh/ATP correspond to those of Electrophorus and Torpedo (Table 1).

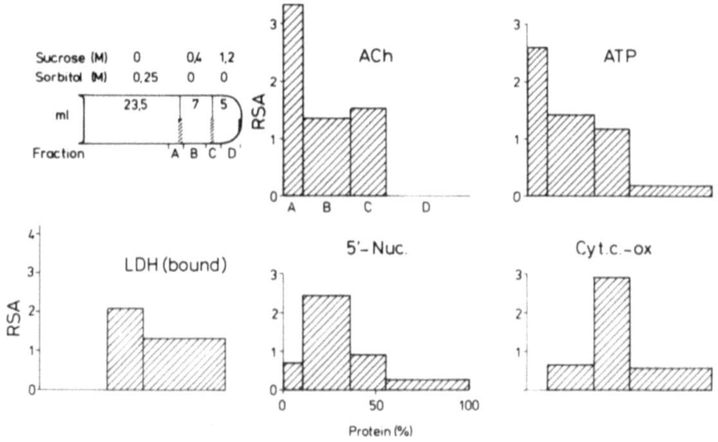

Fig. 5 Refractionation of gradient fraction 3 (Fig. 4) by flotation. The material was layered on a 1.2 M sucrose solution, overlayed by 0.25 M sorbitol solution and centrifuged at 85 000 g-av for 150 min.

Table 1
Molar ratios of ACh/ATP of synaptic vesicles from different electric organs

Electrophorus electricus		
step gradient	10.56 ± 1.54 (4)	this study and ref. 16
zonal gradient	11.09 ± 2.44 (6)	
column chromatography	12.4 (1)	
Malapterurus electricus		
flotation gradient	13.27 ± 1.13 (4)	this study
column chromatography	8.03 ± 0.30 (3)	
Torpedo marmorata		
zonal gradient	5.32 ± 0.45 (21)	ref. 3

Conclusions

Our results suggest that ATP is also a molecular constituent of cholinergic synaptic vesicles in bony fish. It is contained in these fish in a similar molar ratio to ACh as in cholinergic synaptic vesicles isolated from the Torpedo electric organ. Further analysis will have to answer the question whether this is also true for mammalian cholinergic vesicles. We believe that these results are further support for a significant physiological role of ATP in the process of cholinergic signal transmission. It appears likely that ATP may act as a co-transmitter not only in the adrenergic but also in the cholinergic system (ref. 17, this volume and 11). This view is supported by recent findings that ATP has an effect on both the presynaptic nerve terminal (inhibiting transmitter release (5, 10) as well as on the postsynaptic effector membrane (activating cation channels, 1, 7).

References

1. Akasu,T., K. Hirai and K. Koketsu. 1981. Increase of acetylcholine-receptor sensitivity by adenosine triphosphate: a novel action of ATP on ACh-sensitivity. Br.J. Pharmac. 74, 505-507.

2. Dowdall, M.J. and V.P. Whittaker. 1973. Comparative studies in synaptosome formation: the preparation of synaptosomes from the head ganglion of the squid, Loligo paelii. J.Neurochem. 20, 921-935.

3. Dowdall, M.J., A.F. Boyne and V.P. Whittaker. 1974. Adenosine triphosphate, a constituent of cholinergic synaptic vesicles. Biochem.J. 140, 1-12.

4. Dowe, G.H.C., H. Kilbinger and V.P. Whittaker. 1980. Isolation of cholinergic synaptic vesicles from the myenteric plexus of guinea pig small intestine. J.Neurochem. 35, 993-1003.

5. Israël, M. and F.M. Meunier, 1978. The release of ATP triggered by transmitter action and its possible physiological significance: retrograde transmission. J.Physiol. Paris 74, 485-490.

6. Klein, R.L. 1982. In: Neurotransmitter Vesicles (R.L. Klein, H. Lagercrantz and H. Zimmermann, eds.), Academic Press, London. New York. pp.133-174.

7. Kolb, H.A. and M.J.O. Wakelam. 1983. Transmitter-like action of ATP on patched membranes of cultured myoblasts and myotubes. Nature, London, 303, 621-623.

8. Lagercrantz, H. and R.L. Klein. 1982. In: Neurotransmitter Vesicles (R.L. Klein, H. Lagercrantz and H. Zimmermann, eds.), Academic Press, London. New York. pp.89-118.

9. Poisner, A.M. and W. Douglas. 1968. Adenosine triphosphatate and adenosine triphosphatase in hormone containing granules of posterior pituitary gland. Science 160, 203-204.

10. Sawynok, J. and K.H. Jahmadas. 1976. Inhibition of acetylcholine release from cholinergic nerves by adenosine, adenine nucleotides and morphine: antagonism by theophylline. J.Pharmac.exp.Therap. 197, 379-390.

11. Stjärne, L. and P. Åstrand. 1984. Discrete events measure single quanta of adenosine 5'-triphosphate secreted from sympathetic nerves of guinea-pig and mouse vas deferens. Neuroscience 13, 21-28.

12. Tashiro, T. and H. Stadler. 1978. Chemical composition of cholinergic synaptic vesicles from Torpedo marmorata based on improved purification. Eur.J.Biochem. 90, 479-487.

13. Whittaker, V.P., I.A. Michaelson and R.J.A. Kirkland. 1964. The separation of synaptic vesicles from nerve-ending particles (synaptosomes). Biochem.J. 90, 293-303.

14. Wilson, W.S., R.A. Schulz and J.R. Cooper. 1973. The isolation of cholinergic synaptic vesicles from bovine superior cervical ganglion and estimation of their acetylcholine content. J.Neurochem. 20, 659-667.

15. Zimmermann, H. 1982. In: Co-transmission (A.C. Cuello, ed.), The MacMillan Press, London. pp.243-259.

16. Zimmermann, H. and C.R. Denston. 1976. Adenosine triphosphate in cholinergic vesicles isolated from the electric organ of Electrophorus electricus. Brain Res. 111, 365-376.

17. Zimmermann, H. and E.J.M. Grondal. 1985. In: Molecular Basis of Nerve Activity (J.-P. Changeux, F. Hucho, A. Maelicke and E. Neumann, eds.), Walter de Gruyter & Co., Berlin. New York. this volume.

SINGLE CHANNEL FORMATION OF COOPERATIVELY INTERACTING UNITS OF THE HIGHLY-PURIFIED Na-CHANNEL PROTEIN

Günther Boheim and Wolfgang Hanke

Dept. of Cell Physiology, Ruhr-Universität Bochum,
Postfach 102148, D-4630 Bochum, FRG

Jacques Barhanin, David Pauron and Michel Lazdunski

Centre de Biochimie du CNRS, Faculte des Sciences,
Parc Valrose, F-06034 Nice Cedex, France

Summary

The reconstitution of highly purified Na-channels isolated from rat brain into solvent-free planar lipid bilayer membranes was investigated. Series of experiments were carried out with respect to the observation of two characteristic channel types differing in their gating kinetics and conductance values. In order to achieve longer channel state lifetimes the neurotoxin batrachotoxin was added. It is clearly seen that voltage activated channels of the large conductance type spontaneously transform into bursting channels of the low conductance type. Even more evidence comes from the stepwise degradation of large size conductances into smaller ones by the blockade action of tetrodotoxin. The smaller channels exhibit a variety of conductance values which are integral multiples of each other: 14 pS, 28 pS, 42 pS, 56 pS. Generally channel openings and closures proceed in a one-step reaction, irrespective of conductance size. It is suggested that the highly purified Na-channel protein tends to aggregate into clusters. These aggregated protein units interact with very high cooperativity with respect to conformational changes. Activated clusters dissociate and reform spontaneously with some dependence on membrane voltage. Experimental observations

Molecular Basis of Nerve Activity

remained unchanged in series of using completely solvent-free bilayers on the tip of glass pipettes or using decane containing bilayers and/or 500 mM NaCl solutions.

Introduction

Purification and reconstitution of proteins which are responsible for ion translocation in biological membranes represents a reasonable way for elucidating the structural and environmental prerequisites of its native activity patterns. Parameters which influence ion channel activity are composition and physical state of neighbouring lipid molecules, ion composition of the inner and outer electrolyte solutions as well as conformation and state of aggregation of the protein unit. "In-vivo" channel current fluctuations cannot unconditionally be compared with those in reconstituted "in-vitro" systems. Due to the high channel density at the neuromuscular junction or the node of Ranvier it turns out to be impossible to record single channel events directly from these structures by the application of the "patch-clamp" technique (1). Reliable single channel recordings were presented from denervated and cultured embryonic muscle cells (2-4). There is no strict evidence that homogeneously distributed channel proteins in embryonic cells behave similarly to those which are inhomogeneously (5,6) arranged in clusters in adult cells.

Reconstitution of a highly purified channel protein has been successfully demonstrated for the acetylcholine receptor from Torpedo electric organ (7-9) and the neurotoxin-sensitive Na-channel from rat brain (10). With both proteins spontaneous transformations between channel types of different characteristics were observed (10,11). The purpose of this report is to present evidence for an aggregation behavior of the highly purified Na-channel protein after incorporation into planar lipid bilayers. Protein units within these aggregates of variable size interact with very high cooperativity. This is reflected by the appearance of a variety of well-characterized channels which reversibly transform into each other.

Materials and Methods

The Na-channel protein from rat brain was purified as previously described (10,12). Peak fractions of the sucrose gradients were used immediately for reconstitution into phospholipid vesicles. This vesicle suspension was used in reconstitution experiments with planar bilayers directly or after being kept frozen in liquid nitrogen up to several months. Batrachotoxin was the generous gift of Dr. J.W. Daly. Tetrodotoxin was purchased from Sigma and 1-palmitoyl-2-oleoyl-glycero-3-phosphatidylcholine (1,2-POPC) from Avanti. Soybean phospholipids type IIs (asolectin) was obtained from Sigma and partially purified by acetone extraction. Egg-phosphatidylethanolamine (egg-PE) was purified by column chromatography starting from fresh eggs.
Completely solvent-free planar lipid bilayers were formed on the tip of glass pipettes (13,14). Alternatively we obtained virtually solvent-free and decane-containing bilayers by using teflon sandwich septa with a hole of ca. 150 μm diameter (10,15). Vesicle preparations were added to the cis-side. As a consequence of the sidedness of the tetrodotoxin block the cis-side would correspond to the outside of the cell (10). Thus this side is defined as ground and inward current is designated as negative. Temperature: 22 - 23 °C.

Results and Discussion

Completely solvent-free planar bilayers:
The highly-purified Na-channel protein was reconstituted into completely solvent-free asolectin bilayers. We did not find

any significant difference in the observed single channel pattern as compared to bilayers derived from lipid/hydrocarbon solutions. Sonicated asolectin vesicles were mixed with the Na-channel protein containing vesicle suspension and submitted to a freeze-thaw cycle (8,16). The resulting suspension was used as bath solution. After formation of a lipid/protein film at the air/water interface (11) a planar bilayer was established on the tip of a firepolished glass pipette by application of the contact method (14). Fig. 1A shows a schematic drawing of this procedure made under capacitive control.
A typical current fluctuation trace in the absence of any neurotoxin under conditions of alternating (+,-)-voltages is presented in Fig. 1B. At + 50 mV no channel activity is visible. After the voltage jump to -50 mV the two channel types of large and small conductance, respectively, are clearly seen (10). Immediately after the voltage jump the large conductance type appears which then spontaneously transforms into the small one. This channel is characterized by its bursting activity. The step back to + 50 mV again reveals no open channel events suggesting that both channel types adopt resting states (10).
Open state conductances are G_e = 140 pS and G_s = 17 pS, respectively, in symmetrical 150 mM NaCl compartment solutions. Taking into consideration some negatively charged lipids as part of the asolectin lipid mixture the small conductance G_s is about half the value reported recently for a mixture of neutral lipids (10). Under the latter conditions bursting channels of similar conductance have been identified (see following section). Since we could not detect qualitative differences in channel characteristics on the solvent content of the bilayer, most experiments were carried out using virtually solvent-free planar bilayers from neutral lipids.

Virtually solvent-free planar bilayers:
As has been reported in our previous publication (10), channel activity of the highly purified Na-channel protein is induced

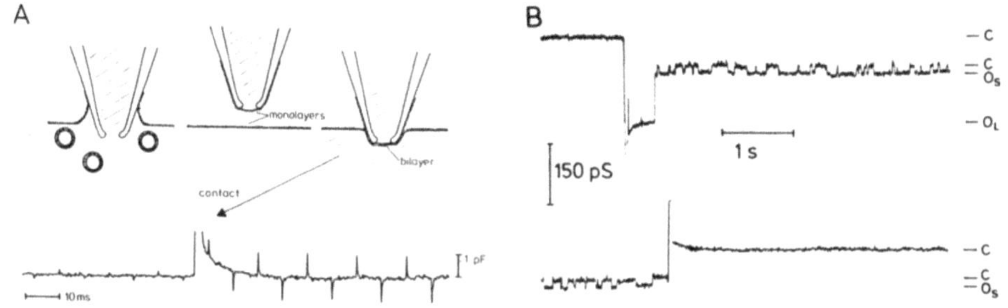

Fig. 1:

A. Upper part - Schematic illustration of the contact method of forming completely solvent-free planar lipid bilayer membranes on the tip of glass pipettes (10).
Lower part - Capacitive transients indicate the apposition of vesicle derived monolayers in form of a bilayer.

B. Current fluctuation trace under the condition of voltage jumps between +50 mV (upper level), -50 mV (downward deflections) and back to +50 mV. In the absence of any neurotoxin a channel type of large conductance (G_l = 140 pS, level difference O_l-C) and a bursting type of small conductance (G_s = 17 pS, level difference Os-C) appear at -50 mV. A spontaneous transformation of the large into the small channel type is visible. The transition occurs from the open state of the large channel type directly into the open state of the small one. At +50 mV both channels adopt corresponding resting states.

Experimental conditions: 60 mg asolectin in 1 ml buffer solution (in mM: 150 NaCl, 4 KCl, 2 $MgCl_2$, 2 $CaCl_2$, 5 Hepes, pH 7.2) were sonicated. Reconstituted vesicles containing the Na-channel protein highly purified by sucrose gradient centrifugation were added to the asolectin vesicle suspension and submitted to a freeze-thaw cycle (16). The lipid/protein film at the surface of the resulting liposome suspension is contacted by the monolayer on the tip of a glass pipette. The pipette was filled with the buffer solution described above. Voltage sign refers to bath as virtual ground.

by the application of appropriate voltage jumps. Fig. 2 shows the current fluctuation pattern which is typically seen during a voltage pulse series. Immediately after application of a negative voltage in most cases a large conductance level appears. This then abruptly falls off either to a lower conductance level of varying size or spontaneously transforms into a different channel type of bursting character. This bursting chan-

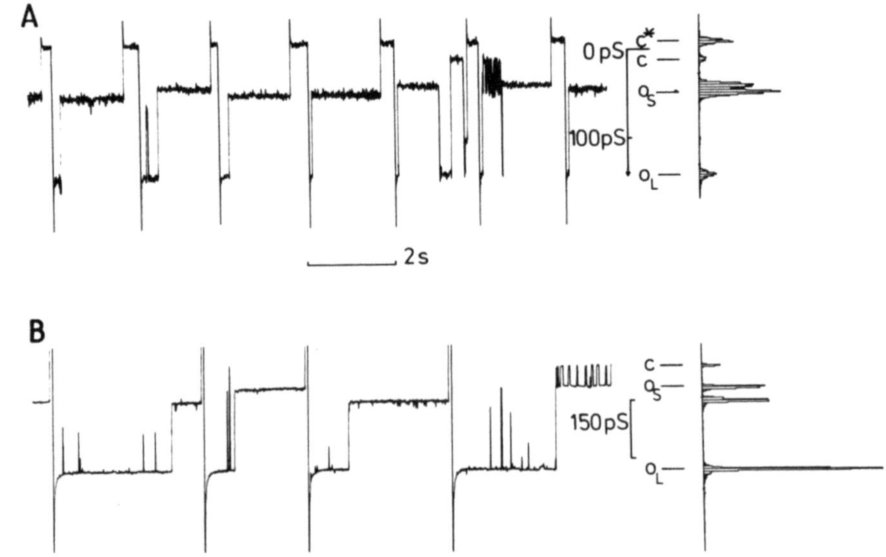

Fig. 2:

Alternative occurrence of the large and the small bursting channel type in the absence (A) or presence (B) of batrachotoxin (BTX).

A. A series of voltage jumps between 0 mV and -70 mV (indicated by the capacitive spikes) reveals the presence of a reproducible large conductance level (O_l) immediately after the jump to -70 mV. This conductance then is abruptly changed to a variety of lower values as demonstrated by the conductance distribution histogram on the right side. After the 6th jump shown on the record the large conductance channel type spontaneously transforms into a bursting channel type of 41 pS conductance (level difference O_S-C). At the end of the burst the large conductance level is adopted again before another lower level different from O_S or C, respectively, is reached. The lowest level observed during this particular experiment is indicated by the 0 pS horizontal bar. Its distance from the C-level amounts 14 pS. A maximum conductance for the large channel type of 141 pS is calculated (level difference O_l - 0 pS). C* means zero current level.

B. In the presence of 0.15 μM BTX a series of voltage jumps between +50 mV and -50 mV is applied. A constant level (O_l) is reached just after the voltage change to -50 mV. Afterwards an abrupt change to a lower conductance level of variable size or a spontaneous transformation into the bursting channel type is seen (level difference O_S-C: G_S = 55 pS). Level C is situated above the bare membrane conductance level. This means that the conductance G_l of the large channel type is at least 270 pS (level difference O_l-C). A comparison of the traces in A and B clearly shows longer lifetimes of the two channel types after the addition of BTX.

Experimental conditions: Virtually solvent-free planar lipid bilayers were formed from a solution of 75 % egg-PE, 20 % 1,2-POPC and 5 % cholesterol in hexane/ethanol (9:1) on a 150 µm diameter hole in a teflon sandwich septum preconditioned with hexane/hexadecane (20:1). Both compartments contained identical buffer solutions (in mM) 140 NaCl, 3 KCl, 0.5 $MgSO_4$, 0.5 $CaCl_2$, 25 Tris-Hepes, pH 7.4.
Protein containing vesicles were resuspended in the buffer solution supplemented with 200 mM sucrose. Suspension was added to the cis-compartment and in B 0.15 µM batrachotoxin to the trans-side.

nel of small conductance is terminated either spontaneously (Fig. 2A) or by a jump to positive voltages. Quite often the small channel exists up to minutes which enables the analysis of its voltage dependent gating behavior at negative voltages (Fig. 3).

The general appearance of the two channel types is similar in the absence or presence of the neurotoxin batrachotoxin (BTX). This toxin is known to prevent inactivation and to prolong state lifetimes of Na-channels in native membrane systems (17, 18). It is clearly seen from Figs. 2A,B and 3A,B that lifetimes of the open state (O_l) of the large channel type and of the open (O_s) and closed (C) states of the small one are significantly increased by BTX. This gives further evidence to the conclusion that both channel types are different expressions of the same highly purified protein.
Whereas the large channel type mainly adopts one state of large conductance and a variety of lower substate conductances during an experimental series, the small channel type is found to express different conductance values which are integral multiples of each other. Conductance values are (in pS ± 10 %): 14, 28, 42 and 56. In the presence of BTX the larger conductances are predominantly adopted. This applies to the small as well as to the large channel type. The surprising and important feature

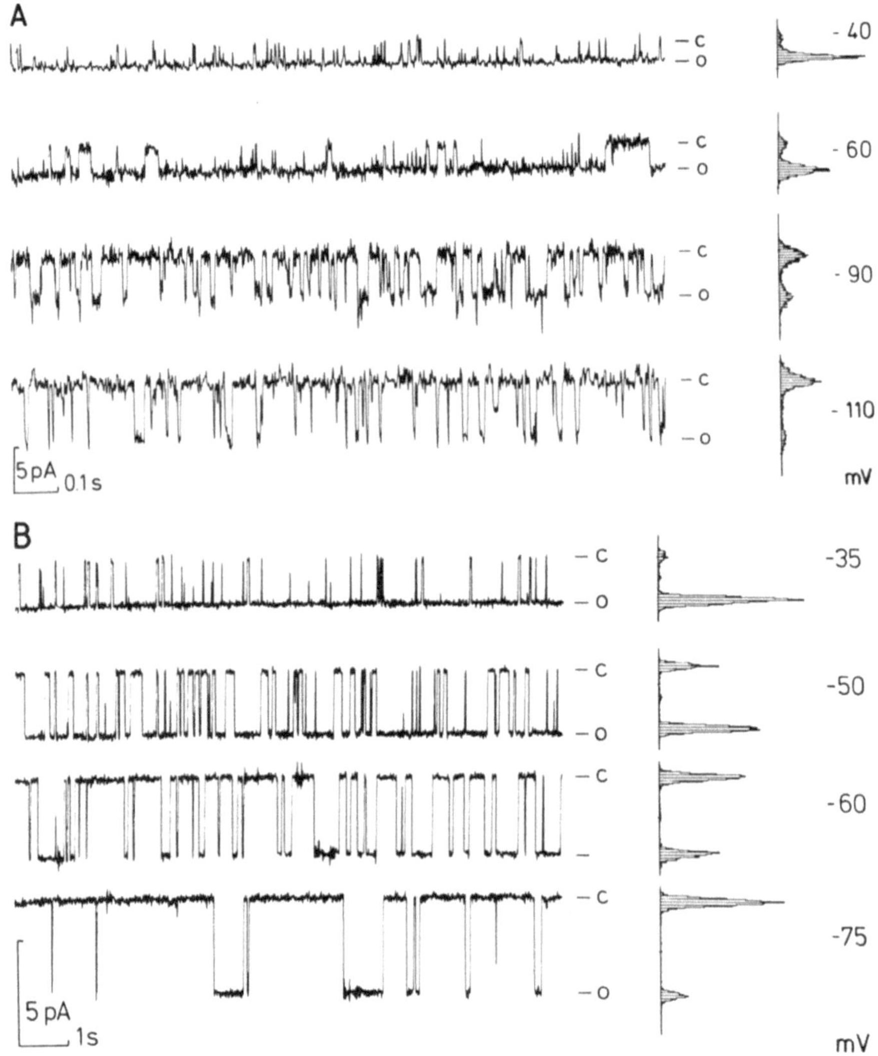

Fig. 3:

Voltage dependence of the distribution between the open and the closed state of the bursting channel type in the absence (A) or presence (B) of 0.15 µM BTX. Experimental conditions are the same as in Fig. 2.

A. The four traces exhibit a shift of the state distribution from mainly being open at -40 mV to mainly being closed at -110 mV. Corresponding level histograms are drawn on the right side of each trace. Conductance difference between the two levels amounts G_S = 56 pS.

B. A similar shift in the distribution between the open and the closed state is seen in the presence of BTX. From mainly being open at -35 mV the channel state changes to mainly being closed at -75 mV. Channel conductance is G_s = 60 pS. It is clearly seen that BTX enlarges channel state lifetimes, e.g. at -60 mV mean lifetimes of the open state τ_o = 50 ms without BTX and τ_o = 240 ms with BTX are found, respectively.

of these different channel expressions is that its gating processes are indistinguishable from "classical" single channel fluctuations (1). If one tries to explain the multitude of channel conformations expressed by a single protein unit (the M_r 270000 Na-channel protein (10)) in terms of aggregate formation, the predominant feature of these clusters would be its highly cooperativity in unit interaction. Then the large channel type would correspond to large clusters and the bursting type to smaller ones, respectively. Spontaneous changes in protein unit interactions would cause the characteristic channel type transformations.
The possible differences in cluster organisation which give rise to the two channel types of different conductance and kinetic properties is completely unknown at present. We cannot even speculate on it. Without BTX the conductance of the large channel type is G_l = 120 - 160 pS and in the presence of BTX at least G_l = 200 - 300 pS, whereas the large low-level conductance in the latter case hints at still larger conductance values. Fig. 2 demonstrates that the mean lifetime of the high conductance state of the large channel type in the absence of BTX (in A: τ_o = 0.39 ms) is significantly shorter than that in the presence of BTX (in B: τ_o = 1.4 s). Similarly the mean lifetime of the open state of the small channel type without BTX (Fig. 3A: τ_o = 50 ms) turns out to be shorter than the corresponding mean lifetime with BTX (Fig. 3B: τ_o = 240 ms). In contrast the conductance sequence (14,28, 42, 56; in pS $\pm$ 10 %) remains unaffected by BTX. Fig. 4A,B presents a half-logarithmic plot of the probability p_o to find the small channel type in the open state. p_o-values were calculated from

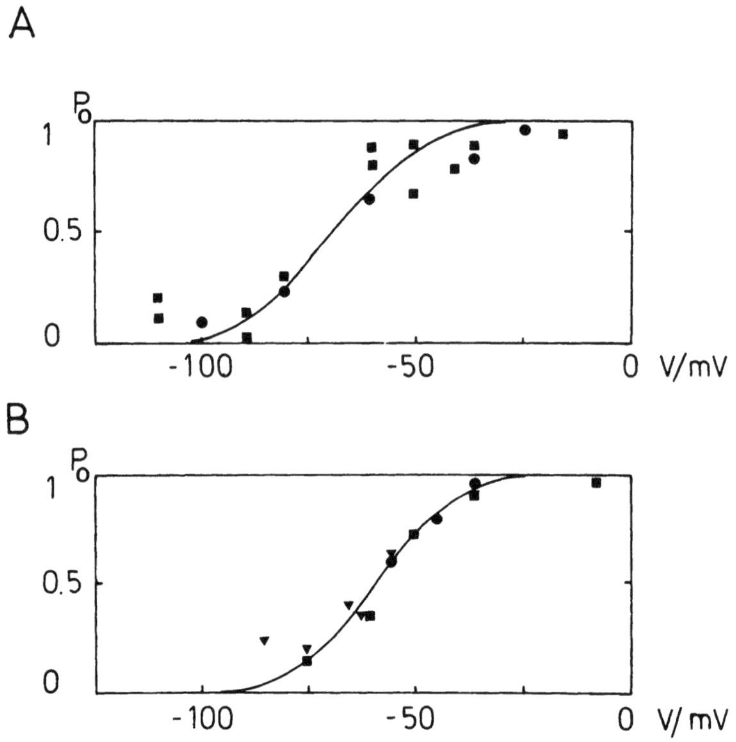

Fig. 4:

Halflogarithmic plot of the probability p_o to find the bursting channel in its open state in the absence (A) or presence (B) of 0.15 µM BTX. Experimental conditions are the same as in Fig. 2.

A. Three "56 pS channels" (■) and one "28 pS channel" (●) are considered. p_o = 0.5 at V_r = -70 mV. This reference voltage V_r is a mean value which applies to the majority of bursting channels observed. Occasionally channels appear with V_r values which differ from the above given mean by at least 50 mV.

B. Corresponding plot of two "56 pS channels" (■,▼) and one "42 pS channel" (●) in the presence of BTX. At p_o = 0.5 a reference voltage V_r = -60 mV is found. A formal gating charge of approximately 3 elementary charges would explain the observed voltage dependence.

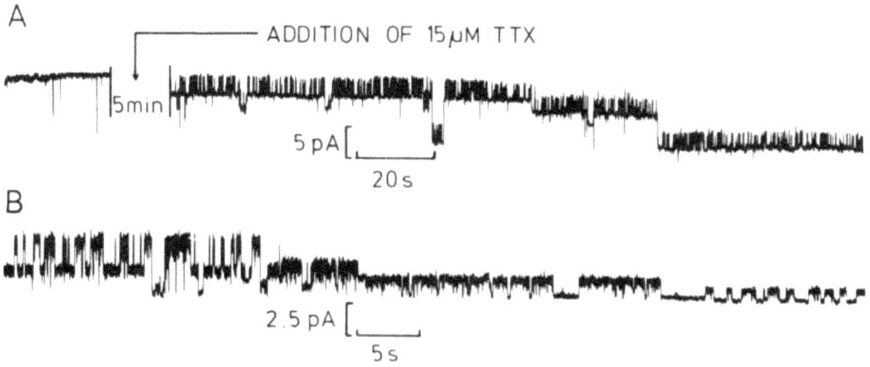

Fig. 5:

Effect of 15 µM tetrodotoxin (TTX) on the conductance induced by the reconstituted highly purified Na-channel protein in the presence of 1 µM BTX. Experimental conditions are the same as in Fig. 2. Applied voltage: -20 mV.

A. After the addition of BTX to the trans-compartment large conductance levels are adopted. Then 15 µM TTX are given to the cis-compartment and after a few minutes a gradual degradation of the large conductance in steps of variable size is observed.

B. The situation at a later time is shown at enlarged current and time scales. Variable step sizes can be clearly distinguished. Broad noise band means bursting channels.

traces similar to those shown in Fig. 3A,B. Channels of different conductance were compared in the absence or presence of BTX. In contrast to in-vivo observations (19) BTX did not shift the p_o/V-curve along the voltage axis. Mean reference voltages (at which $p_o = 0.5$) are $V_r = -70$ mV without BTX and $V_r = -60$ mV with BTX. The slope of the straight line in the $p_o/(1-p_o)$ versus V plot is consistent with a formal gating charge of approximately 3 elementary charges.

The effect of the Na-channel blocking neurotoxin tetrodotoxin (TTX) on the two channel types modified by BTX is shown in Fig. 5. After the addition of 1 µM BTX to the trans-compartment and subsequent channel activation 15 µM TTX had to be added to the cis-side in order to observe the stepwise conductance degradation. Abrupt changes in conductance of variable size and broad noise bands which reflect bursting channels are visible. It is obvious from Fig. 5 that TTX affects both channel types. It does not exclusively act on the single protein unit by closing one 14 pS pathway after the other, but also changes the cooperative interaction between protein units.

Decane-containing planar bilayers:

Since Krueger et al. (20), who reported Na-channel incorporation after addition of native membrane fragments from rat brain into a decane-containing bilayer system, did not present data on a large conductance channel type, we tested our system with decane-containing bilayers in the presence and absence of BTX and furthermore in 140 mM as well as in 500 mM NaCl compartment solutions. We could not detect a general difference to the observations in solvent-free bilayers reported above. The 14 pS channel in 140 mM NaCl seems to correspond to the 30 pS channel in 500 mM salt (20).

The cluster hypothesis:

Classical electrophysiology does not present evidence for highly cooperative gating processes of channel clusters of large and variable size. The recent report of Schindler et al. (11) on cooperative gating of acetylcholine receptor aggregates in planar bilayers does not coincide with corresponding results obtained by the patch-clamp technique on cell membranes (21). However, on electron micrographs monomers and dimer aggregates of the receptor protein can clearly be recognized (22). Similar facts apply to the Na-channel protein. Electron microscopic visualization reveals aggregation into clusters with a unique ribbon-like conformation (23), but no electrophysiological indication of clustered channels exist. This raises the question to whether cluster formation may be an intrinsic property of reconstitution experiments in artificial lipid bilayer membranes. This objection has to be discussed even more seriously if highly purified proteins are reconstituted which are stripped off their native lipid environment. Reconstitution experiments seem to indicate a correlation between the degree of protein purification and the occurrence of channel clusters. The higher the purification step the stronger is the electrochemical evidence for channel cluster formation. In view of these arguments the peculiar lipid environment of a channel protein seems to play a crucial role. Analysis of the lipid environment of the acetylcholine receptor protein at the postsynaptic membrane by Popot et al. (24) revealed a very large amount of lipids with highly unsaturated hydrocarbon chains. On the basis of lipid phase separation phenomena the following explanation might be possible. If the lipid membrane would be structured as a moderate fluid matrix with highly fluid separated domains, then solute molecules would migrate into the more fluid phase. There the highly fluid lipid environment may interrupt cooperative protein interaction which, on the other hand, would be established by the less fluid lipids used for reconstitution experiments. This means that in protein-rich membrane areas less fluid lipids may serve as mechanical transmitter of conformational changes of protein structure, but highly fluid lipids do not.

Acknowledgement

This work was supported by the Minister für Wissenschaft und Forschung des Landes Nordrhein-Westfalen (IV B4 - FA 9273), by the Deutsche Forschungsgemeinschaft (SFB 114), by the Centre National de la Recherche Scientifique, by the Foundation de la Recherche Medicale and by the Ministere de l'Industrie et de la Recherche (grant no. 81.E.0575).

References

1. Hamill, O.P., A. Marty, E. Neher, B. Sakmann, F.J. Sigworth. 1981. Pflügers Arch. 391, 85 - 100.
2. Neher, E., B. Sakmann. 1976. Nature. 260, 799 - 802.
3. Hamill, O.P., B. Sakmann. 1981. Nature. 294, 462 - 464.
4. Sigworth, F.J., E. Neher. 1980. Nature. 287, 447 - 449.

5. Cartaud, J., E.L. Benedetti, A. Sobel, J.P. Changeux. 1978. J. Cell. Sci. 29, 313 - 337.

6. Almers, W., P.R. Stanfield, W. Stühmer. 1983. J.Physiol. 336, 261 - 284.

7. Boheim, G., W. Hanke, F.J. Barrantes, H. Eibl., B. Sakmann, G. Fels, A. Maelicke. 1981. Proc. Natl. Acad. Sci. USA. 78, 3586 - 3590.

8. Tank, D.W., C. Miller. 1983. In: Single-Channel Recording (B. Sakmann, E. Neher, eds.). Plenum Press, London, p. 91 - 105.

9. Labarca, P., J. Lindstrom, M. Montal. 1984. J. Gen. Physiol. 83, 473 - 496.

10. Hanke, W., G. Boheim, J. Barhanin, D. Pauron, M. Lazdunski. 1984. EMBO J. 3, 509 - 515.

11. Schindler, H., F. Spillecke, E. Neumann. 1984. Proc. Natl. Acad. Sci. USA. 81, 6222 - 6226.

12. Barhanin, J., D. Pauron, A. Lombet, R.I. Norman, H.P.M. Vijverberg, J.R. Giglio, M. Lazdunski. 1983. EMBO J. 2, 915 - 920.

13. Wilmsen, U., C. Methfessel, W. Hanke, G. Boheim. 1983. In: Physical Chemistry of Transmembrane Ion Motions (G. Spach, ed.). Elsevier, Amsterdam, p. 479 - 485.

14. Hanke, W., C. Methfessel, U. Wilmsen, G. Boheim. 1984. Bioelectr. Bioenerg. (in press).

15. Boheim, G., W. Hanke, G. Jung. 1983. Biophys. Struct. Mech. 9, 181 - 191.

16. Pick, U. 1981. Arch. Biochem. Biophys. 212, 186 - 194.

17. Albuquerque, E.X., J.W. Daly. 1976. In: Receptors and Recognition Ser. B, Vol. 1 (P. Cuatrecasas, ed.). Chapman & Hall, London, p. 297 - 338.

18. Khodorov, B.I., S.V. Revenko. 1979. Neuroscience. 4, 1315 - 1330.

19. Huang, L.M., N. Moran, G. Ehrenstein. 1982. Proc. Natl. Acad. Sci. USA. 79, 2082 - 2085.

20. Krueger, B.K., J.F. Worley III, R.J. French. 1983. Nature. 303, 172-175.

21. Sachs, F. 1983. In: Single Channel Recording (B. Sakmann, E. Neher, eds.). Plenum Press, London, p. 365 - 376.

22. Cartaud, J., J.L. Popot, J.P. Changeux. 1980. FEBS Lett. 121, 327 - 332.

23. Ellisman, M.H., W.S. Agnew, J.A. Miller, S.R. Levinson. 1982. Proc. Natl. Acad. Sci. USA. 79, 4461 - 4465.

24. Popot, J.L., R.A. Demel, A. Sobel, L.L.M. van Deenen, J.P. Changeux, 1978. Eur. J. Biochem. 85, 27 - 42.

K^+-CHANNELS IN RAT VENTRICULAR CELLS SHOW VOLTAGE-DEPENDENT OUTWARD RECTIFICATION AND NON-LINEAR VOLTAGE TO CURRENT RELATION

W.Schreibmayer, H.Hagauer, H.A.Tritthart

Institute of Medical Physics and Biophysics, University of Graz
Harrachgasse 21/IV, A-8010 Graz

H.Schindler

Biocenter,University of Basel
Klingelbergstr.70, CH-4056 Basel

Introduction

At least two different types of voltage-dependent K^+-conductances exist in cardiac muscle (1): An instantaneous voltage-dependent pathway, which is activated by hyperpolarization (anomalous rectification) and a time-dependent outward current activated on increasing depolarization (delayed rectification). The latter conductance can be described by the Hodgkin-Huxley formulation(2).

Here we describe patch clamp studies on single K^+-selective channels, which show outward rectifying properties. The findings are compared with outward currents of whole cells from the same preparation.

The relation between voltage and single channel current has been found to be non-linear and single channel conductance increased from 24 to about 100 pS with depolarization.

Molecular Basis of Nerve Activity

Results

Whole cell current recordings

Rat ventricular cells were isolated as described by Piper et al.(3) and stored at 37°C in the culture medium M199 (Biochrom KG,Berlin) for up to two days. We used the method described by Marty and Neher(4) to measure whole cell currents at two external $K^{\pm}$concentrations (4 mM and 20 mM, see Figure 1). The electrodes were filled with (in mM): 120 KCl, 0.5 $CaCl_2$and 10 K^{+}/Hepes, pH:7.0 and the external media had the following composition (mM): For 4 mM external K^{+}: 136 NaCl, 4 KCl,0.5 $CaCl_2$,2 $MgCl_2$, 10 Na^{+}/Hepes, pH: 7.0; for 20 mM external K^{+}: 120 NaCl, 20 KCl, 0.5 $CaCl_2$, 2 $MgCl_2$, 10 Na^{+}/Hepes, pH: 7.0. The reversal potentials (measured values: -55 mV at 4 mM and -34 mV at 20 mM) of the outward currents followed the K^{+}-equilibrium potential. Assuming concentrations of 150 mM K^{+} and 15 mM Na^{+} in the cytosol (5), permeability ratios for P_{Na}/P_{K} are calculated to be 0.10 for 4 mM and 0.16 for 20mM K^{+}outside (6). For 20 mM external K^{+} the increase of total cell $K^{\pm}$conductivity with voltage was calculated; the result is plotted in Figure 4 (filled triangles).

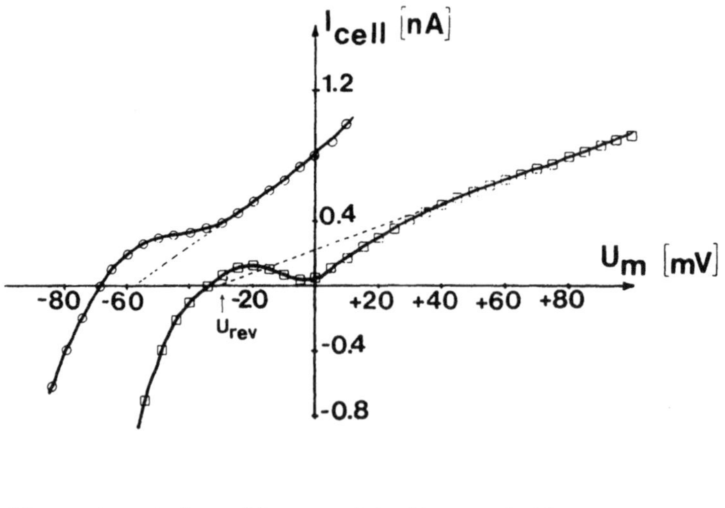

Figure 1 : whole cell current/voltage relation.

Single channel measurements with the patch-clamp method

These studies were performed at room temperature according to the method of Hamill et al.(7) on both cell attached and isolated membrane patches. Single channel currents from patches with a resistance greater than 10 Gohms were low pass filtered at 0.3 or 1 kHz and recorded on PCM tape (Johne and Reilhofer 3K12). Under the ionic condition of 20 mM K^+ outside, pipetts, filled with the external medium, were sealed against the cells and the patch current was measured. The pipett potential was slowly changed to produce a triangular wave (frequency: 0.025 Hz) and the observed single channel currents generated demonstrate the voltage dependence. The measured reversal potential (-34 mV) was taken as the resting potential to calculate the actual voltage across the membrane patch.

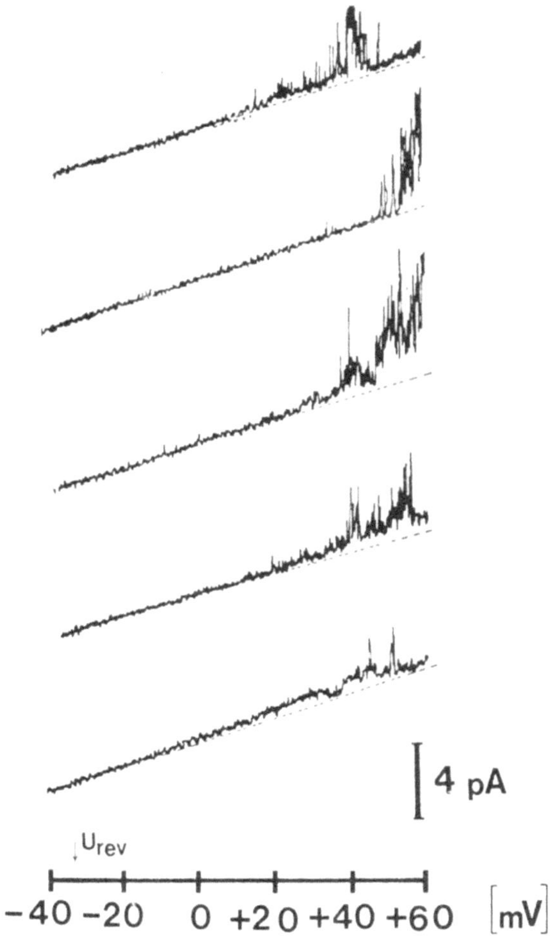

Figure 2

Typical recordings are shown in figure 2. 28 of these single channel traces were averaged on a HP-1000 computer system; the result is shown in Figure 3. After subtraction of the leakage current, the increase in steady state conductivity was calculated and plotted against membrane potential (Figure 4, open triangles); this was then compared with a similar whole cell conductivity plot (Figure 4 , filled triangles).

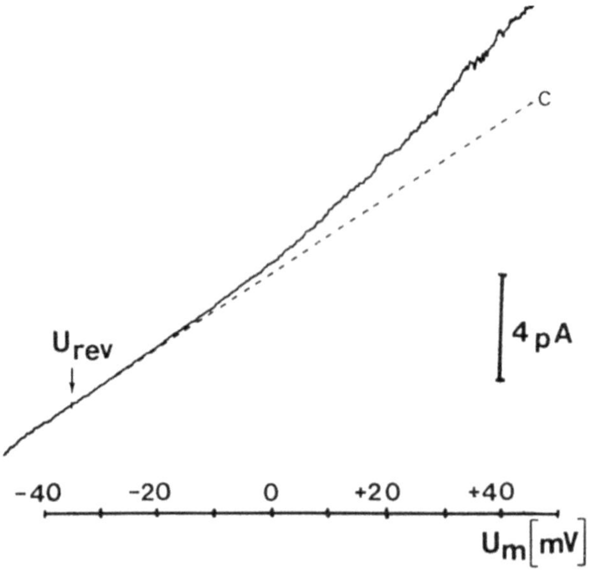

Figure 3 : Averaged single channel traces

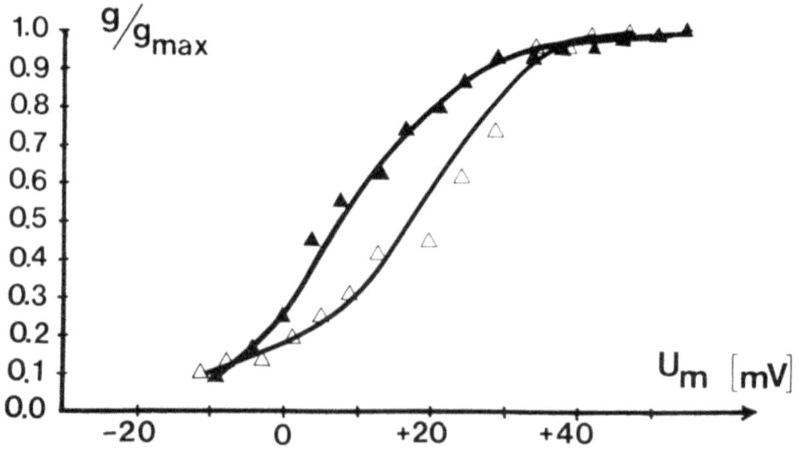

Figure 4 : Comparison whole cell/ patch - clamp

Delayed rectifier currents are known to be blocked specifically by 4-aminopyridine (8,9). With isolated outside-out patches 10 mM 4-aminopyridine blocked single channel currents completely (Figure 5). Pipett medium (mM): 120 KCl, 0.5 $CaCl_2$, 10 K^+/Hepes, pH:7.0; external medium: 120 NaCl, 20 KCl, 0.5 $CaCl_2$, 2 $MgCl_2$, 10 Na^+/Hepes, pH:7.0.

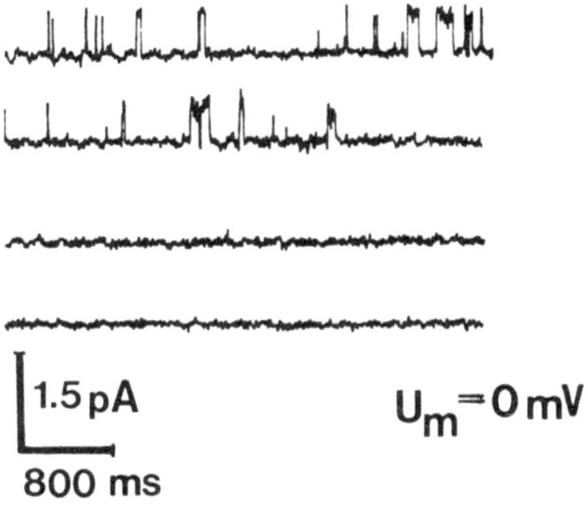

Figure 5 : Block of channels by 4-aminopyridine

On excised outside-out patches, under symmetrical ionic conditions (concentration (mM): 120 KCl, 20 NaCl, 0.5 $CaCl_2$, 10 K^+/Hepes, pH: 7.0), the size of the rectangular jumps in patch current (which we will refer to as an "isolated" rather than a "single channel" event) did not show a constant, unique, conductance over the observed voltage range (+/- 70 mV), but the conductance of the "isolated" event varied between 24 and about 100 pS; the higher values occured more frequently with increasing depolarization. Even at a given potential the "isolated" event current was variable between several discrete levels.

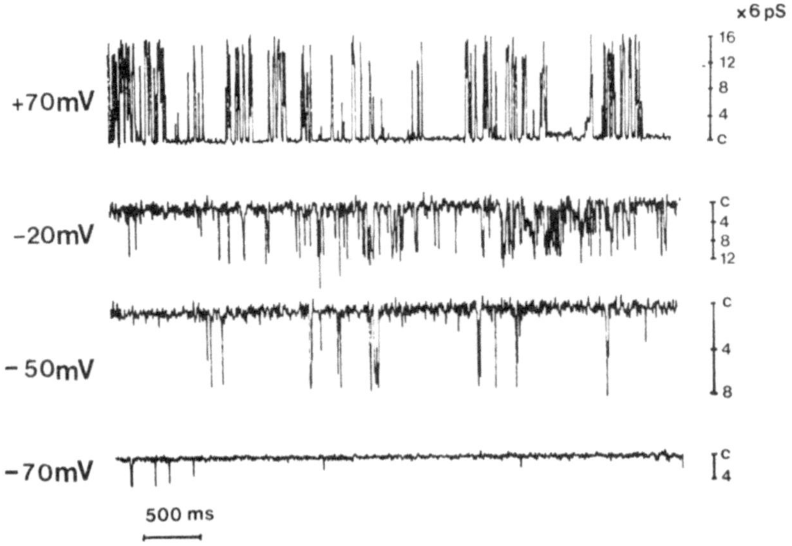

Figure 6 : Isolated event currents at different potentials

Conclusions

The increasing number of active channels with membrane depolarization is evidence of the delayed and outward going rectifyer properties of the channels studied. The block of single channel current by 4-aminopyridine and the selectivity data as well as effects of Cs^+ and Ba^{2+}, reported in a previous paper (10), support this view. The outward currents observed on cell attached membrane patches agree well with outward current measurements across the membrane of whole single cardiac cells (Figure 4), a conductance that is known to follow the Hodgkin-Huxley equations (2).

Different mechanisms have been discussed to explain the steep non-linear relation between membrane-voltage and the size of the "isolated" event (10), i.e. rectangular jumps in patch current. Recently reconstitution experiments (see Schindler and Schreibmayer, this issue) allowed for the assignment of this surprising behaviour to synchronous opening and closing of individual K^+-conductance pathways of 6 pS, indicated here also in Figure 6. The degree of coupling between the individual gating processes has been shown to be governed by voltage and to result in the macroscopic phenomenon known as delayed rectification of membrane current.

Acknowledgements

We thank Prof.H.F.Grundy (Cambridge, England) for reading the manuscript.
This research was supported by the Austrian research fund project No. P4552 and the La-Roche foundation.

References

1.Reuter H. 1984. Ion channels in cardiac cell membranes, Ann. Rev. Physiol. 46, 473 - 484.

2.Noble D. 1962. A modification of the Hodgkin-Huxley equations applicable to Purkinje fibre action and pace-maker potentials, J. Physiol. 160, 317-352.

3.Piper H.M., Probst I., Schwartz P., Hütter F.J. and Spieckermann P.G. 1982. Culturing of calcium stable adult cardiac myocytes, J. Mol. Cell. Cardiol. 14, 397- 412.

4.Marty A. and Neher E. 1983. In: Single-channel recording (Sakmann B. and Neher E., eds.), Plenum Press, New-York and London. pp 107-122.

5.Sperelakis N. 1979. In: Handbook of physiology section 2: The cardiovascular system vol.I: The heart (Berner R.M., Sperelakis N. and Geiger S.R., eds.), Waverly Press, Baltimore and Maryland, pp187-267.

6.Goldman D.E. 1943. Potential, impedance and rectification in membranes, J. Gen. Physiol. 27, 37-60.

7.Hamill O.P., Marty A., Neher E., Sakman B. and Sigworth F.J. 1981. Improved patch-clamp techiques for high resolution current recording from cells and cell-free membrane patches, Pflüg. Arch. 391, 85-100.

8.Kenyon J.L. and Gibbons W.R. 1979. 4-Aminopyridine and the outward current of sheep cardiac Purkinje fibres, J. Gen. Physiol. 73, 139-157.

9.Dubois J.M. 1983. Potassium currents in the frog node of Ranvier, Prog. Biophys. Mol. Biol. 42, 1-20.

10.Schreibmayer W., Tritthart H.A., Zernig G. and Piper H.M. 1984. Single voltage-dependent and outward rectifying K^+-channels in isolated rat heart cells, Europ. Biophys. J. (submitted for publicaton).

A PURIFICATION PROCEDURE OF THE TETRODOTOXIN BINDING COMPONENT CONTAINED IN THE ELECTROPLAXES OF ELECTROPHORUS ELECTRICUS

G. Dandrifosse, Ch. Grandfils, L. Bettendorff, E. Schoffeniels, J. Bontemps
Laboratory of general and comparative Biochemistry, University of Liège,
17, place Delcour, B-4020 Liège, Belgium

Introduction

By using radioactive toxins, purifications of the sodium conducting sites have been achieved from different excitable tissues : electric organ of Electrophorus electricus (1-7), rat skeletal muscle sarcolemma (8,9) and rat brain synaptosomes (10,11).
In the present work the proteins of the Na^+-conducting sites of Eel electroplaxes were purified by a somewhat different and simpler method, using a ^3H-ethylene diamine di-tetrodotoxin derivative (ED-diTTX) recently synthetized in our laboratory (12).
While the solubilization and protection by phosphatidylcholine was similar to described procedures (1,2), the first step of the purification or ion-exchange chromatography in a batch mode was extensively studied to achieve satisfactory recovery of ED-diTTX binding activity. The second purification step was performed using a commercially available gel of wheat germ lectin in contrast to other authors (6,8) who prepared their own gel with a higher degree of substitution. Moreover the second purification step was performed in the absence of stabilizing toxin. The critical parameters of this rapid two-step purification are discussed in detail, as well as the technique of rapid gel filtration used in binding experiments. Precise experimental details are given so that the method is well adapted for new investigations in this field.

Molecular Basis of Nerve Activity

Material and Methods

Binding of (^{3}H)-ethylene diamine di-tetrodotoxin (ED-diTTX)

(^{3}H)-ED-diTTX at 35 Ci/mmol was prepared as described (12). Its binding to solubilized material was assayed according to the fast gel filtration technique of Levinson et al. (13). Small columns of Sephadex-G50 fine were prepared in disposable seringues of 2 ml from Monoject (Brunswick). For the preparation of 100 columns, 15 g of gel were swollen in 50 mM sodium phosphate buffer (pH 6.8) at 90°C for 1 h, and cooled down to room temperature. The gel was transferred into a buchner and washed with 10 bed volumes of buffer. It was finally resuspended to form a 220 ml slurry for degasing. The columns were filled with 2.2 ml of gel suspension retained by a thick filter paper (0.38 mm). Each column was equilibrated in the cold with the incubation buffer by two washings with 1.5 ml of phosphate buffer containing 0.1 % (w/v) Lubrol-PX (Sigma) and phosphatidylcholine (PC) (Lipid Products, Nutfield England) in a Lubrol/PC weight ratio equal to five. Columns were stored in the cold and hydrated with 0.5 ml of buffer before use. In general, they were prepared just before use or the day before.

In a binding experiment the incubation mixture of 0.5 ml was allowed to equilibrate for 5 min. in the presence of 10 nM (^{3}H)-ED-diTTX. Two aliquots of 200 µl were loaded on the columns and centrifuged for 90 sec at full speed on a table-top centrifuge (model Universal Junior S from Heraeus Christ). The void volume fraction of 1.05 ml was directly collected in a small scintillation vial. In these centrifugation experiments the main objective is to obtain a good flow of solvent through the column. For this, a filter paper was found superior over glass wool. In the latter case, a short needle was also added at the end of the seringue.

Unspecific binding was determined by difference. From the total binding measured in the presence of the labelled toxin at 10 nM, the value obtained with a 100 fold excess of cold toxin was substracted. Using this technique, a K_D value of 0.5 nM and a B_m value of 10-12 pmol/mg protein were obtained with the membrane fraction (12).

Membrane suspension from <u>Electrophorus electricus</u>

Membranes were prepared according to Agnew et al.(1) in the presence of 5 mM EDTA and 0.1 mM PMSF (paramethylsulphonyl fluoride, from Sigma). The 48,000 g pellet was suspended at a ratio of 1 ml buffer per g of membrane and stored at -70°C.

Stock solutions for one purification

Lubrol-PX was prepared at 10 % (w/v) and deionized on a mixed-bed resin from Bio-Rad (type AG 501-X8). Aliquots of 10 ml were stored frozen. The Lubrol/PC solution was prepared by the addition of 10 ml Lubrol-PX (10 %) to 200 mg of phospholipid obtained by evaporation by nitrogen in the cold. The phosphate buffer was prepared at pH 6.8 (Na_2HPO_4-H_3PO_4, 0.1 M). It was adjusted to 0.1 mM PMSF before use. The inhibitor was added as powder and dissolved in the buffer at room temperature for 1 h. For DEAE chromatography, a KCl solution at 3 M was used.

Solubilization or 1 % Lubrol extract

A 9 ml membrane suspension was homogenized for several minutes by using a Potter homogenizer at 0°C (teflon pestle). Solubilization was achieved by the addition of 1 ml of detergent (10 %). The mixture was centrifuged at 100,000 g for 60 min. at 2°C to yield the detergent extract. The supernate and the lipidic phase were combined for the purification.

Ion-exchange chromatography on DEAE-Sephadex A-25

The gel was carefully swollen and washed in an excess of 50 mM phosphate buffer. The cooled and degased gel was then equilibrated on a büchner by 10 volumes of buffer containing 0.1 % Lubrol-PX and 0.02 % phosphatidylcholine. A 50 ml-buchner was used to allow a gentle magnetic stirring.

It was thermostatized at 0°C during the whole experiment by immersion in a water-ice mixture.
A gel volume equal to that of the detergent extract was used, i.e. 10 ml. The sample was added to the gel and allowed to interact for 30 min. at 0°C under gentle mixing. After that period of time unretained proteins were recovered from the gel by gravity. Subsequently, the gel was washed three times for 5 min. with the buffer containing 250 mM KCl then three times for 5 min. with the same buffer containing 275 mM KCl.
Aliquots of each washing solution were taken for binding measurements and protein determination using the fluorescamine technique (14) (borate buffer, pH = 9.5) with bovine serum albumin as reference protein.
The elution of proteins was achieved by the addition of 10 ml phosphate buffer containing 800 mM KCl for 30 min, and twice 5 ml of 600 mM KCl for 5 minutes.

Wheat germ lectin affinity chromatography

The support used was the wheat germ lectin linked to macrobeads of Sepharose 6MB from Pharmacia. 10 ml of swollen gel were packed in a column. The whole separation process was performed at a flow rate of 0.8 ml/min.

Results

Ion-exchange chromatography on DEAE-Sephadex A-25

The first purification step initially introduced by Agnew *et al*. (1) consists on an ion-exchange chromatography on DEAE-Sephadex A-25. It has been widely accepted for the channel purification from different organs except in the case of Barchi *et al*. (8) who synthetized their own weak anion exchanger. As the majority of the proteins do not interact with the charged matrix, a procedure in the batch mode is more efficient as regards to the washing of the gel and to the rapidity. Different ionic strengths were assayed both in the washing and eluting steps in order to achieve high binding recovery as well as improved purification yield (Table 1).

Table 1. First purification step using ion-exchange chromatography : determination of the degree of purification and binding recovery for different washing and eluting experimental conditions

	1			2			3		
Scheme for washing and elution steps	Wash : 6 x 200 mM KCl Elution : 10 ml 600 mM 10 ml 400 mM			Wash : 6 x 250 mM Elution : 10 ml 800 mM 2 x 5 ml 600 mM			Wash : 3 x 250 and 3 x 275 mM Elution : 10 ml 800 mM 2 x 5 ml 600 mM		
Initial binding activity (pmol/mg protein)	10.2	10.2	14.5	10.3	8.6	8.3	8.0	10.3	9.0
Eluted binding activity (pmol/mg protein)	103 113 -	102 123 -	192 172 -	118 119 113	126 111 128	102 112 113	197 190 166	179 189 192	183 170 161
Degree of purification	11	12	13	12	15	14	25	19	20
Binding recovery (% of initial activity)	40	41	48	70	60	62	66	59	66

The sample was not adjusted to a particular ionic strength before loading, while the exchanger was equilibrated either with 200 mM or 250 mM KCl before use.

In the experiments reported in Table 1 (scheme 1), the initial protein concentration (P_o) amounted to 7 mg/ml with a mean binding activity of 12 pmol of ^{3}H-ED-diTTX bound per mg protein. Six washes with 10 ml of the same 200 mM salt solution decreased that initial protein concentration down to 0.05 mg/ml. At each step the concentration value was reduced by a factor of 2.5 so that a semi-logarithmic plot of log P (protein concentration) versus eluted volume was found to be linear (results not shown). The good reproducibility of this step was proved when the results of six experiments were plotted, after normalisation of the ordinate to the initial protein concentration, P/P_o. Using this washing scheme, two types of elution were assayed yielding to the following results. With elution by 600 and 400 mM (scheme 1), the best binding recovery was of 48 % for this step with an average purification yield of 12 based on the three experiments. These rather low values could be improved using other salt concentrations.

A first improvement was obtained by increasing the washing salt concentration to 250 mM (scheme 2) and the elution strength to 800, 600 and 600 mM, i.e., an additional elution step. Under these conditions, the binding recovery increased up to 60-70 % but the purification factor was found to be unchanged.

For increasing further the degree of purification, a biphasic washing process was assayed (scheme 2). Instead of 6 washes at constant 250 mM concentration, the washing concentration was raised to 275 mM in the three final washes, while maintaining the same elution scheme. The main advantage of this system was a better purification factor of more than 20 and a good binding recovery. It should be mentioned that the application of the 275 mM salt concentration did not modify the linarity of the elution plot (see above) but released some binding activity at very low protein concentration. While the highest eluted activities were about half that of the original sample in schemes 1 and 2, it amounted to 13, 28 and 43 pmol/mg prot in scheme 3 where 275 mM KCl was used. The corresponding values of protein concentrations were 0.18, 0.08 and 0.04 mg/ml.

The high value of the purification factor obtained here is in good agreement with the data of the literature (1,6) with the advantages that the critical parameters have been clearly pointed out.

Affinity chromatography on wheat germ lectin Sepharose-6 MB

This purification step adopted by various authors (6,8,10) appears as a faster step as compared to gel filtration chromatography after concentration of the DEAE pooled fractions on ultrafilters (1). In this work, we aimed to use a commercial lectin gel without protection of the receptor by the toxin, i.e. a more simple procedure using a column system.
As shown in Fig. 1, a 5 ml-column was loaded with the pooled fractions of the DEAE step at a specific activity of 186 pmol/mg protein. During the load of the column and its washing, a significant binding activity was measured, slightly inferior to that of the loaded sample. As soon as the gradient of N-acetylglucosamine was applied a sharp peak of binding activity was observed. The majority of the binding was located in the leading part of the protein elution profile (Fig. 1b). At the peak the degree of purification for this step is 6.5 fold, which is comparable to the value observed by Norman et al. (6) in the presence of the stabilizing TiTxγ toxin and using a home-made gel with a higher degree of lectin-substitution. When comparing the protein and activity patterns, it appears that this purification steps is partially unsatisfactory. Indeed, an important binding activity was measured before the gradient elution. Moreover, the binding activity was associated to only one minor part of the removed proteins suggesting non specific binding of protein to the affinity support.

Conclusions

The purification of the Na^+-channel proteins appears as a simple procedure as usual chromatographic techniques are used. This is not the case as no more significant progresses have been described since the earlier works of Agnew et al. (1). In addition before the developmental study of the purification process, different steps have to be accomplished.
First, a Na^+-channel probe has to be available, i.e. a labelled form of a neurotoxin. Such a marker has been previously synthetized and purified in

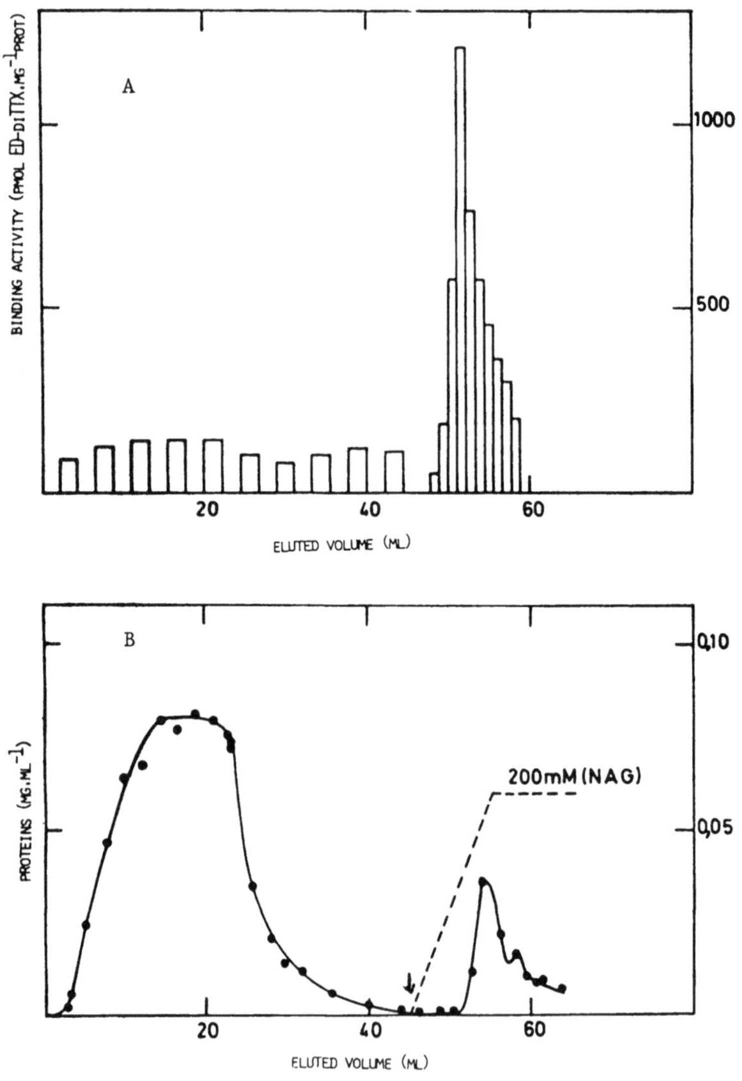

Fig. 1 : Binding (part A) and Protein (part B) profiles for the second purification step consisting of a chromatography on a 5 ml wheat germ lectin Sepharose-6MB column.
The 0-200 mM N-acetylglucosamine gradient of 10 ml was applied as indicated by the arrow.

our laboratory (12). The derivative obtained was a new labelled form of tetrodotoxin prepared with a high overall yield and a high specific activity.

Secondly, the binding properties of this probe have to be studied. The experiments are usually carried out first with a membrane suspension for which the free/bound separation is performed on glass fiber filters. The binding parameters of the solubilized material have to be determined by rapid gel filtration, as described by Levinson et al. (13). In our case, the method has been fully optimized with respect to the following parameters : design of the bottom part of the column for obtaining a good flow, centrifugation time and speed for a total protein recovery (bound toxin) without contamination of the free toxin and, delay time between sample applications and column centrifugation (15).

The third step before attempting the purification is to check the stabilization of the solubilized TTX receptor by exogenous phospholipids. This problem has been analyzed previously (2,15).

After these considerable efforts, the purification may be investigated as reported in this work. Our aim was to present a simple and economic purification procedure, which did not require the synthesis of chromatographic media nor the use of an expensive labelled toxin which has to be added to the buffers to stabilize the receptor.

Moreover the reproducibility of the presented experiments may be appreciated. This is not the case of numerous publications where typical experiments are usually described. The purification yields obtained with our method ranged between 20 and 25 for the first purification step and 5 and 6 for the second column affinity chromatography. In a separate publication (16), we described the rapid and sensitive monitoring of this purification by high-performance gel filtration, i.e. using a parameter different from charge and affinity for the characterization. This technique clearly showed the evolution of the purification towards the enrichment in one component of high-molecular weight. As reported here, the affinity step using a lectin column appeared inferior to the ion-exchange process.

The results obtained show that the method we described in this paper, allows to purify 120 times the sodium conducting sites, a value close to the one reported by others for the most purified fractions containing these proteins.

Acknowledgements

This work was supported by a grant nr 2.4518.80 from the Fonds de la Recherche Fondamentale Collective (Belgium).

References

1. Agnew, W.S., S.R. Levinson, J.S. Brabson, M.A. Raftery. 1978. Proc. Natl. Acad. Sci. USA 75, 2006-2010.

2. Agnew, W.S., A.M. Raftery. 1979. Biochemistry 18, 1912-1918.

3. Agnew, W.S., A.C. Moore, S.R. Levinson, A.M. Raftery. 1980. Biochem. Biophys. Res. Comm. 92, 860-866.

4. Ellisman, M.M., W.S. Agnew, J.A. Miller, S.R. Levinson. 1982. Proc. Natl. Acad. Sci. USA 79, 4461-4465.

5. Miller, J.A., W.S. Agnew, S.R. Levinson. 1983. Biochemistry 22, 462-470.

6. Norman, R.I., A. Schmid, A. Lombet, J. Barhanin, M. Lazdunski. 1983. Proc. Natl. Acad. Sci. USA 80, 4164-4168.

7. Lombet, A., R.I. Norman, M. Lazdunski. 1983. Biochem. Biophys. Res. Comm. 114, 126-130.

8. Barchi, R.L., S.A. Cohen, L.E. Murphy.1980. Proc. Natl. Acad. Sci. USA 77, 1306-1310.

9. Barchi, R.L. 1983. J. Neurochem. 40, 1377-1385.

10. Catterall, W.A., C.S. Morow, R.P. Hartshorne. 1979. J. Biol. Chem. 254, 11379-11387.

11. Hartshorne, R.P., W.A. Catterall. 1981. Proc. Natl. Acad. Sci. USA 78, 4620-4624.

12. Bontemps, J., R. Cantineau, C. Grandfils, P. Leprince, G. Dandrifosse, E. Schoffeniels. 1984. Anal. Biochem. 139, 149-157.

13. Levinson, S.R., C.J. Curatalo, J. Reed, M.A. Raftery. 1979. Anal. Biochem. 99, 72-84.

14. Udenfriend, S., S. Stein, P. Bohlen, W. Daerman, W. Leimgruber, M. Weigele. 1972. Science 178, 871-873.

15. Bontemps, J., C. Grandfils, G. Dandrifosse, E. Schoffeniels.1984. Arch. Internat. Physiol. Biochim. 92, 39-45.

16. Bontemps, J., C. Grandfils, L. Bettendorff, G. Dandrifosse, E. Schoffeniels. 1984, submitted to J. Chromatog.

HIGH-YIELD SYNTHESIS OF A NEW Na-CHANNEL MARKER : (^{3}H)ETHYLENEDIAMINE DI-TETRODOTOXIN OR ED-diTTX

Ch. Grandfils, L. Bettendorff, E. Schoffeniels, J. Bontemps and G. Dandrifosse

Laboratoire de Biochimie Générale et Comparée, Université de Liège, Place Delcour, 17 - B4020 Liège - Belgique

Introduction

Among the variety of biological marker used for testing the purification degree of the sodium channel, tetrodotoxin (TTX) is one of the most widely used. This neuroactive agent exerts its highly specific action on sodium channel (1). To make use of this toxin for biochemical studies, different tritiation techniques have been attempted (1). The first methods, consisting in an unspecific tritium exchange, gave very low specific activities and caused an important breakdown of the molecule. Higher specific radioactivities were obtained more recently by chemical modification of TTX (2,3) but the overall yield of the synthesis only reached 1-3 %. We have reinvestigated the procedure described by Chicheportiche at al (4). The first reaction step, consisting in oxidation of the TTX-C11 alcohol group has been optimized with a yield of reaction of 30-50 % determined by high performance liquid chromatography. This modification allowed a more efficient coupling of (^{3}H) ethylenediamine with oxidized TTX, producing a derivative composed of one TTX molecule bound to each of the two amino group of (^{3}H) ethylenediamine. The TTX derivative, produced with a specific activity of 45 Ci/mole and a global yield of 15 % shows excellent binding properties towards a membrane extract prepared from electroplax membranes (Kd = 0,8 nM). Competition curve with unlabelled TTX on the same material shows the existence of two saturable sites whose affinities are quite different.

Molecular Basis of Nerve Activity

Results

Initial synthesis following the method of Chicheportiche et al (3) led to low overall yields. The TTX oxidation appears as the limiting step of the synthesis since the characterization of the formed carbonyl function revealed by diphenylhydrazine test is reaching the limit of detection.

The original results presented by Pfitzner and Moffatt (4) suggests that this limit step could be improved by raising the reagents concentration after a careful dehydration of the dimethyl sulfoxide (DMSO) and H_3PO_4 solution.

To assess this hypothese, we studied the quality of the reagents and the influence of them on the oxidation yield using (^{14}C) testosterone as model molecule.

The results presented in fig. 1a show clearly a linear increase of the oxidation yield within the concentration range tested. If one assume a similar reactivity for TTX in identical reaction conditions, a yield of only 6-7 % for a TTX concentration of 20 mM as reported by Chicheportiche et al (3) was not surprising.

By increasing the concentration of the reagents to 160 mM for TTX (fig. 1b), a comparable oxidation yield for TTX is thus obtained after 90 min. The kinetic of this reaction was monitored by HPLC. Native and oxidized forms of TTX were separated isocratically on an ultrasphere-ODS column from Altex (5 µm, 0,42 x 15 cm). The retention of the two molecules was modulated by adjustement of the pH of the aqueous mobile phase by H_3PO_4. An U.V. monitor operating at 195 nm was used for detection.

The formation of dicyclohexylurea cristals in the reaction mixture doesn't prove that an oxidation is occuring. Their formation depends solely on the residual water content in all the reagent used. The oxidation yield value of 53 % is not improved by longer reaction times up to 24 h, but constitutes a considerable improvement when compared to the litterature (3).

The coupling reaction of oxidized TTX with (^{3}H) ethylenediamine is performed in presence of $NaCNBH_3$ as reductive agent of the aldimine initially

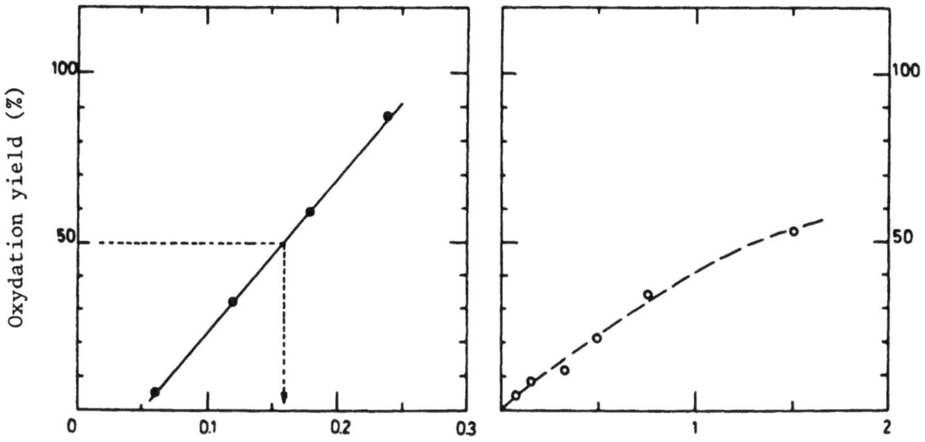

Fig. 1 : oxidation of testosterone (a) and tetrodotoxin (b) by DMSO and dicyclohexylcarbodiimide in presence of phosphoric acid
a : influence of testosterone concentration on its oxidation yield
b : kinetic of the tetrodotoxin oxidation (0, 16 M, 40°C)

formed. 7,5 mCi of the radioactive amine, corresponding to 0,17 µmole of of ethylenediamine were added to 2 µmoles of oxidized TTX. In these conditions, namely a large excess of the carbonyl compound, the formation of a di-tetrodotoxin derivative : TTX-NH-CH_2-CH_2-NH-TTX was favoured. The purification of the TTX derivative was achieved in two steps.
The first separation consists in a chromatography on a cation-exchanger : Bio-Rex 70. In order to avoid the loss of material, we performed only one chromatographic passage on this resin. The gradient elution profile was designed to get a total separation of the tritiated amine free from (^3H) ED-diTTX released at higher ionic strength. Fig. 2 shows the elution profile of radioactivity. The main peak 3 appearing between 0,12 and

0,60 M NH_4HCO_3 includes 53 % of the recovered tritium. Only the fractions corresponding to this peak show highly specific binding with excitable membranes. Compare to the results of Chicheportiche et al, the shape of the main peak suggests a better homogeneity of the TTX derivative and a higher coupling yield. Peak 2 is assigned to (^{3}H)ethylenediamine.

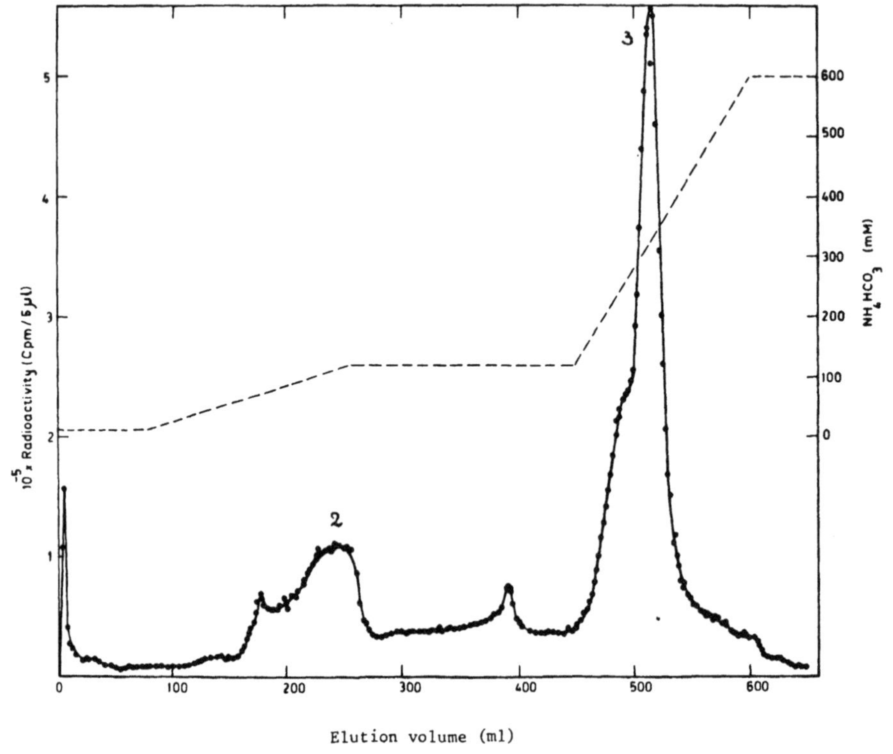

Fig. 2 : Purification of (^{3}H) ethylenediamine ditetrodotoxin by cation-exchange chromatography.
Full and dotted liner refer to radioactivity measurement and to the elution profile respectively.

The fractions that make peak 3 were pooled and lyophilised. Bio-gel P2 is then used to repurify and to desalt the radioactive TTX derivative with a 10 mM pyridine-formate buffer at pH 4. At approximately two column volumes, a main symetrical peak appears. The corresponding fractions were then lyophilized for further caracterization.
The physico-chemical analysis was performed by determination of three parameters. Firstly, the presence of free amino group has been tested using the sensitive fluorescamine technique (5).
In spite of a limited detection lowered to 50 pmol per test, no reactivity was found, which means that in our derivatives both amine groups are engaged with carbonyl group(s) of oxidized TTX. Secondly, the fluorescence produced by our derivative was measured after hydrolysis in NaOH 3,5 M (80°C, 30') (6). The tritiated derivative gives a fluorescent compound whose excitation and emission characteristics are identical to those of the C9-base arising from TTX alkaline hydrolysis.
Keeping in mind, the conditions of the coupling step of the synthesis, these above results support the view that our radioactive ligand is constituted of two oxidized TTX per (^{3}H) ethylenediamine molecule.
This assumption was confirmed by the comparison of native TTX and tritiated derivative molecular weight using a gel filtration medium : Sephadex G10.
The volume ratio (VE/Vo) represents the ratio of the eluated volume on void volume.
ED-diTTX and TTX eluated at VE/Vo = 1,27 and 1,61 respectively with a difference in capacity factor of 0,34. This difference is similar with that obtained with the oxidized and reduced forms of glutathione used to calibrated the column. This result clearly confirmed the composition of the final product.
Finally, a mouse survival assay revealed that the toxic activity of our derivative was fully retained. Binding properties were studied using membranes prepared from the electric organ of Electrophorus electricus as starting material.

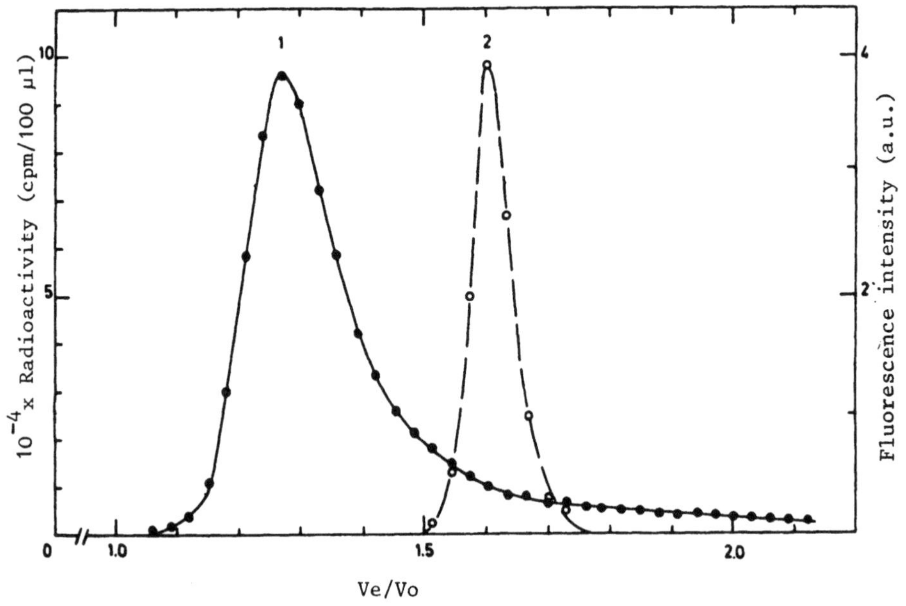

Fig. 3 : Separation of ED-diTTX (peak 1) from native TTX (peak 2) on a Sephadex G10 column. ED-diTTX was detected by radioactivity (left ordinate), while TTX was measured fluorimetrically (right ordinate) after alkalinehydrolysis.

Fig. 4 shows two saturation curves obtained with a lubrol-PX extract of the membranes prepared according to the method of Miller et al (7). The binding results obtained by rapid gel filtration method (10) were compared with a more reliable equilibrium analysis technique (above curve). These results show that our derivative displays a high affinity for the sodium conductance site (Kd = 0,79 nM) comparable to that of the native toxin. This parameter doesn't significantly vary whatever the method used for the separation of bound and free toxin. However, B_{max}, the number of receptor per protein mg, is underestimated by 15-20 % using gel filtration method. This result observed also by other authors (8,9) could be explained by taking into account the relative dissociation of the ligand-receptor complex. Bottom figure (Fig. 4b) represents the Scatchard plot transformation of the above data.

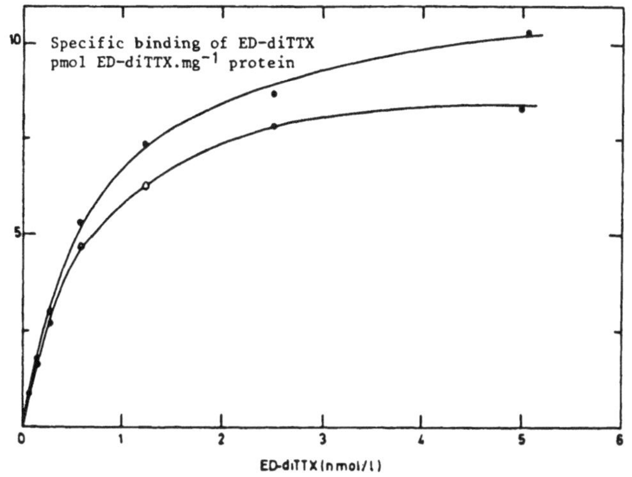

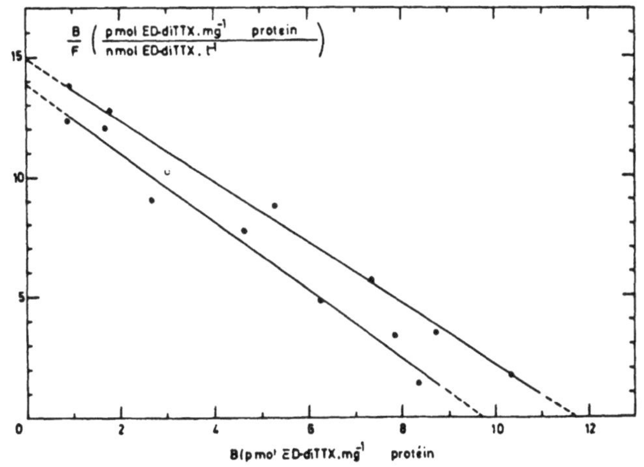

Fig. 4a : Binding of ED-diTTX on a Lubrol-PX extract of membranes prepared following the method of Miller (7). Top and bottom curves represent respectively the results given by an equilibrium dialysis technique and a gel filtration method.

Fig. 4b : Scatchard plot of the data from fig. 4a.

The relative affinities of ED-diTTX and native TTX for the sodium conductance site were measured by a competitive assay represented in fig. 5.

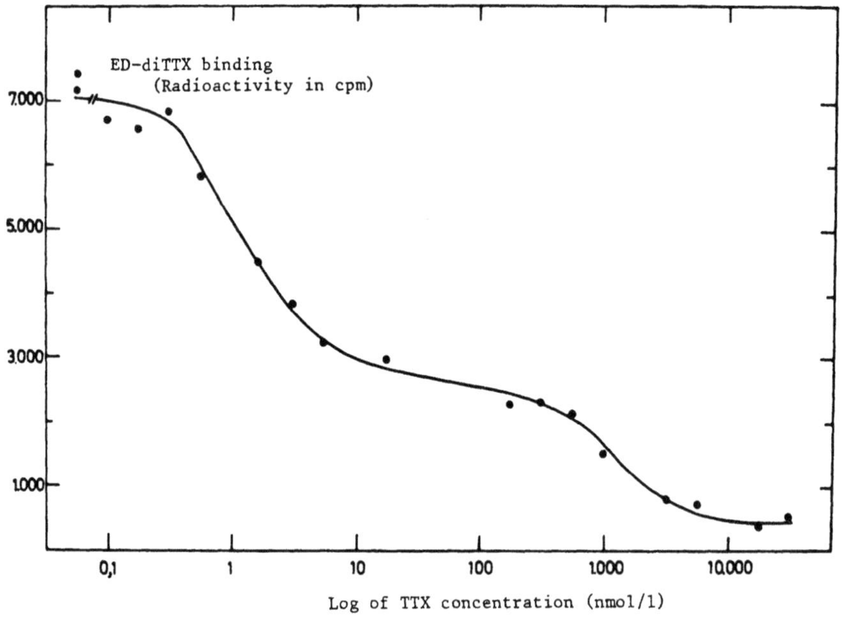

Fig. 5 : Competition curve binding between ED-diTTX and TTX observed on a Lubrol-PX extract of membrane prepared from the main electric organ of Electrophorus electricus.

This competition curve appears as clear a biphasic one. The results suggest the existence of two saturable sites whose affinities are well distinct with IC50 values equal to 1 nM and 1 µM respectively. The existence of impurities in ED-diTTX solution has been rejected as a possible reason for this result since the derivative appeared as homogeneous by HPLC. The same effect was also obtained with two different TTX solutions.

Two major explanations could be proposed. The first supposes the existence of an extra-site associated directly or not with the excitable membranes extract.
The bifunctional caracter of our TTX derivative could also explain the existence of the second site.

Acknowledgement

We would like to acknowledge the excellent technical assistance of Mr. Louis Roland. This work received the financial support of the FRFC, Belgium.

References

1. Ritchie, J.M., R.B. Rogart. 1977. The Binding of Saxitoxin and Tetrodotoxin to Excitable Tissue. Rev. Physiol. Biochem. Pharmacol. 79, 1-50.

2. Grünhagen, H.H., M. Rack, R. Stämpfli, H. Fasold, P. Retter. 1981. Chemically Tritiated Tetrodotoxin : Physiology Activity and Binding to Na-Channels. Archives of Biochemistry and Biophysics 206, 198-204.

3. Chicheportiche, R., M. Balerna, A. Lombet, G. Romey, M. Lazdunski. 1980. Synthesis of New, Highly Radioactive Tetrodotoxin Derivatives and Their Binding Properties to the Sodium Channel. Eur. J. Biochem. 104, 617-625.

4. Pfitzner, K.E., J.G. Moffatt. 1965. Sulfoxide Carbodiimide Reactions. I. Facile oxidation of Alcohols. J. Am. Chem. Soc. 87, 5661-5670.

5. Udenfriend, S., S. Stein, P. Böhlen, W. Dairman, W. Leimgruber, M. Weigele. 1972. FLUORESCAMINE : A Reagent for Assay of Amino Acids, Peptides, proteins, and Primary Amines in the Picomole Range. Science 178, 871-872.

6. Nunez, M.T., S. Fischer, E. Jaimovich. 1976. A Fluorimetric method to determine Tetrodotoxin. Anal. Biochem. 72, 321-325.

7. Miller, J.A., W.S. Agnew, S.R. Levinson. 1983. Principal Glycopeptide of the Tetrodotoxin/Saxitoxin Binding Protein from Electrophorus electricus : Isolation and Partial Chemical and Physical Characterization. Biochemistry 22, 462-470.

8. Levinson, S.R., C.J. Curatalo, J. Reed, M.A. Raftery. 1979. A Rapid and Precise Assay for Tetrodotoxin Binding to Detergent Extracts of Excitable Tissues. Analytical Biochemistry 99, 72-84.

9. Hartshorne, R.P., W.A. Catterall. 1984. The Sodium Channel from Rat Brain. The Journal of Biological Chemistry. 259, 1667-1675.

INSECT OLFACTORY CELLS: ELECTROPHYSIOLOGICAL AND BIOCHEMICAL STUDIES

K.E. Kaissling, U. Klein, J.J. de Kramer, T.A. Keil, S. Kanaujia, J. Hemberger
Max-Planck-Institut für Verhaltensphysiologie, D-8131 Seewiesen, West-Germany

1. Introduction and Summary

Male moths recognize airborne pheromone molecules by means of extremely sensitive and highly specific receptor cells, which are located in olfactory hair sensilla on the antennae. One pheromone molecule may be sufficient to elicit a nerve impulse (1). Smallest alterations of chemical structure lead to strongly reduced effectiveness of the stimulus compound (2).

The morphological structure of insect olfactory sensilla offers several opportunities for different experimental approaches. 1. Electrical responses of single receptor cells in single sensilla can be recorded extracellularly. 2. Odour sensitive dendritic membranes can be isolated out of hairs collected from antennal branches. 3. "Receptor lymph" fluid surrounding the receptor cell dendrites can be collected from tips of opened hairs for biochemical analysis.

This report summarizes structural, electrophysiological, and biochemical studies of chemical transduction in olfactory sensilla of the silk moth Bombyx mori, Antheraea polyphemus, and Antheraea pernyi. Small transient potential fluctuations precede the nerve impulses elicited by weak pheromone stimulation (3). These so-called "elementary receptor potentials" may reflect single ion channel openings in the dendritic membrane (2). Responses to pheromone derivatives suggest that the ion channels open in different kinds depending on the chemical structure of the stimulus molecule. Isolated dendritic membrane shows structural units in electron micrographs, which may indicate membrane integrated proteins (4). The receptor lymph has surprisingly high potassium and sulfur concentrations (2). A pheromone binding protein and a pheromone degrading enzyme are found in the receptor lymph (4, 5).

Molecular Basis of Nerve Activity

2. The structure of olfactory hair sensilla

Olfactory hair sensilla (sensilla trichodea) in moths are epithelial sense organs and consist of usually two bipolar receptor neurons responding to two different pheromone components and three auxiliary cells (fig. 1). The outer segments of the receptor cell dendrites run through the hair lumen of a cuticular hair which is filled with an extracellular fluid, the so-called "receptor lymph" or "sensillum liquor" (6). The odour molecules adsorbed on the hair surface penetrate the hair wall most likely via the numerous pore tubules which extend into the hair lumen, often contacting the dendritic membrane in electron microscopical preparations (fig. 2) (7, 8). It remains an open question wether the stimulus molecules reach the dendritic cell membrane via these contacts or via the receptor lymph.

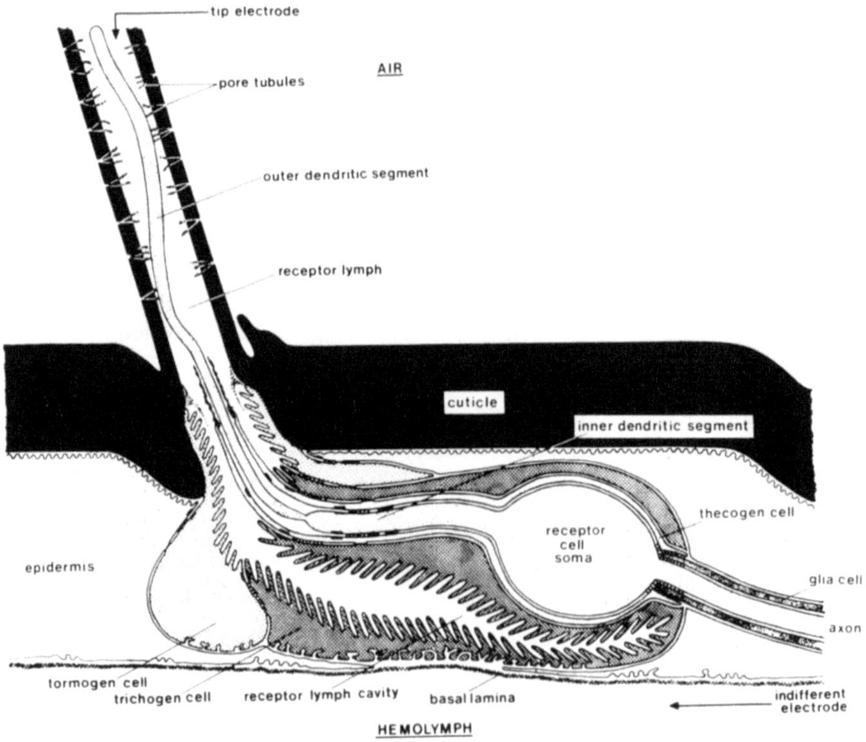

Fig. 1. Diagram of an olfactory sensillum trichodeum showing one receptor cell and three auxiliary cells.

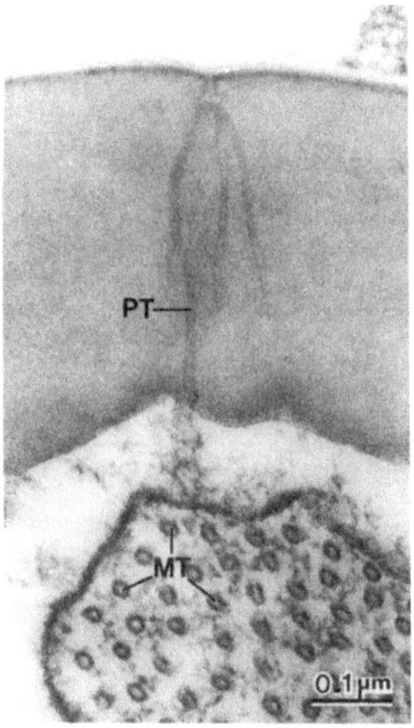

Each antenna of the male moth of Antheraea polyphemus bears 55,000 sensilla trichodea (fig.1) besides several thousand smaller olfactory sensilla of other types and sensilla of other stimulus modalities (9). Biochemical analyses are limited by the small amount of material obtained from an olfactory sensillum: about 1 pl of receptor lymph and up to 600 μm^2 of membrane area of outer dendritic segments within the hair (10, 11).

Fig. 2. Cross section of an olfactory hair with pore tubule (PT) contacting the dendritic membrane. MT, cytoplasmic microtubules. Transmission electron micrograph (From ref. 8).

3. Responses to single pheromone molecules

The extracellularly recorded elementary receptor potentials from a single olfactory cell are negatively directed potential fluctuations of an amplitude of about 0.5 mV and a duration of 10 - 50 ms (Fig. 3) (2). Corresponding current fluctuations could be observed under voltage clamp conditions. Most likely, these events reflect conductance changes in the dendritic membrane of the receptor cell. In order to estimate the size of the conductance change per elementary event the electrical properties of the olfactory sensillum were analyzed. An equivalent network diagram based on the geometry of the sensillum (10, 11) was developed. Several electrical parameters were measured besides the amplitudes of the elementary voltage and current fluctuations: the transepithelial voltage (+40 mV) and resistance ($2 \cdot 10^8 \Omega$) of the sensillum, the maximum change of voltage (-20 mV) and resistance (-15%) due to pheromone stimulation, and the specific resistance of the receptor lymph ($\sim 40\ \Omega$cm) (2, 12). Other parameters as the resistance of the dendritic membrane could not be measured directly.

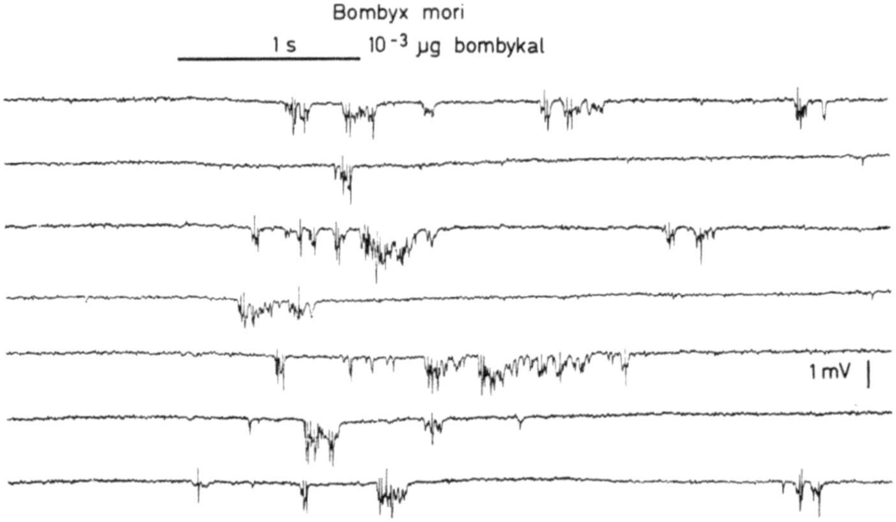

Fig. 3. Elementary receptor potentials and nerve impulses as recorded extracellularly from an olfactory hair. Each trace shows the response to a one second stimulus of the pheromone component E-10, Z-12 hexadecadienal (µg on the odour source).

A computer simulation of the equivalent circuit diagram revealed a conductance increase of the dendritic membrane in the range of 30 pS for an elementary receptor potential. This value corresponds to the conductance of single ion channel openings induced by acetylcholine in muscle cells (13).

The time course of the elementary receptor potentials as recorded under current clamp conditions was, usually, more rounded if compared with the single channel fluctuations observed in patch clamp measurements in other receptor systems (13). A rounded shape is expected from the capacitances of 30 pF in parallel to the main portion of the resistance across the sensillum ($1.5 \cdot 10^{8} \Omega$) which has been ascribed to the apical membranes of receptor and auxiliary cells (14).

These studies, together with nerve impulse measurements at weak pheromone stimulation, led to the working hypothesis that one pheromone molecule induces the opening of a single ion channel and causes a depolarization of the receptor cell which is able to elicit a nerve impulse (2).

4. Specificity of olfactory receptor cells

Minimal alterations of pheromone molecules -- for example removal, shifting or change of configuration of a double bond -- reduce the stimulus effectiveness 10 to 1000 fold if measured as the mean amplitude of the receptor potential. In addition, certain pheromone derivatives produce altered fluctuations of the receptor potential (fig. 4) (2). Typically, the receptor potential fluctuates as expected from superimposition of a number of elementary receptor potentials. Shortening of the pheromone molecule by one or two CH_2-groups or removal of a double bond can produce much smoother signals. This response type may be produced by a large number of comparatively small elementary responses. The single events may have a much smaller duration beyond the limits of resolution of the recording method. Ligand-dependent life times of the open state of ion channels are known from the acetylcholine receptor (13).

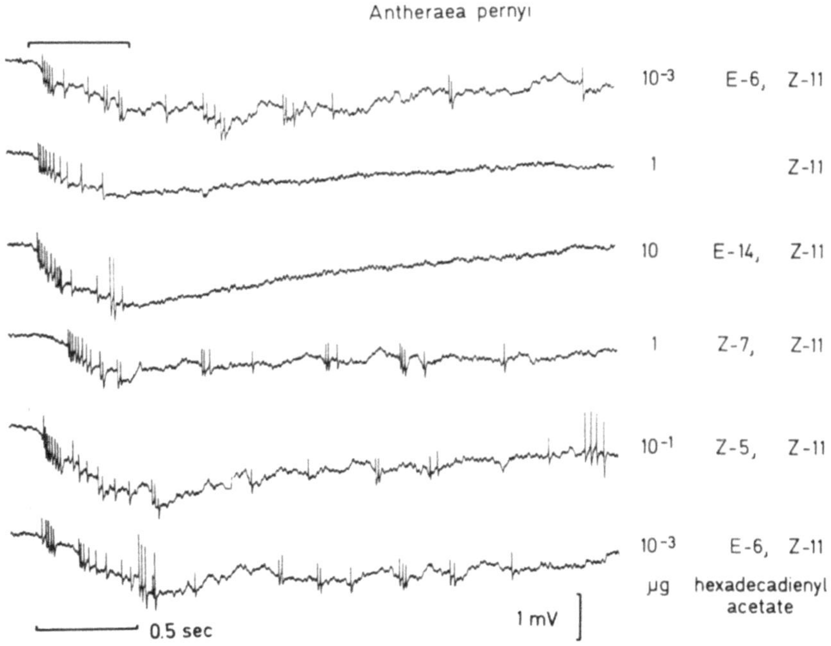

Fig. 4. Responses of a single receptor cell to the pheromone component E-6, Z-11 hexadecadienyl acetate and to 4 derivatives with reduced effectiveness (cf. amounts in µg loaded on the odour source). Fluctuating (traces 1, 4-6) and smooth (traces 2, 3) receptor potentials with nerve impulses. (From ref. 2).

5. Isolated dendritic membrane

The dendritic membranes were separated from other epithelial membranes by breaking off the hairs from frozen antennae (fig. 5) (4). This treatment caused the dendrites to disintegrate and to form vesicles which could be washed out of the hair fragments by suspension in buffer. The membrane vesicles were separated from the hair cuticle by several filtration and centrifugation steps. Vesicles treated with cationized ferritin as an electron microscopical marker for negative charges often showed depositions of the label on the outside. This suggests preservation of the cell coat, already known from dendrites studied in situ (15). Unlabelled vesicles ocurred as well which may have a reversed membrane orientation. After negative staining of the membrane vesicles, sometimes structural units became apparent which had a diameter of about 3 nm (fig. 6). They were irregularly distributed with a maximal density of about 30,000 per μm^2. These units may indicate proteins, perhaps receptor molecules, integrated in the dendritic membrane (4).

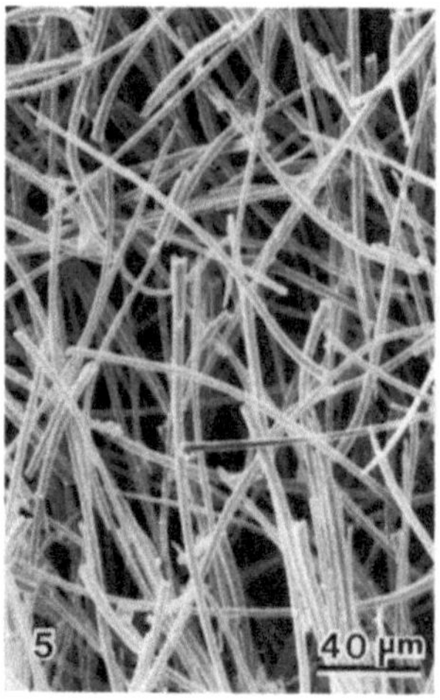

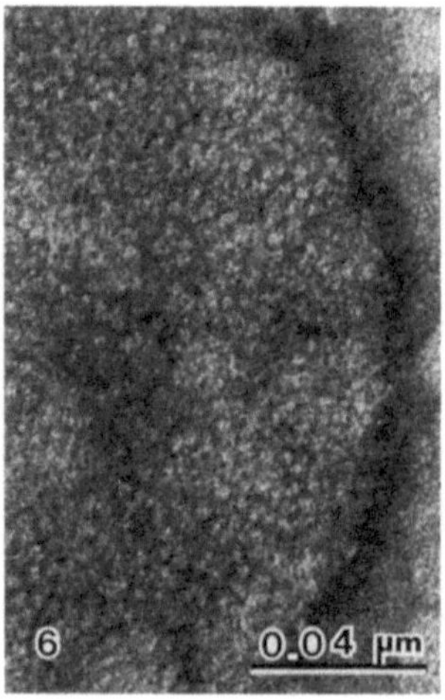

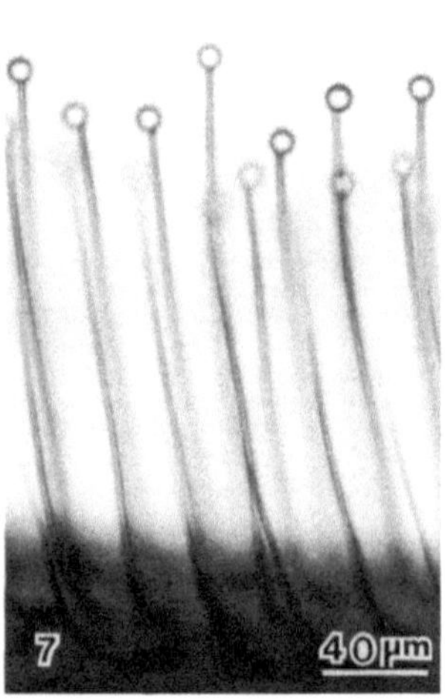

Fig. 5. Isolated olfactory hairs of *Antheraea polyphemus*. Scanning electron micrograph.

Fig. 6. Negatively stained vesicle of dendritic membrane from isolated olfactory hairs of *Antheraea polyphemus*. Transmission electron micrograph.

Fig. 7. Receptor lymph droplets extruded through the tip of opened olfactory hairs of *Antheraea polyphemus*. Light microscopical preparation.

6. Ionic composition of the receptor lymph

Isolated receptor lymph (fig. 7) subjected to flame photometry revealed a concentration of about 200 mM K and about 25 mM Na. In comparison hemolymph contained only about 40 mM K and 12 mM Na. The high K-value, also found in the receptor lymph of insect mechanoreceptors, may be produced by an electrogenic K^+-pump located in the folded membranes of the auxiliary cells (16). These Na/K ratios were confirmed by energy dispersive X-ray microanalysis, which allowed in addition rough estimates of Mg, P, S, Cl, and Ca contents in two species (fig. 8). The high content of sulfur (about 100 mM) in the receptor lymph may indicate the presence of acidic mucopolysaccharides. Such organic anions may compensate for the lack of anions, suggested by the low ratio of Cl/K. Ten seconds after strong pheromone stimulation, the receptor lymph showed a further reduced Cl/K ratio whereas the S/K ratio increased. These effects, however, cannot be explained to date (2).

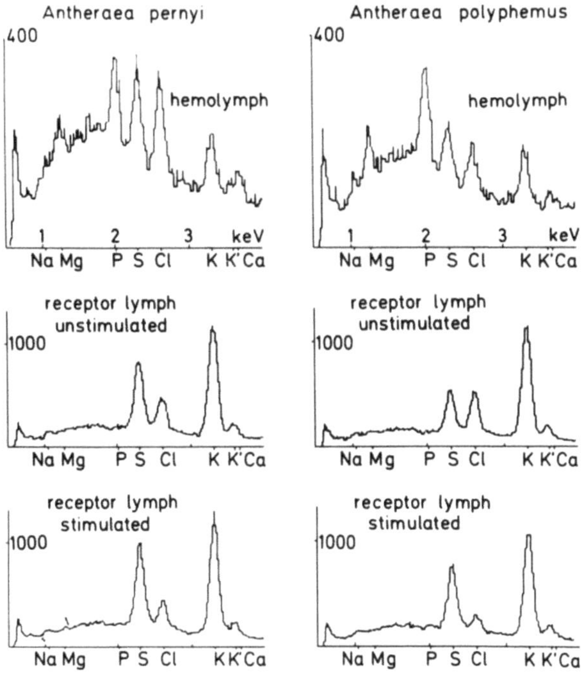

Fig. 8. Energy dispersive X-ray spectra obtained from receptor lymph droplets of single olfactory hairs and from hemolymph. (From ref. 2).

7. Pheromone degrading enzymes

Enzymatic degradation of the pheromone component bombykol (E-10, Z-12 hexadecadienol) has been shown in antennae of male and female Bombyx mori and also in other parts of the body (17). At least three esterases can be demonstrated electrophoretically in homogenates of antennae of male Antheraea polyphemus, using enzymatic staining (fig. 9a) (5). The molecular weight of the esterases ranges from 55,000 to 90,000 D, according to gradient gel electrophoresis. Only the smallest esterase was found to be located in the receptor lymph of the sensilla trichodea of males and was, therefore, called "sensillar esterase" (5). The electrophoretically purified enzyme degraded one of the female pheromone components, E-6, Z-11 hexadecadienyl acetate (18), with an apparent K_M of $6 \cdot 10^{-7}$M (19).

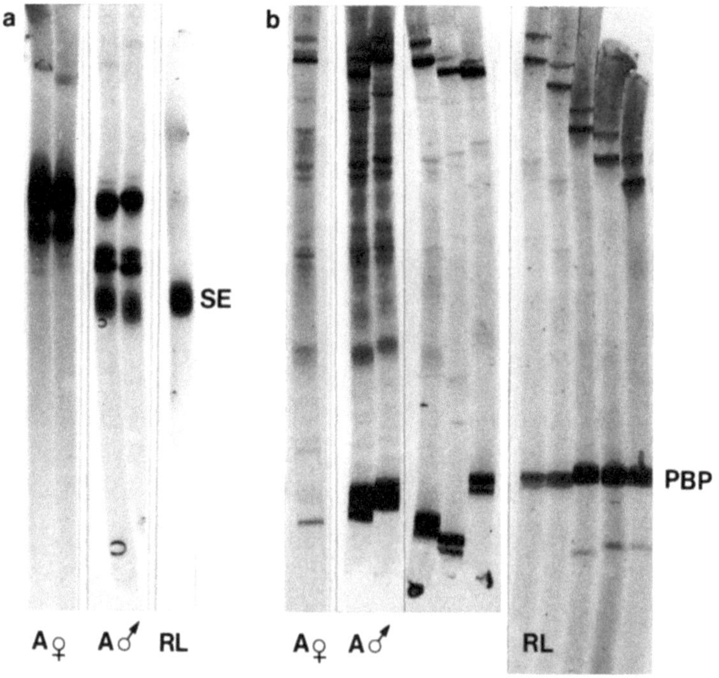

Fig. 9. Polyacrylamide microgradient gels of soluble proteins from homogenized antennae (A) and isolated receptor lymph (RL) of Antheraea polyphemus.
a) Gels stained using naphthyl acetate as a substrate for esterases. SE, sensillar esterase. b) Gels stained with Coomassie blue. PBP, pheromone binding protein.

8. Pheromone binding protein

The most prominent soluble protein in the antennae of male Antheraea polyphemus (fig. 9b) binds to radiolabelled pheromone as found by fluorography after gel electrophoresis (5). This protein has been called "pheromone binding protein". It occurs in the receptor lymph but could also be detected in minute amounts in the hemolymph (2) and in antennae of females. The binding protein has a molecular weight in the range of 30,000 D, according to non-denaturing gel electrophoresis and gel filtration. In SDS gels it shows a molecular weight of about 15,000 D, suggesting the presence of two subunits. The concentration of the binding protein can be calculated as about 10 mM assuming that all of the protein (up to 15 μg per antenna (5)) is located in the receptor lymph. This would mean a several hundred fold excess of binding protein molecules over the maximum number of receptors expected in the dendritic membrane. The purified binding protein shows saturable and specific binding of pheromone with a dissociation constant of about $6 \cdot 10^{-8}$ M (fig. 10). Saturation was reached with about 1 pheromone molecule per binding protein. In competition experiments with hexadecanyl acetate, or E-6, Z-11 hexadecadienol, the same amount of inhibition was reached at a 10 or 300 fold higher concentration, respectively, as compared to the unlabelled pheromone.

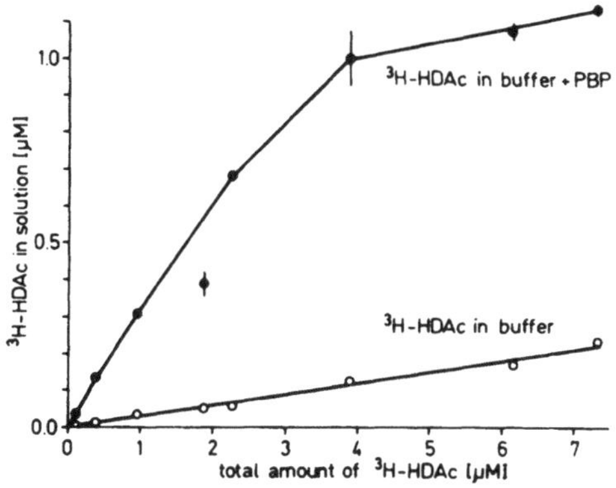

Fig. 10. Binding of E-6, Z-11 ^{3}H-hexadecadienyl acetate (^{3}H-HDAc) to purified pheromone binding protein (PBP). Lower curve: 3% of pheromone are free in buffer solution, 97% remain at the glass wall. Upper curve: pheromone in buffer + binding protein. The curve saturates at about 25% of pheromone in solution.

9. Possible functions of proteins in the receptor lymph

Proteins in the receptor lymph may be involved in the transport of pheromone towards the receptor cell dendrite and in the "rapid inactivation" of pheromone. Both processes are postulated from electrophysiological data and experiments with ^{3}H-labelled pheromone (2, 3, 17, 20). Transport of pheromone from the hair surface towards the dendritic membrane seems to be a prerequisite for the recognition at the receptor cell. The pheromone molecules may reach the dendritic membrane via the pore tubules and enter the receptor lymph after the stimulatory effect at the cell membrane. In an alternative model, the pheromone binding protein may serve as a carrier for the pheromone (20), if the pore tubules do not contact the dendritic cell membrane in vivo (15). It is not yet possible to decide between the tubule model and the carrier model. In any case, considerable amounts of pheromone move into the hair lumen. Receptor lymph collected 1-2 min after stimulation with ^{3}H-labelled pheromone contained 40% of the radioactivity associated with the hairs (19).

A "rapid inactivation" (2) of pheromone has been postulated because the receptor potential declines within seconds after the end of a pheromone stimulus whereas the pheromone molecules remain in much longer association with the hairs. In Bombyx mori the "rapid inactivation" may be a non-enzymatical process because the metabolic half-life of the pheromone component bombykol was 3 min on intact antennae of males (3, 17). In preliminary experiments, the metabolic half-life of pheromone associated with intact Antheraea polyphemus antennae was in a similar range (21). It is conceivable that binding of pheromone to the binding protein can reduce the concentration of pheromone at the receptors and, thus, act as a rapid non-enzymatic inactivation in the receptor lymph (5). This proposed mechanism would exclude a carrier function of the binding protein. Another rapid inactivation mechanism is required for the pheromone molecules associated with the hair wall. Surprisingly, the purified sensillar esterase degraded the pheromone much faster than expected from metabolic half-life of pheromone in intact antennae (20). The role of this enzyme in vivo remains to be clarified. Further studies on transport and inactivation of pheromone molecules may help to understand the dynamics of the receptor cell response and, consequently, the transduction process in these cells.

References

1. Kaissling, K.E., E. Priesner. 1970. Naturwiss. 57, 23.

2. Kaissling, K.E., J. Thorson. 1980. In: Receptors for Neurotransmitters, Hormones, and Pheromones in Insects (D.B. Sattelle et al., eds.). Elsevier North Holland, p. 261.

3. Kaissling, K.E. 1974. In: Biochemistry of Sensory Functions (L. Jaenicke, ed.). Springer Verlag, Berlin, p. 243.

4. Klein, U., T.A. Keil. Cellular and Molecular Neurobiol. (in press).

5. Vogt, R.G., L. Riddiford. 1981. Nature (London) 293, 161.

6. Ernst, K.D. 1969. Z. Zellforsch. mikr. Anat. 94, 72.

7. Steinbrecht, R.A., B. Müller. 1971. Z. Zellforsch. mikr. Anat. 117, 570.

8. Keil, T.A. 1982. Tissue and Cell 14, 451.

9. Boeckh, J., K.E. Kaissling, D. Schneider. 1960. Zool. Jb. (Anat.) 117, 559.

10. Gnatzy, W., W. Mohren, R.A. Steinbrecht. 1984. Cell Tissue Res. 235, 35.

11. Keil, T.A. 1984. Zoomorphology 104, 147.

12. Kaissling, K.E. 1980. In: Int. Symp. Olfaction and Taste VII (H. van der Starre, ed.). IRL Press, London, p. 81.

13. Neher, E., B. Sakmann. 1976. Nature (London) 260, 779.

14. De Kramer, J.J., K.E. Kaissling, T.A. Keil. 1984. Chem. Senses 8, 289.

15. Keil, T.A. 1984. Tissue and Cell 16 (5) (in press).

16. Thurm, U., J. Küppers. 1980. In: Insect Biology in the Future (M. Locke and D.S. Smith, eds.). Academic Press, New York, p. 753.

17. Kasang, G., K.E. Kaissling. 1972. In: Olfaction and Taste IV. (D. Schneider, ed.). Wissensch. Verlagsges., Stuttgart, p. 200.

18. Kochansky, J., J. Tette, E.F. Taschenberg, R.T. Cardé, K.E. Kaissling, W.L. Roelofs. 1975. J. Insect Physiol. 21, 1977.

19. Kanaujia, S., K.E. Kaissling: J. Insect Physiol. (in press).

20. Vogt, R.G. 1984. Thesis. University of Washington.

21. Kasang, G. (personal communication).

EFFECTS OF EXTRACELLULAR ATP ON ^{86}Rb INFLUX IN CHICK MYOTUBES; INDICATIONS OF A COTRANSMITTER ROLE IN NEUROMUSCULAR TRANSMISSION.

J. Häggblad, H. Eriksson, E. Heilbronn
Unit of Neurochemistry and Neurotoxicology, University of Stockholm, S-172 46 Sundbyberg, Sweden.

Introduction

Adenosinetriphosphate (ATP) has previously been proposed to have a potentiating effect on the postsynaptic responses to acetylcholine (ACh) in striated muscle (1). This effect has been suggested to be caused by interaction of ATP with an allosteric site on the nicotinic acetylcholine receptor (nAChR) and to occur without changes in the binding affinity of ACh to the nAChR while ATP alone had no effect on the resting membrane potential (2). Recently, ATP was suggested to have transmitter-like actions on patch-clamped membranes of cultured myotubes (3). This paper describes further experimental evidence for an ion-translocating effect by extracellular ATP on cultured myotubes as measured by biochemical techniques. In the present experiments influx of ^{86}Rb in chick myotubes was studied as a consequence of extracellular ATP application.

Methods

Thigh and breast muscles from 11-12 days old chick embryos were dissected and placed in culture medium (Ham´s F10-medium, 4 % fetal calf serum, 9 % newborn calf serum, 2 % chick embryo extract, 50 U/ml penicillin and 50 µg/ml streptomycin). The tissue was transferred to a glass-teflon homogenizer and care-

Molecular Basis of Nerve Activity

fully disrupted (5 strokes). The dissociated cells were seeded on non-coated plastic tissue-culture dishes (Falcon, 5 cm in diameter) to dispose of the majority of muscle fibroblasts. After plating of fibroblasts the resultant cell slurry was diluted and seeded on collagen-coated dishes at a density of 1.0×10^6 cells/dish. At day 2 after preparation the cells were treated with 10^{-5}M cytosine arabinoside to dispose of remaining fibroblasts. The medium was changed every 2-3 days. The cultures were used for experiments 7-9 days after preparation. At this time the cells had a typical myotube morphology and contracted spontaneously.

ATP-induced ^{86}Rb-influx was measured in the myotubes with a technique previously described by Catterall (4). Uptake of ^{86}Rb was studied during 30 s periods at room temperature. Routinely, 20 µCi ^{86}Rb was used per dish. The uptake medium (130 mM NaCl, 5.4 mM KCl, 1.8 mM $CaCl_2$, 0.8 mM $MgCl_2$, 5.5 mM D-glucose, 1 mM NaH_2PO_4, 50 mM HEPES, pH 7.4) was supplemented with 5 mM ouabain to keep the background uptake low. At the end of the uptake period the medium was drawn off and the cells were washed three times with ice-cold medium (164 mM NaCl, 5.4 mM KCl, 1.8 mM $CaCl_2$, 0.8 mM $MgCl_2$, 5.0 mM HEPES, pH 7.4). Subsequent to washing cells were detached from the plates with 1 M NaOH and radioactivity determined in a Packard 500C γ-counter at 15-500 keV energy-window settings.

^{86}Rb (1-12 mCi/mg) was obtained from the Radiochemical Centre, Amersham, England. β,γ-methylene ATP, ATP (disodium salt), ADP, AMP, adenosine, cytosine arabinoside, ouabain and carbachol were from Sigma, St Louis, MO, USA. β,γ-imido ATP was purchased from Serva, Heidelberg, FRG and α-bungarotoxin from Miami Serpentarium, Miami, FA, USA. Cell culture medium and sera were obtained from Flow Laboratories Svenska AB, Solna, Sweden. All other chemicals used were of analytical grade. Buffers and media were made in double deionized water.

Results

A dose dependant influx of ^{86}Rb into the myotubes was observed due to extracellular application of ATP. The observed dose-response curve is discontinuous (figure 1). The first rising phase of ^{86}Rb-influx has an EC_{50}-value of approximately 10 µM. Above 50 µM a plateau is reached. At concentrations above 1 mM a second non-saturable increase in ^{86}Rb-uptake is observed.

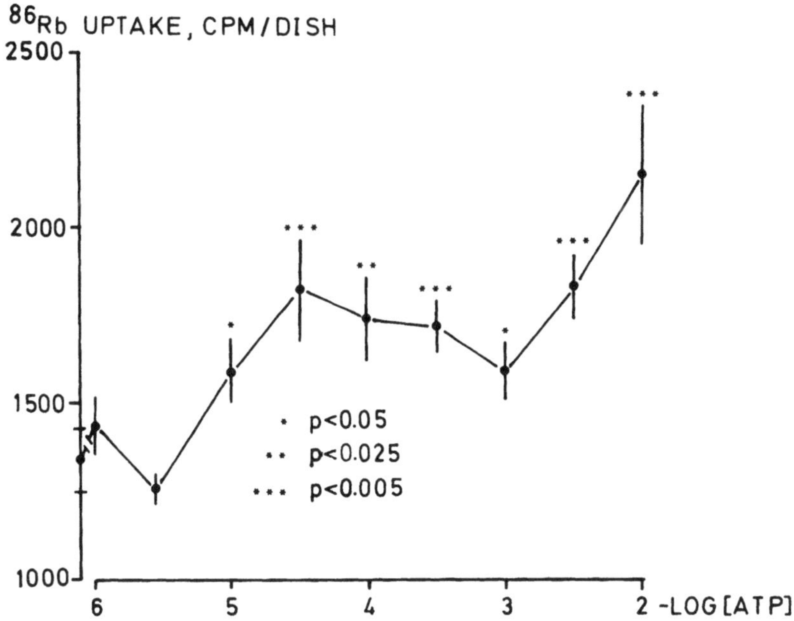

Figure 1. Dose-response curve for ATP-induced ^{86}Rb-uptake in chick myotubes.
Values are expressed as mean±SEM (n=4). p-values were determined with a two-tailed t-test.
Throughout the concentration range an equimolar Mg·ATP-complex was used as effector.

The action of ATP is inhibited by preincubation with β,γ-imido ATP and β,γ-methylene ATP. These ATP-analogues on their own had no effect (table 1). None of the tested ATP-metabolites were as effective as ATP in producing influx (table 2). In experiments were ouabain was excluded from the uptake medium the effect of ATP on ^{86}Rb-influx was still present, though the background uptake was higher (table 1).

Table 1. ATP-induced influx of ^{86}Rb in chick myotubes; Effects of β,γ-imido and β,γ-methylene ATP.

Conditions	^{86}Rb-uptake % of control	Statistics	
no drug	100±4	-	(n=4)
0.5 mM ATP	121±7	p<0.05	(n=4)
0.5 mM β,γ-imido ATP	102±4	n.s.	(n=4)
0.3 mM β,γ-methylene ATP	99±7	n.s.	(n=3)
1.0 mM carbachol	134±3	p<0.001	(n=4)
no drug 1)	100±4	-	(n=4)
0.5 mM ATP 1)	100±2	n.s.	(n=4)
0.5 mM β,γ-imido ATP 1)	105±9	n.s.	(n=4)
1.0 mM carbachol 1)	131±3	p<0.005	(n=4)
no drug 2)	100±8	-	(n=3)
0.3 mM ATP 2)	106±3	n.s.	(n=3)
0.3 mM β,γ-methylene ATP 2)	103	-	(n=2)
1.0 mM carbachol 2)	192±9	p<0.005	(n=3)
no drug 3)	100±3	-	(n=4)
0.5 mM ATP 3)	122±4	p<0.005	(n=3)
1.0 mM carbachol 3)	142±2	p<0.001	(n=3)

Values are expressed as mean±SEM. p-values were determined with a two-tailed t-test by comparison to the no-drug situation in each experimental condition. 100 % basal uptake corresponds to approximately 1400 cpm.
1) and 2); preincubated (5 min) with 0.5 mM β,γ-imido or 0.3 mM β,γ-methylene ATP, respectively.
3); no ouabain added in uptake medium.

In order to check if alteration of extracellular concentrations of divalent cations, due to the chelating action of ATP, was responsible for the increase in uptake of ^{86}Rb a series of control experiments were performed (table 3). Exclusion of Ca^{2+} slightly increased both basal and ATP-evoked uptake of ^{86}Rb. Omission of Mg^{2+} did not significantly alter ATP-induced uptake. Basal uptake in the absence of Mg^{2+} was slightly increased. Addition of equimolar amounts (to ATP) of either Ca^{2+} or Mg^{2+}

to the uptake medium had no significant effect on ATP-induced uptake of ^{86}Rb.

Table 2. Ion-translocation effect of carbachol, ATP and metabolites.

Conditions	^{86}Rb-uptake % of control	Statistics	
no drug	100±7	-	(n=6)
1.0 mM ATP	128±3	$p<0.05$	(n=6)
1.0 mM ADP	116±5	$p<0.1$	(n=6)
1.0 mM AMP	101±3	$p<0.5$	(n=6)
1.0 mM adenosine	112±5	$p<0.2$	(n=6)
1.0 mM carbachol	150±8	$p<0.001$	(n=6)

Values are expressed as mean±SEM. p-values were determined with a two-tailed t-test. 100 % basal uptake corresponds to 1510±105 cpm (mean±SEM).

Table 3. ATP-induced ^{86}Rb-uptake in chick myotubes; Effects of calcium and magnesium ions.

Conditions	^{86}Rb-uptake % of control	Statistics	
Standard uptake-medium	100±4	-	(n=4)
0.5 mM ATP	120±2	$p<0.005$	(n=4)
0.5 mM ATP + 0.5 mM $CaCl_2$	123±5	$p<0.01$	(n=4)
0.5 mM ATP + 0.5 mM $MgCl_2$	126±12	$p<0.05$	(n=3)
0.5 mM ATP no $CaCl_2$ added	131±6	$p<0.005$	(n=3)
0.5 mM ATP no $MgCl_2$ added	126±3	$p<0.005$	(n=3)
no $CaCl_2$ added	117±6	$p<0.1$	(n=4)
no $MgCl_2$ added	117±9	$p<0.2$	(n=4)

Values are expressed as mean±SEM. p-values were determined with a two-tailed t-test. 100 % basal uptake corresponds to 1126±41 cpm (mean±SEM).

In some experiments the action of ATP was compared to activation of the nAChR-coupled ion-channel by a cholinergic agonist. The maximal ion-translocation effect of ATP was not of the same magnitude as that elicited by the cholinergic agonist carbachol (figure 2, eg table 1). Furthermore, at maximal response concentration of ATP, the effect of ATP was additive to the actions elicited by carbachol (figure 2). Preincubation with α-bungarotoxin did not affect the ATP-induced uptake of ^{86}Rb while carbachol-mediated influx was totally abolished (table 4). Preincubation with β,γ-imido ATP or β,γ-methylene ATP did not affect carbachol-induced uptake of ^{86}Rb (table 1).

Fibroblasts cultured from chick embryos showed no sensitivity to ATP (data not shown).

Table 4. Effect of α-bungarotoxin on ATP- and carbachol-induced uptake of ^{86}Rb in chick myotubes.

Conditions	^{86}Rb-uptake % of control	Statistics	
no drug	100±8	-	(n=6)
0.3 mM ATP	137±9	p<0.025	(n=6)
1.0 mM carbachol	159±11	p<0.005	(n=6)
0.3 mM ATP[1)]	142±17	p<0.05	(n=6)
1.0 mM carbachol[1)]	100±6	n.s.	(n=6)

Values are expressed as mean±SEM. p-values were determined with a two-tailed t-test. 100 % basal uptake corresponds to 1760±150 cpm (mean±SEM).
1); preincubated 30 min with 200 nM α-bungarotoxin.

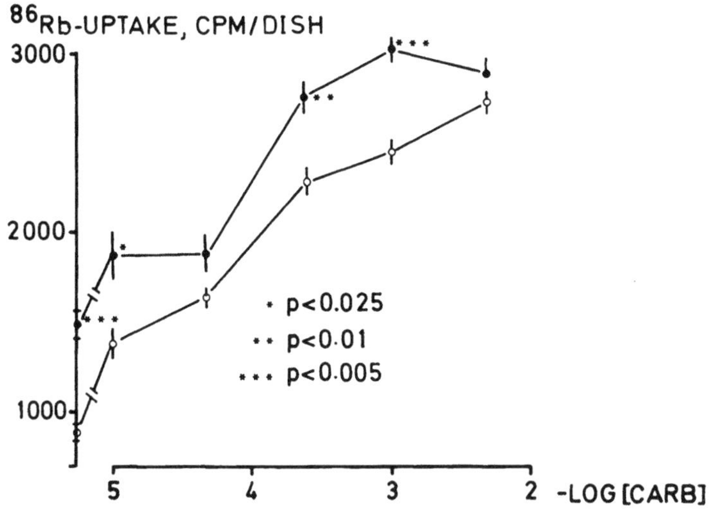

Figure 2. Effect of ATP on carbachol-induced uptake of ^{86}Rb in chick myotubes.
Values are expressed as mean±SEM, n=3. p-values were determined with a two-tailed t-test.
o-o; carbachol ●-●; carbachol + 0.3 mM ATP

Discussion

The presented results may be interpreted as further evidence for a transmitter action of ATP on cultured muscle cells. Our results are in accordance with observations reported by Kolb and Wakelam (3) who suggested a transmitter like action of ATP in the same tissue with a patch-clamp technique. These authors proposed that the action of ATP is mediated via an ATP-receptor that could be an embryonic precursor to the myotube nAChR. To strengthen this view they compared similarities in charge distribution between ATP and ACh (5). The ion-translocation effects of ATP we observe in the low concentration range (1-500 µM) of the discontinuous dose-response curve probably corresponds to an action similar to that reported by Kolb and Wakelam (3), while the second rise in ^{86}Rb-influx (>1 mM) most probably is due to cell damage caused by the high concentrations of ATP. The increase in basal and ATP-induced ^{86}Rb-uptake in the absence of Ca^{2+}-ions is probably due to "leak"-

transport. It is known that the permeability for monovalent cations is controlled by extracellular Ca^{2+} in many eucaryotic cells (cf 6). The finding that omission of Mg^{2+} did not affect ATP-induced influx is more obscure as a Mg-ATP complex often is necessary to elicit biological effects with ATP. Enough Mg^{2+} to complex ATP may, however, have been present.

Our results suggest that ATP and cholinergic agonists act on different mechanisms. The maximal effects elicited by ATP and carbachol were additive, α-bungarotoxin did not inhibit ATP-induced ^{86}Rb-uptake and preincubation with β,γ-imido ATP or β,γ-methylene ATP had no effect on uptake elicited by carbachol. It is interesting to note that the β,γ-derivatives of ATP had no ion-translocation effect in themselves. This suggests that the β,γ-high energy bond of ATP is needed to elicit the ATP-induced uptake of ^{86}Rb. β,γ-imido and β,γ-methylene ATP also seem to bind to the receptive site for ATP as they inhibit the action of ATP.

In frog sartorius muscles ATP has previously been reported to have a modulatory action on the nAChR by increasing its sensitivity to ACh (2) as measured by electrophysiological techniques. To exert this action, ATP was suggested to bind to an allosteric on the nAChR and increase either channel life-time or conductance or both at the same time without affecting the binding affinity for ACh. In our experiments as well as in (3) ATP, in itself, had an effect on ion-transport, thus a transmitter action may be proposed.

Finally, the embryonic nature of cells in cultures must be taken into consideration. It could well be that ATP serves a transmitter role early in development while at later stages this role is completely taken over by ACh and "only" the modulatory actions of ATP remain.

Acknowlegement

This research was supported by a grant from the Swedish Medical Research Council (grant no B85-13X-03907-13B).

References

1. Ewald, D.A. 1976. J. Membr. Biol. 29, 47.

2. Akasu, T., Hirai, K., Koketsu, K. 1981. Br. J. Pharmac. 74, 505.

3. Kolb, H.-A., Wakelam, M.J.O. 1983. Nature (Lond) 303, 621.

4. Catterall, W.A. 1975. J. Biol. Chem. 250, 1776.

5. Boyd, H.J. 1973. J. Theor. Biol. 42, 49.

6. Latzkowits, L., Fajszi, C. 1982. In: Handbook of Neurochemistry (A. Lajtha, ed.) Plenum Press, New York, p 1.

II. The Acetylcholine Receptor: Structural Aspects

A MONOCLONAL ANTIBODY SPECIFIC FOR ONE OF THE TWO AGONIST BINDING SITES AT THE ACETYLCHOLINE RECEPTOR

G. Fels, A. Maelicke
Max-Planck-Institut für Ernährungsphysiologie,
Rheinlanddamm 201, D-4600 Dortmund, FRG

D. Schäfer
Max-Planck-Institut für Systemphysiologie,
Rheinlanddamm 201, D-4600 Dortmund, FRG

J.P. Walrond, Th.S. Reese
Laboratory of Neurobiology, NIH-NINCDS
Marine Biological Laboratory, Woods Hole, MA 02543 / USA

Abstract

We have raised against the acetylcholine receptor from Torpedo marmorata a monoclonal antibody (WF6) with the following characteristics: The antibody binds to an extracellular region of the native receptor but does not bind to the denatured receptor. Its target region is not a very common immunogenic site at the receptor surface. Binding to the receptor is inhibited by representative agonists and α-neurotoxins but not by low molecular weight antagonists. Under saturating conditions, WF6 blocks half of the binding sites for acetylcholine at the receptor; with these sites blocked, acetylcholine or its agonists cannot induce ion flux through the receptor-integrated channel.

Introduction

Until recently most studies of the structure-function relationship of the nicotinic acetylcholine receptor have been performed

Molecular Basis of Nerve Activity

by means of ligands with well defined pharmacology and target sites: cholinergic agonists and antagonists, local anesthetics and direct channel blockers (1,2). In contrast, monoclonal antibodies can be raised, in principle, against any site at the surface of the receptor. An appropriate bank of antibodies (3) therefore should provide a much wider selection of marker specificities than is available with the classical ligands. The antibody approach is limited, however, by the fact that most proteins including the acetylcholine receptor (4) possess only few main immunogenic regions. Using appropriate preparations and immunization protocols (3,5), these limitations can be overcome and selective markers of widely varying specificity can be obtained (3).

Here we summarize our studies with an antibody (WF6) obtained by subcloning the hybridoma cell-line XR-6-G10 (3). Antibody WF6 is similar but not identical in its basic properties to mAb6 of Watters and Maelicke (3). The properties of this antibody as detailed here show the unique advantages of these probes over other ligands in structural and functional studies of the receptor.

Materials and Methods

The preparation and assay procedures for monoclonal antibodies and the receptor have been described previously (3). The receptor was iodinated to 35 Ci/mmol employing Bolton-Hunter reagent (Amersham, FRG). Gold-labeled anti-mouse IgG was obtained from Jenssen (Belgium). The ultrastructural studies were performed together with Dr. T.S. Reese and Dr. J. Walrond at the Marine Biological Laboratory at Woods Hole, MA, and Dr. D. Schäfer from the Max-Planck-Institut für Systemphysiologie, Dortmund. The binding assay for ^{3}H-acetylcholine has been described previously (6) and was carried out in collaboration with Dr. E.K.

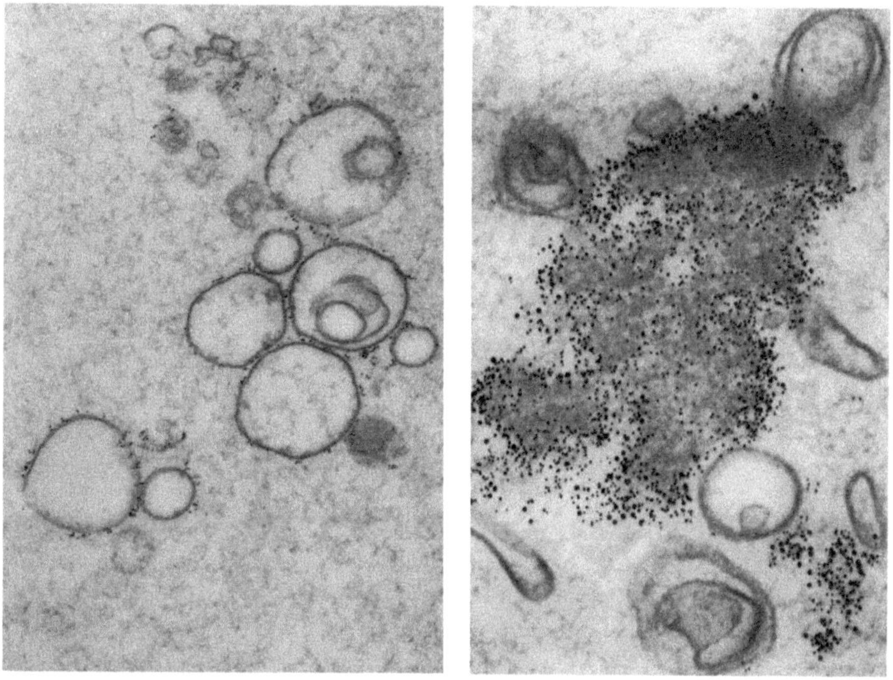

Fig. 1. Thin section of rapidly frozen and freeze-substituted Torpedo membrane fragments after successive labeling with monoclonal antibody WF6 and gold(5 nm)-labeled anti-mouse IgG (Fig. 1a, left). In Fig. 1b (right) additional labeling with protein A (10 nm) was employed.

Wolff from our laboratory. The ion-flux measurements were performed by Dr. M. Covarrubias and R. Plümer-Wilk from our laboratory employing the fluorescence quenching of 1,2,6,8-pyrene tetrasulfonat by Tl^+ as monitoring signal (7). Immunoblot procedures were performed together with M. Schreiber from our laboratory by means of the Biorad Assay kit (170-6502) or peroxidase-conjugated protein A (Sigma) following the Biorad instructions. The dilutions were as follows: mAb WF6 0.01 mg/ml, anti Torpedo acetylcholine receptor serum 1:300, peroxidase labeled reagents 1:1000.

Results

Antibody WF6 was derived from XR6-G10 (3) and, as its parent clone, also reacted with both receptor-rich membranes and purified receptor from Torpedo marmorata. Furthermore, the cross-reactivity pattern with Electrophorus receptor and rat muscle receptor was also comparable to XR6-G10 while in contrast to the latter WF6 did not bind to ganglionic tissue (3,8). Similar to the parent clone, WF6 binds with rather high affinity to the Torpedo receptor ($K_D < 1 \times 10^{-9}$M (9)); this binding is competitively inhibited by α-cobratoxin, acetylcholine, carbamoylcholine and succinyldicholine (3).

To further explore the properties of WF6 the following studies were performed:
(i) Micrographs were taken from thin sections of rapidly frozen and freeze-substituted Torpedo membrane vesicles successively labeled with WF6 and gold-labeled anti-mouse IgG in order to test whether WF6 is bound to the extra-cellular part of these membranes (Fig. 1). Independent of whether the membrane vesicles were treated with saponin or not, the majority of labeled molecules appeared on their outside indicating extracellular labeling. Fig. 1b is unusual in that large parts of it show top views rather than cross fractures (Fig. 1a); thus, the high density of the label is demonstrated. As the employed membrane preparation was not further enriched by gradient centrifugation (6), membrane vesicles without label are rather frequently observed (Fig. 1b).

(ii) The subunit specificity of antibody WF6 was tested in immunoblot experiments. For this purpose, nitrocellulose filter paper blots of polyacrylamid gels of SDS extracts of Torpedo membranes were incubated with WF6 or, for comparison, with rabbit antisera raised against the whole receptor protein. After washing, the blotted proteins specifically reacting with

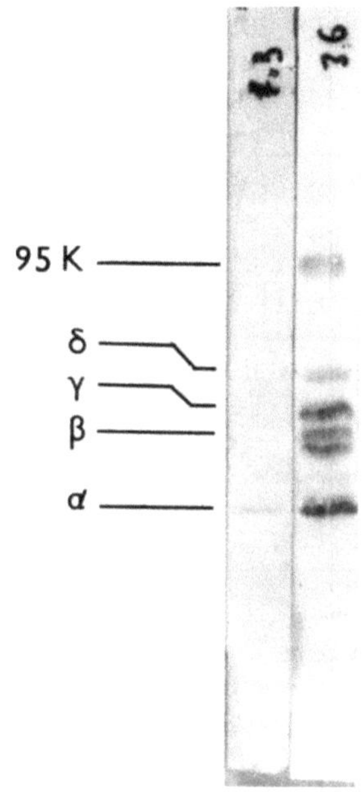

Fig. 2. Subunit specificity of mAb WF6. SDS-page separated proteins from _Torpedo_ membrane fragments were transferred to nitrocellulose sheet and reacted with the following antibodies: mAb WF6 at 0.01 mg/ml (left lane); rabbit anti _Torpedo_ acetylcholine receptor serum (1:300 in PBS) (right lane). Labeled protein bands were visualized via peroxidase-labeled anti-mouse IgG or protein A.

WF6 or the rabbit serum were visualized with enzyme-labeled protein A or anti mouse IgG (see Methods). As shown in Fig. 2, WF6 labeled only one protein band which by its apparent molecular weight appeared to be the α-polypeptide of the receptor. In contrast and as expected, all polypeptides of the receptor were labeled by the rabbit serum.

(iii) To compare with and extend the previously reported ELISAs for XR6-G10 and the _Torpedo_ receptor, radioimmuno assays with WF6 and ^{125}I-labeled receptor were performed. As shown in the Table, WF6 did not bind to heat-denatured receptor, or to native receptor after saturation with α-cobratoxin. In contrast, the interaction of an immune serum from a _Torpedo_ receptor-immunized rabbit with the receptor was little if at all affected by the presence of α-cobratoxin or denaturation of the receptor. Furthermore, when the binding of WF6 and another monoclonal antibody (XR33-E7) (mAb 10 of (3)) of similar concentration was

Table Binding of mAb WF6 and anti-acetylcholine receptor serum to ^{125}I-acetylcholine receptor

antigen employed	amount of antigen (in cpm) precipitated as complex with	
	mAb WF6	anti-AChR serum
receptor	13161	15820
receptor + toxin	1439	16429
denatur. receptor	2172	15963
denatur. receptor + toxin	1672	n.d.

50 µl of purified mAb WF6 (initial concentration 1 mg/ml) or 50 µl of antiserum (1:10 in PBS) were incubated with 5 µl ^{125}I-AChR (50.000 cpm, Bolton Hunter labeled to 35 Ci/mmol) in 150 µl PBS 0.05% Tween 80 containing BSA at 1 mg/ml concentration. The complex was precipitated after 12 h at 4°C with 100 µl of Staphylococcus aureus (Immunoprecipin, BRL). Radioactivity of the precipitate was measured after two washes with each 1 ml PBS Tween. n.d. = not determined.

measured after saturation of the receptor with a rabbit anti-receptor immune serum (Fig. 3), only little influence of the immune serum on WF6 binding but a strong competitive effect on XR33-E7 binding was observed. Consequently, WF6 is directed against antigenic determinants that require the intact three-dimensional structure of the receptor, and these determinants do not define a main immunogenic region at the receptor surface.

(iv) Competition binding studies (by ELISAs) of antibody WF6 and cholinergic ligands to the *Torpedo* receptor (Fig. 4) agreed with the finding for the parent clone (3) in that only α-cobratoxin, acetylcholine, carbamoylcholine and suberyldicholine acted as effective competitors. Extending the list of cholinergic ligands employed, not only tubocurarine, hexamethonium and decamethonium but also the local anesthetics dibucaine and quinacrine did not compete with WF6 for receptor binding.

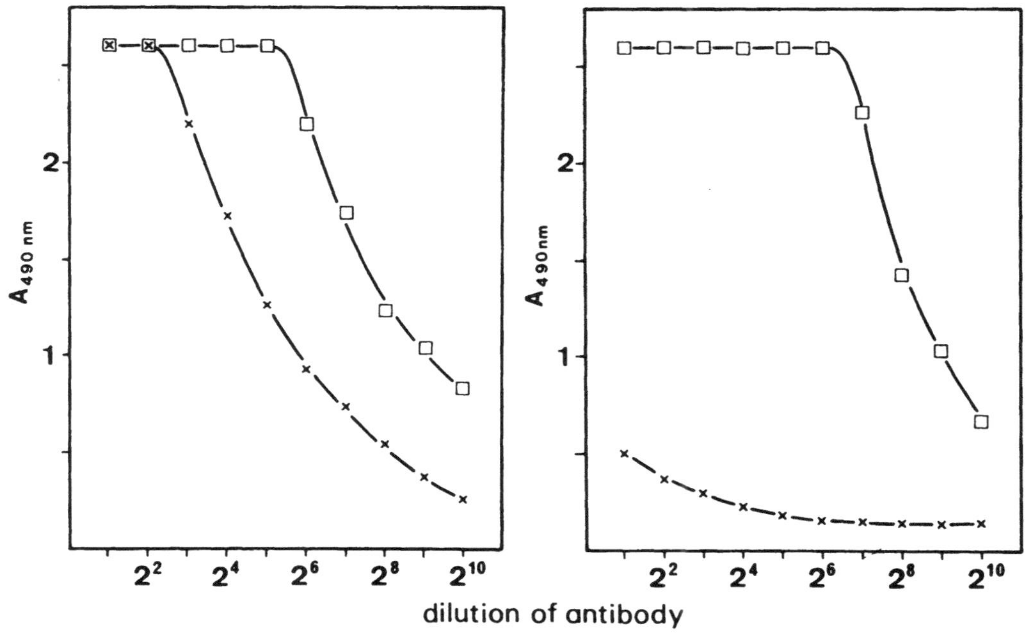

Fig. 3. Competition ELISA employing purified acetylcholine receptor from *Torpedo marmorata* and identical concentrations of (left) mAb WF6 or (right) mAb W10 (3): serial dilution of the antibody (1:10 diluted in PBS) (□) or in anti-*Torpedo* acetylcholine receptor serum (x) (1:100 in PBS) after preincubation (2 h, 37°) with the antiserum.

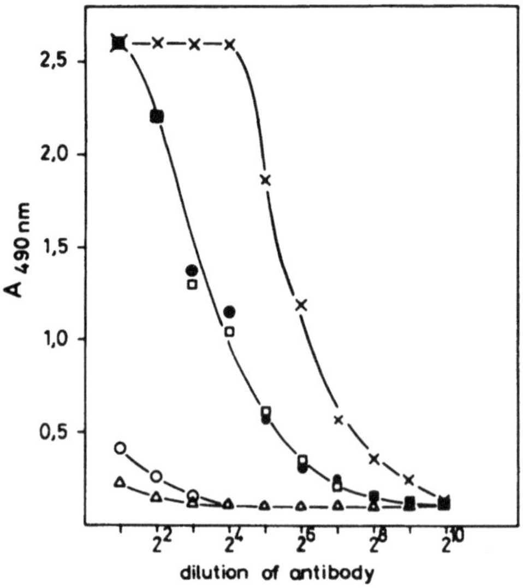

Fig. 4. Competition ELISA with purified Torpedo acetylcholine receptor and mAb WF6: serial dilution of the antibody in buffer (x), 10^{-3}M suberyldicholine (□), 10^{-3}M carbamoylcholine (●), 10^{-3}M acetylcholine (O), or 10^{-6}M α-cobratoxin (Δ). Dilution curves performed in 10^{-3}M solutions of tubocurarine, hexamethonium, decamethonium, quinacrine or dibucaine coincided, within the limits of experimental error, with the control curve in the absence of any ligand (x). Ligands were dissolved in PBS and incubated with the immobilized receptor for 15 min (toxin for 2 h) prior to serial dilution with the antibody.

(v) The competition of antibody WF6 and acetylcholine for receptor binding was further explored employing a rapid centrifugation assay (6). As shown in Fig. 5, saturating concentrations of WF6 apparently block one half of the binding sites for acetylcholine at the membrane-bound receptor. Under these conditions of saturating blockade, the class of high affinity binding sites for acetylcholine disappears.

(vi) In view of the ongoing discussions as to whether one or two molecules of acetylcholine per receptor monomer are required for channel activation (10,11), it was interesting to test whe-

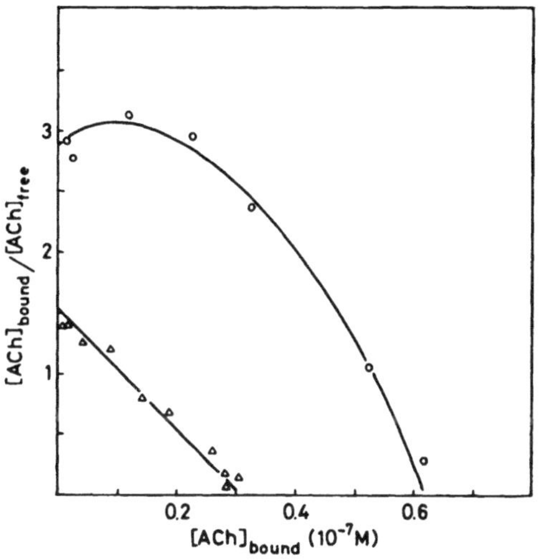

Fig. 5. Scatchard plot of ^{3}H-acetylcholine binding to membrane-bound acetylcholine receptor from Torpedo marmorata. Experimental details see (6,9). Initial concentration of receptor sites in both experiments R_{oo} = 32 nM. Initial concentration of antibody $C_{oo} = 10^{-6}$ M. Acetylcholine binding sites R_o and equilibrium dissociation constant K_D in the absence (o) of antibody R_o = 64 nM, K_{D1} = 21 nM, K_{D2} = 7 nM, in the presence (Δ) of antibody R_o = 31 nM, K_D = 19 nM.

ther channel activation by acetylcholine occurs in the presence of saturating concentrations of antibody WF6. For this purpose, the kinetics of Tl^{+}-influx into receptor-rich membrane vesicles were studied in the absence and presence of saturating concentrations of WF6. As shown in Fig. 6, the remaining Tl^{+}-influx in the presence of WF6 (upper trace) was similar if not identical to the background "leakage" also observed in the presence of saturating concentrations of α-toxin. We conclude that both binding sites per receptor monomer must be saturated with acetylcholine or agonist to activate the receptor-integral channel.

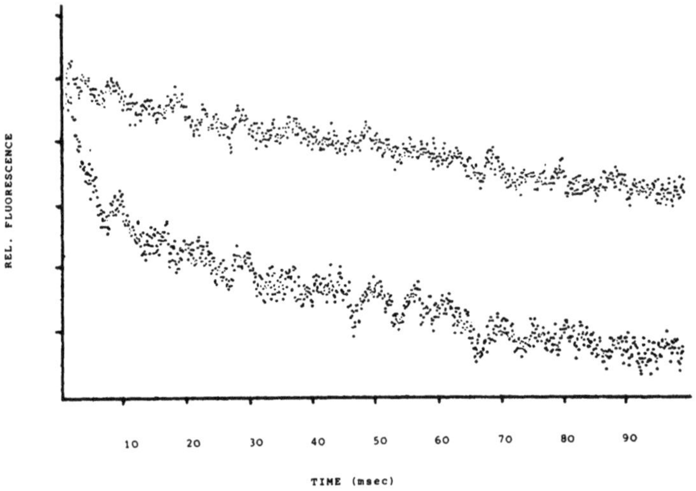

Fig. 6. Effect of mAb WF6 on the kinetics of Tl^+-influx into receptor-rich membrane vesicle from Torpedo marmorata (initial concentration 2 x 10^{-7}M toxin binding sites). Vesicles, containing the fluorophore 1,2,6,8-pyrene tetrasulfonat and suspended in 160 mM $NaNO_3$, 10 mM HEPES, 0.1 mM Tetram (6), pH 7.4, were rapidly mixed with equal volumes of a solution containing 0.1 mM acetylcholine, 130 mM $NaNO_3$, 30 mM $TlNO_3$, 10 mM HEPES, pH 7.4, and the change in fluorescence was recorded. The experiments were carried out with a Durrum D-110 stopped-flow fluorimeter monitoring the fluorescence quenching of the fluorophore by Tl^+. Lower trace: Time course of Tl^+ influx into vesicles after receptor activation by acetylcholine. Upper trace: Tl^+ influx into vesicles preincubated for 180 min with mAb WF6 (10^{-6}M). The upper trace coincides with the control determined in the absence of acetylcholine.

Conclusions

Monoclonal antibody WF6 recognizes a minor immunogenic region within the ligand binding region of the acetylcholine receptor from Torpedo marmorata. From its binding sites at the receptor, WF6 can block at most half of the binding sites for acetylcholine indicating that these either differ in structure or are differently linked to the antibody binding sites. Blockade of one half of the binding sites for acetylcholine at the receptor prohibits activation of the receptor channel by acetylcholine indicating that the binding of two transmitter molecules is a necessary requirement for channel activation.

Acknowledgement

We thank H. Rimpel, G. Wehmeier, H. Poloczek and J. Michaud for excellent technical assistance. The electronmicroscopic studies in collaboration with T.S. Reese were supported by the NATO grant SA 5-205 RG (0409/83). The biochemical studies were supported by grants from the Deutsche Forschungsgemeinschaft (SFB 168) and the Fonds der chemischen Industrie.

References

1. Maelicke, A.. 1984. Angew.Chem. 96, 193-219.
2. Changeux, J.-P., A., Devillers-Thiery, P. Chemouilli. 1984. Science 225, 1335-1345.
3. Watters, D., A. Maelicke. 1983. Biochemistry 22, 1811-1819.
4. Lindstrom, J.. 1983. Neurosci.Comm. 1, 139-156.
5. Maelicke, A., D. Watters, G. Fels, R. Plümer. 1984. J.Rec.Res., in press.
6. Fels, G., E.K. Wolff, A. Maelicke. 1982. Eur.J.Biochem. 127, 31-38.
7. Gonzalez-Ros, J.M., J.A. Ferragut, M. Martinez-Carion. 1984. Biochem.Biophys.Res.Comm. 120, 368-375.
8. Fels, G., H. Breer, A. Maelicke. 1983. In: Toxins as Tools in Neurochemistry (F. Hucho and Y.A. Ovchinnikov, eds.). Walter de Gruyter & Co., Berlin, pp. 127-140.
9. Fels, G., et al. in preparation.

10. Sheridan, R.E., H.A. Lester. 1975. Proc.Natl.Acad.Sci.USA 72, 3496-3500.

11. Maelicke, A., H. Prinz. 1983. In: Modern Cell Biology (B.H. Satir, ed.). Alan R. Liss, Inc., New York, 1, pp. 171-197.

THE MOLECULAR BIOLOGY OF ACETYLCHOLINE RECEPTORS.

A. Mauron, P. Nef, and M. Ballivet.
Department of Biochemistry, University of Geneva,
1211 Geneva 4, Switzerland.

Introduction

In the last few years, interest in molecular and genetic approaches to neurobiology has waxed high. Although great successes have already been registered in the cloning of genes important in nerve cells (mostly neuropeptides so far) we will restrict this discussion to the use of recombinant DNA methods in the study of neurotransmitter receptor proteins.

What benefits can be expected from applying cDNA cloning to a receptor of neurobiological interest? Firstly, this approach will yield the complete primary sequence of the protein. If a full-length cDNA clone is isolated, then its nucleotide sequence can easily be established and the amino-acid sequence of the protein deduced. The seemingly roundabout way of getting this information by cloning and sequencing DNA is likely to be the easiest, at least if a complete sequence is required. Knowledge of the primary sequence in turn allows one to make predictions about the secondary structure and the membrane domains of the receptor polypeptide(s). Secondly, the cloning procedure will yield specific nucleic acid hybridization probes, which can be put to several uses: They are needed to detect and characterize the mRNA species encoding the receptor with a view to assessing their cell-type specificity and their regulation. Such probes are also necessary to explore the structure of the gene, (in

Molecular Basis of Nerve Activity

particular how it is divided into exons and introns); whether the genome contains other genes of related sequence, the whole group forming one gene family; what the evolutionary relationships are within such a gene family. Such probes can also be used to screen cDNA or genomic libraries of other organisms, so that the initial investment of effort in getting the first cDNA clone in one particular species need not be repeated. This is based on the fact that cognate genes of different organisms often show enough DNA sequence homology for specific probes to cross-react in a nucleic acid hybridization reaction.

The nicotinic acetylcholine receptor (AChR) present at the neuromuscular junction of vertebrates was the first neurotransmitter receptor to be studied with the tools of recombinant DNA technology. A review of the work undertaken in this field by this and other laboratories should illustrate several of the principles alluded to above.

Cloning the AChR

The discovery that the electrical organ of Torpedo was exceptionally rich in AChR was decisive in starting the biochemical work on this molecule (1). By 1980, work in many laboratories had led to the purification of the Torpedo receptor and the elucidation of many of its biochemical properties (for a review, see Conti-Tronconi and Raftery, (2). The nicotinic acetylcholine receptor is a ligand-gated, cation-specific ionophore. Binding of acetylcholine leads to the transient opening of the channel. Certain snake toxins such as a-bungarotoxin have a very high affinity for the acetylcholine-binding site and block it specifically, a property widely used to purify and assay the AChR. The receptor was found to be an oligomeric protein containing four different kinds of subunits in the stoichiometry $\alpha_2\beta\gamma\delta$. Both the capability of binding acetyl-choline or toxins and of functioning as a cation channel reside in this pentameric glycoprotein. The two α subunits contain the ligand binding sites whereas the channel function seems more or less delocalized among all subunits. Furthermore, it was recognized that many of the insights coming from work with the Torpedo receptor would probably hold for the junctional receptor of vertebrates generally.

As the methods for cloning DNA complementary to mRNA became available, it was realized that the electric organ would be a rich source of the mRNA species encoding the AChR (3). Poly(A)-RNA extracted from this tissue was therefore used by many groups to produce cDNA libraries, which were screened using several methods, including colony hybridization to oligonucleotide probes designed from known N-terminal partial protein sequencing data (4). cDNA for all four subunits of the Torpedo californica receptor were eventually obtained, their nucleotide sequence determined and the amino-acid sequence of the polypeptides deduced. References to this work are included in Table 1, which compiles all reports containing cloning and/or sequencing data for the AChR subunits published so far.

Table 1. Reports of AChR Cloning and Sequencing Data

Organism	Nature of clones and AChR subunit encoded	First author, date and reference
Torpedo californica	cDNA γ	Ballivet 1982 (5)
		Claudio 1983 (6)
" "	cDNA α	Noda 1982 (7)
" "	cDNA β, δ	Noda 1983 (8)
" "	cDNA γ	Noda 1983 (9)
" "	genomic δ	Hershey 1983 (10)
Torpedo marmorata	cDNA α	Giraudat 1982 (11) Sumikawa 1982 (12) Devillers-Thiery 1983 (13)
Mouse	cDNA α	Merlie 1983 (14)
Calf	cDNA α	Noda 1983 (15)
"	cDNA γ	Takai 1984 (16)
Human	genomic α	Noda 1983 (15)
Chicken	genomic α	Ballivet 1983 (17) and unpublished results
"	genomic γ, δ	Nef (18, in press)

Analysis of AChR protein sequence

As soon as the primary sequences of all four subunits of the Torpedo AChR were known, several features became evident (Noda et al.,9). It was clear that the four sequences showed significant homology over their full length and could easily be aligned on each other. Furthermore, various computer methods for the analysis of local hydrophobicity (19, 20) and prediction of secondary structure (21, 22) were applied to these sequences and showed that the four subunits are very similar in these respects as well. Each subunit has four hydrophobic stretches, which are thought to constitute αhelical transmembrane domains. The striking overall similarity in primary sequence and secondary structure predictions provides an underlying rationale for the pseudosymmetrical arrangement of the subunits in the receptor oligomer which was already suggested by physico-chemical studies (reviewed by Kistler et al. 23). Also, as several lines of evidence strongly suggest that the N-termini of the sub-units lie on the extracellular aspect of the membrane, the logical conclusion is a model in which each AChR polypeptide traverses the membrane four times, with the C-termini ending up on the outside as well (6,9,13). The large N-terminal hydrophilic domain contains the carbohydrate residues and, in the case of the α subunits, the acetylcholine binding site. Another relatively large hydrophilic domain, containing the 100 to 140 amino acids between transmembrane regions 3 and 4, is present on the cytoplasmic side.

As to the precise arrangement of the poylypeptides in the membrane, an alternative model is nevertheless possible. Guy (24) and Finer and Stroud (25) have proposed that a stretch of amino-acids immediatly upstream of the fourth hydrophobic domain constitute an additional transmembrane α-helix. This area, which in the previous model would be part of the large cytoplasmic domain mentioned earlier, is not hydrophobic. Instead, it has the potential of forming an amphiphilic α-helix, namely a transmembrane α-helical domain in which hydrophobic amino-acids would be clustered on one side, interacting with membrane lipids (and/or hydrophobic regions of the other subunits), whereas hydrophilic residues would be similarly clustered, forming a hydrophilic aspect of the helix: the latter would join with other similar domains of the other subunits to form the polar lining of the ion channel, which would contain roughly equal numbers of positive and negative residues. As a consequence, each polypeptide would now go through the membrane five times, and the C-termini therefore end up on the cytoplasmic side. In principle, one could discriminate between the two models on the basis of experiments showing the location of the C-termini relative to the membrane. Results published so far are conflicting, as discussed elsewhere in this volume (Barkas and Juillerat).

The conspicuous sequence homology between the various subunits of Torpedo AChR suggests that the corresponding genes must have evolved from a common, single ancestral gene by succesive duplications, as proposed originally by Raftery et al. (4). As the $\alpha\beta$ pair and the $\gamma\delta$ pair show closer kinship than other combinations, the phylogenetic tree describing these relationships was designed by Noda et al.(9) to show a first duplication of the ancestral gene a_0 to give two genes named a_1 and a_2. Then, at some later time, a_1 was duplicated to yield α and β and a_2, to give γ and δ.

On the basis of cross-hybridization with Torpedo probes, it has been possible to obtain clones - both cDNA and genomic - from other vertebrate species (table 1), so that some information is available about AChR sequences of human, mouse, calf and chicken. Where available, the deduced amino-acid sequences of the AChR subunits in other organisms show a striking resemblance to those of Torpedo. Moreover, there usually is more sequence homology between the same subunits in different species than between different subunits of one given organism, a finding consistent with the idea that subunits of like name are truly cognate in an evolutionary sense. This is particularly evident when comparing α subunits of different species, which are most highly conserved. This, in a way, is not too surprising as the α subunit contains the acetylcholine binding site and therefore is presumably subject to additional structural constraints as compared to the other subunits.

Inspection of the deduced primary sequences of the AChR polypeptides from these other sources reveals the same secondary structure elements already described in the Torpedo sequences, particularly the four hydrophobic domains thought to be transmembrane segments. It is also apparent that these segments (at least the first three) are especially well conserved in sequence among various organisms, thus justifying a posteriori the choice of probes corresponding to these areas to screen libraries of different organisms (18). Another area of high sequence conservation is evident within the N-terminal extracellular domain. Table 2 shows the core of this conserved area.

Table 2. Comparison of Sequences around the Area Thought to Be the Acetylcholine Binding Site in Subunits.

	128 … 142	
Torpedo α	PPAIFKSYCEIIVTHFPFDQQNCTMKL	(7)
Human α	PPAIFKSYCEIIVTHFPFDEQNCSMKL	(15)
Calf α	PPAIFKSYCEIIVTHFPFDEQNCSMKL	(15)
Chick α	PPAIFKSYCEIIVTYFPFDQQNCSMKL	(17)
Torpedo β	PSAIYRSSCTIKVMYFPFDWQNCTMVF	(8)
Torpedo γ	PPAIYRSTCPIAVTYFPFDWQNCSLVF	(6,9)
Human γ	PPAIFRSSCPISVTYFPFDWQNCSLKF	(Ballivet, unpubl.)
Calf γ	PPAIFRSSCPVSVTFFPFDWQNCSLIF	(16)
Chick γ	PPAIYRSSCSIHVTYFPFDWQNCTMVF	(18)
Torpedo δ	PPAIFRSSCPINVLYFPFDWQNCSLKF	(8)
Chick δ	PPAIFRSACPINVNFFPFDWQNCTLKF	(18)

Among the features conserved in all subunits, not just α's, are the two cysteine residues at positions 128 and 142 (numbers refer to Torpedo α) and the N-linked glycosylation site QNCT/S around cysteine 142. In all known α sequences, this is the only possible such glycosylation site, whereas more of these potentially exist on the other subunits. That this glycosylation is somehow involved with acetylcholine binding is suggested by experiments in which tunicamycin, an inhibitor of core glycosylation, prevents the maturation of newly synthesized subunits to the α-toxin binding state (26). More interesting in view of the unique role played by the α subunit are α-specific features, i.e. sequence elements found in all α subunits and none of the others: Table 2 shows these in boxes. Of particular interest are amino-acid positions 125 and 132: At position 125, all α subunits have a lysine whereas all non-α have an arginine. Similarly at position 132, all α sequences show a glutamic acid or glutamine where all non-α have a tryptophan. In addition, there is a pair of cysteines at positions 192-193 (not shown on the figure), which is also α-specific. It is tempting to speculate that, in α subunits, these cysteines make disulfide bridges with the ones at 128 and 142, so that all major α-specific elements would be brought together in close proximity. Indeed, experiments by Karlin et al. (27) using the sulfhydryl-dependent affinity ligand MBTA (4-(N-maleimido-phenyltrimethyl ammonium)) suggest that the binding site is close to a disulfide bridge.

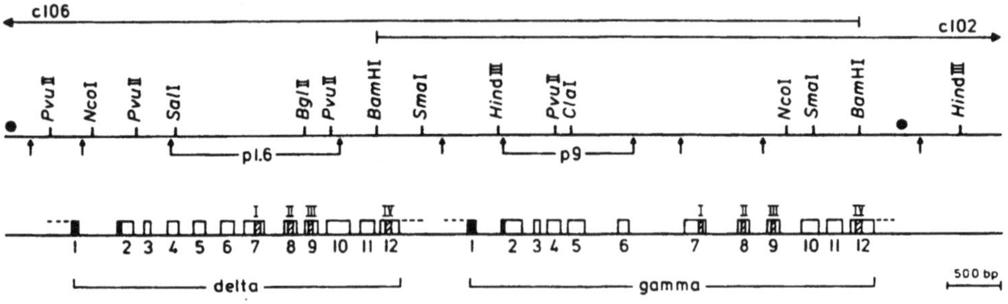

<u>Fig. 1.</u> Structure of genes coding for the γ and δ subunits in chicken. The numbered boxes represent exons (data from ref. 18). Transmembrane domains are also indicated (I-IV).

Gene Structure and Evolution

Several laboratories have screened genomic libraries of Torpedo and other organisms in order to study the structure of the genes encoding the AChR subunits. This laboratory screened a lambda-genomic library of chicken and obtained clones encoding the α, γ and δ subunits (17, 18). Interestingly, it was found that the γ and δ genes are very closely linked (fig. 1). They are coded by the same strand of DNA and there are only 740 base pairs between the end of the δ coding sequence and the initiator codon of γ (18). Genomic blot experiments showed that the clone containing both γ and δ sequences did not result from artifactual recombination but faithfully reflected the situation in the chicken genome. The DNA sequence of this and an overlapping clone were determined and the extent of exons and introns could be determined by matching with the Torpedo cDNA sequences. The result, shown on figure 1 shows that the two genes have the same number of exons. In addition, all exon-intron borders occur at homologous positions between or within codons. Also, individual exons seem to encode specific structural modules : this is quite clear in the case of exons 7,8,9 and 12 which encode the four putative hydrophobic transmembrane domains, whereas the amphiphilic helix proposed by Finer and Stroud would be encoded by exon 11. This is yet another instance of the rule that structural and/or functional domains of a polypeptide chain tend to be encoded by distinct exons, a finding related to Gilbert's hypothesis of protein evolution by "exon-shuffling" (28). More recently, Go's calculations (29) suggested that, for globular proteins like lysozyme or globin, modules defined by purely structural criteria tend to be coded for by separate exons. The work with the AChR would suggest that this may hold for integral membrane proteins as well. In any event, it seems clear that the exon-intron structure of these genes has been faithfully conserved over a considerable period, even as the sequence itself was diverging significantly (the two deduced primary sequences under consideration show only 49% conservation). This laboratory has also isolated and partially sequenced clones for the α-subunit of chicken (17 and unpublished results). The distribution of exons and introns there is similar, but not identical. Nevertheless, even though the α gene has 9 exons instead of 12, it appears that exon 7 of α corresponds to the sum of exons 8 and 9 in γ and δ. A similar relationship exists for exon 8 of α and exons 10 + 11 of the γδ pair. As shown in fig.5, the structure of exon 7 of α can be derived by "clean" intron excision from the corresponding exons in the γ and δ genes, vz. exons 8 and 9. The structure of the human α gene, cloned and sequenced by Noda et al. (15), is very similar. Again, There are nine exons and the transmembrane domains are encoded by the same exons as in the chicken α gene.

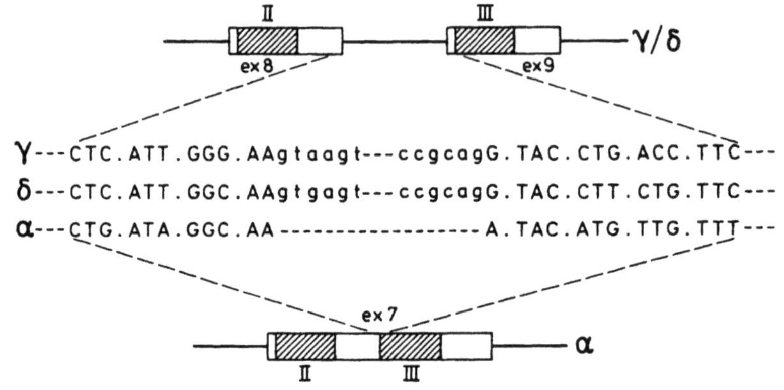

Fig. 2. Derivation of α gene exon-intron structure by precise intron excision. Aligning the nucleotide sequences of the chicken α, γ, and δ genes in the area shown demonstrates that α exon 7 can be derived from γδ exons 8 and 9 by precise excision of an intron.

The fact that the $\gamma\delta$ pair of genes is closely linked is rather striking in view of the proposed phylogeny of the AChR genes discussed above. The latter would indeed predict the generation of a linked $\gamma\delta$ pair by gene duplication. On the other hand, the fact that this linkage survived many hundreds of million years to present-day chicken means that for a long time, the γ and δ gene had the potential for concerted evolution, i.e. the two genes could be corrected against each other by crossover fixation (or gene conversion), as is most probably the case in many multi-gene gene families (30,31,32,33). If this did happen, then the time of gene duplication giving rise to the γ and δ gene may not correspond to their actual evolutionary separation, something to be borne in mind in any attempts to evaluate rates of evolution for the AChR genes.

From the phylogenetic tree under consideration, we would conclude that in all vertebrates, the AChR gene form a small gene family comprising four non-allelic members, α,β,γ, and δ, and that throughout vertebrate evolution, genes encoding the same subunit are really cognate, as mentioned before. This implies that duplication and evolutionary separation among the four subunit genes had been completed by the time of the vertebrate radiation, more than 700 million years ago.

The question of wether the AChR multi-gene family contains additional members has been raised by several investigators. If one considers only such AChR subunits as have been purified (i.e. electroplax and neuromuscular junction), no sequence heterogeneity has ever been detected, contrary to initial expectations that, for instance, the two α subunits could by encoded by distinct genes (see the discussion in Klarsfeld et al., 34). On the other hand, it has been suggested that the CNS contains a nicotinic AChR related to, but not identical to the classical AChR (35,36) and there, the issue of separate but related genes being expressed is more open. Genomic blot analyses using α and δ-specific probes in Torpedo (34, 10)and α,γ,δ probes in chicken (18 and our unpublished results) suggest that the genes encoding the AchR subunits are unique by conventional hybridization criteria. This, at least would seem to exclude the existence in these genomes of closely related additional genes but negative hybridization data are rarely completely conclusive as they depend critically on the choice of probes and reaction conditions. Work is in progress in this and other laboratories to answer these questions by screening brain cDNA libraries for putative brain-specific AChR transcripts.

References

1. Nachmanson, D. 1959. Chemical and molecular basis of nerve activity. Academic Press, New York.

2. Conti-Tronconi, B.M. and Raftery, M.A. 1982. The Nicotinic Cholinergic Receptor: Correlation of Molecular Structure with Functional Properties. Ann. Rev. Biochem. 51,491.

3. Mendez, B., Valenzuela, P., Martial, J.A. and Baxter, J.D. 1980. Science 209, 695.

4. Raftery, M.A., Hunkapiller, M., Strader, C. and Hood, L.E. 1980. Science 208, 1454.

5. Ballivet, M., Patrick, J., Lee, J. and Heinemann, S. 1982. Proc. Natl. Acad. Sci. USA 79, 4466.

6. Claudio, T., Ballivet, M., Patrick, J. and Heinemann, S. 1983. Proc. Natl. Acad. Sci. U.S.A. 80, 1111.

7. Noda, M., Takahashi, H., Tanabe, T., Toyosato, M., Furutani, Y., Hirose, T., Asai, M., Inayama, S., Miyata, T. and Numa, S. 1982. Nature 299, 793.

8. Noda, M., Takahashi, H., Tanabe, T., Toyosato, M., Kikyotani, S., Hirose, T., Asai, M., Takashima, H., Inayama, S., Miyata, T. and Numa, S. 1983. Nature 301, 251.

9. Noda, M., Takahashi, H., Tanabe, T., Toyosato, M., Kikyotani, S., Furutani, Y., Hirose, T., Takashima, H., Inayama, S., Miyata, T. and Numa, S. Nature 302, 528.

10. Hershey, N.D., Noonan, D.J., Mixter, K.S., Claudio, T. and Davidson, N. 1983. Cold Spring Harbor Symp. Quant. Biol. 48, 79.

11. Giraudat, J., Devillers-Thiery, A., Auffray, C., Rougeon, F. and Changeux, J.P. 1982. The EMBO Journal 1, 713.

12. Sumikawa, K., Houghton, M., Smith, J.C., Bell, L., Richards, B.M. and Barnard, E.A. 1982. Nucl.Ac.Res. 10,5809

13. Devillers-Thiery, A., Giraudat, J., Bentaboulet, M. and Changeux, J.P. 1983. Proc. Natl. Acad. Sci. U.S.A. 80, 2067.

14. Merlie, J.P., Sebbane,R., Gardner, S. and Lindstrom, J. 1983. Proc. Natl. Acad. Sci. U.S.A. 80, 3845.

15. Noda, M., Furutani, Y., Takahashi, H., , Toyosato, M., Tanabe, T., Shimizu, S., Kikyotani, S., Kayano, T., Hirose, T., Inayama, S. and Numa, S. Nature 305, 818.

16. Takai, T., Noda, M., Furutani, Y., Takahashi, H., Notake, M., Shimizu, S., Kayano, T., Tanabe, T., Tanaka, K., Hirose, T., Inayama, S., and Numa, S. 1984. Eur. J. Biochem. 143, 109.

17. Ballivet, M., Nef. P., Stalder, R. and Fulpius B. 1983. Cold Spring Harbor Symp. Quant. Biol. 48, 83.

18. Nef. P., Mauron, A., Stalder, R., Alliod, C. and Ballivet, M. Proc. Natl. Acad. Sci. U.S.A. (in press).

19. Hopp, T.P. and Woods, K.R. 1981. Proc. Natl. Acad. Sci. U.S.A. 78, 3824.

20. Kyte, J. and Doolittle, R.F. 1982. J. Mol. Biol. 157, 105.

21. Chou, P.Y. and Fasman, G.D. 1978. Ann. Rev. Biochem. 47, 251.

22. Garnier, J., Osguthorpe, D.J. and Robson, B. 1978. J. Mol. Biol. 120, 97.

23. Kistler, J., Stroud, R.M., Klymkowsky, M.W., Lalancette, R.A. and Fairclough, R.H. 1982. Biophys. J. 37, 371.

24. Guy, H.R. 1984. Biophys. J. 45, 249.

25. Finer-Moore, J. and Stroud, R.M. 1984. Proc. Natl. Acad. Sci. U.S.A. 81, 155.

26. Merlie, J.P. 1984. Cell 36, 573.

27. Karlin, A., Cox, R., Kaldany, R.R., Lobel, P. and Holtzmann, E. Cold Spring Harbor Symp. Quant. Biol. 48, 1.

28. Gilbert, W. 1978. Nature, 271, 501.

29. Go, M. 1983. Proc. Natl. Acad. Sci. U.S.A. 80, 1964.

30. Smith, G.P. 1973. Cold Spring Harbor Symp. Quant. Biol. U.S.A. 38, 507.

31. Hood, L. Cambell, J.H. and Elgin, S.C.R. 1975. Annu. Rev. Genet. 9, 305.

32. Smith, G.P. 1976. Science 191, 528.

33. Baltimore, D. 1981. Cell 24, 592.

34. Klarsfeld, A., Devillers-Thiery, A., Giraudat, J. and Changeux, J.P. 1984. EMBO J. 3, 35.

35. Norman, R.I., Mehraban, F., Barnard, E.A. and Dolly, J.O. 1982. Proc. Natl. Acad. Sci. U.S.A. 79, 1321.

36. Swanson, L.W., Lindstrom, J., Tzartos, S., Schmued, L.C., O'Leary, D.D.M. and Cowan, W.M. 1983. Proc. Natl. Acad. Sci. U.S.A. 80, 4532.

Acknowledgements

This work was supported by grants of the Swiss National Fund for Scientific Research to Bernard Fulpius (3.685.0.80) and Marc Ballivet (3.432.0.83). We thank Bernard Fulpius for his constant support and Thomas Barkas for fruitful discussions.

ANTIBODIES TO SYNTHETIC PEPTIDES AS PROBES OF NICOTINIC ACETYLCHOLINE RECEPTOR STRUCTURE

T. Barkas, M. Juillerat
Département de Biochimie, Université de Genève,
CH 1211-GENEVA 4, Switzerland

Introduction

Due to the recent explosion in information on the primary structure of the nAChR from Torpedo (1-6) and other species (7-10), we have an unrivalled opportunity of exploring receptor structure and function. Of prime interest are the structure of the molecule as a receptor protein per se (how the chains are arranged in the membrane and relative one to another, where the binding sites are located, where the ion channel is located and how it operates, etc.) and as the major antigen in the human neuromuscular disease, myasthenia gravis.

One approach now feasible is the use of antibodies directed against synthetic peptides chosen from the primary sequence of the receptor. Using this method, we might hope to localise ligand-binding sites, to test models predicted from the primary structure, and to identify and characterise antigenic sites. This approach has considerable advantages over the use of polyclonal or monoclonal antibodies to intact or denatured receptor:-

(i) The site of binding of the anti-peptide antibodies can be precisely defined (given certain limitations discussed below).

(ii) It allows the production of antibodies to sites which would not normally be immunogenic.

In the last four years, it has been repeatedly demonstrated that it is possible to raise antibodies to a wide variety of peptides taken from a protein sequence and that these antibodies will often recognise not only the denatured protein but also the native structure (see reviews 11, 12; 13). In fact, it has been claimed that any peptide might be used to raise such anti-protein antibodies, provided only that the peptide is exposed on the surface of the protein (14). This contrasts

Molecular Basis of Nerve Activity

dramatically with the low number of epitopes recognised by the immune system when the intact protein is used as the immunogen, and leads to the concept of immunodominant regions.

In this paper, we will review what is known regarding the method and draw attention to some pit-falls which are not always taken into account during experimental design or interpretation of results, and will illustrate the general approach with examples using peptides derived from the sequence of the alpha-chain of the nAChR.

Methodology

1. Choice of peptide

(i) Exposure on the native protein

If the purpose is to obtain antibodies reactive with the native protein, it is obviously necessary to choose a peptide which is at least partially exposed on the surface (14). In the case of proteins of known structure, this is a simple matter (see 12). In other cases, for example the nAChR, this information is not available and less precise methods must be used. Hydrophilic areas in the primary sequence are good candidates for exposed sites, and, in the case of small, globular, soluble proteins, a correlation between the major peak of hydrophilicity and an antigenic site has been made (15). The presence of a possible glycosylation site might also be indicative of an exposed region; this does not necessarily correlate with a hydrophilic region as, in the alpha chain of the receptor, the proposed binding site (Asn 141) is contained in a markedly hydrophobic segment. N- and C- termini are often antigenic and exposed; this has proven to be the case for the nAChR (16, 17).

(ii) Other peptides

Peptides might be of interest because of the results of chemical modification studies, e.g. the cysteine residues of the nAChR which are reactive with affinity reagents. If the purpose is simply to raise antibodies which will recognise denatured protein, e.g. for use in immunoblotting, it might be thought that almost any peptide would suffice. However, hydrophobic peptides pose problems of solubility (which might be overcome by the addition of extra charged residues). Similarly, peptides containing several cysteine residues create difficulties due to uncontrolled formation of disulphide bridges, resulting in complex, ill-defined mixtures of aggregates, again with attendant solubility problems.

(iii) Length of the peptide

A minimum length of six amino acids has been suggested for the elicitation of an anti-peptide response (11, 12). It might be relevant that this is approximately the size of an antibody binding site. It should be noted, however, that a long peptide containing a sequence A might present A in a different conformation than if peptide A was used alone. This effect has been observed for the binding of monoclonal anti-virus antibodies to peptides (18).

(iv) Final design

Having made the basic choice of peptide, several refinements can be made to the final product. For labelling purposes, it is useful to add a terminal Tyr residue if one is not already present. To increase solubility and to facilitate the use of glutaraldehyde as the coupling agent for the binding of the peptide to carrier protein, Lys residues can be added. Some groups prefer to couple the peptide to the carrier via cysteine residues; these can also be added at this stage. We ourselves prefer to make such additions to the N-terminus of the specific peptide, as we have observed for a range of peptides derived from the nAChR, HLA antigens and $beta_2$-microglobulin that the response to the synthetic peptide was restricted almost exclusively to the C-terminal six amino acids.

2. Coupling to carrier protein and immunisation

Coupling using glutaraldehyde (19), m-maleimidobenzyl N-hydroxysuccinimide (MBS) (14, 20) or N-succinimidyl 3-(2-pyridyldithio) propionate (SPDP) (21) are most commonly used. Various carrier proteins (KLH, ovalbumin, albumin, tetanus toxoid) have been used with no obvious advantages of any particular protein. The degree of substitution with peptide is conveniently estimated using radiolabelled tracer peptide or amino acid analysis and is normally of the order of 10-20% (w/w) peptide in the conjugate. Rabbits have routinely been used for raising the antisera. No apparent advantage of the intradermal over the intramuscular route of injection in terms of titre or affinity of the antibodies was found (our unpublished results).

3. Detection of antibodies and determination of titres and affinities

Affinity and titre of antisera can be measured using crude antisera by a modification of the method of Steward (22). Briefly, a fixed amount of labelled antigen is mixed with increasing amounts of unlabelled antigen, then reacted with a fixed amount of antiserum. The bound antigen is separated from free antigen using Staphylococcus aureus or a second antibody (23). The reciprocal of the final antibody concentration is determined from the y intercept of a graph of 1/bound antigen against 1/free antigen and the affinity constant is derived from this value and the slope of the graph. Ammonium sulphate precipitation, as used in the original method, was found to overestimate titres (23).

Titres of anti-peptide sera, in general, are similar to or slightly lower than titres to normal protein antigens. The affinities of anti-peptide antibodies have not been commonly measured. Using the above method, we obtain values of 10^6 -10^7 M^{-1} for anti-peptide antibodies, with either peptide or receptor as the antigen, compared with 10^7 -10^8 M^{-1} for normal anti-nAChR antibodies, using receptor as the antigen (23; Barkas and Roth, unpublished results). This is similar to the values reported for the binding of natural anti-hepatitis B antibodies to synthetic peptides (25), for anti-hapten antibodies (26-28) and for anti-HSA antibodies (29), but is considerably lower than the value of 10^8 -10^9 M^{-1} estimated for anti-peptide antibodies binding to calmodulin (21). It would be reasonable to assume that higher affinity antibodies might result from immunisation with a lower dose of conjugates or with conjugates containing a lower density of peptide. Nevertheless, in a systematic study, using rabbits and rats, we have found no effect of varying the dose of a given conjugate or decreasing the content of peptide over a hundred-fold range (Barkas, unpublished results). However, this work was performed using KLH as carrier. KLH is itself a multimer and the density of peptide presented to the lymphocyte is high. It is possible that low level conjugation to smaller monomeric carriers might be advantageous. Measurement of antibody affinity is of obvious importance in deciding the type of detection system to use and in interpreting results of binding assays.

The above method is useful, if a little tedious, in allowing the measurement of both titres and affinities on a quantitative basis using crude antisera. Simple qualitative detection systems can also be used, such as ELISA, agglutination, viral neutralisation, etc.

4. Precautionary notes

(i) Affinity-purification of the antibodies

It is strongly recommended that all anti-peptide antisera are subjected to affinity-purification. Using crude antisera to a variety of nAChR peptide-conjugates or KLH alone, we have observed strong "specific" binding on immunoblots to a variety of standard proteins i.e. the same antisera reproducibly recognise the same standard protein. Using affinity purified antibodies, this is not the case (Barkas and Kistler, unpublished results).

(ii) Reaction with "native" protein

The discrimination which we wish to make is between truly native protein and protein which has not been deliberately denatured but which might contain some denatured or degraded protein. Simply, for example, showing a reactivity of anti-peptide antibodies with ^{125}I-labelled receptor or with receptor coated on to plastic does not exclude the possibility that the binding is to denatured protein. We ourselves have observed a reactivity of antibodies to peptide 151-169 of the alpha-chain (see below) with "native" nAChR. However, on closer investigation, it was found that the "native" receptor recognised by the antibodies was a subpopulation which did not bind labelled toxin, nor could it bind to immobilised toxin (30), and presumably represents denatured receptor in the preparation. A minimal requirement for showing binding to the native structure should be the recognition of toxin-labelled receptor (assuming the antibodies are not directed to this site) and/or the receptor in the membrane.

(iii) Possibility of reaction with other peptides

It cannot be assumed that because a reaction is observed between antibodies (raised against a certain peptide A) and the whole protein that binding to the protein occurs via peptide A. If the peptide has the same configuration as in the native protein and if the peptide is exposed, it is reasonable to suppose that the antibodies would bind at least to site A. However, it is also possible that they bind to another similar site somewhere in the protein. If the above conditions are not met, it is possible that the antibodies bind solely to other sites in the molecule. An example of similar peptides occurring at other regions of the protein can be taken from the nAChR work. If we consider the C-terminus of the alpha-chain FAGRLIELSQEG, we find an identical pentapeptide GRLIE and the tripeptide RLI in the N-terminus of the gamma and delta chains respectively. In the alpha chain itself, we find several pentapeptides which have residues identical to the sequence FAGRLIELSQEG in three out of five positions, which could conceivably react with the antibodies. Fortunately for our purpose, almost all the antibodies produced recognised the subpeptide ELSQEG which is unique to the C-terminus of alpha (16).

(iv) Lack of reactivity of anti-peptide antibodies with the protein

In certain cases, this might be readily explained by the presence of carbohydrate on the appropriate sequence in the protein. In other cases, lack of reactivity might be due to an inappropriate conformation, due e.g. to disulphide-bridges present in the protein and absent in the peptide or vice-versa, or to other steric constraints.

Anti-peptide antibodies and the nicotinic acetylcholine receptor

The peptides described below are all taken from the sequence of the alpha chain of Torpedo (1).

1. C-terminal

This peptide FAGRLIELSQEG (P3), described above, was of interest from two points of view. Firstly, the marked discrepancy between the molecular weights of the alpha chain as determined from the cDNA data (greater than 50K daltons) or by SDS gels (40K daltons) might result from post-translational cleavage of the protein. The mature alpha chain was already known to contain the proposed N-terminus (31), therefore if processing did occur, the proposed C-terminal sequence should be absent from the mature alpha chain. Secondly, two models have been proposed for the mode of insertion of the chain in the membrane, one of which places the C-terminal intracellular (32, 33) and the other extracellular (3, 4, 6). Using such an antiserum, we hoped to resolve these questions.

A double lysine derivative of P3 was coupled to KLH using glutaraldehyde and antisera were raised in rabbits (16). The peptide itself did not bind to antibodies to native nAChR, but did bind detectably to antibodies to denatured nAChR. Anti-peptide antibodies bound to the peptide, to non-denatured and SDS or acid-denatured ^{125}I-receptor and to toxin-labelled receptor. The titre of the antisera were 0.1-1.7 μM for peptide and 0.1-0.4 μM for receptor. Between 40 and 100% of the anti-peptide antibodies recognised the receptor. The affinities for either antigen were $1\text{-}6.10^{6}\ M^{-1}$. (23, unpublished results). Immunoblots showed binding was exclusively to the alpha chain and this could be further localised, using tryptic digestion of vesicles (34), to the C-terminal 100 amino-acids. Antibodies purified on the unique peptide ELSQEG, which shows no homology whatsoever to other regions of the α-chain, were equally effective in binding to the receptor in solution and the alpha band on immunoblots, showing that cleavage does not occur. Ratnam and Lindstrom (24) have also recently demonstrated that antibodies to a similar peptide recognize the mature alpha subunit. Using micro-sequencing of peptides, Conti-Tronconi and Raftery reached the same conclusions (35).

The anti-P3 antibodies also bound to the receptor in Torpedo membranes. However, marked difficulties were encountered in trying to localise the "sideness" of the binding. No binding was detectable on electron microscopy, using protein A gold. Agents such as saponin gave equivocal results; such procedures for the production of "leaky" membranes were of dubious value as any increase in binding might be explained either by the C-terminus being internal or by increasing the extracellular exposure of the C-terminus, which is expected to protrude only a short distance from the lipid bilayer. Experiments using chick muscle cells in culture (Harvey, Roth, Juillerat and Barkas, submitted) showed inhibition of the depolarisation response to iontophoretically-applied acetylcholine. This effect was observed with either crude antisera or affinity-purified antibodies and was blocked by preincubating the antibodies with free peptide. This suggests either that the C-terminus is extracellular or that the antibodies bind to a similar site elsewhere in the chick receptor. Ratnam and Lindstrom (24) have recently demonstrated absorption of antipeptide antibodies, raised against C-terminal peptides of the α, β, γ, and δ chains of Torpedo, to LIS-treated vesicles and not to native vesicles. However, the approach is open to the criticism mentioned above, namely that permeabilisation of membranes might simply make an external site more accessible. These authors also find that a thousand-fold excess of soluble receptor over added antibodies was required to produce 50% block of the antibody, suggesting that their antibodies also are of low affinity. Affinity-purified antibodies were not used, and no proof was given that the binding was not to other sites of the appropriate chain or to membrane proteins other than the receptor.

These antibodies have been used as affinity ligands to isolate the receptor under mild conditions, using free peptides as eluting agent (Falquet, Juillerat and Barkas, unpublished results).

2. N-terminal

Using a conjugate of residues 1-10 (SEHETRLVAN) with C-terminally added CYG coupled to KLH using MBS, we have been unable to raise antibodies which recognise the receptor. Similar results have been reported by Ratnam and Lindstrom (24). However, another group (17) report that, using a conjugate of residues 1-20 coupled to BSA via a carbodiimide, antibodies have been produced which specifically recognise the alpha chain on immunoblots. The antibodies bind both to iodinated-reduced carboxymethylated receptor and to ^{125}I-labelled receptor: however, no proof was given that this binding to "native" receptor is not due to the presence of denatured receptor in the preparation. These antibodies are reported not to bind to membranes.

3. Peptide 151-169 ("Main immunogenic region")

It has been suggested that, based on the hydrophilicity profile, peptide 161-166 might be an important antigenic site (1, 6, 7). This idea results from an overinterpretation of the observation of Hopp and Woods (15) that, for a small, soluble, globular protein, the major peak of hydrophilicity correlates with an antigenic site. Firstly, on a hydrophilicity plot, it is not at all clear whether this peptide is the main peak of hydrophilicity. Secondly, the receptor is not a small, soluble, globular protein. However, we have synthesised peptide 151-169, YDGTKVSISPE-SDRPDLST, (P1), containing this region, and shown that it is not recognised by antibodies to the native receptor, whereas it is bound by 25-33% of the antibodies raised against acid-denatured receptor (30). Antibodies to the conjugate bind the peptide, denatured receptor and "native" receptor. However, this last binding was to a subpopulation of ^{125}I-nAChR which could not bind iodinated toxin and did not bind to immobilised toxin, suggesting that it is denatured material in the preparation.

This region contains two Pro residues and has been predicted to be a beta-turn section. Our results could be explained on a model where this peptide is contained in a cleft and only becomes exposed on denaturation.

It is also of interest to note that many of the peaks of hydrophilicity of the receptor chains correlate with the presence of introns in the genomic sequence (see data 7, 10), suggesting that hydrophilic sites might simply correspond to exposed regions between functional domains of the receptor.

4. Cysteine or "toxin-binding site" peptides

It has been known for a long time that the receptor contains a disulphide bridge which is important in maintaining the integrity of the ligand-binding site (reviewed in 36). Four cysteine residues occur in the proposed extracellular region of the alpha chain at positions 128, 142, 192 and 193. Numa's group suggests a disulphide bridge between the first two Cys residues as a possible candidate (1). However, this area is highly conserved across all four chains of Torpedo, whereas Cys 192, 193 are unique to alpha. Moreover, if we postulate a pocket containing peptide P1, residues 128 and/or 143 could easily be brought into contact with residues 192 and 193 to generate one or two disulphide bridges, further stabilising the pocket. It should be noted that essentially all of the sequence C-terminal to Cys 142 as far as the proposed first transmembrane section (residues 210-240), which contains fewer aminoacids than the other chains, is unique to alpha, as are the two Pro residues in peptide P1 which contribute to the proposed beta-turn structure. This area is, therefore, a prime candidate for a ligand-binding region.

We are currently working with peptides from this region, in particular residues 127-143 (P2) and 189-202 (P4) and variants thereon. Peptide P2 is of further interest as that thought to carry the carbohydrate of the alpha chain. However, this region presents difficulties, as the peptides are quite hydrophobic, creating solubility problems. Once solubilised and coupled to carriers (via Lys using glutaraldehyde or via Cys using MBS or SPDP), they stimulate the production of antipeptide antibodies which do not react even with denatured receptor. In the case of P2, this does not appear to result from the presence of carbohydrate on the receptor as treatment with endoglycosidase H, which removes 70% of the reactivity with concanavalin A, has no effect on the binding. However, although these peptides are not recognised by antibodies to native receptor, they do bind to antibodies to denatured receptor. In the case of P2, 10-20% of such antibodies recognise the peptide. Using these antisera (which also bind to native receptor), it is possible to show that P2 blocks the binding of equimolar amounts of receptor and acid-denatured receptor, suggesting that the binding site is exposed on the native protein, but, as in the case of the C-terminal peptide P3, is not an immunogenic site. We are currently using this approach to isolate specific antibodies to these regions.

Conclusion

As can be readily seen from the current scientific literature, anti-peptide antibodies are proving to be extremely useful tools. Such antibodies have advantages over polyclonal and monoclonal antibodies produced by immunisation with a protein as the site of binding can be more precisely determined and anti-peptide antibodies can be produced to many sites which would not be immunogenic in the whole protein. The disadvantages of the method are that antibodies can only be produced to linear sequences and the possible low affinity of the antibodies obtained.

Synthetic peptides and anti-peptide antibodies can be expected to yield valuable information regarding the structure, function and antigenicity of the nAChR. They could be used to localise important antigenic and ligand-binding sites, either by direct binding of the peptides to characterised monoclonal antibodies or by using the antibodies to localise the cleavage site of proteolytic enzymes which generate fragments, already characterised in terms of toxin-or antibody-binding or in carbohydrate content. The antibodies could also be used directly to determine the ligand-binding sites, to test predicted models for the structure of the receptor, and to isolate peptides of interest from proteolytic digests of the nAChR.

Acknowledgements

We would like to thank our colleagues and collaborators, Professor B.W. Fulpius, Drs. M. Ballivet, B. Roth, J. Kistler, B. Schwendimann, A.L. Harvey, J. Moody, P. Nef, and J. Falquet for their help in this work, which was supported by the Swiss National Foundation for Scientific Research. The secretarial assistance of Mrs. M. Pittet is gratefully acknowledged.

References

1. Noda, M., Takahashi, H., Tanabe, T., Toyosato, M., Furutani, Y., Hirose, T., Asai, M., Inayama, S., Miyata, T. and Numa, S. (1982). Primary structure of α-subunit precursor of Torpedo californica acetylcholine receptor deduced from cDNA sequence. Nature 299, 793-797

2. Sumikawa, K., Houghton, M., Smith, J.C., Richards, B.M. and Barnard, E.A. (1982). The molecular cloning and characterisation of cDNA coding for the α-subunit of the acetylcholine receptor. Nucl. Acids Res. 10, 5809-5822

3. Claudio, T., Ballivet, M., Patrick, J. and Heinemann, S. (1983). Nucleotide and deduced amino acid sequences of Torpedo californica acetylcholine receptor γ-subunit. Proc.Nat.Acad.Sci., USA 80, 1111-1115

4. Devillers-Thiery, A., Giraudat, J., Bentaboulet, M. and Changeux, J-P. (1983) Complete mRNA coding sequence of the acetylcholine binding α-subunit of Torpedo marmorata acetylcholine receptor: a model for the transmembrane organization of the polypeptide chain. Proc.Nat.Acad.Sci., USA 80, 2067-2071

5. Noda, M., Takahashi, H., Tanabe, T., Toyosato, M., Kikyotani, S., Hirose, T., Asai, M., Takashima, H., Inayama, S., Miyata, T. and Numa, S. (1983). Primary structures of β and δ-subunit precursors of Torpedo californica acetylcholine receptor deduced from cDNA sequences. Nature 301, 251-255

6. Noda, M., Takahashi, H., Tanabe, T., Toyosato, M., Kikyotani, S., Furutani, Y., Hirose, T., Takashima, H., Inayama, S., Miyata, T. and Numa, S. (1983). Structural homology of Torpedo californica acetylcholine receptor subunits. Nature 302, 528-532

7. Noda, M., Furutani, Y., Takahashi, H., Toyosato, M., Tanabe, T., Shimizu, S., Kikyotani, S., Kayano, T., Hirose, T., Inayama, S. and Numa, S. (1983). Cloning and sequence analysis of calf cDNA and human genomic DNA encoding α-subunit precursor of muscle acetylcholine receptor. Nature 305, 818-823

8. Ballivet, M., Nef, P., Stalder, R. and Fulpius, B. (1983). Genomic sequences encoding the α-subunit of acetylcholine receptor are conserved in evolution. Cold Spring Harb. Symp.Quant. Biol. XLVIII, 83-87

9. Takai, T., Noda, M., Furutani, Y., Takahashi, H., Notake, M., Shimizu, S., Kayano, T., Tanabe, T., Tanaka, K-I., Hirose, T., Inayama, S. and Numa, S. (1984). Primary structure of γ-subunit precursor of calf-muscle acetylcholine receptor deduced from the cDNA sequence. Eur.J.Biochem. 143, 109-115

10. Nef, P., Mauron, A., Stalder, R., Alliod, C. and Ballivet, M. Structure, linkage and sequence of the two genes encoding the γ and δ subunits of the nicotinic acetylcholine receptor. Proc.Nat.Acad.Sci., USA (in press)

11. Lerner, R.A. (1982). Tapping the immunological repertoire to produce antibodies of predetermined specificity. Nature 299, 592-596

12. Lerner, R.A. (1983). Synthetic vaccines. Scientific American 249, 48-56

13. Niman, H.L., Houghten, R.A., Walker, L.E., Reisfeld, R.A., Wilson, I.A., Hogle, J.M. and Lerner, R.A. (1983). Generation of protein-reactive antibodies by short peptides is an event of high frequency: implications for the structural basis of immune recognition. Proc.Nat.Acad.Sci., USA 80, 4949-4953

14. Green, N., Alexander, H., Olson, A., Alexander, S., Shinnick, T.M., Sutcliffe, J.G. and Lerner, R.A. (1982). Immunogenic structure of the influenza virus hemagglutinin. Cell 28, 477-487

15. Hopp, T.P. and Woods, K.R. (1981). Prediction of protein antigenic determinants from amino acid sequences.. Proc.Nat.Acad.Sci, USA 78, 3824-3828

16. Barkas, T., Juillerat, M., Kistler, J., Schwendimann, B. and Moody, J. (1984) Antibodies to synthetic peptides as probes of acetylcholine receptor structure. Eur.J.Biochem. 143, 309-314

17. Neumann, D., Fridkin, M. and Fuchs, S. (1984). Anti-acetylcholine receptor response achieved by immunisation with a synthetic peptide from the receptor sequence. Biochem.Biophys.Res.Comm. 121, 673-679

18. Emini, E.A., Jameson, B.A. and Wimmer, E. (1983). Priming for and induction of anti-poliovirus neutralizing antibodies by synthetic peptides. Nature 304, 699-703

19. Reichlin, M. 1980. In:Methods in Enzymology, Vol. 70 (H.V. Vunakis and J.J. Langone, eds.) Academic Press, New York. pp.159-165

20. Lerner, R.A., Green, N., Alexander, H., Liu, F-T., Sutcliffe, J.G. and Shinnick, T.M. (1981). Chemically synthesised peptides predicted from the nucleotide sequence of the hepatitis B virus genome elicit antibodies reactive with the native envelope protein of Dane particles. Proc.Nat.Acad.Sci., USA 78, 3403-3407

21. Van Eldik, L.J., Fok, K-F., Erickson, B.W. and Watterson, D.M. (1983). Engineering of site-directed antisera against vertebrate calmodulin by using synthetic peptide immunogens containing an immunoreactive site. Proc.Nat.Acad. Sci.,USA 80, 6775-6779

22. Steward, M.W. 1978: In: Handbook of Experimental Immunology, 3rd. ed., Vol. 1 (D.M. Weir, ed.) Blackwell Scientific Publications, Oxford. pp.16.1-16.20

23. Roth, B. and Barkas, T. Ammonium sulphate precipitation overestimates titres of anti-peptide antisera. Immunol.Lett. (in press)

24. Ratnam, M. and Lindstrom, J. (1984). Structural features of the nicotinic acetylcholine receptor revealed by antibodies to synthetic peptides. Biochem.Biophys.Res.Comm. 122, 1225-1233

25. Brown, S.E., Zuckerman, A.J., Howard, C.R. and Steward, M.W. (1984). Affinity of antibody responses in man to hepatitis B vaccine determined with synthetic peptides. Lancet ii, 184-187

26.Ronco, J.R., Sciutto, E., Leoni, J., Margni, R.A. and Binaghi, R.A. (1984). Interaction of purified precipitating and non-precipitating (coprecipitating) antibodies with hapten and with haptenated protein. Evidence of an asymmetric antibody molecule. Immunology 52, 449-456

27. Eisen, H.N. and Siskind, G.W. (1964). Variations in affinities of antibodies during the immune response. Biochemistry 3, 996-1007

28. Stupp, Y., Yoshida, T. and Paul, W.E. (1969). Determination of antibody-hapten equilibrium constants by an ammonium sulphate precipitation technique. J. Immunol. 103, 625-627

29. Steward, M.W. and Petty, R.E. (1972). The antigen-binding characteristics of antibody pools of different relative affinity. Immunology 23, 881-887

30. Juillerat, M., Barkas, T. and Tzartos, S. (1984). Antigenic sites of the nicotinic acetylcholine receptor cannot be predicted from the hydrophilicity profile. FEBS Lett. 168, 143-148

31. Raftery, M.A., Hunkapiller, M.W., Strader, C.D. and Hood, L.E. (1980). Acetylcholine receptor: complex of homologous subunits. Science 208, 1454-1457

32. Guy, H.R. (1984). A structural model of the acetylcholine receptor channel based on partition energy and helix packing calculations. Biophys.J. 45, 249-261

33. Finer-Moore, J. and Stroud, R.M. (1984). Amphipathic analysis and possible formation of the ion channel in an acetylcholine receptor. Proc.Nat.Acad.Sci., USA 81, 155-159

34. Wennogle, L.P. and Changeux, J-P. (1980). Transmembrane orientation of proteins in acetylcholine receptor-rich membranes from Torpedo marmorata studied by selective proteolysis. Biochemistry 106, 381-393

35. Conti-Tronconi, B.M., Hunkapiller, M.W. and Raftery, M.A. (1984). Molecular weight and structural nonequivalence of the mature subunits of Torpedo californica acetylcholine receptor. Proc.Nat.Acad.Sci, USA 81, 2631-2634

36. Karlin, A. 1980. In: Cell Surface Reviews, Vol. 6 (C.W. Cotman, G. Poste and G.L. Nicolson, eds.) North-Holland Publishing Company, Amsterdam. pp.192-260

REGULATION OF THE NICOTINIC ACETYLCHOLINE RECEPTOR BY PHOSPHORYLATION

E. Heilbronn, H. Eriksson, R. Salmonsson
Unit of Neurochemistry and Neurotoxicology, University of Stockholm, Sundbyberg 172 46, Sweden

Introduction

The understanding of the molecular mechanism by which neurotransmitters produce biological responses in their target cells is important. Among the modulatory reactions involved in the function of the pre- and post-synapse are the phosphorylation and desphosphorylation of specific target proteins (1). Several pathways are known along which these protein phosphorylations may proceed. This paper concerns studies on one such target protein, the nicotinic acetylcholine receptor (nAChR) of the postsynaptic cell (skeletal muscle, electroplax), the target of the motor and the electromotor neurones. Protein kinases are a group of enzymes that transfer a phosphate group from ATP to proteins. Usually the amino acids serine, threonine or tyrosine in the protein are phosphorylated. A number of substances can act as activators of these protein kinases; table 1 shows substances suggested to serve this purpose in the phosphorylation of nAChR. Many of the studies on nAChR phosphorylation, however, have not been carried out under conditions existing *in vivo* or even with endogenous kinases and their functional relevance is thus not clear. Further, work demonstrating regulation *in vivo* of levels of presumptive protein kinase activators are missing. There is little information about the chemical nature of the protein kinases involved in nAChR phosphorylation or of their localization within the postsynaptic cell. Finally, receptor

Molecular Basis of Nerve Activity

Table 1. Literature on the Phosphorylation of the Nicotinic Acetylcholine Receptor from Electric Fish

Source	Phosphorylated Subunits	Kinase Activator	References
Torpedo californica	β, γ, δ	-----	1
Electrophorus electricus	48 000	-----	2
T. marmorata	-----	-----	15
T. californica	-----	not cAMP	16
T. marmorata	γ, δ	-----	9
T. californica	(α) β, γ, δ	Ca/calmodulin	17
T. californica	(α) γ, δ,	-----	18
T. fuscomaculata	-----	cGMP, not cAMP	8
T. californica	γ, δ	cAMP	6
T. californica	-----	not, Ca/calmodulin	19
T. californica	δ	Ca/phospholipid	5
T. marmorata	γ, δ	cAMP	20
T. californica	γ, δ	cAMP, not Ca/calmodulin	21
T. marmorata	γ, δ	cAMP and ?	this paper

phosphorylation should be a reversible process, but dephosphorylation has hardly been studied at all (see however 2, 3). In the present paper we describe endogenous protein kinase activities present in microsacs obtained from the electric organ of Torpedo marmorata and resulting in nAChR phosphorylation. We further describe some properties of a cAMP-dependent protein kinase present there and phosphorylating the γ and ∂ subunits of nAChR.

Materials and Methods

Phosphorylation of nAChR was studied on membranes or microsacs obtained from the electric organ of Torpedo marmorata, on sucrose gradient fractions of the monomeric (9S) and the dimeric (13S) form of nAChR or on affinity chromatography purified receptor (4). Sucrose gradient fractions were obtained after solubilisation of microsacs with the detergent octa ethyleneglycol mono n-dodecylether ($C_{12}E_8$, Nikko Chemicals Co., Japan). Most work was done with deepfrozen electric organs, stored in liquid nitrogen or at -90°C. When it became obvious that the protein kinases under study changed their properties upon storage of the organs, material from live fish or recently excised organs was used. The fish was obtained from the Marine Biological Station of Arcachon, France. All solutions used contained the protease inhibitors EDTA, EGTA, PMSF, aprotinin, leupeptin and pepstatin. Buffers were either Tris-HCl pH 7.4 or MES, pH 6.9. Ouabain was added in most experiments and shown to improve the degree of protein phosphorylation considerably. The separation of the protein kinase activity from detergent solubilised microsacs from nAChR was carried out by sedimentation velocity centrifugation, using 5-20% sucrose containing 0.1% $C_{12}E_8$ and, when required, 10 mM DTT, 10 μM cAMP or both. In the gradient, nAChR was localised by an α-bungarotoxin binding assay. Marker proteins were used to estimate the S-values of the nAChR fractions. The kinase activity was localised by checking the ability of the various fractions to phosphorylate the 13S receptor fraction in the presence of γP^{32}-ATP.

Endogenous protein kinase activities were studied with solubilised microsacs or with electroplax membranes. Microsacs were solubilised because the protein kinase activity is localised on their inside and ATP only poorly penetrates over the membranes. Phosphorylation experiments were started, after preincubation of nAChR with the various components, by the addition of γP^{32}-ATP. Analysis of results was done by semiquantitative or quantitative analysis of autoradiograms of SDS-polyacrylamide gels after electrophoresis, in the presence of purified nAChR.

Sequencing of phosphorylated peptides from nAChR was performed on subunits isolated from affinity chromatography purified nAChR. Phosphorylation was performed with the catalytic subunit of cAMP-dependent protein kinase. Subunits were separated, and isolated by preparative gelelectrophoresis. Phosphorylated subunits were then treated with proteases and phosphorylated peptides were purified by HPLC before amino acid analysis and sequencing.

Results

Endogenous phosphorylation of nAChR

A basic phosphorylation of nAChR and other proteins, not completely extinguished by any of the protein kinase inhibitors tested, occurs in solubilized microsacs. Some such phosphorylation may occur at the amino acid tyrosine within the β, γ and ∂ subunits of nAChR (1). Some or all phosphorylating enzymes are localised inside the microsac as phosphorylation of proteins is much lower when intact microsacs are used. It is known that microsacs have right side out membranes.

Basic phosphorylation is increased when Ca^{2+} is added to the medium. The phosphorylation of nAChR does not seem to increase further if calmodulin of phosphatidylserine are added together with Ca^{2+}. A Ca^{2+}/phosphatidylserine activated protein kinase has been shown to phosphorylate the ∂ subunit of nAChR in an entirely artificial system (5).
Endogenous cAMP-dependent protein kinase was found to phosphorylate nAChR subunits γ and ∂, as previously described (6). We found that the degree of stimulation by cAMP was dependent on experimental conditions. It was usually about 2 fold when microsacs from reasonably fresh electric organs were used, but disappeared upon storage of the sacs or the organ. Stimulation increased several fold in the presence of

ouabain. The protein inhibitor of this kinase inhibited 80-90% of the phosphorylation of the receptor subunits (fig 1).

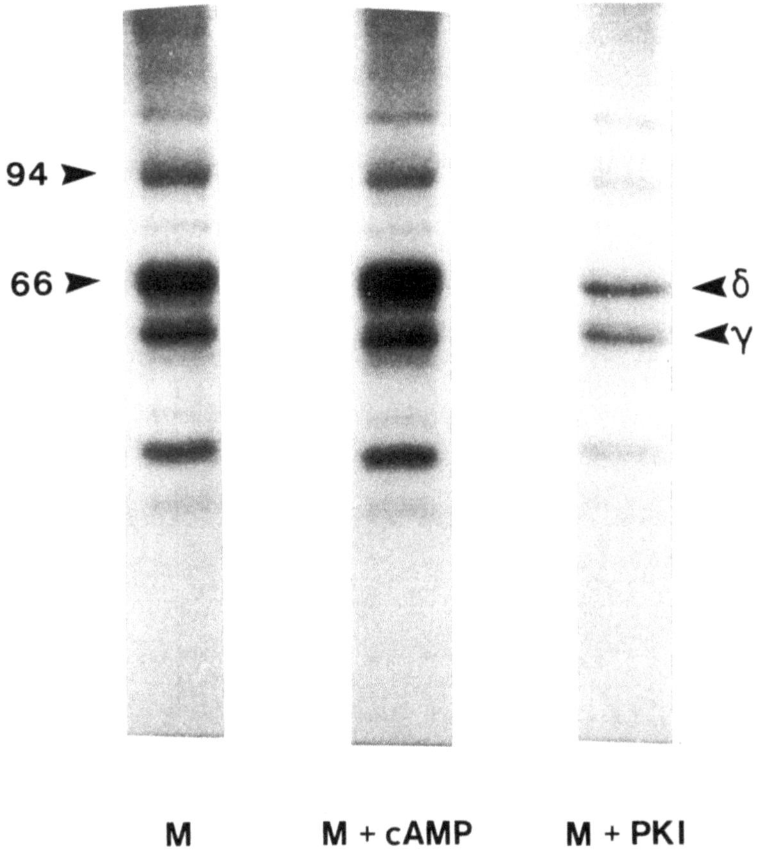

Fig 1. Autoradiograms of gelectrophoreses showing endogenous cAMP-dependent phosphorylation of nAChR-containing membranes. Proteins were phosphorylated in the presence of membranes (M) only, of membrane + cAMP, of membranes + protein kinase inhibitor (PKI).

Separation of protein kinase(s) from nAChR

After solubilisation of microsacs and sedimentation velocity centrifugation on a sucrose density gradient nAChR was found in fractions corresponding to sedimentation coefficients 9S and 13S. The phosphorylating activity, as tested either on histone or on the 13S nAChR, was found in a 7S fraction. This fraction may contain more than one protein kinase. The 13S nAChR was well separated from the 7S kinase while the 9S nAChR fraction was somewhat contaminated by the kinase(s) (fig 2). Both receptor fractions were phosphorylated by 7S; no comparisions of kinetics have yet been performed. The 7S kinase activity is stimulated by cAMP and inhibited by the cAMP-dependent kinase inhibitor.

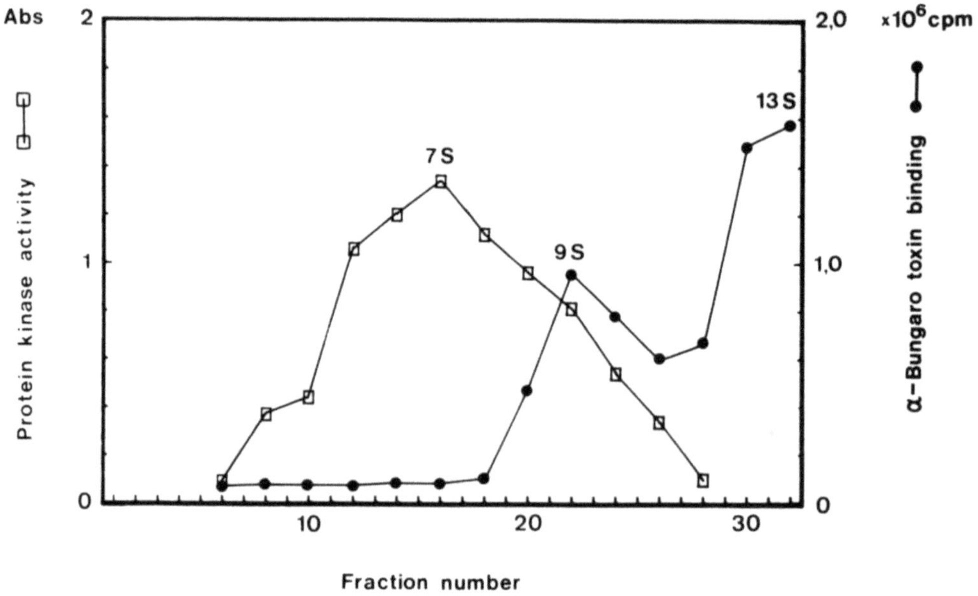

Fig 2. Diagram from sedimentation velocity centrifugation on sucrose density gradient showing separation of nAChR peak 13S from and overlap of 9S peak with the 7S protein kinase peak. Details of analysis, see paper.

Properties of the cAMP-dependent protein kinase

As mentioned, the top of the protein kinase activity peak, estimated with the aid of reference proteins from a similar sucrose gradient, was shown to correspond to a sedimentation coefficient of about 7S, which roughly corresponds to a M_r

of 160 000. Addition of cAMP to solubilised microsacs and to the gradient did not move the protein kinase activity from its 7S position. To test the possibility that a reductive environment is needed to split the solubilised protein kinase into its regulatory and catalytic activities, dithiothreitol (DTT) was added. In the absence of cAMP this addition partly moved the nAChR phosphorylating capacity of the 7S fraction to a position corresponding to a sedimentation coefficient of 3S, corresponding to a M_r of 40 - 50 000. Addition of both cAMP and DTT totally transferred the protein kinase activity to the 3S fraction (fig 3). Both the activity of the 7S fraction and of the 3S fraction was inhibited by the cAMP dependent protein kinase inhibitor. The observations suggest that the 7S fraction contains the protein kinase holoenzyme and the 3S fraction the catalytic subunit of the enzyme.

Biochemical studies on nAChR phosphorylation

Both the monomeric and the dimeric form of nAChR were phosphorylated. The presence of carbachol or α-bungarotoxin did not prevent phosphorylation. Kinetic comparisions of phosphorylation of nAChR alone, in the presence of agonist or in that of antagonist have not yet been made.

Determination of phosphorylated amino acids

Peptides obtained from receptor subunits phosphorylated by cAMP-dependent protein kinase are currently isolated and sequenced in collaboration with D. Eaker, Uppsala. Looking at the primary receptor structure as given by Noda et al (7), one sees that suitable sites of phosphorylation might be the serine in position 254 of the γ-chain and that of position 361 of the ∂-chain. These serines are in sequences arg-X-serine, suitable for phosphorylation, and are located on the inside of the postsynaptic cell.

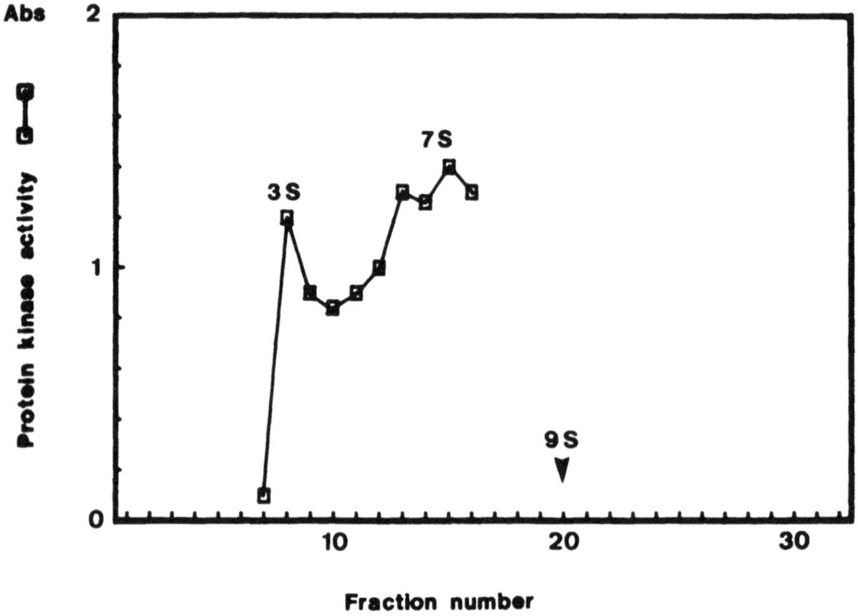

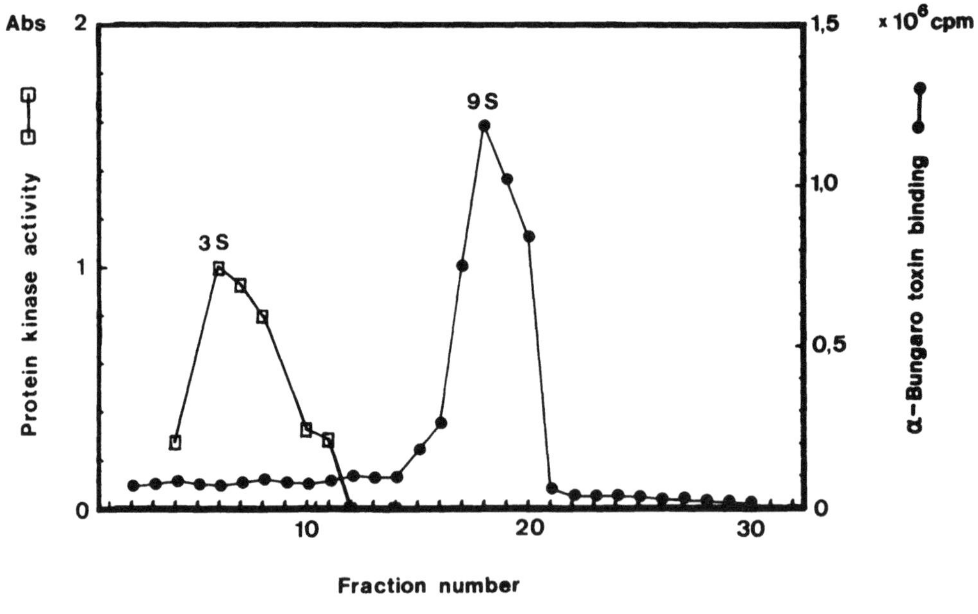

Fig 3. Diagrams showing a) partial transfer of a 7S protein kinase activity peak to a 3S activity peak in the presence of DTT, and b) total transfer of 7S protein kinase activity to a 3S activity peak in the presence of cAMP and DTT.

Discussion

Our studies have shown that endogenous phosphorylation of nAChR and a number of other proteins occurs in the electroplax. The phosphorylating enzyme(s) is (are) largely localised inside the microsac as phosphorylation is higher when it is carried out with solubilised microsacs than with intact structures. The localisation of the protein kinase(s) in the postsynaptic cell of the electric organ is very important and could be carried out by the use of antibodies. We have so far not been able to isolate enough kinase protein from the microsacs to prepare such antibodies. The protein content of the sedimentation velocity gradient fraction is extremely small though the kinase activity is easily demonstrable. This observation leaves open the possibility that the protein kinase activities we are demonstrating come from e g presynaptic impurities present, e g via inclusions into the microsacs during their preparation and are not truly postsynaptic.

We have in our work clearly demonstrated that the cAMP-dependent protein kinase activity present in a solubilized preparation of microsacs phosphorylates a number of proteins, among them nAChR. In the latter molecule the subunits γ and ∂ are phosphorylated. We have further shown that this cAMP-dependent kinase is subject to ageing in stored electric organs or microsacs. The kinase is unable to dissociate in the presence of cAMP unless a reductive environment is provided. _In vivo_ such an environment may be provided by the presence of e g glutathione in the cell. This reductive environment may be necessary to prevent the SH-groups of the protein kinase subunits from reacting with each other to form disulfide bonds which would prevent a complete separation of the regulatory from the catalytic subunits, yet allow the opening up of the catalytic areas of the holoenzyme (fig 4).

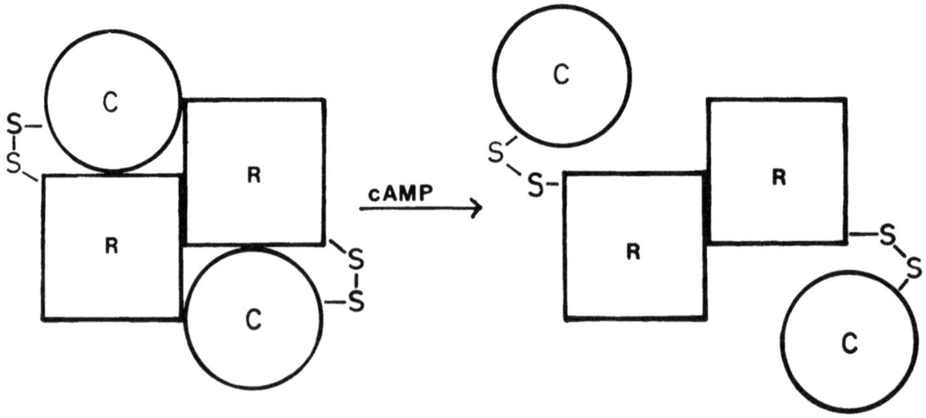

Fig 4. Scheme illustrating possible noncomplete activation of cAMP-dependent protein kinase in the absence of a reductive environment.

Our work so far tells little about the functional meaning of receptor phosphorylation. Such work is presently going on. It has been suggested that nAChR phosphorylation might be connected with the availability, affinity or the function of the neurotransmitter binding site of nAChR (8). The embryonic receptor has been shown to differ from the adult one in its degree of phosphorylation (9) perhaps in different turnover time, in different immunological properties or in different capability to move within the membrane. The properties of the ion channel have been suggested to depend on the degree of phosphorylation (10). The immediate transmitter-triggered opening of the cation channel of nAChR should, however, be too quick a reaction to be dependent on a phosphorylation - desphosphorylation process. The measuring of channel opening and closing by patch clamp techniques would be interesting, provided patches containing nonphosphorylated or phosphorylated nAChR can be obtained. Desensitation is a somewhat slower process and could be affected by phosphorylation (10), though it is obvious from work with purified receptor in reconstructed membranes (11) that changes from nonphosphorylated to phosphorylated receptor or vice versa are not involved in desensitation per se. Presently, it seems most likely that receptor phosphorylation and dephosphorylation is associated with receptor localisation, movement and perhaps turnover. Studies have shown that the level of cAMP is of importance for the synthesis (and localisation?) of nAChR in myotubes (12).

There is no information about regulation of the levels of the substances known to stimulate the protein kinase activity within the electroplax cell. We would like to finish this paper by discussing a possibility concerning cAMP. The levels of this substance might be controlled by the Ca^{2+} moving into the cell simultaneously with neurotransmitter action. A transmitter-induced Ca^{2+} entry across the postsynaptic membrane has been described (13) for frog endplates. Ca^{2+}-influx over postsynaptic membranes, especially when transmitter action is prolonged (e g artificially by the use of anticholinesterase) has also been suggested (14). In our final figure (fig 5) we present a scheme suggesting a function of Ca^{2+} or Ca^{2+}/calmodulin as activators of an adenylate cyclase. This scheme must be tested by correlating transmitter-induced Ca^{2+} influxes into microsacs with their levels of cAMP, We are aware of other possibilities, one of them being a direct action of Ca^{2+} on a protein kinase. Some of our experiments suggest such a possibility.

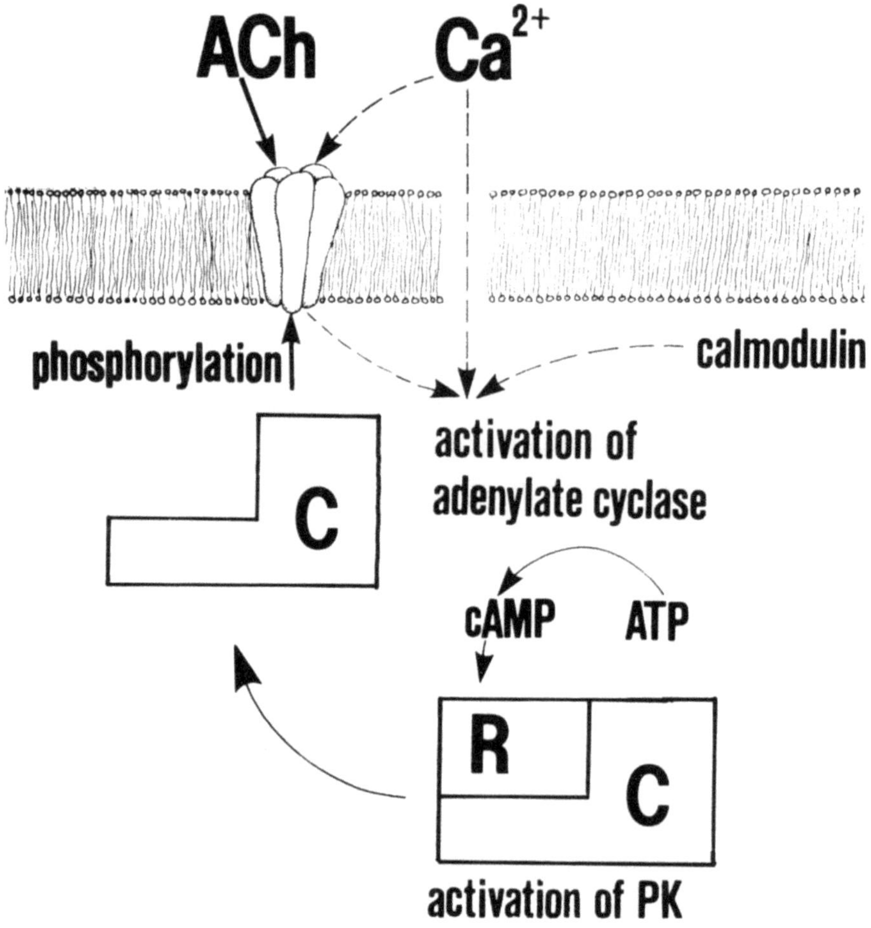

Fig 5. Scheme suggesting regulation of cAMP levels in the electrocyte of Torpedo by Ca^{2+} entering together with neurotransmitter ACh. Ca^{2+} entry either through the nAChR channel or through another channel, perhaps a Ca^{2+}-channel. A Ca^{2+}/calmodulin dependent activation of an adenylate cyclase could occur.

Acknowledgement

This research was supported by a grant from the Swedish Natural Science Research Council.

References

1. Nestler, S.J., Walaas, S.J., Greengard, P. 1984. Science 225, 1357.

2. Teichberg, V.I., Changeux, J.-P. 1977. FEBS Lett. 74, 71.

3. Gordon, A.S., Milfray, D., Davis, C.G., Diamond, I. 1979. Biochem.Biophys.Res.Comm. 87, 876.

4. Heilbronn, E., Mattsson, Ch. 1974. J.Neurochem. 22, 315.

5. Huganir, R.L., Albert, K.A., Greengard, P. 1983. Soc. Neurosci. Abstr. 9, 578.

6. Huganir, R.L., Greengard, P. 1983. Proc.Natl.Acad.Sci. 80, 1130.

7. Noda, M., Takahashi, H., Tanabe, T. et al. 1983. Nature 302, 528.

8. Carstens, M.E., Weller, M., Neethking, A.C., Taljaard, J.J.F. 1982. Mol.Cell.Biochem. 42, 161.

9. Saitoh, T., Changeux, J.-P. 1981. Proc.Natl.Acad.Sci. 78, 4430.

10. Gordon, A.S., Davis, C.G., Milfay, D., Diamond, I. 1977. Nature 267, 539.

11. Anholt, R., Fredkin, D.R., Deerinck, T. et al. 1982. J.Biol.Chem. 257, 7122.

12. Blosser, J.C., Appel, St.H. 1980. J.Biol.Chem. 255, 1235.

13. Miledi, R., Parker, I., Schalow, G. 1980. J.Physiol. 300, 197.

14. Kawabuchi, M. 1982. J. Neuropathol.Exptl.Neurol. 41, 298.

15. Heilbronn, E., Björk, C., Elfman, L., Hartman, A., Mattsson, C. 1979. Advances in Cytopharmacology (H. Ceccarelli and F. Clementi, eds.). Raven Press, p 151.

16. Gordon, A.S., Diamond, I. 1980. J.Supramol.Struct. 14, 163.

17. Smilowitz, H., Hadjian, R.A., Dwyer, J., Feinstein, M.B. 1981. Proc.Natl.Acad.Sci. 78, 4708.

18. Davis, C.G., Gordon, A.S., Diamond, I. 1982. Proc.Natl. Acad.Sci. 79, 3666.

19. Palfrey, H.C., Rothlein, J.E., Greengard, P. 1983. J.Biol. Chem. 258, 9496.

20. Eriksson, H., Salmonsson, R., Liljeqvist, G., Heilbronn, E. 1984. Dynamics of Cholinergic Function (I. Hanin, ed.). Raven Press (in press).

21. Zavoico, G.B., Comerci, C., Subers, E., Egan, J.J., Huang, C.-K., Feinstein, M., Smilowitz, H. 1984. Biochim.Biophys. Acta. 770, 225.

MOLECULAR INTERACTIONS BETWEEN THE MAJOR PROTEINS OF THE POST-SYNAPTIC DOMAIN IN TORPEDO MARMORATA ELECTROMOTOR SYNAPSE.

J. Cartaud, C. Kordeli

Microscopie Electronique, Institut J. Monod C.N.R.S., Université Paris VII, 2 Place Jussieu 75005 PARIS, France.

H.O. Nghiêm, J.P. Changeux

Neurobiologie Moléculaire, Institut Pasteur, 28 Rue du Docteur Roux, 75015 PARIS, France.

Physiological, biochemical and morphological evidence has stressed that the postsynaptic membranes of chemical synapses possess some unique features. In the adult neuromuscular junction of vertebrates as well as in the closely related electromotor synapse of electric fishes, the postsynaptic membrane contains a high density of acetylcholine receptor molecules (Ach-R). These molecules form regular arrays in the plane of the juxtaneural part of the membrane (1, 2). As a consequence of this long-range supramolecular organization, the receptors are strongly immobilized (3). In addition, the postsynaptic membrane is usually associated with elaborate specializations, the basal lamina and the cytoplasmic densities that underline the receptor clusters. Cytoplasmic filaments that contact the densities are particularly abundant in the subneural sarcoplasm (2, 4).

The postsynaptic structure, in particular the Ach-R clusters, is extremely stable and is maintained for a long period after denervation (5 , 6). It is thought that many aspects of the morphogenesis and stabilization of the synapse could be accounted for by macromolecular interactions between the membrane, the basal lamina and the cytoskeleton. However, little is known about the molecules which could be involved in such interactions.

Molecular Basis of Nerve Activity

The fish electric organs have provided a uniquely rich source of cholinergic synapses permitting extensive biochemical and structural characterization of the major functional components (7). For this reason the electric organ constitutes an excellent model system to study the molecules involved in the immobilization of the Ach-R in the mature postsynaptic domain.
In this review, we present recent data on the molecular architecture of the postsynaptic domain, particularly on aspects that may be relevant to the stabilization of the Ach-R.

The In-Plane and Transmembrane Disposition of the Ach-R in the Postsynaptic Membrane.

From the first electrophysiological evidences demonstrating high Ach-sensitivity at the motor end-plate, to the most recent structural analysis, the postsynaptic membrane emerges as an extremely specialized domain in which the "receptive substance", is concentrated. To day's electron micrographs, particularly those obtained using the freeze-etching technique, disclose an original arrangement of the rosettes of receptor in double rows in the plane of the juxtaneural part of the membrane (1,2,8,9). Within these rows, dimeric structural units can be recognized. Since the native, heavy form of the Ach-R represents a dimer formed by the covalent association of two coplanar Ach-R rosettes (reviewed in 10), the supramolecular organization of the receptors in the membrane likely relies upon a linear assembly of the Ach-R doublets. However, the fact that stable clusters of receptors exist in the absence of covalent bonding between doublets implies the existence of additional factors involved in cluster formation.
Selective proteolysis of membrane fragments shows that the Ach-R polypeptides have a transmembrane orientation (11). The recent deciphening of the amino-acid sequence of the four subunits of the receptor has further extended our knowledge on the transmembrane distribution of the protein in regard to the

bilayer (12, 14). Models of the transmembrane organization based on sequence data are in agreement with the biophysical data derived from X-ray diffraction and electron microscopy. The picture that emerges from these studies is that all the receptor subunits exposed at both surfaces of the membrane span the bilayer several times.

The Membrane-Bound 43K Protein : a Major Component of the Postsynaptic Densities that Binds to the Ach-R.

The postsynaptic membranes purified from adult Torpedo electric tissue retain most of their functional and structural properties, including the ability to respond to externally applied cholinergic agonists by an increase in ionic permeability. The major structural features associated with the membrane in situ, such as the Ach-R tracks and the associated postsynaptic densities, are also conserved in the isolated membrane fragments (figure 1). These membranes are comprised of the Ach-R subunits

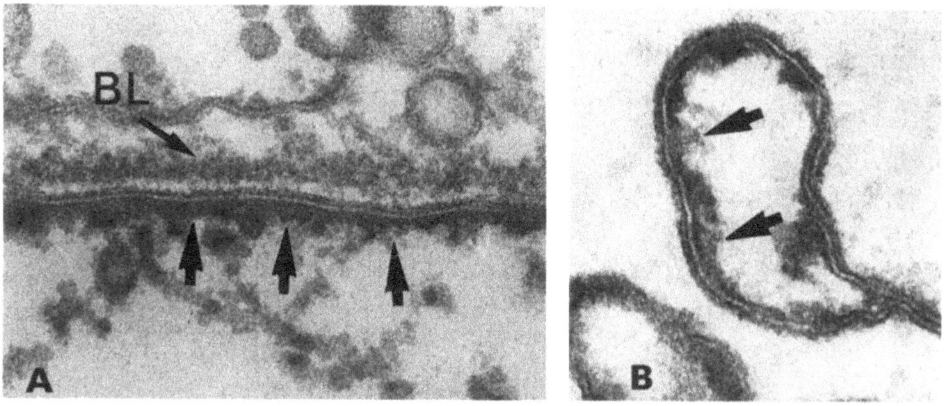

Figure 1. The Torpedo postsynaptic membrane in situ (A) and after isolation (B). Electron dense material - or densities - is observed at the cytoplasmic side of the Ach-R - rich membrane (arrows). BL = basal lamina (x 120,000)

(up to 40% of the total protein mass) and of one major peripheral membrane protein of apparent molecular mass 43,000 daltons (referred to as 43K or ν1 protein) which is distinct from actin (15).
The peripheral, non receptor proteins can be extracted from the membrane by treatment at alkaline pH (16) or by extraction with low concentration of lithium diiodosalicylate (17). Extraction of the peripheral proteins has essentially no effect on the functional properties (ligand binding and ionic permeability) of the membrane. However, important "structural" alterations, such as increases in rotational (18, 21) and translational (22, 23) mobility, sensitivity to proteolysis (24) and to heat denaturation (25) of the receptors have been reported. Morphological studies have demonstrated that cytoplasmic condensations, coextensive with the receptor-rich areas of the membrane are no longer visible following alkaline extraction (figure 2)(23,26). These results suggest that the main molecular constituent of the submembraneous condensations, the 43K protein, occupy the same region of the postsynaptic membrane as the receptor and may contribute to its immobilization via a direct interaction. The demonstration of the codistribution of the Ach-R and 43K proteins would constitute an important step in the understand-

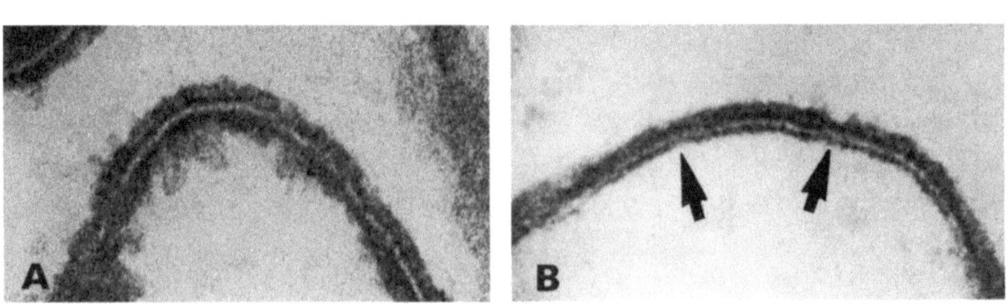

Figure 2 : Following alkaline extraction, the postsynaptic densities are no longer visible. A - untreated membrane fragment, B - membrane fragment extracted for 30 min. at pH 11 at 4°C. (x 290,000)

ing of the membrane architecture in this system. For this purpose, a direct immunocytochemical localization of the 43K protein at the E.M. level was undertaken. This was achieved by preparing monoclonal antibodies against the membrane-bound 43K protein and by immunolabelling of electric organ fragments with colloïdal gold-conjugated immunoreagents (figure 3). As expected the localization of the 43K protein is strictly limited to the cytoplasmic surface of the Ach-R-rich domains of the membrane (27). Double labelling experiments further show that immunostaining is totally absent from the bottom of the synaptic folds, areas which are also devoid of receptors (28).
High resolution immunocytochemical studies of the tangential distribution of the 43K protein in isolated sheets derived from the innervated face of the electrocytes disclose evidences for a distribution in double rows quite similar to that of the receptors (29). Additional evidence for a topographical proximity of the two proteins in the membrane was recently achieved by crosslinking experiments in which a specific crosslink between the 43K protein and the β subunit of the receptor was demonstrated (30).
In conclusion, the peripheral, 43K protein, is set in register with respect to the Ach-R and is likely to participate in the structural stabilization of the postsynaptic membrane domain. Its localization at the cytoplasmic face of the membrane raises interesting possibilities for an interaction with the cytoskeleton.

Cytoskeleton - Membrane Interaction in the Electroplaque.

Cytoskeleton - membrane interactions are thought to play a decisive role in the development and maintenance of cell shape and behaviour. Indeed, specialized membrane domains are often characterized by a close association with well-defined cytoskeletal components.

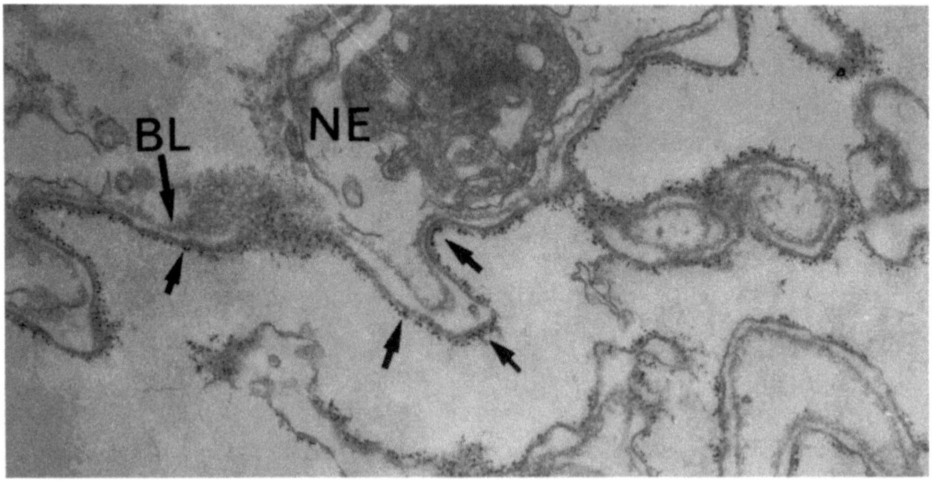

Figure 3 : Immunogold labelling of an innervated sheet with anti-43K monoclonal antibody. A nearly continuous labelling is observed at the cytoplasmic face of the postsynaptic membrane (arrows). NE = nerve ending; BL = basal lamina; (x 30,000)

The synaptic area is an example of such a specialized membrane domain where Ach-R clusters are associated with differenciations of the underlying cytoplasm : the synaptic densities and attached filaments (4, 31, 32).

The postsynaptic folds that characterize the mature neuromuscular junction are indeed associated with typical cytoskeletal proteins, including a cytoplasmic form of actin (33), several actin-binding proteins (α-actinin, filamin, vinculin (34)), and an intermediate-sized filament-related 51K protein (35). Torpedo electromotor synapses lack most of the postsynaptic folds. However, sufficient structural similarities exist between the two systems that justify the study of the molecular interactions that take place at the Torpedo postsynaptic membrane, as a simplified model.

Though purified postsynaptic membrane fragments usually retain some cytoskeletal material, such as the postsynaptic densities

(23, 26) or actin (36), most of the attached filaments observed in situ are lost. Furthermore, since these vesicles reseal with almost total conservation of their initial polarity (right-side out), the cytoplasmic side of the membrane is not readily accessible. Studies of cytoskeleton - membrane interactions were thus carried-out on large sheets derived from the innervated face of electrocytes by gentle homogeneization and low-speed centrifugation (1). Electron micrographs of deep-etched, rotary-shadowed (1) or negatively-stained (figure 4) samples disclosed a dense cytoskeletal network attached to the postsynaptic membrane. Ultrastructural and immunocytochemical analyses of this cytoskeleton reveal that it consists essentially of intermediate-sized filaments, probably desmin filaments, and actin (29). As noted previously, the 43K is restricted to the cytoplasmic surface of the postsynaptic Ach-R-rich membrane areas.

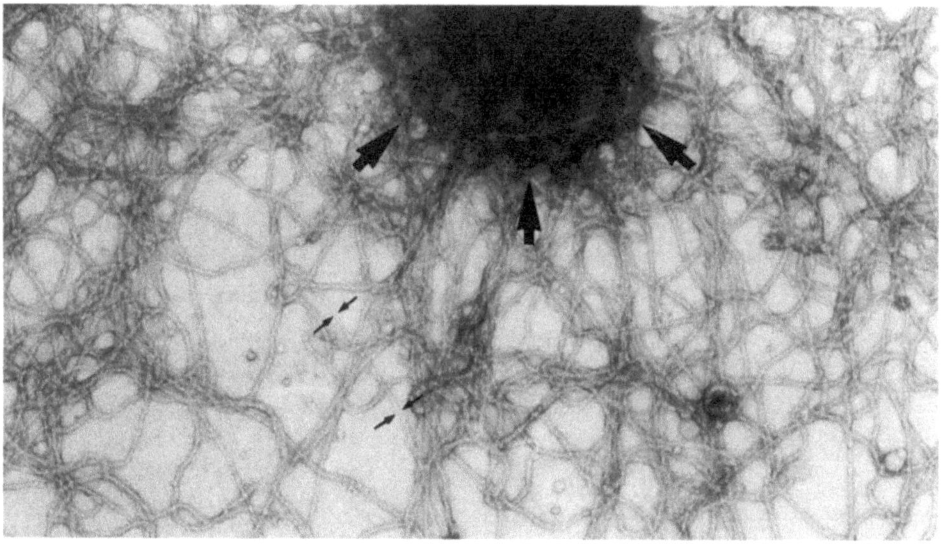

Figure 4 : The postsynaptic cytoskeleton revealed on a negatively stained innervated membrane sheet. As in Figure 3, the material was labelled with the anti-43K antibody. Gold particles are restricted to the membrane area (large arrows). The cytoskeleton is characterized by a dense meshwork of intermediate-sized filaments. Some thin (actin ?) filaments are also observed (small arrows) (x 50,000).

The filaments observed in contact with the membrane are never labelled on their lengths with the anti-43K antibody. Interestingly, the free-ends of detached filaments are sometimes observed with a punctual labelling (37). These images suggest that the intermediate-sized filaments might attach to the postsynaptic densities, possibly via the 43K protein. This protein that is associated on one hand with the Ach-R clusters in the membrane is, on the other hand likely to be implicated in the anchorage to the cytoskeleton. It would thus constitutes the intermediate link that insure the anchorage of Ach-R clusters to the underlying cytoskeleton.
The presence of the 43K - or a closely related - protein at the rat neuromuscular junction has been recently established (38, 39). Unlike the other cytoskeletal components present at the junction, the 43K protein appears to be a specific Ach-R associated protein. It is thus plausible that the 43K protein serves as an anchoring protein at the neuromuscular junction as well. However, since several cytoskeletal proteins are also found in this area, it is necessary to define how these proteins are associated to maintain the postsynaptic architecture. Preliminary results show, quite surprisingly that several of the cytoskeletal proteins that are present at the rat neuromuscular junction (in particular the actin-binding proteins) are not detected at the <u>Torpedo</u> postsynaptic domain (29). This would indicate that actin and the actin-binding proteins rather participate in the maintenance of the complex-architecture of the synaptic folds, the anchorage of the Ach-R clusters being mediated in both cases by the 43K protein.

Acknowledgements

This work was supported by grants from the Muscular Dystrophy Association of America, the College de France, the Ministere de la Recherche et de la Technologie, the Centre National de la Recherche Scientifique, the Institut National de la Santé et de la Recherche Médicale, the Commissariat à l'Energie Atomique and the Université Paris VII. C. Kordeli was supported by grants from the Association des Myopathes de France.

References

1. Heuser, J.E. and S.R. Salpeter. 1979. Organization of acetylcholine receptors in quick-frozen, deep-etched, and rotary-replicated Torpedo postsynaptic membrane. J. Cell Biol. 82, 150-173.

2. Hirokawa, N. and J.E. Heuser. 1982. Internal and external differenciations of the postsynaptic membrane at the neuromuscular junction. J. Neurocytol. 11, 487-510.

3. Rousselet, A. and P.F. Devaux. 1977. Saturation transfer electron paramagnetic resonance on membrane-bound proteins. II. Absence of rotational diffusion of the cholinergic receptor protein in Torpedo marmorata membrane fragments. Biochem. Biophys. Res. Commun. 78, 448-454.

4. Couteaux, M.R.. 1981. Structure of the subsynaptic sarcoplasm in the interfolds of the frog neuromuscular junction. J. Neurocytol. 10, 947-962.

5. Frank, E., Gautvik, K. and M. Sommersehild. 1975. Persistence of junctional acetylcholine receptors following denervation. Cold Srping Harbor Symposium on Quantitative Biology 40, 275-281.

6. Bourgeois, J.P., J.L. Popot, A. Ryter and J.P. Changeux. 1978. Quantitative studies on the localization of the cholinergic receptor protein in the normal and denervated electroplaque from Electrophorus electricus. J. Cell Biol. 79, 200-216.

7. Changeux, J.P. 1981. The acetylcholine receptor: an "allosteric" membrane protein. Harvey Lect. 75, 85-254.

8. Cartaud, J., E.L. Benedetti, A. Sobel and J.P. Changeux. 1978. A morphological study of the cholinergic receptor protein from Torpedo marmorata in its membrane environment and in its detergent extracted form. J. Cell Sci. 29, 313-337.

9. Cartaud, J. 1980. A critical re-evaluation of the structural organization of the excitable membrane in Torpedo marmorata electric organ. In Ontogenesis and Functional Mechanisms of Peripheral Synapses. J. Taxi, editor, Elsevier, North Holland-Amsterdam-New York 199-210.

10. Popot, J.L. and J.P. Changeux. 1984. The nicotinic receptor of acetylcholine : structure of an oligomeric integral membrane protein. Physiological Rev. (in press).

11. Wennogle, L.P. and J.P. Changeux. 1980. Transmembrane orientation of proteins present in acetylcholine receptor-rich membranes from Torpedo studied by selective proteolysis. Eur. J. Biochem. 106, 381-393.

12. Devillers-Thiery, A, J. Giraudat, M. Bentaboulet and J.P. Changeux. 1983. Complete mRNA coding sequence of the acetylcholine binding α subunit of Torpedo marmorata acetylcholine receptor : a model for the transmembrane organization of the polypeptide chain. Proc. Natl. Acad. Sci. USA 80, 2067-2071.

13. Claudio, T., M. Ballivet, J. Patrick and S. Heineman.1983. Nucleotide and deduced amino acide sequences of Torpedo californica acetylcholine receptor γ subunit. Proc. Natl. Acad. Sci. USA 80, 1111-1115.

14. Noda, M., H. Takahashi, T. Tanabe, M. Toyosato, S. Kikyotani, Y. Furutani, T. Hirose, H. Takashima, S. Inayama, T. Miyata and S. Numa. 1983. Structural homology of Torpedo californica acetylcholine receptor subunits. Nature (Lond.). 302, 528-532.

15. Sobel, A., M. Weber and J.P. Changeux. 1977. Large-scale purification of the acetylcholine-receptor protein in its membrane bound and detergent extracted forms from Torpedo marmorata electric organ. Eur. J. Biochem. 80, 215-224.

16. Neubig, R.R., E.K. Krodel, N.D. Boyd and J.B. Cohen. 1979. Acetylcholine and local anesthetic binding to Torpedo nicotinic post-synaptic membranes after removal of non-receptor peptides. Proc. Natl. Acad. Sci. USA 76, 690-694.

17. Porter, S. and S.C. Froehner. 1983. Characterization and localization of the Mr = 43,000 proteins associated with acetylcholine receptor-rich membranes. J. Biol. Chem. 258, 10034-10040.

18. LO, M. M. S., P.B. Garland, J. Lamprecht, and E.A. Barnard 1980. Rotational mobility of the membrane-bound acetylcholine receptor of Torpedo electric organ measured by phosphorescence depolarization. FEBS (Fed. Eur. Biochem. Soc.) Lett. 111, 407-412.

19. Rousselet, A., J. Cartaud and P.F. Devaux. 1979. Importance des interactions protéine-protéine dans le maintien de la structure des fragments excitables de l'organe électrique de Torpedo marmorata. C.R. Seances Acad. Sci. 289, 461-463.

20. Rousselet, A., J. Cartaud and P.F. Devaux. 1981. Effects of temperature, lipid modification and pH on the mobility of the major proteins of the receptor-rich membranes from Torpedo marmorata. Biochim. Biophys. Acta 648, 169-185.

21. Rousselet, A., J. Cartaud, P.F. Devaux and J.P. Changeux. 1982. The rotational diffusion of the acetylcholine receptor in T. marmorata membrane fragments studied with a spin-labelled α-toxin. Importance of the 43K protein. EMBO J. 1, 439-445.

22. Barrantes, F.J., W. Ch. Neugebauer and H.P. Zingsheim.1980. Peptide extraction by alkaline treatment is accompanied by rearrangement of the membrane-bound acetylcholine receptor from Torpedo marmorata. FEBS (Fed. Eur. Biochem. Soc.)Lett. 112, 73-78.

23. Cartaud, J., A. Sobel, A. Rousselet, P.F. Devaux and J.P. Changeux. 1981. Consequences of alkaline treatment for the ultrastructure of the acetylcholine receptor-rich membranes from Torpedo marmorata electric organ. J. Cell Biol. 90, 418-426.

24. Klymkowsky, M., J.E. Heuser and R.M. Stroud. 1980. Protease effects on the structure of acetylcholine receptor membranes from Torpedo californica. J. Cell Biol. 85, 823-838.

25. Saitoh, T., L.P. Wennogle and J.P. Changeux. 1979. Factors regulating the susceptibility of the acetylcholine receptor protein to heat inactivation. FEBS (Fed. Eur. Biochem. Soc.) Lett. 108, 489-494.

26. Sealock, R. 1982. Cytoplasmic surface structure in postsynaptic membranes from electric tissue visualized by tannic acid-mediated negative contrasting. J. Cell Biol. 92, 514-522.

27. Nghiêm, H.O., J. Cartaud, C. Dubreuil, C. Kordeli, G. Buttin and J.P. Changeux. 1983. Production and characterization of a monoclonal antibody directed against the 43,000 ν1 polypeptide from Torpedo marmorata electric organ. Proc. Natl. Acad. Sci. USA 80, 6403-6407.

28. Sealock, R., B. Wray and S.C. Froehner.1984. Simultaneous ultrastructural localization of the acetylcholine receptor and the Mr 43,000 protein in Torpedo electroplax using monoclonal antibodies. J. Cell Biol. 98, 2239-2244.

29. Kordeli, C., J. Cartaud, H.O. Nghiêm, D. Paulin, L. Pradel and J.P. Changeux. (in preparation).

30. Burden, S.J., R.L. DePalma and G.S. Gottesman. 1983. Crosslinking of proteins in acetylcholine receptor-rich membranes : association between the β-subunit and the 43K subsynaptic protein. Cell 35, 687-692.

31. Birks, R., H.E. Huxley and B. Katz. 1960. The fine structure of the neuro-muscular junction of the frog. J. Physiol. (Lond.). 150, 134-144.

32. Ellisman, M.H., J.E. Rash, L.A. Staehelin and K.R. Porter. 1976. Studies of excitable membranes. II A comparison of specialization at neuromuscular junctions and non junctional sarcolemmas of mammalian fast and slow twitch muscle fibers. J. Cell Biol. 68, 752-774.

33. Hall, Z.W., B.W. Lubit, and J.H. Schwartz. 1981. Cytoplasmic actin in postsynaptic structures at the neuromuscular junctions. J. Cell Biol. 90, 789-792.

34. Bioch, R.J. and Z.W. Hall. 1983. Cytoskeletal components of the vertebrate neuromuscular junction : vinculin, α-actinin and filamin. J. Cell Biol. 97, 217-223.

35. Burden, S.J. 1982. Identification of an intracellular postsynaptic antigen at the frog neuromuscular junction. J. Cell Biol. 94, 521-530.

36. Strader, C.O., E. Lazarides and M.A. Raftery. 1980. The characterization of actin associated with postsynaptic membranes from Torpedo californica. Biochem. Biophys. Res. Commun. 92, 365-373.

37. Cartaud, J., C. Kordeli, H.O. Nghiêm and J.P. Changeux. 1983. La protéine ν1 de 43000 daltons : pièce intermédiaire assurant l'ancrage du récepteur cholinergique au cytosquelette sous-neural ?. C.R. Acad. Sci. Paris 297, 285-289.

38. Froehner, S.C., V. Gulbrandsen, C. Hyman, A.Y. Jeng, R.R. Neubig and B. Cohen. 1981. Immunofluorescence localization at the mammalian neuromuscular junction of the Mr 43,000 protein of Torpedo postsynaptic membranes. Proc. Natl. Acad. Sci. USA 78, 5230-5234.

39. Froehner, S.C. 1984. Peripheral proteins of postsynaptic membranes from Torpedo electric organ identified with monoclonal antibodies. J. Cell Biol. 99, 88-96.

THE GLYCINE RECEPTOR OF RAT SPINAL CORD: PROGRESS IN THE INVESTIGATION OF A NEURONAL ION CHANNEL PROTEIN

Heinrich Betz, Gabriele Grenningloh, Bertram Schmitt
Institute for Neurobiology, ZMBH, Center for Molecular Biology, University of Heidelberg, D-6900 Heidelberg FRG

Abstract

The amino acids gammaaminobutyric acid (GABA) and glycine are the major inhibitory neurotransmitters of the vertebrate central nervous system (CNS). Both amino acids increase the chloride conductance of the postsynaptic membrane via specific receptor proteins. This report reviews our recent studies of the glycine receptor (GlyR) of rat spinal cord.

Introduction

David Nachmansohn devoted his life to exploring the neurobiology of acetylcholine. This initiative has been extraordinarily fruitful: Today, the enzymes involved in the synthesis of this neurotransmitter, the receptor proteins involved in its recognition, the enzyme species causing its inactivation/degradation and the respective high-affinity uptake system responsible for transmitter/precursor reutilization have all been characterized in sometimes astonishing detail (see this volume). Acetylcholine is, however, only a minor neurotransmitter in the CNS. Here, a plethora of other compounds including catecholamines, peptides and amino acids have been identified as bona fide or candidate neurotransmitter substances. Amongst the best established CNS transmitters are the amino acids glycine and GABA; together they represent the major inhibitory

Molecular Basis of Nerve Activity

neurotransmitters in the vertebrate brain and spinal cord. Both GABA and glycine act by increasing the chloride conductance of the neuronal membrane, i.e., they hyperpolarize the postsynaptic cell. The receptors for these amino acids have been investigated by anatomical, electrophysiological and pharmacological techniques and shown to be separate molecules on the basis of the following criteria:

i) Glycine and GABA receptors are differently distributed in the CNS. Glycine is the major transmitter in spinal cord and brain stem, whereas GABA predominates in the brain (1);
ii) although certain amino acids, such as taurine, can activate both the GABA receptor and the GlyR, the agonist and, in particular, the antagonist pharmacology of both receptors is clearly different (see below);
iii) photoaffinity labeling experiments have shown that the drug-binding polypeptides of the GlyR and the GABA receptor have different molecular weights (2).

Despite these various differences, the glycine- and GABA-activated chloride channels show striking similarities in their overall gating behavior (3,4). Therefore, a common ion channel component for this class of receptor proteins has recently been proposed (4). Here, we present new data on the polypeptide composition and immunology of the GlyR and suggest a structural model for this chemically gated ion channel protein.

Pharmacology of the GlyR

The number of currently known ligands of the GlyR is comparatively small. Besides glycine, the amino acids ß-alanine and taurine are also efficient agonists of this receptor ($K_D \sim$ 5-20 µM; see ref. 5); GABA, in contrast, has little effect below 1 mM.

Strychnine, a heterocyclic alkaloid from Strychnos nux vomica, is an extremely potent antagonist of the glycine receptor (K_D 5-10 nM). All the toxic effects of this alkaloid can be attributed to blocking glycinergic transmission in different regions of the CNS (6). Already in 1973, Young and Snyder demonstrated high-affinity binding of [^{3}H]strychnine to membrane preparations of rat spinal cord (7). This binding was blocked not only by glycine and its agonists, but also by chloride and other anions capable of mediating glycine responses, and it has thus been postulated to occur at a site associated with the ion channel of the receptor (8). Different investigators have used [^{3}H]strychnine binding to monitor GlyR sites in a simple radioreceptor assay. We have found strychnine and its derivatives powerful tools in analyzing and purifying the GlyR protein.

Avermectin B1a, an antihelmintic macrocyclic lactone disaccharide, is known to increase GABA responses by binding to an allosteric site of the GABA receptor (9). Interestingly, this compound is a potent inhibitor ($K_D < 1$ µM) of [^{3}H]strychnine binding to spinal cord membranes or detergent extracts (10). This observation supports the previously noted similarity between GlyR and GABA receptor proteins.

Biochemistry of the GlyR

Our first step in the biochemical characterization of the GlyR was the solubilization of [^{3}H]strychnine binding sites from synaptic membrane fractions of rat spinal cord (11). These experiments revealed that the purified GlyR behaved as a glycoprotein of $M_r \sim 250{,}000$ whose native conformation was stringently dependent on the presence of phospholipids. All experiments with solubilized GlyR thus had to be performed in buffers containing mixed phosphatidylcholine/detergent micelles.

Further investigation of GlyR was greatly facilitated by the discovery that strychnine is a natural photoaffinity ligand of this receptor protein (12). Upon UV illumination in the presence of spinal cord membranes, [^{3}H]strychnine became irreversibly incorporated into its high-affinity receptor binding site.

SDS polyacrylamide gel electrophoresis of the illuminated membranes identified a polypeptide of M_r = 48,000 (48K) as the strychnine binding subunit of the GlyR. During the past few years, this photoaffinity labeling reaction has been characterized in more detail. The kinetics and the pharmacology of the UV-induced incorporation of [^{3}H]strychnine into membranes were both consistent with the hypothesis that strychnine is cross-linked at or close to the antagonist binding site of GlyR (2). Also, photoaffinity labeling proved useful for exploring the topology in the membrane of the 48K subunit: incubation of [^{3}H]-strychnine-labeled membranes with different proteases led to a degradation of the 48K band to polypeptides of lower molecular weight (2). Even extensive proteolysis failed, however, to remove the radioactive label from the membranes. Strychnine binding was therefore postulated to occur in a receptor domain which is protected by the hydrophobic environment of the neuronal membrane (2). This interpretation is consistent with the previously proposed association of strychnine binding with the ionic channel of the GlyR (8).

An important class of tranquilizer drugs, the benzodiazepines, affect GABA-ergic transmission and can be incorporated into a membrane polypeptide of ∿ 50K upon UV illumination (13). This GABA-receptor subunit was shown to be different from the strychnine binding polypeptide of GlyR (2).

For affinity purification of GlyR, a physiologically active derivative of strychnine, 2-aminostrychnine, was coupled to agarose beads via a long hydrophilic spacer (5). This affinity resin bound 80-90% of the GlyR sites present in detergent extracts of spinal cord membranes. Furthermore, about 20% of the strychnine binding activity could be recovered after extensive washing of the resin and biospecific elution with the receptor agonist glycine. This simple protocol provided a standard procedure for isolating GlyR to near homogeneity in one step: Currently, about 50 µg of purified receptor protein can be obtained within 2-3 days. Also, this affinity chromatography was successfully employed in the purification of the GlyR of pig spinal cord (14).

Upon SDS polyacrylamide gel electrophoresis, purified GlyR from rat or pig spinal cord was separated into three polypeptides of 48K, 58K and 93K (Fig. 1). The stoichiometry of these polypep-

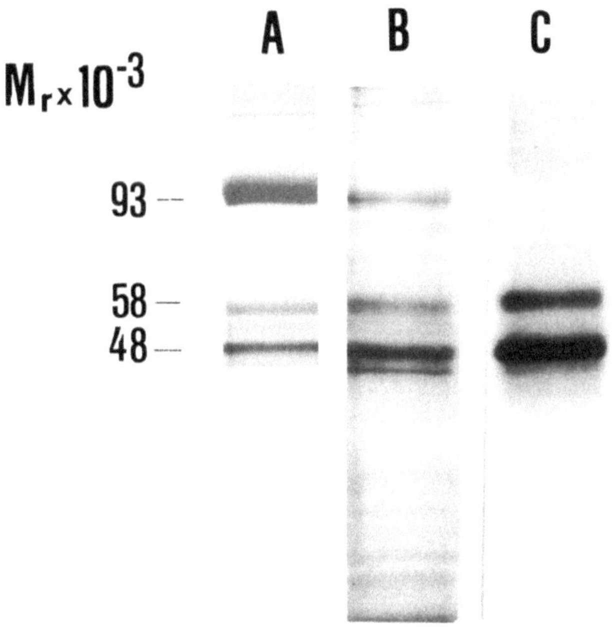

Fig. 1. SDS polyacrylamide gel electrophoresis of affinity-purified GlyR. A: GlyR purified from rat spinal cord according to (5); B: GlyR purified from pig spinal cord using Triton X100, protease inhibitors and iodoacetamide (14); C: same as B, but purification in the absence of iodoacetamide and using cholate as a detergent. Note loss of 93K polypeptide in preparation C.

tides has not yet been clarified; however, usually the 48K polypeptide appeared to be more abundant than the other two. Also, the 48K and 58K polypeptides were shown to bind concanavalin A, and thus are glycopeptides (14).

UV illumination in the presence of [^{3}H]strychnine labeled the 48K and, to a much lesser extent, the 58K polypeptide. These polypeptides thus participate in the formation of the extracellular ligand binding domain of the glycine receptor (5,14). The function of the 93K polypeptide so far remains mysterious. Recent experiments have showed that, under certain conditions, it may be separated from the 48K and 58K GlyR subunits. Also, the 93K polypeptide was rapidly lost in the absence of protease inhibitors (Fig. 1; see ref. 14). An interesting speculation relates the 93K polypeptide to the ion channel function of GlyR. Its lack of glycosylation and its intracellular accessibility (see below) are both consistent with this hypothesis.

Peptide mapping experiments on the individual GlyR polypeptides produced some peptide fragments of identical size from all three receptor subunits. The 48K and 58K polypeptides in particular gave rather similar proteolytic finger prints (15). These results suggested that the different subunits of the GlyR may contain homologous domains and thus evolved from a common ancestor polypeptide, as has been shown for the nicotinic acetylcholine receptor of fish electric organ (16).

In preliminary experiments, Drs. D. Graham and H.G. Schindler were able to show that the affinity-purified GlyR can produce glycine-induced conductance changes upon reconstitution into planar lipid bilayers. This conductance change was, moreover, glycine specific, strychnine sensitive and chloride dependent. It therefore appeared that affinity-purified GlyR contained both the ligand binding and ion transport functions of this membrane-bound signal transducer protein. This finding strongly argues against the previously postulated "floating receptor/common ion channel" hypothesis of GABA- and glycine-mediated postsynaptic inhibition (3,4).

Immunology of GlyR

In order to develop new ligands for different functional sites of the GlyR, we produced polyclonal and monoclonal antibodies against the affinity-purified GlyR protein. Using standard procedures of hybridoma production and soft agar cloning, nine different cell lines were established which secrete monoclonal antibodies (mAbs) capable of binding the ^{125}I-labeled GlyR protein (15). In Western blots four of these antibodies bound to one of the different denatured GlyR subunits; three mAbs did not bind to blots and are thus presumably conformation dependent. Most interestingly, however, two of the antibodies recognized more than one of the GlyR polypeptides: GlyR mAb 4a bound to the 48 and 58K polypeptides and GlyR mAb 7a to the 93K and, to a lesser extent, the 48K polypeptides. These immunological data extend the abovementioned structural similarities of the different GlyR polypeptides (15).

In collaboration with Drs. J. Barker and P. St. John at NIH and Drs. A. Triller and H. Korn at the Pasteur Institute, most of our GlyR mAbs were tested in immunocytochemical experiments on primary cultures and tissue sections of spinal cord and brain stem. These preliminary experiments revealed a) a co-extensive staining of synaptic regions and neuronal processes by mAbs directed against the different GlyR polypeptides; b) differences in the accessibility of the binding domains of the mAbs specific for different subunits of the receptor. Antibodies recognizing the 48K and 58K polypeptides all stained intact cultured neurons, whereas staining with an mAb specific for the 93K polypeptide was observed only with sectioned material. Furthermore, by electron microscopy this antibody was shown to bind to the inner leaflet of the neuronal membrane (A. Triller et al., unpublished). A major immunodominant domain of the 93K polypeptide thus appears to be cytoplasmic.

A structural model of the GlyR

On the basis of our previous data, we proposed a structural

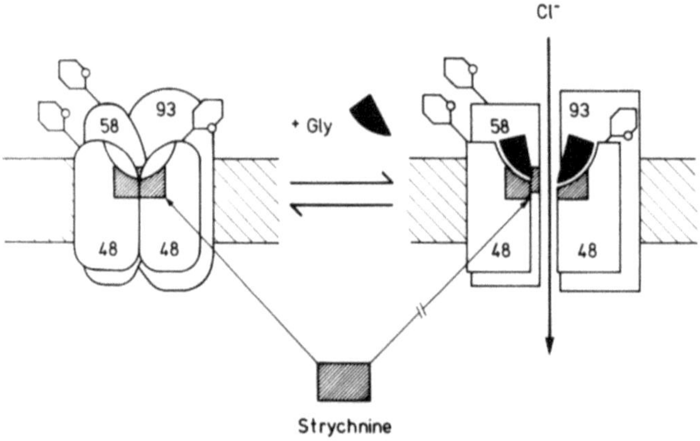

Fig. 2. A hypothetical model of the glycine receptor in its closed and open conformations.

model of the GlyR as shown in Fig. 2 (16). According to this model, the glycine receptor is a heteropolymer of two copies of the 48K polypeptide and one copy each of the 58K and the 93K polypeptides. From competition experiments, the agonist binding sites of the receptor were assigned to the 48K polypeptide chains. In agreement with this proposal, electrophysiological data suggest that more than one glycine molecule is required for chloride channel opening (4). The number of strychnine binding sites is not known; however, our data are consistent with a single strychnine binding site which overlaps the agonist site and includes not only the (two) 48K chain(s), but also the 58K polypeptide (14).

So far this model of the GlyR has not been clearly proven. However, certain of its features can now be refined: i) our antibody binding experiments suggest that the major portions of the 48K and the 58K polypeptides are extracellular; also, these polypeptides have been shown to be glycosylated (14); ii) the 93K polypeptide differs from the other glycine receptor subunits in that it does not bind lectin and is accessible for mAb binding only from the cytoplasmic site of the membrane. Whether the 93K polypeptide has a function in ion translocation remains to be determined.

It may be noted here that the structural model proposed for the GlyR also may be valid for the GABA receptor protein: i) upon affinity chromatography, two GABA receptor polypeptides of 53K and 57K were identified which can both be labeled by benzodiazepines (17);ii) from irradiation inactivation studies, a GABA receptor polypeptide of >100K has been proposed. In view of the astonishing similarity of glycine- and GABA-mediated chloride conductance changes a high degree of homology between these two receptor systems may be expected. The molecular cloning and sequencing of cDNA coding for the different GlyR and GABA receptor polypeptides should provide an answer to these questions.

Acknowledgements

We thank C. Schröder and H. Krischke for expert technical assistance and Rosemary Franklin for help with the preparation of the manuscript. This work was supported by grants from the Deutsche Forschungsgemeinschaft and the Bundesministerium für Forschung und Technologie.

References

1. Snyder, S.H., J.P. Bennett. 1976. Ann.Rev.Physiol. 38, 153.

2. Graham, D., F. Pfeiffer, H. Betz. 1983. Eur.J.Biochem. 131, 519.

3. Barker, J.L., R.N. McBurney. 1979. Nature 277, 234.

4. Hamill, O.P., J. Bormann, B. Sakmann. 1983. Nature 305, 805.

5. Pfeiffer, F., D. Graham, H. Betz. 1982. J.Biol.Chem. 257, 9389.

6. Zarbin, M.A., J.K. Wamsley, M.J. Kuhar. 1981. J. Neurosci. 1, 532.

7. Young, A.B., S.H. Snyder. 1973. Proc.Natl.Acad.Sci.USA 70, 2832.

8. Young, A.B., S.H. Snyder. 1974. Proc.Natl.Acad.Sci.USA 71, 4002.

9. Fritz, L.C., C.C. Wang, A. Gorio. 1979. Proc.Natl.Acad.Sci. USA 76, 2062.

10. Graham, D., F. Pfeiffer, H. Betz. 1982. Neurosci.Lett. 29, 173.

11. Pfeiffer, F., H. Betz. 1981. Brain Res. 226, 273.

12. Graham, D., F. Pfeiffer, H. Betz. 1981. Biochem.Biophys.Res. Commun. 102, 1330.

13. Möhler, H., M.K. Battersby, J.G. Richards. 1980. Proc.Natl. Acad.Sci.USA 77, 1666.

14. Graham, D., F. Pfeiffer, R. Simler, H. Betz: Biochemistry (in press)

15. Pfeiffer, F., R. Simler, G. Grenningloh, H. Betz: Proc.Natl. Acad.Sci.USA (in press)

16. Betz, H., D. Graham, F. Pfeiffer, H. Rehm. 1983. In: Toxins as Tools in Neurochemistry (F. Hucho and Y.A. Ovchimikov, eds.). Walter de Gruyter, Berlin New York, p. 245.

17. Sigel, E., C. Mamalaki, E.A. Barnard. 1983. J.Biol.Chem. 258, 6965.

ANTIBODIES TO SYNTHETIC PEPTIDES AS PROBES FOR THE α-SUBUNIT AND FOR THE CHOLINERGIC BINDING SITE OF THE ACETYLCHOLINE RECEPTOR

Drorit Neumann[1], Jonathan M. Gershoni[2], Mati Fridkin[3] and Sara Fuchs[1]

Departments of [1]Chemical Immunology, [2]Biophysics and [3]Organic Chemistry, The Weizmann Institute of Science, Rehovot 76100, Israel

Introduction

In an attempt to identify the binding site of the nicotinic acetylcholine receptor (AChR), a number of experimental approaches have been adopted. Affinity labels such as maleimido benzyltrimethylammonium iodide (MBTA;1) and bromoacetylcholine (BAC;2), have been employed and were shown to specifically alkylate a sulfhydryl group in the α-subunit, present in close proximity (1 nm) to the cholinergic binding site.

The analysis of α-bungarotoxin (α-BTX) binding to proteolysed or deglycosylated α-subunit has also been employed in search of the acetylcholine (ACh) binding site. Protein blotting has been found to be particularly useful in this respect. From such experiments it can be concluded that the isolated immobilized α-subunit can bind α-BTX (3-6) and that this binding is to the physiologically relevant ACh binding site as it is MBTA sensitive (7). The capacity for toxin binding can be maintained in fragments as small as 8 kd. Noteworthy is the fact that α-BTX binding exists whether Asn 141 is glycosylated or not, as demonstrated by the fact that this α-BTX binding is endoglycosidase H (Endo H) susceptible (4).

Molecular Basis of Nerve Activity

In the present study we have used synthetic peptides and their antibodies to investigate the ACh binding site. According to the sequence data and affinity labelling experiments Noda et al. (8) have proposed that the first two cysteine residues (128 and 142) are within the binding site of AChR. Smythies (9) had proposed that the first 17 amino acids of the α-subunit could accommodate a cholinergic binding site. His prediction was based on secondary structure analysis by Chou and Fasman (10).

In the following we will report on our experiments with two synthetic peptides (1-20 and 126-143) and their respective antibodies. The antipeptide antibodies were used to test if one can correlate the presence of the corresponding peptide sequences with the cholinergic binding site as probed by α-BTX binding. Our results indicate that the cholinergic binding site lies beyond Glu 152 which is the first *Staphylococus aureus V8* protease cleavage site after Asn 141.

Results and Discussion

We have synthesized peptides, 1-20 and 126-143, from the α-subunit of Torpedo AChR according to the available sequence data (8,11,12). Synthesis of the peptides was performed by the Merrifield solid phase procedure (13). The amino acid sequence for the two peptides was as follows: Peptide 1-20; H-Ser-Glu-His-Glu-Thr-Arg-Leu-Val-Ala-Asn-Leu-Leu-Glu-Asn-Tyr-Asn-Lys-Val-Ile-Arg-OH. Peptide 126-143; H-Ser-Tyr-Cys-Glu-Ile-Ile-Val-Thr-His-Phe-Pro-Phe-Asp-Gln-Gln-Asn-Cys-Thr-OH

The synthetic peptides were first used to establish whether and to what extent antibodies against these determinants are elicited in rabbits immunized with intact receptor or denatured AChR (RCM-AChR;14). We have demonstrated that both peptides reacted significantly with anti RCM-AChR antibodies by solid phase radioimmunoassay, whereas their binding to anti intact AChR antibodies was very low (Neumann et al., in preparation).

The synthetic peptides were conjugated to bovine serum albumin (BSA) as a carrier protein. The conjugates were injected into rabbits in order to elicit antipeptide antibodies, as previously described (15). Both antipeptide antibodies were shown to bind the respective peptide as well as Torpedo AChR in its intact or denatured form. In order to test to what extent the antipeptide antibodies recognized the native and denatured receptor inhibition experiments were performed. Such experiments indicated that the binding of antipeptide antibodies to AChR could be inhibited either by the peptide, by AChR or by RCM-AChR (Fig. 1). The peptides and RCM-AChR inhibited the binding of ^{125}I-radiolabelled AChR to the respective antibodies to the same extent. AChR was a weaker inhibitor, and 10 fold higher concentrations were required to achieve a 50% inhibition. These results suggest that in the intact receptor the sequences which correspond to the synthetic peptides (1-20; 126-143) are either not fully exposed or that they are maintained in a configuration which is recognized to a lower degree by the antibodies.

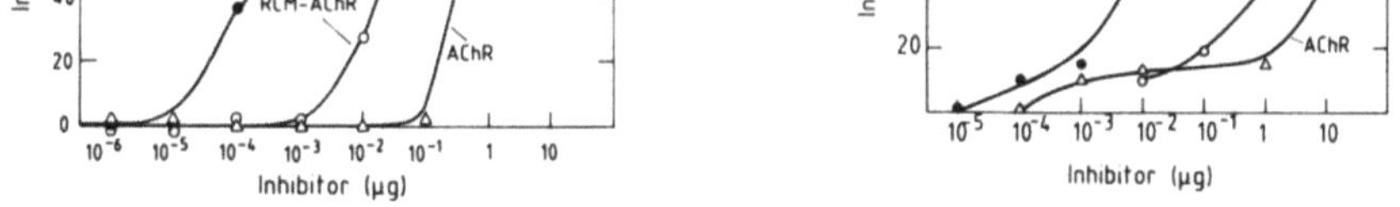

Figure 1. Anti-peptide antisera react better with denatured AChR (RCM-AChR) than with native AChR.
The binding of anti-peptide antibodies to AChR could be inhibited either by the corresponding peptide (●), by AChR (Δ) or by RCM-AChR (o).

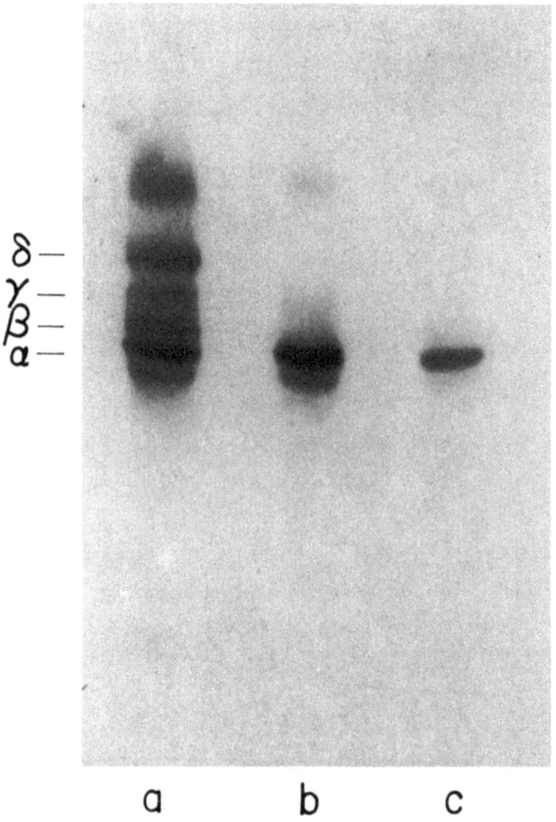

Figure 2. Binding of anti-peptide antibodies to the α-subunit as detected by immunoblotting. Strips were reacted with anti-AChR antibodies (a), anti-peptide 1-20 antibodies (b), and anti-peptide 126-143 antibodies (c).

In order to assess the subunit specificity of both antipeptide antibodies the immunoblotting method was employed (16). In spite of the sequence homology between the receptor subunits, particularly in the region between the first two cysteine residues (128-142), the antibodies reacted only with the α-subunit of AChR as shown in Fig. 2. An identical pattern was observed with a monoclonal antibody against the receptor α-subunit (15).

It was previously shown that the isolated α-subunit can bind α-BTX (3-6) and that when the α-subunit is subjected to V8 digestion, an 18 kd toxin binding fragment is obtained. It has also been demonstrated that MBTA labelling of AChR inhibits α-BTX binding to this 18 kd fragment (7). The antipeptide antibodies were used to test whether proteolytic fragments of the α-subunit which contained the corresponding amino acid sequences could also bind α-BTX. Protein blots of the α-subunit digested with S. aureus V8 were prepared as described (7). These were then probed with antipeptide 1-20 antibodies, antipeptide 126-143 antibodies and ^{125}I-labelled α-BTX. The results of such an experiment are shown in Fig. 3.

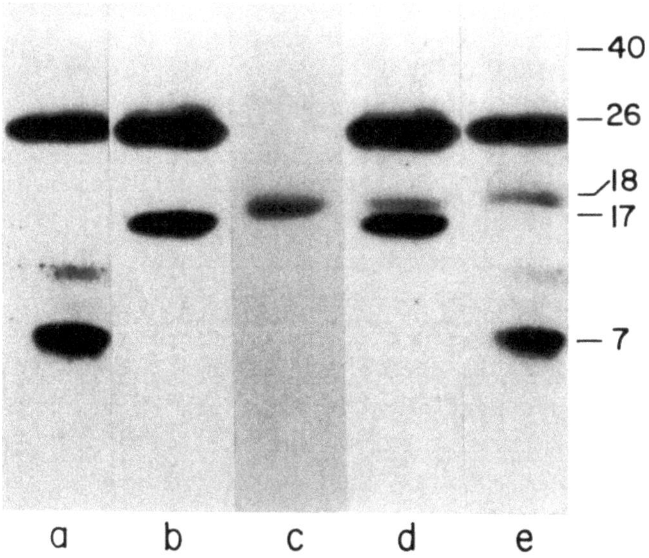

Figure 3. V8 fragments of AChR α-subunit probed with α-BTX and antipeptide antibodies. Blots of the proteolytic digests were incubated with antipeptide 1-20 (a), antipeptide 126-143 (b) and α-BTX (c). Further analysis was accomplished by probing the toxin labelled blots a second time with antipeptide 1-20 (e) and with antipeptide 126-143 (d).

A 26 kd fragment reacted with the antibodies but did not bind toxin. Other fragments exclusively bound antipeptide 1-20 antibodies (9 kd) and antipeptide 127-143 antibodies (17 kd). Only one fragment, 18 kd, bound α-BTX. In order to unequivocally establish that the α-BTX binding fragment was distinct from the 17 kd fragment detected with the antipeptide 126-143 antibodies, the blots that had been initially probed with the antipeptide antibodies were probed for a second time with α-BTX. In both cases the α-BTX bound to the 18 kd fragment which was conclusively different from the other previously labelled fragments (Fig. 3,d & e). It should be noted that a monoclonal antibody (5.5) which is site specific (17) reacted in a specific manner with the 18 kd fragment (data not shown). These experiments indicate that the α-BTX binding site is present in a fragment which contains neither of the synthetic peptide sequences.

Further characterization of the various peptide fragments was accomplished by deglycosylation experiments using Endo H (4). Asn 141 is considered to be the only asparagine residue in the entire α-subunit that fulfills the substrate requirements for N-glycosylation (8). Thus, the N-glycosidically linked oligosaccharide side chain serves as a marker for the presence of Asn 141. Therefore, the α-subunit was isolated from AChR samples that were incubated in the presence or absence of Endo H. Protein blots of these α-subunits, digested with S. aureus V8 protease were then prepared and probed with anti-peptide 1-20 antibodies, anti-peptide 126-143 antibodies and α-BTX (Fig. 4). Only those fragments that bound antipeptide 126-143 antibodies were found to be susceptible to the Endo H treatment as indicated by the shift in Mr (about 2 k) corresponding to the removal of the sugar side chain. The α-BTX binding fragment was unaltered in the Endo H treated sample. We therefore conclude that the α-BTX binding site and MBTA modified cysteine should at least be beyond the first V8 cleavage site after asn 141, i.e. Glu 152. As MBTA labels a sulphydryl group in or very close to the receptor binding site (1), and since the only two cysteines present beyond residue 152 of the postulated extracellular portion (up to Leu 210) are in position 192 and 193, we propose that MBTA labels one or both of these two cysteines as has been reported (18).

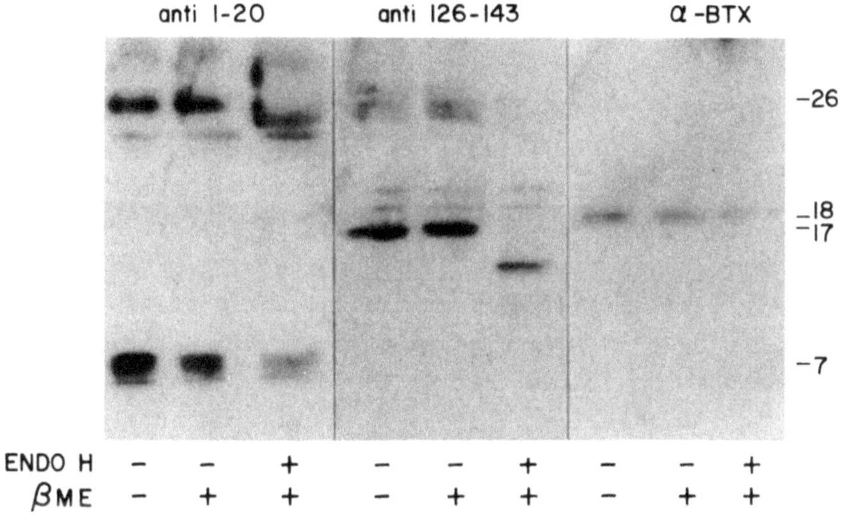

Figure 4. The effect of Endo H and β-mercaptoethanol (βME) on the V8 fragments of the α-subunit. The various receptor samples were resolved on polyacrylamide gels and digested with S. aureus V8 protease. The fragments were then blotted and probed with α-BTX or antibodies as indicated.

Cahill and Schmidt (19) employed antibodies against MBTA to isolate the MBTA labelled peptides obtained following pronase digestion. From statistical analysis of amino acid compositions of the peptides they proposed that cysteine residue 142 is the labelled one. On the other hand, Kao et al. (18) have recently sequenced the MBTA labelled, CNBr cleaved peptides of the α-subunit, and concluded that the labelling is either in cysteine 192 and/or 193. This is in agreement with our results. The possibility that cys 192 and/or cys 193 are involved in the receptor binding site seems especially attractive as these tandem two cysteine residues are unique to the α-subunit and reside in a portion of the subunit where there is no sequence homology with the other receptor subunits (18,20).

Analysis of proteolytic fragments obtained following digestion of the α-subunit with S. aureus V8 protease have been previously performed (7,21,22). Gullick et al. (21) have also demonstrated that an MBTA labelled peptide (19 kd) is distinct from a

(17 kd) peptide which contains carbohydrate (as demonstrated by staining with periodate-Schiff reagent). These results are in agreement with ours, if indeed these two fragments correspond respectively to our 18 kd and 17 kd fragments. Conti-Tronconi et al. (22) reported also on a carbohydrate containing V8 fragment of 17 kd which is most probably the same 17 kd fragment described by us (Figs. 3,4) and by Gullick et al. (21). Conti-Tronconi et al. (22) also reported on another higher molecular weight fragment (19 kd) which stains for carbohydrate only after prolonged staining with periodate-Schiff reagent and claims that it is similar to their 17 kd fragment in sequence, yet differs in the degree of glycosylation. If this is the case, then the 19 kd described by these authors (22) does not correspond to our 18 kd fragment.

In summary, in this study we have been able to localize the cholinergic binding site to a region of the AChR α-subunit which is no larger than 60 amino acids (152 to 210). This region contains sequences which are relatively unique for the α-subunit and might therefore explain the selective binding of toxin to the α-subunit inspite of the considerable overall homology amongst the four receptor subunits. The proposed binding sequence is in close proximity and may be contiguous with cysteines 192, 193. These conclusions are currently being tested.

Acknowledgements

We wish to thank Ms. Dora Barchan for excellent technical assistance. This work was supported by grants from the Muscular Dystrophy Association of America, the Los Angeles Chapter of the Myasthenia Gravis Foundation and the United States-Israel Binational Science Foundation (BSF) to S.F. J.M.G. is a fellow of the Batsheva de Rothschild Foundation.

References

1. Karlin, A., D. Cowburn. 1973. Proc. Natl. Acad. Sci. USA 70, 3636.

2. Damle, V.N., M. McLaughlin, A. Karlin. 1978. Biochem. Biophys. Res. Commun. 84, 845.

3. Haggerty, J.G., S.C. Froehner. 1981. J. Biol. Chem. 255, 8294.

4. Gershoni, J.M., E. Hawrot, T.L. Lentz. 1983. Proc. Natl. Acad. Sci. USA 80, 4973.

5. Tzartos, S.J., J.P. Changeux. 1983. The EMBO J. 2, 381.

6. Oblas, B., N.D. Boyd, R.H. Singer. 1983. Anal. Biochem. 130, 1.

7. Wilson, P.T., J.M. Gershoni, E. Hawrot, T.L. Lentz. 1984. Proc. Natl. Acad. Sci. USA 81, 2553.

8. Noda, M., H. Takahashi, T. Tanabe, M. Toyosato, Y. Furutani, T. Hirose, M. Asai, S. Inayama, T. Miyata, S. Numa. 1982. Nature 299, 793.

9. Smythies, J.R. 1980. Medical Hypotheses 6, 943.

10. Chou, P.Y., G.D. Fasman. 1978. Ann. Rev. Biochem. 47, 251.

11. Sumikawa, K., M. Moughton, J.C. Smith, L. Bell, B.M. Richards, E.A. Barnard. 1982. Nucl. Acids Res. 10, 5809.

12. Devillers-Thiery, A., J. Giraudat, M. Bentaboulet, J.-P. Changeux. 1983. Proc. Natl. Acad. Sci. USA 80, 2067.

13. Merrifield, R.B. 1965. Science 150, 178.

14. Bartfeld, D., S. Fuchs. 1977. FEBS Lett. 77, 214.

15. Neumann, D., M. Fridkin, S. Fuchs. 1984. Biochem. Biophys. Res. Commun. 121, 673.

16. Gershoni, J.M., G.E. Palade. 1982. Anal. Biochem. 124, 396.

17. Mochly-Rosen, D., S. Fuchs. 1981. Biochemistry 20, 5920.

18. Kao, P.N., A.J. Dwork, R.J. Kaldany, M.L. Silver, J. Wideman, S. Stein, A. Karlin. 1984. J. Biol. Chem. 259, 11662.

19. Cahill, S., J. Schmidt. 1984. Biochem. Biophys. Res. Commun. 122, 602.

20. Changeux, J.-P., A. Devillers-Thiery, P. Chemouilli. 1984. Science 225, 1335.

21. Gullick, W.J., S. Tzartos, J. Lindstrom. 1981. Biochemistry 20, 2173.

22. Conti-Tronconi, B.M., M.W. Hunkapiller, M.A. Raffery. 1984. Proc. Natl. Acad. Sci. USA 81, 2631.

POSTSYNAPTIC MODULATION OF NEURONAL FIRING PATTERN BY ADENOSINE

P. Schubert, K.S. Lee, W. Tetzlaff, G.W. Kreutzberg

Max Planck Institute for Psychiatry
Am Klopferspitz 18a, D-8033 Planegg-Martinsried, F.R.Germany

In addition to neurotransmitters, a second group of neuronal signals has been more recently characterized, the neuromodulators. They are able to modify the complex activity pattern which is generated in the brain as the result of a multitude of transmitter mediated events. They do so by either interfering with the process of synaptic transmission, affecting e.g. transmitter release, or they modify directly the properties of the target neuron by changing the nerve cell excitability. The group of neuromodulators is rather heterogeneous including also the classical transmitter acetylcholine, the function of which was the central topic at this symposium. In sympathetic neurons acetylcholine was found to block via muscarinic receptors a voltage sensitive potassium current, the M-current. Normally, the M-current serves to reduce the tendency for repetitive discharge in sympathetic neurons. In the presence of acetylcholine, however, single stimuli are capable of eliciting repetitive cell discharge (see 1).
In contrast, a suppression of the tendency for repetitive firing is apparently exerted by another neuromodulator, the nucleoside adenosine. As a general cell constituent adenosine is ubiquitously present; it is released in an activity-related manner from neurons (2,3) and presumably also from glial cells (4). If low micromolar concentrations are reached in the extracellular space, then local nerve cell activity will be affected. Both sponaneous firing as well as evoked discharges elicited by the stimulation of afferent fibers were found

Molecular Basis of Nerve Activity

to be depressed in the presence of adenosine (5,6,7,8), an effect which is receptor-mediated and can be blocked by the adenosine receptor antagonist theophylline. In particular, the evoked synaptic potentials, the EPSPs as well as the IPSPs, were reduced in size by adenosine which pointed to a synaptic site of action. This was supported by studies, mainly performed on peripheral nerves, demonstrating that adenosine reduces transmitter release (for review see 9). Accordingly, the electrophysiologically observed depressive effect of adenosine on evoked potentials could be easily explained as resulting from an inhibitory effect on the release of neurotransmitters.

In view of such a plausible presynaptic mode of action the possibility of an additional postsynaptic adenosine action has received little attention (10, 11). But this possibility that membrane properties and nerve cell excitability are directly affected by adenosine would considerably increase its power and significance as a neuromodulator which is able to control the firing pattern of a neuron. The present studies were therefore undertaken to examine the possible contribution of postsynaptic mechanisms to the neuromodulatory action of adenosine. These studies utilized the rat hippocampus which shows a relatively simple and anatomically well investigated architecture and which allows simple electrophysiological investigations with an in vitro slice preparation.

In a first series of experiments we approached the localization of the receptors which mediate the modulatory effect of adenosine on nerve cell function. We knew from previous studies that the adenosine induced depression of hippocampal evoked synaptic potentials is mediated by an A1 adenosine receptor (12). The pharmacological profile of this high affinity receptor has been well characterized and N6 substituted adenosine derivatives such as cyclohexyladenosine (CHA) have been shown to be rather specific agonists for this receptor type (13). Therefore, tritium labelled CHA has been used to localize the distribution of the A1 receptors in brain tissue by means of autoradiography. In the hippocampus, CHA-binding shows a highly selective distribution with a particularly high density of A1 receptors in the stratum radiatum and stratum oriens of the CA1-area, the regions where the apical and the basal dendrites of the pyramidal neurons are located together with their respective afferent input fibers (see fig. 1 and 2).

In order to differentiate between a pre- and or postsynaptic localization of the A1-receptors, lesions of the major input systems to the hippocampus were performed and the effect on the autoradiographically demonstrable ^{3}H-CHA-binding was examined (14). A semiquantitative evaluation of the autoradiographs was performed by means of microdensitometry using a computerized texture analysis system (Leitz TAS). Differences in the extinctions between experimental and control animals measured in homologous areas were expressed as percent of control. Removal of the contralateral hippocampus, which leads to a degeneration of the powerful commissural input, resulted in a considerable loss of CHA-binding in the terminal fields of these fibers, i.e. in the stratum radiatum and oriens of CA1 and CA3 and also in the stratum moleculare of the dentate gyrus. Compared to controls, the reduction in ^{3}H-CHA-labelling amounted to 20 t0 25 percent. This suggests that a considerable number of A1 adenosine receptors are associated with the commissural input fibers, although transsynaptic influences cannot be entirely ruled out. If the Schaffer collaterals, the intrahippocampal projection from CA3 to CA1 were disrupted by destroying the CA3 area, the resulting reduction of CHA-binding in their terminal field was comparably low, only ca. 10 percent. Since small electrolytic lesions placed in the CA3 area necessarily involve a concomitant damage of at least some of the commissural fibers, the effect resulting from the degeneration of the Schaffer collaterals might be even less. Apparently, this strong intrinsic hippocampal input system has less A1 adenosine receptors. This suggests that the different intrahippocampal fiber connections

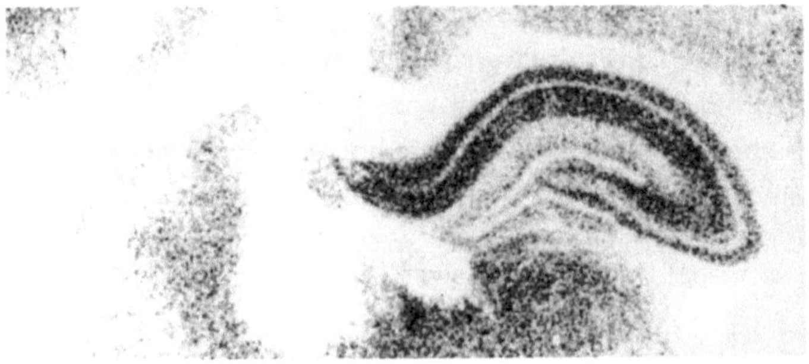

Fig.1 Evidence for a presynaptic location of A1 adenosine receptors. Removal of the commissural and entorhinal inputs to the hippocampus decreases the autoradiographically demonstrable binding of 3H-cyclohexyladenosine (3H-CHA). This decrease is clearly seen in the molecular layer of the dentate gyrus (compare to the normal hippocampus on the left side in fig.2).

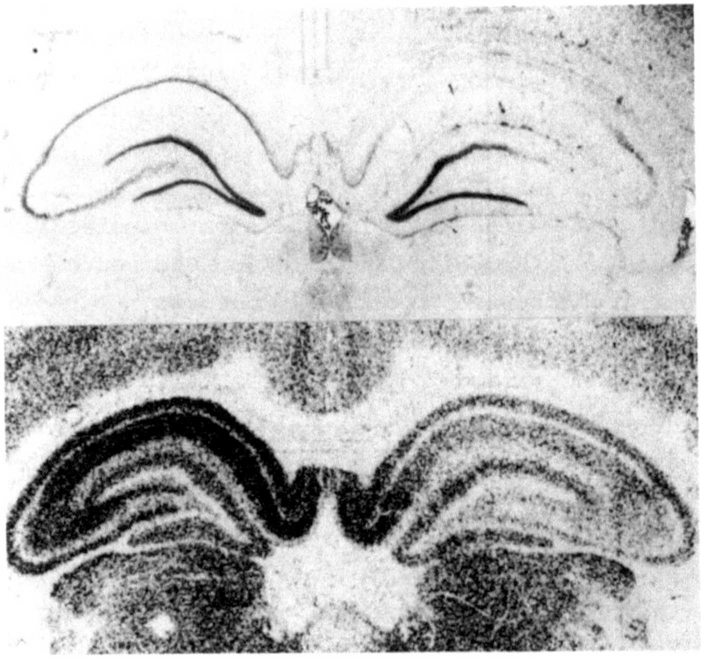

Fig.2 Evidence for a postsynaptic localization of A1 adenosine receptors. Kainic acid was injected into the right hippocampus resulting in a degeneration of hippocampal neurons (see above, Nissl-stained sclice). The autoradiography reveals a dramatic decrease of 3H-CHA-binding on the injected side (right) compared to the control (left).

differ in their content of A1 receptors. Also the fiber system connecting the dentate gyrus with the hippocampal proper, the mossy fibers, seems to have only small amounts of ^{3}H-CHA binding sites. Destroying the granule cells of the dentate gyrus from which the mossy fibers originate does not affect CHA-binding in the CA3 terminal field.

There are apparently also differences between the various extrinsic afferent systems to the hippocampus. Murray and Cheney (15) have reported that elimination of the cholinergic, serotonergic and noradrenergic afferents does not alter CHA-binding in the hippocampus. On the other hand, destroying the largest extrinsic hippocampal input, which originates in the entorhinal cortex, results in a considerable loss of CHA-binding. The disappearance of CHA-binding is particularly obvious in the dentate gyrus molecular layer where the entorhinal fibers terminate on the granule cell dendrites in a sharply confined layer and which is distinctly positive for CHA binding in controls (see fig. 1).

In addition to a presynaptic localization, evidence was also obtained for a postsynaptic localization of the A1 receptors. This was demonstrated by destroying the hippocampal target neurons themselves either by injection of kainic acid or by local electrocoagulation. The degeneration of the CA1 pyramidal neurons following kainic acid injection was accompanied by a dramatic loss of CHA-binding of 40 percent (see fig. 2). Similarly, if a circumscribed lesion was produced by electrolysis in the cell body layer of CA1 neurons, a localized decrease of CHA-binding was found in the stratum radiatum where the dendrites of the lesioned neurons are located. These findings strongly suggest that a considerable amount of adenosine A1 receptors is associated with the postsynaptic hippocampal neurons, particularly with the membranes of their dendritic processes.

Whether these postsynaptically-located A1 adenosine receptors exert a direct modulatory effect on nerve cell function was tested electrophysiologically in a second series of experiments (see 16,17). The electrophysiological investigations were performed on an in vitro hippocampal slice preparation (see fig.3). Evoked neuronal potentials were recorded extracellularly from the soma layer in the CA1 pyramidal cells. One stimulation electrode waş located in the stratum radiatum, (Sr), allowing an orthodromic synaptic activation of the CA1-neurons via the commissural fibers and Schaffer collaterals. A second stimulation electrode was located in the alveus,(Sa), allowing a direct non-synaptic activation of the neurons by antidromic stimulation of their axons. During the course of the experiment slices were constantly superfused with a standard medium to which adenosine and other compounds or drugs to be tested were added. Technical details of this in vitro technique are described elsewhere (18).

In order to ensure that an observed adenosine effect was indeed postsynaptic, the possibility of synaptic effects was eliminated by performing the experiments under conditions in which synaptic function was blocked. This was achieved by lowering the level of calcium ions in the superfused medium from 2.2 mM to 0.2 mM while the concentration of magnesium ions was raised from 2.4 mM to 4 mM. At such calcium concentrations synaptically evoked population spikes can no longer be elicited;upon stimulation of the afferent fibers in the stratum radiatum only a presynaptic fiber potential is recorded (fig. 3, first row). The postsynaptic pyramidal neurons are still excitable,however, and can be activated by direct antidromic stimulation of their axons. But, instead of responding to antidromic stimulation with a single spike, the neurons respond with a series of afterpotentials in low calcium medium (fig.3, 2.row). Thus, the tendency of a neuron to exhibit repetitive discharges is dramatically increased in low calcium. This is also indicated by the development of spontaneous interictal spiking resembling epileptiform activity (see also 19,20). Adenosine, if added in micromolar concentrations to the low calcium medium, blocks the generation of afterpotentials in a dose dependent manner (see fi.3).

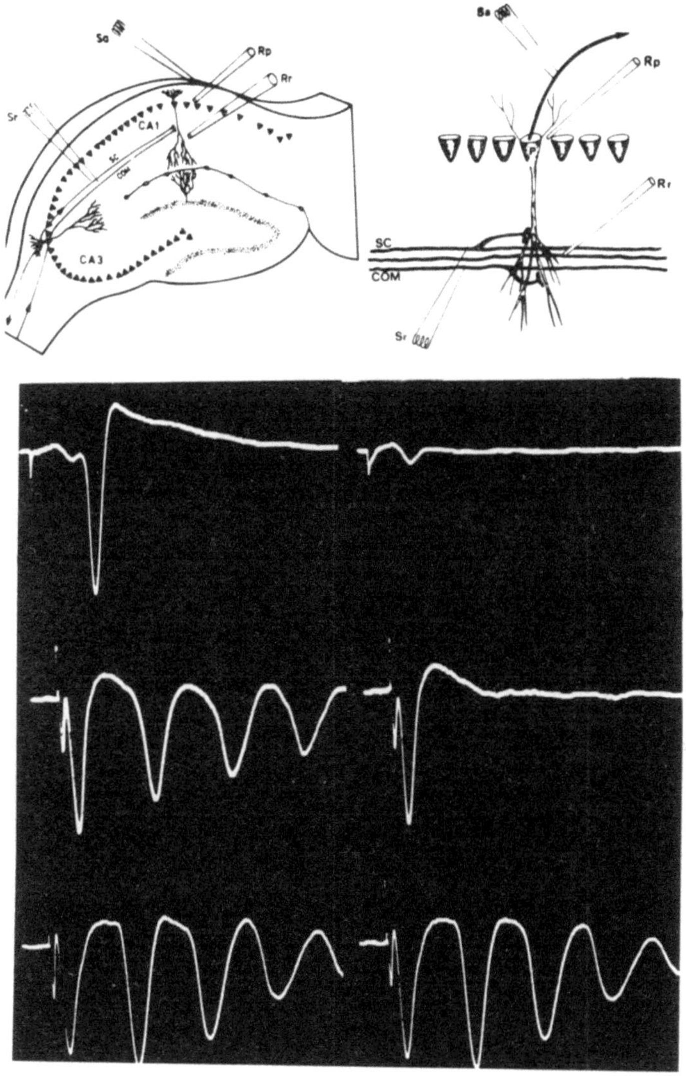

Fig.3 Location of stimulating- (S) and recording (R)-electrodes in the hippocampal slice (above) and somatic potentials evoked in CA1 neurons. 1.row: blockade of synaptic transmission. The population spike evoked by orthodromic stimulation in normal medium (left) can no more be elicited in low calcium medium (right). 2.row: Generation of afterpotentials elicited in low calcium medium by antidromic stimulation (left) is blocked by 0.3 µM L-PIA (right). 3.row: in the presence of 50 µM 4-aminopyridine the addition of even 5 µM l-PIA (right) is ineffective.

This means that adenosine is able to modify the firing pattern of a neuron, in the absence of synaptic transmission, by a direct postsynaptic action. When testing the action of the adenosine analogue phenylisopropyl adenosine (PIA), the l-isomer was found to be by far more effective than the d-isomer. Such stereospecificity of action by PIA is characteristic of an A1 site indicating that this postsynaptic modulatory effect is mediated by an A1 adenosine receptor.

In an attempt to localize the site of this depressive effect of adenosine on the generation of afterpotentials, the nucleoside was pneumatically injected on defined areas along the extension of the CA1 neurons. Surprisingly, injections into the somatic layer, where the antidromic spike is presumably generated, were less effective than injections into the apical dendritic region. This is consistent with the autoradiographic observation that the CHA-binding sites are enriched in the dendritic areas.

The observed capability of adenosine to modulate the neuronal firing pattern may not only be relevant for nerve cell function in the normal brain. It may be of particular importance under pathological conditions in which the tendency of neurons for abnormal repetitive discharge is increased. Such an anticonvulsant effect of adenosine on pathological cell discharge has been described previously by Dunwiddie (21). Interictal spiking generated experimentally by the application of penicillin was found to be abolished by adenosine. Although in this case a synaptic action of adenosine cannot be excluded, the observed effect resembles closely the adenosine effect demonstrated in the present study and a similar postsynaptic site of action might be assumed.

Concerning the mechanism of action it is interesting that adenosine has been shown by intracellular recordings to lead to a membrane hyperpolarization (10, 11). Among several other possibilities this might be explained by a change in the potassium conductance of the neuronal membrane. Such an activation of a potassium conductance by adenosine has been in fact shown in heart muscle cells (22). Also the finding by Perkins and Stone (23) that 4-aminopyridine, a potassium channel blocker, diminished the adenosine produced depression of cortical nerve cell firing in 50 % of the neurons studied suggested a possible influence of adenosine on transmembrane potassium fluxes. 4-aminopyridine has been studied particularly in respect to its presynaptic action. In contrast to the action of adenosine this drug causes a dramatic increase of transmitter release. This effect is thought to result from a prolongation of the presynaptic spike duration leading to an increased influx of calcium into the axon terminal (19). We therefore tested whether the adenosine depression of afterpotentials generated upon antidromic stimulation in low calcium medium can be antagonized by 4-aminopyridine. While this K -channel blocker had a small excitatory effect of its own on the amplitude of the afterpotentials, it had a powerfulantagonistic effect on the adenosine mediated depression. In the presence of 50 uM 4-aminopyridine the potent A1 receptor agonist, l-phenyl-isopropyladenosine (l-PIA), was ineffective (see fig. 3). The amplitude of the afterpotentials remained completely unchanged by 5 uM l-PIA. This is indeed a pronounced effect since in the absence of 4-aminopyridine, l-PIA abolished the generation of afterpotentials completely at a concentration of 0.3 uM (see fig. 3).

In conclusion, evidence was presented for a modulatory effect of adenosine on the pattern of nerve cell firing reducing the tendency for repetitive cell discharge. In addition to the well established presynaptic action of adenosine on transmitter release adenosine appears to affect directly the membrane properties of the postsynaptic neuron. An adenosine induced increase of the potassium permeability seems to be the most likely mechanism of action. The effect is receptor mediated, i.e. by the A1 receptor. Evidence is presented that these receptors are localized also on the postsynaptic site where they seem associated predominantly with the dendritic membranes of the hippocampal pyramidal neurons studied.

References

1. Adams,P.R., D.A.Brown, A.Constanti.1981. J.Physiol.33o,537.

2. McIlwain,H. 1972. Biochem.Soc. Symp.36,69.

3. Schubert,P., K.Lee, M.West, S.Deadwyler, G.Lynch. 1976. Nature 260,541.

4. Lewin,E., V.Bleck. 1979. J.Neurochem. 33,365.

5. Phillis,J.W., G.K.Kostopoulos, J.Limacher. 1974. Can.J.Physiol.Pharmacol. 52,1226.

6. Okada,Y., Y.Kuroda. 1975. Proc.Jap.Acad. 51,491.

7. Schubert,P., U.Mitzdorf. 1979. Brain Res. 172,186.

8. Dunwiddie,T.V., B.J.Hoffer. 1980. Br.J.Pharmacol. 69,59.

9. Fredholm,B.B., P.Hedqvist. 1980. Biochem.Pharmacol. 29,1635.

10. Phillis,J.W., J.P.Edstrom, G.K.Kostopoulos, J.R.Kirkpatrick. 1979. Canad.J.Physiol.Pharmacol. 57,1289.

11. Siggins,G., P.Schubert. 1981. Neurosci.Lett. 23,55.

12. Reddington,M.,K.S.Lee,P.Schubert.1982.Neurosci.Lett.28,275.

13. Bruns,R.F., J.W.Daly, S.H.Snyder. 1980. Proc.Natl.Acad.Sci. USA 77,5547.

14. Schubert,P.,K.S.Lee,W.Tetzlaff,G.W.Kreutzberg: in prep.

15. Murray,T., D.L.Cheney. 1982. Neuropharmacol. 21,575.

16. Lee,K.S., P.Schubert, U.Heinemann. 1984. Brain Res. 321,160.

17. Schubert,P.,K.Lee. 1984. Clin.Neuropharmacol.7,Suppl.1,410.

18. Lynch,G.,P.Schubert. 1980. Ann.Rev.Neurosci. 3, 1.

19. Jefferys, J.G.R., H.L.Haas. 1982. Nature 300,448.

20. Yaari,Y., A.Konnerth, U.Heinemann. 1983. Exp.Brain Res. 51, 151.

21. Dunwiddie,T.V.. 1980. Epilepsia 21,541.

22. Jochem,G., H.Nawrath. 1983. Experientia 39,1347.

23. Perkins,M.N., T.W.Stone. 1980. Br.J.Pharmacol. 70, 425.

MAJOR PROTEINS OF THE ELECTRIC ORGAN FROM TORPEDO MARMORATA - POSSIBLE ASSOCIATIONS WITH THE POSTSYNAPTIC MEMBRANE

V. Witzemann, C.M. Boustead, and J.H. Walker
Abteilung Neurochemie, Max-Planck-Institut für biophysikalische Chemie, D-3400 Göttingen

Introduction

In the process of chemical neurotransmission receptors play a central role in that they enable cells to respond to extracellular chemical stimuli. By far the best understood neurotransmitter receptor is the nicotinic acetylcholine receptor (AChR) due to its fortuitous occurrence in high concentrations in the electric organs of certain electric fish, which have been used by many research groups as source material and an ideal model system of neuromuscular synapses. During development the AChR molecules become concentrated and stabilized at sites of synaptic contacts. Only recently growing interest has been shown in asking in how far proteins associated in a peripheral manner with the receptor-containing membrane are involved in regulating AChR function and the formation of the postsynaptic membrane (1-6). Morphological studies revealed that filaments (5-12 nm in diameter) are attached to this membrane (7,8) and biochemical analysis of AChR enriched membranes obtained upon subcellular fractionation indicated a number of polypeptides (M_r 43 000, 50 - 55 000, 90 - 100 000) being associated in a peripheral manner to the receptor-containing membrane.

Molecular Basis of Nerve Activity

Results and Discussion

Over the years the properties of membrane bound AChR have been studied using membrane fragments purified on sucrose density gradients (9,10). These membranes are highly enriched in receptor protein and Fig. 1 shows a typical polypeptide pattern obtained upon separation on SDS-denaturing polyacrylamide gels. Besides the four AChR subunits, additional polypeptides are consistently found at M_r 43 000, 55 000 and at 90 - 100 000. Brief treatment of these fragments with pH 11 has been shown to efficiently remove 'peripherally' associated proteins (2,3) reducing the protein composition of the membrane fragments to the characteristic pattern of the four subunits of the AChR complex (Fig. 1b). The material solubilized by pH 11 treatment contains a major band at M_r 43 000 (Fig. 1e). It was shown (6, 11) that this band consists of a number of different polypep-

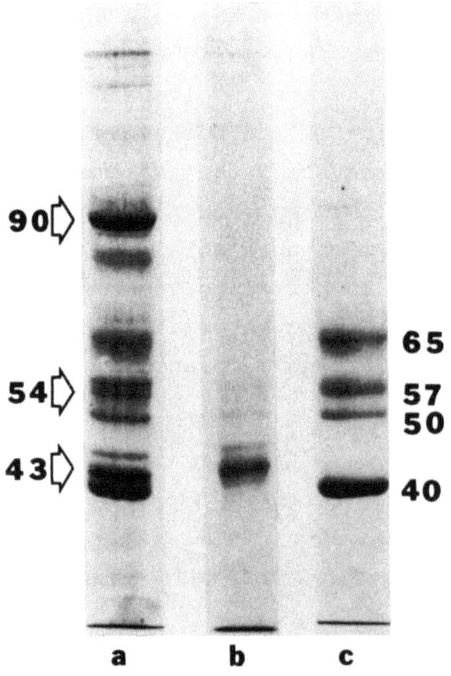

Fig. 1: SDS-Polyacrylamide gel electrophoresis of a) membrane fragments enriched in acetylcholine receptor protein. Major non-receptor polypeptides are indicated by arrows and $M_r \times 10^{-3}$. The M_r 90 000 component occurs in high concentrations in the non-innervated membrane of electrocytes (29) and may represent a subunit of the Na^+/K^+-ATPase (19, 20); b) proteins solubilized at pH 11.0 treatment; c) membrane fragments after pH 11.0 treatment; the receptor subunits α (40), β (50), γ (57), and δ (65) are indicated ($M_r \times 10^{-3}$).

tides, initially named ν_1, ν_2 and ν_3 (11) which can be distinguished due to characteristic differences in their isoelectric properties. Fig 2 shows the pattern of these polypeptides obtained upon two-dimensional gel electrophoresis.

The association of actin with the receptor membranes had been shown by Strader et al (12). Of the polypeptides separated upon two-dimensional gel electrophoresis it was shown that the polypeptide at pH 5.4 represents actin (6,13, and Fig. 3). The minor polypeptides, ranging from pH 6.2 - 6.5 have been iden-

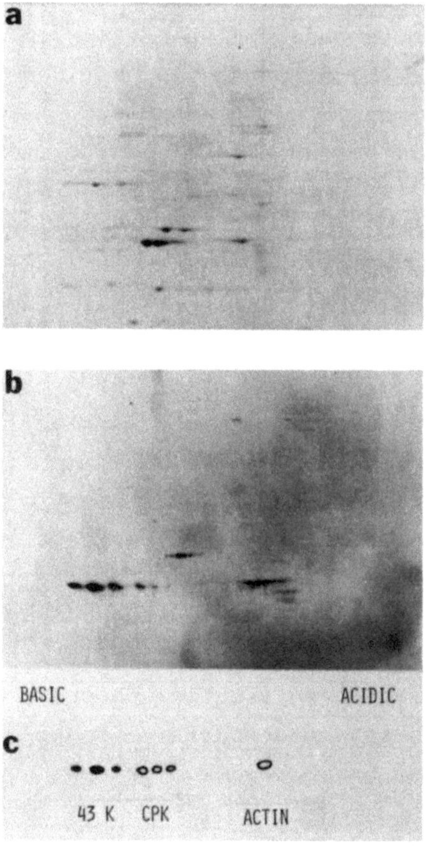

Fig. 2: Two-dimensional analysis of a) proteins of the electric organ, soluble at low ionic strength (≙ major cytoplasmic proteins)

b) proteins peripherally associated with the acetylcholine receptor-enriched membranes (as shown also in Fig. 1b)

c) schematic representation of the M_r 43 000 polypeptides (I.P.~7.0) creatine phosphokinase (CPK;~I.P. 6.5) and actin (I.P. 5.4).

tified as creatine phosphokinase (13,14). Since the majority of these polypeptides is found in the cytoplasm and represent up to 10% of the total cytoplasmic protein content (Fig. 2 shows the polypeptides soluble at low ionic strength), they may be merely entrapped in vesiculated membranes during homogenization.

The electrocytes are derived embryologically from muscle tissue (7); it is therefore of interest whether the creatine phosphokinase is of the muscle-specific (MM) type as found by Gysin (13), or whether it actually represents the brain (BB) type (14). The isoenzymes from Torpedo muscle electric organ as well as from Torpedo brain were purified by $(NH_4)_2SO_4$ fractionation followed by DEAE-cellulose chromatography (15). Analysis of the enzymes revealed distinct differences in isoèlectric properties: the BB-type enzyme was focusing at more basic pH values ($\sim$7.0) than the MM-type enzyme and the electric organ enzyme was identical with the enzyme purified from muscle tissue (not shown). Antibodies raised against the electric organ enzyme readily reacted with the muscle enzyme but not with the enzyme from brain (Fig. 3) demonstrating that the electric organ enzyme is indeed of the MM-type. Furthermore there is no difference between the cytoplasmic creatine phosphokinase and the ν_2 polypeptides associated with the receptor enriched membranes with regard to isoelectric and immunochemical properties. No immunochemical cross-reactivity can be detected between creatine phosphokinase, actin, and the M_r 43 000 (ν_1) polypeptides associated with the postsynaptic membrane. Immunohistochemical staining of electrocytes shows distribution of creatine phosphokinase throughout the cell. Additional experiments will be conducted to decide whether there is a specific functional association between this enzyme and the postsynaptic membrane. In this respect it is of interest to note that a close association of creatine phosphokinase with cellular structures such as mitochondria, cytoskeletal components and plasma membranes has been demonstrated (16, and references therein).

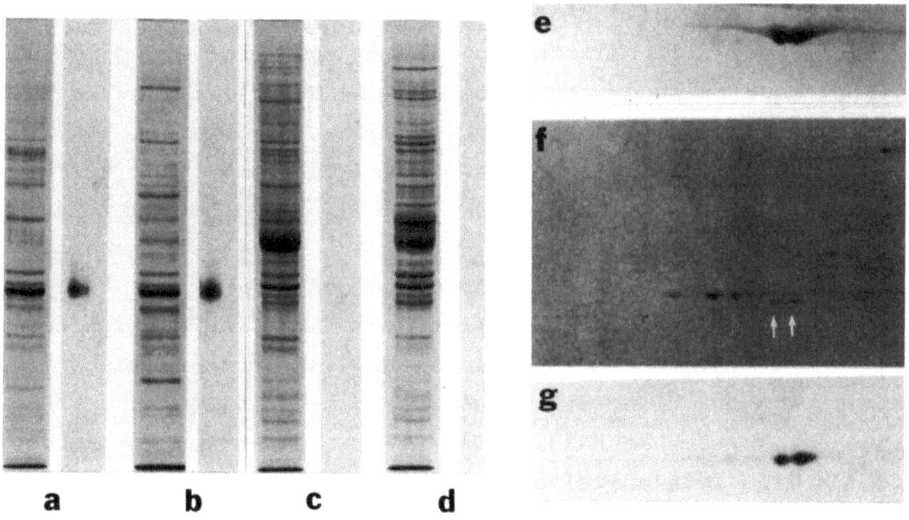

Fig. 3: SDS-polyacrylamide gel electrophoresis of cytoplasmic protein (= soluble at low ionic strength) from Torpedo marmorata and immunochemical identification of creatine phosphokinase (MM-type). a) Electric organ, b) back muscle, c) brain, d) electric lobe. The left lanes show the polypeptides after Coomassie Brilliant Blue staining and the right lanes the corresponding 'immune blots'. The antibodies used were raised to purified electric organ creatine phosphokinase. Immune complexes were visualized using peroxidase-conjugated second antibody. The experiment demonstrates that the creatine phosphokinase is of the MM-type. e) Two-dimensional analysis of electric organ creatine phosphokinase, f) two-dimensional analysis of pH 11.0 soluble proteins associated with the acetylcholine receptor enriched membranes; the minor polypeptides indicated by arrows have identical isoelectric properties as the creatine phosphokinase subunits. g) Immunochemical identification of creatine phosphokinase subunits (MM-type) using anti-creatine phosphokinase antibodies as described above.

The major M_r 43 000 polypeptides (ν_1) have been shown to be localized on the cytoplasmic side of the postsynaptic membrane (17,18, and references therein). It has been suggested that these polypeptides play a role in controlling the mobility and stability of the AChR molecules (21-24). This view was supported by experiments where the M_r 43 000 polypeptides could be crosslinked to the β-subunit (M_r 50 000) of the AChR (25).

We have recently used an actin-overlay technique similar to one described in the literature (26) to show that the M_r 43 000 polypeptides bind under these conditions actin (27). The presence of actin and actin-binding proteins was demonstrated immunochemically after the proteins were separated on polyacrylamide gels and transferred onto nitrocellulose sheets by capillary diffusion. With monoclonal anti-actin antibodies and peroxidase conjugated second antibody one can demonstrate the presence of actin in the AChR-enriched membrane fragments. After pH 11.0 extraction no detectable amount of actin remained in the membrane fraction but was found in the pH 11.0 wash. The actin-overlay technique (27), was used to detect actin-binding proteins. The actin-binding buffer routinely contained calcium (0.2 mM) although no difference was seen when actin incubation was performed in the presence of EGTA or in the presence of calcium and calmodulin. In these experiments actin is labelled as before but in addition several other proteins are stained, most notably the M_r 43 000 polypeptides. The pH 11.0 washed membranes contain little if any of these actin-binding proteins. That the major actin-binding protein, associated with the receptor-containing membranes, actually is the M_r 43 000 polypeptide is demonstrated in Fig. 4. The characteristic isoelectric forms of this protein were not labelled by direct incubation with monoclonal antibody and only after pre-incubating the nitrocellulose blot with purified actin were these polypeptides stained by the monoclonal antibody, thus supporting the notion that the M_r 43 000 polypeptides can interact with actin.

Analyzing the protein composition of the developing electric organ from Torpedo marmorata (6) it was found that the significant increase of the receptor-associated M_r 43 000 polypeptides coincided with the exponential increase of the electric organ discharge measured upon stimulation, an increase in electrocyte surface contacted by nerves and an increase in hydrophobic '6 S' acetylcholinesterase. These parameters are

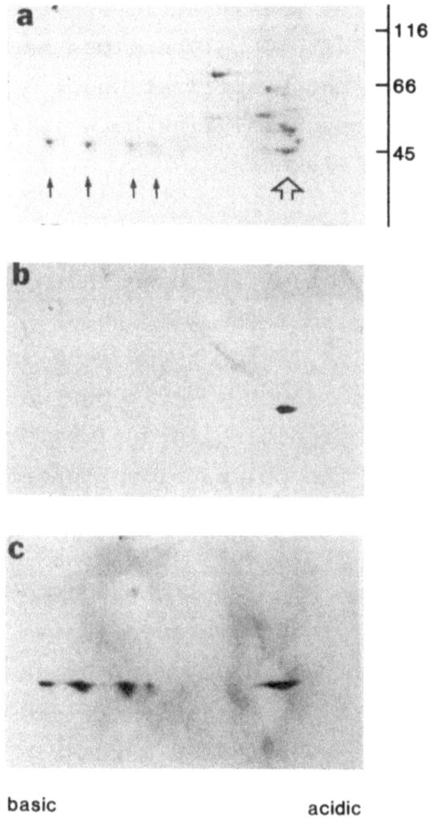

Fig. 4: Two-dimensional analysis of proteins eluted by pH 11.0 treatment from acetylcholine receptor-enriched membrane fragments.
a) Proteins stained with Coomassie Brilliant Blue. The open arrow indicates the position of actin, the filled arrows indicate the heterogeneous M_r 43 000 polypeptides.
b) Immunochemical demonstration of actin.
c) Immunochemical demonstration of actin and actin-binding proteins. (27)

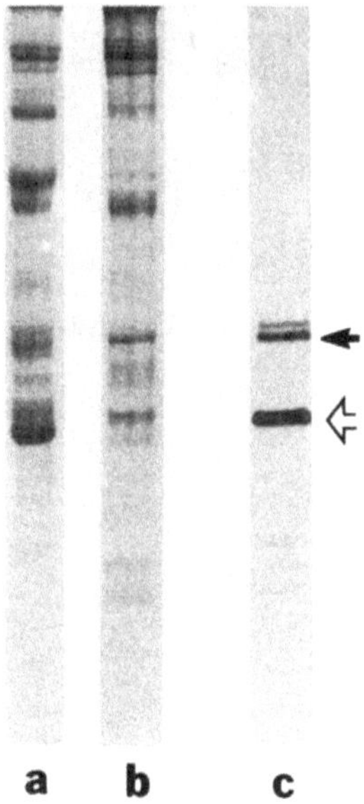

Fig. 5: SDS-polyacrylamide gel electrophoresis of
a) acetylcholine receptor-enriched membranes;
b) Triton X-100 insoluble proteins = cytoskeleton fraction derived from membrane fragments in a);
c) immunochemical identification of actin (open arrows) and a desmin-related M_r 54 000 polypeptide (closed arrow) using monoclonal anti-actin and anti-desmin antibodies.

indicative of the functional maturation of the synaptic structures during synaptogenesis (6). The M_r 43 000 polypeptides may be directly involved in the aggregation and stabilization of AChR molecules in that they provide an essential link between the membrane-bound AChR to cytoskeletal elements.

Several authors have identified a polypeptide of M_r 50-55 000 with the AChR membrane (28,29). Walker et al. (30) have shown a M_r 54 000 polypeptide being associated with these membranes (Fig. 5). After detergent extraction using Triton X-100, the insoluble fraction was found to be enriched in this M_r 54 000 polypeptide. Two-dimensional gel electrophoresis revealed isoelectric properties similar to desmin, i.e. the polypeptide focused slightly to the basic side of actin, also present in this fraction. Immunochemical staining using monoclonal anti-desmin antibodies demonstrated that the M_r 54 000 polypeptide in the cytoskeletal fraction obtained from AChR-enriched membranes is immunologically related to desmin (Fig. 5). The desmin immunoreactivity extends throughout the electrocyte and is not only associated with the AChR containing membrane. Like actin this polypeptide occurs also in large amounts in the electric organ and could represent part of the intracellular framework which is necessary to maintain the unusual structure of electrocytes which can measure up to 1 cm in diameter but are only about 15 μm thick.

Acknowledgements

The research described here was supported by the Max-Planck-Gesellschaft and the Deutsche Forschungsgemeinschaft, Project grant SFB 33.

References

1. Sobel, A., M. Weber, J.-P. Changeux. 1977. Eur.J.Biochem. 80, 215.

2. Neubig, R.R., E.K. Krodel, N.D. Boyd, J.B. Cohen. 1979. Proc.Natl.Acad.Sci.USA 76, 690.

3. Elliot, J., S.G. Blanchard, W. Wu, J. Miller, C.D. Strader, P. Hartig, H.-P. Moore, M.A. Raftery. 1980. Biochem.J. 185, 667.

4. Sealock, R. 1982. J.Cell Biol. 92, 514.

5. Barrantes, F.J. 1982. J. Cell Biol. 92, 60.

6. Witzemann, V., D. Schmid, C. Boustead. 1983. Eur.J.Biochem. 131, 235.

7. Fox, G.Q., G.P. Richardson. 1979. J.Comp.Neurol. 185, 293.

8. Heuser, J.E., S.P. Salpeter. 1979. J.Cell Biol. 82, 150.

9. Changeux, H.-P. 1981. Harvey Lect. 75, 85.

10. Conti-Tronconi, B.M., M.A. Raftery. 1981. Ann.Rev.Biochem. 51, 491.

11. Gysin, R., M. Wirth, S.D. Flanagan. 1981. J.Biol.Chem. 256, 11373.

12. Strader, C.D., E. Lazarides, M.A. Raftery. 1980. Biochem. Biophys.Res.Commun. 92, 365.

13. Gysin, R., B. Yost, S.D. Flanagan. 1983. Biochemistry 22, 5781.

14. Barrantes, F.J., G. Mieskes, T. Wallimann. 1983. Proc.Natl. Acad.Sci.USA 80, 5440.

15. Witzemann, V., C. Boustead. (submitted)

16. Erickson-Viitanen, S., P. Viitanen, P.J. Geiger, W.C.T. Yang, S.P. Bessman. 1982. J.Biol.Chem. 257, 14395.

17. Nghiem, H.-O., J. Cartaud, C. Dubreuil, C. Kordeli, G. Buttin, J.-P. Changeux. 1983. Proc.Natl.Acad.Sci.USA 80, 6403.

18. Sealock, R., B.E. Wray, J. Froehner. 1984. J.Cell Biol. 98, 2239.

19. Elfman, L., E. Heilbronn, P.L. Jorgensen. 1982. Biochim. Biophys.Acta 693, 273.

20. Lindstrom, J., B. Walter (nave), B. Einarson. 1979. Biochemistry 18, 4470.

21. Lo, M.M.S., P.B. Garland, E. Lamprecht, E.A. Barnard. 1980. FEBS Lett. 111, 407.

22. Saitoh, T.H., L.P. Wennogle, J.-P. Changeux. 1979. FEBS Lett. 108, 489.

23. Barrantes, F.J., D.Ch. Neugebauer, H.P. Zingsheim. 1980. FEBS Lett. 111, 73.

24. Rousselet, A., J. Cartaud, P.F. Devaux, J.-P. Changeux. 1982. EMBO J. 1, 439.

25. Burden, S.J., R.A. De Palma, G.S. Gottesmann. 1983. Cell 35, 687.

26. Snabes, M.C., A.E. Boyd, J. Boyan. 1981. J.Cell Biol. 90, 809.

27. Walker, J.H., C.M. Boustead, V. Witzemann. 1984. EMBO J. 3, 2287.

28. Burden, S. 1982. J.Cell Biol. 94, 521.

29. Froehner, S.C. 1984. J.Cell Biol. 99, 88.

30. Walker, J.H., C.M. Boustead, V. Witzemann, J. Shaw, K. Weber, M. Osborn. (submitted)

PROTEIN-BLOT ANALYSIS OF THE NICOTINIC ACETYLCHOLINE RECEPTOR

Jonathan M. Gershoni

Department of Biophysics, The Weizmann Institute of Science, Rehovot, Israel

Edward Hawrot, Paul T. Wilson and Thomas L. Lentz

Yale University School of Medicine, 333 Cedar Street, New Haven, CT 06510, USA

Introduction

Occupancy of the cholinergic binding site of the nicotinic acetylcholine receptor (AChR) elicits allosteric changes in this protein complex which result in membrane depolarization (1). The molecular details of these events are of great interest and it is believed that by identifying and characterizing the precise peptide domain responsible for ligand binding one will gain better understanding of receptor function and the mechanisms involved in neurotransmission. For the past few years we have been studying the nicotinic cholinergic binding site by employing a novel methodology, namely protein blotting (2).

In principle, purified AChR or crude membrane preparations, both derived from <u>Torpedo californica</u> electric organ, are resolved into their constituents by polyacrylamide gel electrophoresis (PAGE), then electrophoretically transferred to positively charged nylon membrane filters (PCM), which are subsequently incubated with ^{125}I-labeled α-bungarotoxin (BTX).

Molecular Basis of Nerve Activity

The bound toxin is detected autoradiographically. This approach has allowed us to demonstrate that the α-subunit of AChR harbors the binding site for BTX, that BTX binding does not require the N-glycosidically linked oligosaccharide side chain of the α-subunit, and that proteolytic fragments of the α-subunit, as small as 8kd, can still specifically bind BTX. Moreover, we have used the protein-blotting assay to reveal toxin-binding polypeptides in muscle, as well as in brain tissue, in a variety of animals (3-5). This paper summarises our findings.

Methods

Purification of Torpedo AChR

Frozen electric organ tissue was homogenized in 40 mM Tris.HCl pH 7.4/1 mM Na_2EDTA/0.1 mM PMSF/0.05% NaN_3. The debris was removed by filtration through cheese cloth, followed by low speed centrifugation (10 min at 4000xg). Membranes rich in AChR (about 15% of the total protein) were pelleted from the supernatant by centrifugation for 45 min at 100000xg. These membranes were either used directly for protein-blot analyses or solubilized in 1% Triton-X-100 and applied to a cobra toxin-Sepharose column. Purified AChR was eluted from the column with 1 M carbamylcholine.

Protein-blotting and toxin overlays

Electric organ membranes or purified AChR were solubilized at room temperature in 50 mM Tris.HCl pH 6.8/2% lithium dodecyl sulfate (LDS)/3% 2-mercaptoethanol/0.1% bromophenol blue. The samples were then electrophoresed (at 4°C) on 10% polyacrylamide gels which contained no detergent. LDS (0.1%) was present only in the upper reservoir buffer. At the end of electrophoresis the gels were either stained in Coomassie brilliant blue or used immediately for protein blotting. Protein blotting was achieved electrophoretically (200 mA for 2 h at 4-8°C) in an apparatus equipped with an electrode array which provided a uniform electric field (6). PCM (7) was used as the immobilizing matrix; however, nitrocellulose membrane filters have also been found to be quite adequate. After transfer, the blots were quenched for 1 hr at 45°C in 1% hemoglobin in 10 mM phosphate buffered saline pH 7.4 (PBS). The quenched blots were then incubated at room temperature with ^{125}I-BTX (1-5 nM, 1-5 x 10^6 cpm) overnight, or as indicated, in fresh hemoglobin-containing PBS. Afterwards the blots were washed in PBS (4°C, 5 x 5 min). The blots were then placed between pieces of saran wrap and used to expose Kodak X-AR-5 film at -70°C using a DuPont Lightning Plus intensifying screen. Exposures were routinely carried out for 12-48 hr.

Proteolytic digestion of the α-subunit

The α-subunit of AChR was excised out of gels and subjected to proteolytic digestion according to Cleveland et al. (8). The fragments were separated on 15% acrylamide gels which were then blotted and probed with ^{125}I-BTX (4).

Other analyses

Toxin overlays of a variety of AChR preparations were also performed. AChR treated with different concentrations of maleimido-benzyltrimethylammonium iodide (MBTA) (9) provided evidence that the toxin binding site assayed on the blots corresponded to the MBTA sensitive cholinergic binding site. Endoglycosidase-H treatment of the AChR (3) provided deglycosylated α-subunits. Muscle tissue from chick embryos, denervated rat legs, and BC3H-1 cells grown in culture as well as duck and fish brains were used for affinity purification of AChR proteins. The homogenized tissues were solubilized and applied to cobra-toxin columns or BTX columns. The bound material was released from the columns with LDS sample buffer, and run on LDS gels as described above. Blots were then prepared and toxin-overlayed.

Results and Discussion

Figure 1 demonstrates the toxin overlay of blots prepared from crude membrane samples of Torpedo electric organ and purified Torpedo AChR. The binding of toxin could be obtained even when samples were subjected to boiling in SDS containing buffer, electrophoresis at room temperature and subsequent blotting onto nitrocellulose membrane filters. The intensity of the signal appeared to be significantly higher, however, when the procedure was carried out under mild denaturing conditions as described in Methods.

The formation of the toxin/α-subunit complex was directly dependent on the amount of toxin (not shown) or α-subunit present (Figure 2A). Quantification of the bound toxin demonstrated that a linear response was obtained using crude membrane samples ranging between 25 and 100 µg. In view of this result, MBTA-treated membranes were analysed.*

AChR containing membranes were treated with various concentrations of MBTA (0-40 µM). Then the membranes were either directly assayed for toxin binding or used to prepare blots which were subsequently toxin-overlayed. As can be seen in Figure 2B, the extent of MBTA inhibition of toxin binding was the same in both the intact AChR and the blotted α-subunit, thus proving that the toxin binding site of the isolated immobilized α-subunit is equated with the MBTA binding site.

In order to determine the affinity of BTX for the blotted

---

* One does not expect to observe MBTA inhibition of toxin binding greater than 50-70% (9).

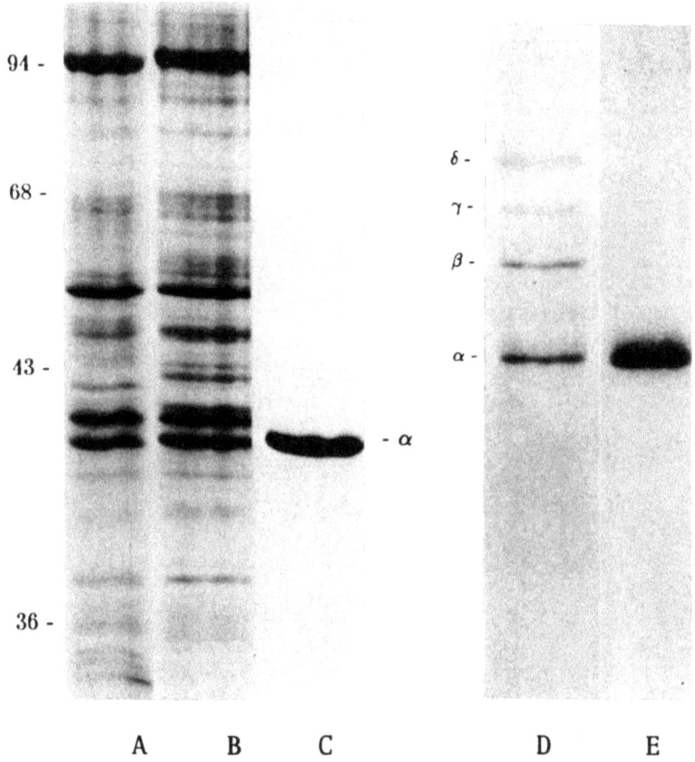

Figure 1. Crude membranes from Torpedo electric organ (A,B,C) or purified AChR (D,E) were separated by LDS-PAGE at room temperature (A) or at 4°C (B-E). The gels were stained with Coomassie brilliant blue (A,B,D) or blotted, probed with ^{125}I-BTX and autoradiographed (C,E).

α-subunit, the following experiments were performed. Identical samples of AChR were blotted and incubated overnight with ^{125}I-BTX, washed in cold PBS and then washed further at room temperature for various periods of time. Quantification of the toxin bound for each time point was plotted and is shown in Figure 2C. The half-life of the complex was calculated.

When blots of AChR were incubated with ^{125}I-BTX for different periods of time, it became evident that the association of BTX

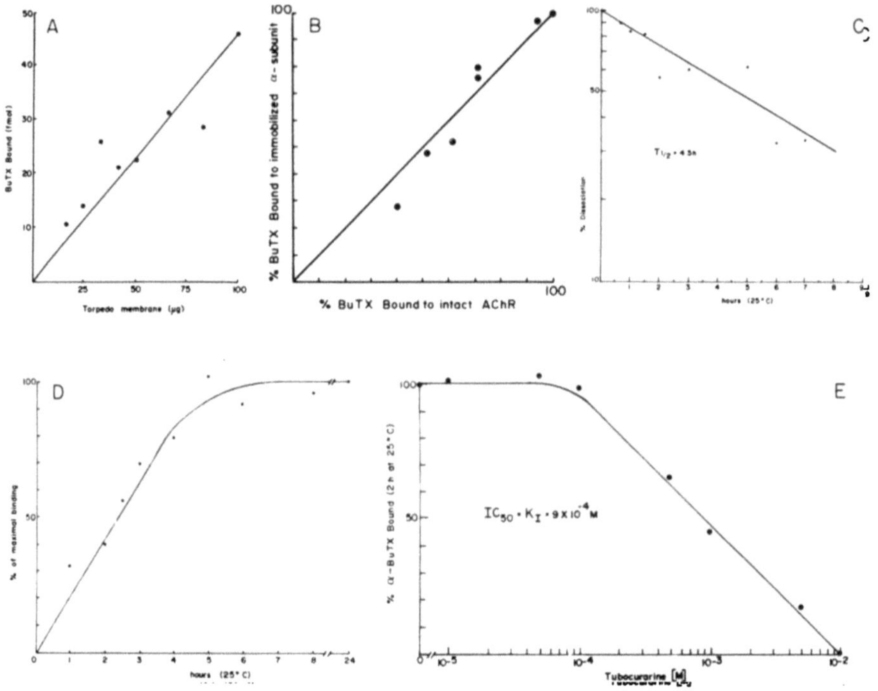

Figure 2. Quantification of ^{125}I-BTX binding to the AChR α-subunit (see text). Note that each point in (B) represents a different concentration of MBTA used.

with the α-subunit reached equilibrium only after 6 h at room temperature (Figure 2D).

This association could be competed with other antagonists as is demonstrated in Figure 2E. In such experiments, blots of AChR were coincubated (for 2 h) with various concentrations of d-tubocurarine and a constant concentration of ^{125}I-BTX (2 nM). The blots were washed and the signals quantified.

From the above it appears that the binding of toxin to immobilized α-subunit is specific and physiologically relevant and so the method of toxin-overlay of blotted AChR was used to further analyse the toxin binding site. For example, purified AChR was subjected to endoglycosidase-H digestion. This treatment selectively removes N-glycosidically linked oligosaccharide side chains. The treated AChR was used for the preparation of blots which were then toxin overlayed and compared with a control sample of AChR.

The migration of deglycosylated α-subunit was altered due to the endoglycosidase-H treatment (as can be seen by a shift in mobility corresponding to about 2k in molecular weight). Both the glycosylated and deglycosylated α-subunits bound ^{125}I-BTX with equal efficiencies (Figure 3). This result is of particular importance not only because it demonstrates that the sugar is not required for the toxin binding observed, but also that the toxin-binding α-subunit is indeed N-glycosylated. This would infer that should differences in the two α-subunits of AChR exist (10), they can not be ascribed to the glycosylation state of Asn 141.

Toxin-overlays of blots prepared from proteolysed α-subunits have proved to be extremely useful (Figure 4). The α-subunit was isolated from LDS-gels and subjected to proteolysis with S. aureus V8 protease (Fig.4B), papain (Fig.4C), bromelain (Fig.4D), and proteinase K (Fig.4E) in situ in the stacking gels of 15% polyacrylamide gel, as described by Cleveland et al. (8). Toxin overlays of blots prepared from these gels revealed that for the different proteases, unique fragments varying from 30-8kd maintained the capacity for binding of ^{125}I-BTX.

This binding to proteolytic fragments could be shown to be sensitive to MBTA treatment of the intact AChR from which they were derived (Figure 5). This approach should allow one eventually to dissect out and further characterize the cholinergic binding site. Results pertaining to the identification of the ligand binding site obtained by employing this blot procedure in conjunction with antibodies generated against synthetic peptides of the α-subunit are described by Neumann et al. (these proceedings).

Finally, the toxin overlay technique has been used to reveal BTX binding proteins of AChRs derived from a variety of tis-

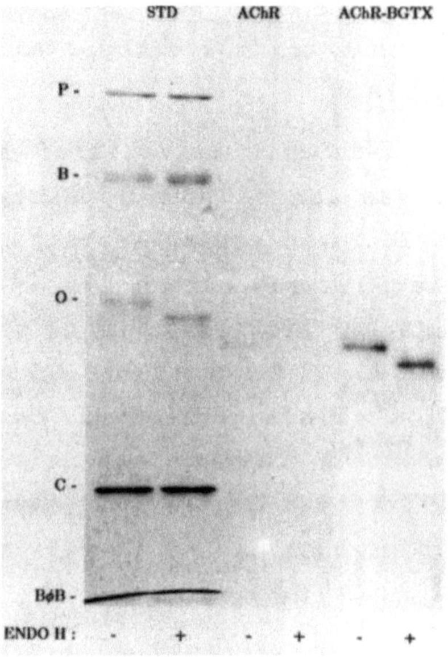

Figure 3

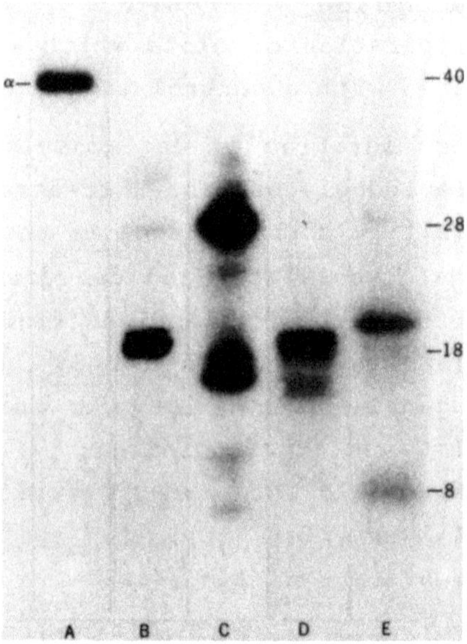

Figure 4

Figure 3. A mixture of protein standards (P, phosphorylase B; B, bovine serum albumin; O, ovalbumin; and C, carbonic anhydrase) and purified AChR were incubated in the presence or absence of endoglycosidase H (ENDO H). The samples were then separated by LDS-PAGE and the gels were either stained with Coomassie briliant blue (first four lanes) or blotted, overlayed with ^{125}I-BTX and autoradiographed (last two lanes). Note that only ovalbumin was affected by the ENDO-H treatment, indicating that the shift in migration is due to removal of sugar rather than proteolysis.

Figure 4. An autoradiogram of blotted proteolytic fragments of the α-subunit probed with ^{125}I-BTX (see text). Lane A is α-subunit non-proteolysed.

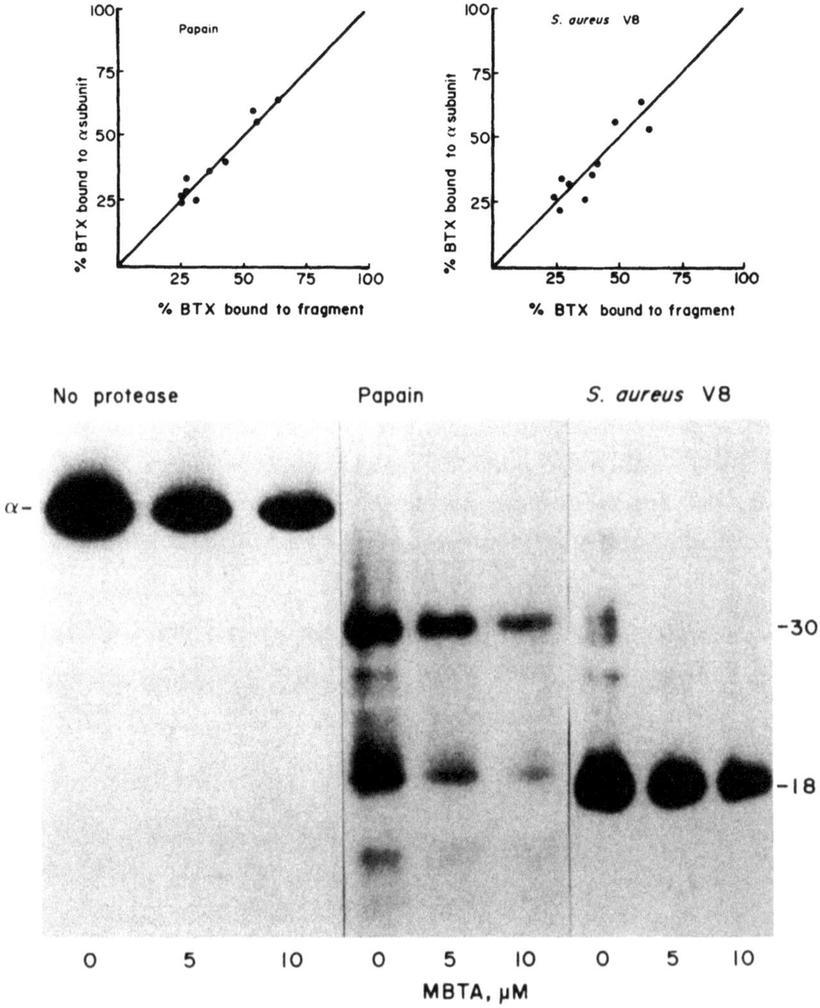

Figure 5. Autoradiographs of intact α-subunits or proteolysed α-subunits derived from AChR treated with various concentrations of MBTA. Quantification of such experiments demonstrates that the MBTA treatment inhibits the toxin binding to the major proteolytic fragments, or to the intact α-subunit, to the same extent (each point represents a different concentration of MBTA used).

sues. In electric organ tissues from Torpedo and Electrophorus, where AChR is relatively abundant, crude membrane preparations could be used for the detection of BTX binding proteins. In order to detect BTX binding proteins in tissues less plentiful in AChR, an enrichment step had to be performed prior to LDS-PAGE. Muscle tissue from chick embryos, for example, was first applied to a toxin column. The material that adhered to the column was used for blotting, and resulted in only a 40k polypeptide that bound the ^{125}I-BTX (silver staining of the proteins resolved on the gel showed many additional bands, none of which were recognized by the toxin). Similar experiments were performed on denervated rat muscle and BC3H-1 cells. Whereas signals are also obtained at 40k-41k, it appears that these experiments approach the limit of detection of the technique (11 and unpublished observations).

Most interesting is the fact that in using toxin overlays of blots we have been able to detect BTX-binding components in the central nervous system. Blots from brain tissue obtained from chicks, ducks and a variety of fish, were prepared in the manner used in the analysis of chick embryonic muscle. Toxin-overlays revealed high molecular weight (approx. 220k) polypeptides that bound BTX. This binding was extremely sensitive to boiling of the sample, could be competed with d-tubocurarine or nicotine, and was found to be unaffected by the presence of atropine. As yet it is not clear whether or not the high molecular weight protein that binds BTX represents a single polypeptide or an aggregate of a number of receptor components. No toxin binding was detected in experiments in which rat and guinea pig brains were analysed (5 and unpublished observations).

In summary, protein blotting has been adapted to and subsequently perfected for the analysis of the nicotinic AChR. It has been found to be extremely convenient and useful as it has allowed the identification and characterization of sub-receptor elements without having to isolate and purify them.

By modifying the experimental conditions, it is hoped to be able to determine the requirements needed to reconstitute high affinity binding of BTX to the AChR. By this effort to elucidate the structures involved in ligand binding and receptor assembly, it is believed that a more comprehensive understanding of the processes involved in neurotransmission will be realized.

Acknowledgements

We thank Donna Klimowicz and Andrea Boissevain for their excellent technical assistance, Pamela Ossorio for the preparation of the figures and Dvorah Ochert for preparing this manuscript.

References

1. Changeux, J.-P. 1981. The Harvey Lectures 75, 85-254.
2. Gershoni, J.M. and G.E. Palade. 1983. Anal.Biochem. 131, 1-15.
3. Gershoni, J.M., E. Hawrot, and T.L. Lentz. 1983. Proc.Natl.Acad.Sci.USA 80, 4973-4977.
4. Wilson, P.T., J.M. Gershoni, E. Hawrot, and T.L. Lentz. 1984. Proc.Natl.Acad.Sci.USA 81, 2553-2557.
5. Hawrot, E., P.T. Wilson, J.M. Gershoni, A.L. Boissevain, and T.L. Lentz. 1984. Neurosci.Abs. 10, 936 (#275.8).
6. Gershoni, J.M., F.E. Davis, and G.E. Palade. 1984. Anal.Biochem. (in press).
7. Gershoni, J.M. and G.E. Palade. 1982. Anal.Biochem. 124, 396-405.
8. Cleveland, D.W., S.G. Fisher, M.W. Kirschner, and U.K. Laemmli. 1977. J.Biol.Chem. 252, 1102-1106.
9. Karlin, A. 1977. Meth. Enzymol. 46, 582-590.
10. Conti-Tronconi, B.M., M.W. Hunkapiller, and M.A. Raftery. 1984. Proc.Natl.Acad.Sci. USA 81, 2631-2634.
11. Lentz, T.L., J.M. Gershoni, P. Wilson, D. Klimowicz, L.M. Hall, and E. Hawrot. 1983. Neurosci.Abs. 9, 159 (#48.5).

III. The Acetylcholine Receptor: Functional Aspects

ACETYLCHOLINE RECEPTOR (FROM *ELECTROPHORUS ELECTRICUS*): A COMPARISON OF SINGLE-CHANNEL CURRENT RECORDINGS AND CHEMICAL KINETIC MEASUREMENTS

George P. Hess*, Hans-Albert Kolb[1], Peter Läuger[1], Ernest Schoffeniels[2], Werner Schwarze[1], Jayant B. Udgaonkar, and Elena B. Pasquale
Section of Biochemistry, 270 Clark Hall, Cornell University, Ithaca, NY 14853 U.S.A.

Summary

Here are described some fast reaction techniques that have been adapted for chemical kinetic investigations of transmembrane processes. First are described the results obtained in studies of the acetylcholine receptor in membrane vesicles prepared from the electroplax of *E. electricus*. It is then shown how the chemical kinetic measurements made with vesicles can be compared with single-channel current measurements made with the electroplax from which the vesicles were prepared. The biological significance of the results obtained so far is summarized. Finally, a new approach, which allows chemical kinetic investigations to be made of receptor function in the cells of the nervous system, is described.

Introduction

As recently as 20 years ago only one biochemical laboratory was interested in the cholinergic system, that of Professor David Nachmansohn at Columbia University. He had discovered the enzyme choline acetylase and was the first to demonstrate that the hydrolysis of ATP can be utilized in the biosynthesis of biologically important compounds like acetylcholine (1). His group then isolated acetylcholinesterase in an

1. Fakultät für Biologie, Universität Konstanz, Germany.
2. Institut Léon Fredericq, Universite de Liège, Belgium.

Molecular Basis of Nerve Activity

highly purified form (2,3). Understanding the mechanism by which this enzyme catalyzes reactions, investigation of its inhibition by organic phosphates (used in nerve gases and insecticides), and the reactivation of the phosphorylated enzyme by designed site-specific reactivators, were major contributions from his laboratory that were of scientific and clinical interest (4,5). In 1953, Nachmansohn suggested that the acetylcholine receptor is a protein different from acetylcholinesterase (6), and subsequently Changeux succeeded in isolating the receptor protein from the electric organ of <u>Electrophorus electricus</u> (7).

Nachmansohn's laboratory was the obvious choice for a biochemist interested in neurobiology, and in 1970 I (GPH) made my first visit there. I had just returned from a sabbatical year at the MRC Laboratory of Molecular Biology in Cambridge where we had worked out the crystal structure of a derivative of a proteolytic enzyme, α-chymotrypsin, (8) that was enzymatically inactive but that still had an intact binding site for small molecules (9). Our hope of using this derivative to investigate the crystal structure of an enzyme:substrate complex was defeated by the packing of the crystal which blocked the substrate-binding site. While I was in Cambridge, David Blow and I completed our respective reviews on the structure and function of this enzyme (10,11) and I felt it was a good time to leave this area of research.

On the way back to Ithaca I talked to Professor Nachmansohn. He was familiar with the fast reaction techniques that had been developed for chemical kinetic studies of protein-mediated reactions in solution (12-15). He, like I, was also impressed by the major contribution these techniques had made to our understanding of the mechanism by which proteins catalyze and regulate the multitudes of reactions involved in the synthesis and degradation of cellular constituents (16-20). Was it possible to use the same approach in investigations of membrane-bound proteins, particularly those that regulate the transmission of signals between cells? Professor Nachmansohn immediately drew my attention to the elegant experiments done by his group with the acetylcholine receptor-containing electroplax of <u>E. electricus</u> (21-24). He then mentioned that Jean-Pierre Changeux had succeeded in preparing acetylcholine receptor-containing vesicles from the electroplax of E. electricus, which allowed one to measure the acetylcholine-induced flux of tracer ions (25).

After many stimulating discussions with David Nachmansohn on neurobiology, and interesting discourses on philosophy, art and politics, we abandoned chemical kinetic measurements made with soluble proteins in solution, and started to work on the receptor-rich vesicles. One of the pleasures we received from this new direction of research was our continuing contact with Professor Nachmansohn and his laboratory, with David's infectious enthusiasm for science and his intellectual approach to the world in general.

I will now briefly describe the approaches we used to adapt fast reaction techniques for chemical kinetic investigations of transmembrane processes, using receptor-rich membrane vesicles prepared from the electroplax of E. electricus or Torpedo californica.

Methods and Techniques

Preparation of vesicles of uniform size and receptor content: Chemical kinetic measurements of receptor-controlled ion translocation in vesicles depend on a preparation in which the receptor site distribution and vesicle diameter are uniform (26,27). We developed a purification procedure that allowed us to separate the vesicles containing acetylcholine receptors from the remaining vesicles in the Kasai and Changeux preparation (25). The principle of the procedure is the equilibration of all the vesicles in the preparation with Cs^+ and then the exchange of Cs^+ for Na^+ in those vesicles that respond to acetylcholine. We could then separate the vesicles that contain receptors and are equilibrated with Na^+ from the vesicles that are filled with Cs^+ and do not contain receptors. A cesium density gradient centrifugation was the last step in this procedure (26,28). The mean diameter and receptor content of the vesicles used in our experiments are 3600 ± 600 Å and 12 receptors respectively.

Development of flow techniques with a time resolution of 5 ms: The Kasai and Changeux preparation (25) has been used for measurements of receptor-controlled flux in many laboratories (for a recent review see (29)) but it was not possible to make measurements much earlier than a few seconds after mixing the vesicles, acetylcholine, and tracer ions. In 1979, we adapted the pulsed quench flow technique, developed for reactions in solutions (30), to measure ion translocation in vesicles using tracer (31,32) and non-radioactive ions (33). The time resolution of these techniques is 5 ms, which is an improvement by a factor of 1000 over previously used methods. The results are shown in Figure 1 a and b. An optical technique was subsequently developed that allows one to measure Cs^+ flux in vesicles (34) (Figure 1c). The use of fluorescence quenching to measure translocation of Tl^+ in vesicles by stopped flow techniques had been introduced earlier by Moore and Raftery (35a). Cs^+ has an advantage over Tl^+ in that in the absence of acetylcholine, influx occurs in the hr time region, whereas Tl^+ influx occurs in the sec time region.

Results

The minimum mechanism in Figure 2 is a summary, in pictorial form, of the results that we obtained using the various fast reaction techniques. In agreement with the electrophysiological measurements of Katz and Thesleff (35b) made over 25 years ago, the minimum model consists of two conformations of the receptor: an active form that can give rise to an open transmembrane channel (A) and an inactive (desensitized) form (I) that cannot.

In the model shown, K_1 and K_2 are the dissociation constants for the binding of acetylcholine to the active and the inactive form of the receptor respectively. Φ is the equilibrium constant for the channel-opening process, and K_R is the dissociation constant of the regulatory site for acetylcholine. The influx rate coefficient contains the intrinsic constant $\bar{J}$. $\bar{J}$ is the specific reaction rate for the receptor-controlled ion translocation process and is directly related to the single-channel conductance, γ (45). R_0 represents the moles of receptor sites per liter internal vesicle volume and is characteristic for a given vesicle preparation.

Now will follow a brief comparison of the conclusions that are based on electrophysiological measurements made with cells to those that are based on chemical kinetic measurements made with vesicles prepared from the same tissue. We observed (37,38) that the affinity of acetylcholine for the active form of the receptor is about 100 times lower than the value

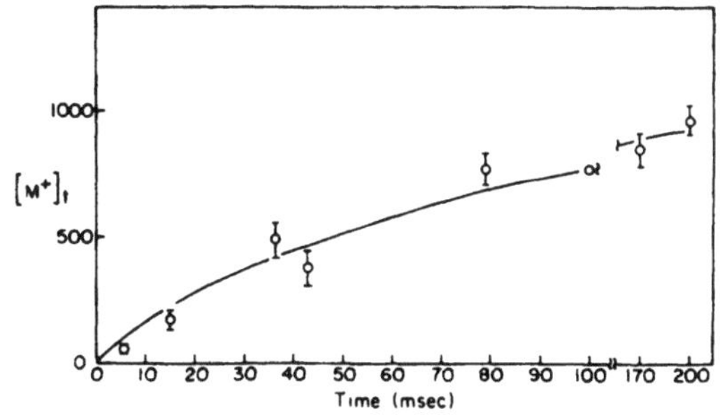

a

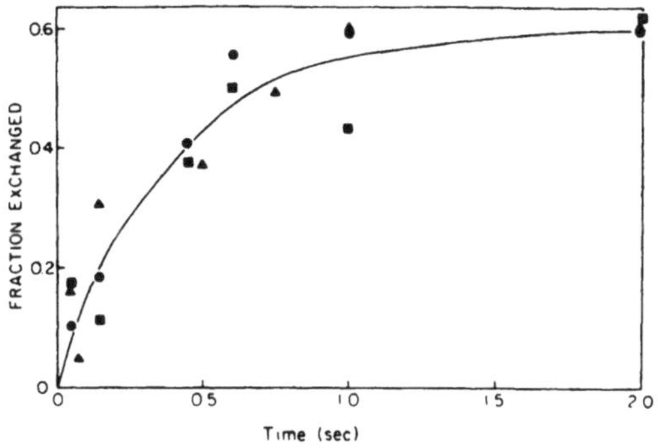

b

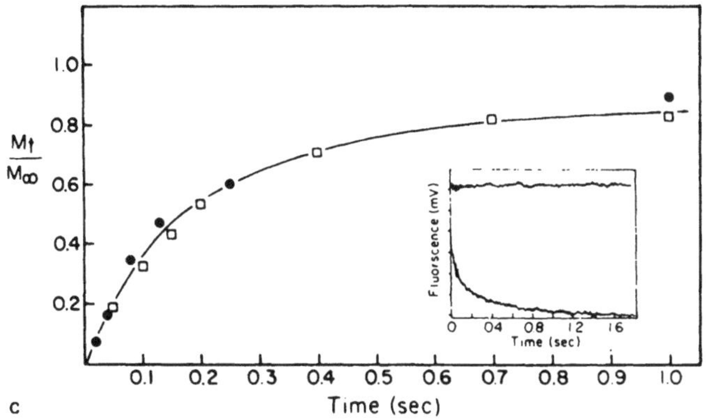

c

Figure 1: Different techniques for measuring ion translocation processes in vesicles using (a) a quench-flow technique and tracer ions (31,32), (b) a quench flow technique and vidicon flame emission spectroscopy to detect Na^+ (O), Li^+ (Δ) and K^+ (□) (33), and (c) a spectrophotometric stopped flow technique that measures Cs^+ flux (34), is shown in the inset. A comparison between results obtained using the stopped flow (□) and quench flow (○) techniques is shown in the main part of this Figure.

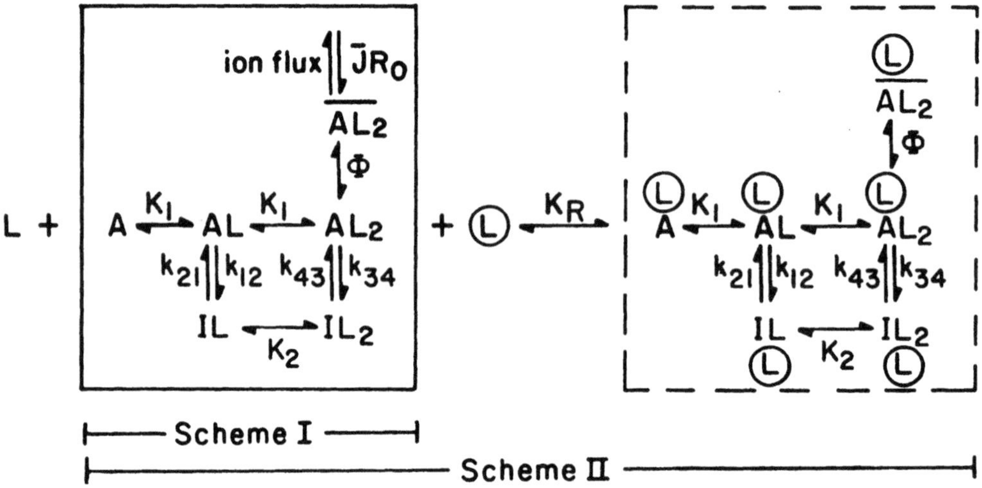

Figure 2: The minimum mechanism for receptor-controlled ion translocation based on chemical kinetic measurements with a time resolution of 5 ms. Membrane vesicles were used and the concentration of acetylcholine, carbamoylcholine, and suberyldicholine were varied over at least a 2000-fold concentration range. L represents ligand and (L) represents ligand bound to the inhibitory site of the receptor, A, AL_1 and AL_2 the receptor species with none, one or two ligand molecules bound, and $\overline{AL}_2$ represents the open-channel form of the receptor. k is the rate constant for interconversions between active A and desensitized I receptor forms. IL and IL_2 represent the inactive form of the receptor with one or two ligand molecules bound. All other constants are defined in the text.

obtained from physiological experiments (39) and comparable to the amount of acetylcholine released between cells. Changes in the concentration of acetylcholine can, therefore, have major effects on the number of receptor sites involved in channel opening and on the transmission of signals between cells. Similarly, chemical kinetic measurements indicate that inactivation of the receptor occurs almost 100 times faster (40) than was deduced from electrophysiological experiments (41,42).

Reactivation occurs in the 200 ms time region. The inactivation-reactivation of the receptor could have an important physiological function, namely the storage of information for short periods of time (36). Chemical kinetic investigations have led to the discovery of a voltage-dependent regulatory site for acetylcholine (43). Depending on the amount of acetylcholine released from the nerve terminal and on the resting potential of the cell, a slight change in transmembrane potential can have a dramatic effect on the acetylcholine receptor molecule that can form transmembrane channels. A change in the transmembrane potential of 10 millivolts can change the number of receptor molecules that can form channels by a factor of two, which may well be sufficient to determine whether or not a signal will be transmitted.

We now know, from single-channel current measurements made with _E. electricus_ electroplax, that the electrophysiological measurements that led to the general belief that if there is any control of the transmission of signals between cells the receptor probably plays no part in it were made with the desensitized receptor.

Single-channel current measurements made with the electroplax: It is possible to compare the results obtained in chemical kinetic measurement made with vesicles with single-channel current measurements made with the cells from which the vesicles were prepared. The relationship between $\bar{J}$, the specific reaction rate obtained from chemical kinetic measurements (44), and γ, the single-channel current conductance, has recently been derived (45). $\bar{J} = \frac{RT}{F^2} \frac{\gamma}{[M]} N_A$, where R, T, and F represent the molar gas constant, absolute temperature, and Faraday constant respectively. γ, N_A and [M] represent the single-channel conductance, Avogadro's Number, and the molar concentration of inorganic ions that can pass through the channel, respectively.

The evaluation of γ has been described in detail (46,47). The evaluation of $\bar{J}$ is based on measurements of the effect of ligand concentration on J_A, the rate coefficient for the receptor-controlled ion translocation in the absence of inactivation (44). The effect of four activating ligands on J_A is shown in Figure 3. At least a 2000-fold concentration range was used with each ligand. It can be seen that at saturating concentrations of activating ligand the maximum observed J_A value, $J_{A(max)}$(obs), for suberyldicholine is about twice that for carbamoylcholine and 3 times larger than that for trans-Bis Q(48). We suggested that the reason for these differences in $J_{A(max)}$(obs) values is due to the equilibrium between the receptor in the closed-channel form with two ligand molecules bound to it, AL_2, and in the open-channel form, also with two ligand molecules bound to it, $\overline{AL}_2$ (Figure 2) (44). It is well known that equilibria between protein conformations can be strongly influenced by the ligand molecules that are bound. We assumed, therefore, that the number of ions that pass through the open receptor channel is independent of the ligand used, but that the equilibrium between the two receptor conformations, corresponding to the closed-channel (AL_2) and open-channel form ($\overline{AL}_2$) (Figure 2), characterized by the equilibrium constant Φ (Figure 3), depends on the type of ligand bound to the receptor. This assumption is in agreement with the results obtained by Colquhoun and colleagues (51), who demonstrated that the single-channel conductance of the receptor in embryonic rat muscle is the same for 10 ligands.

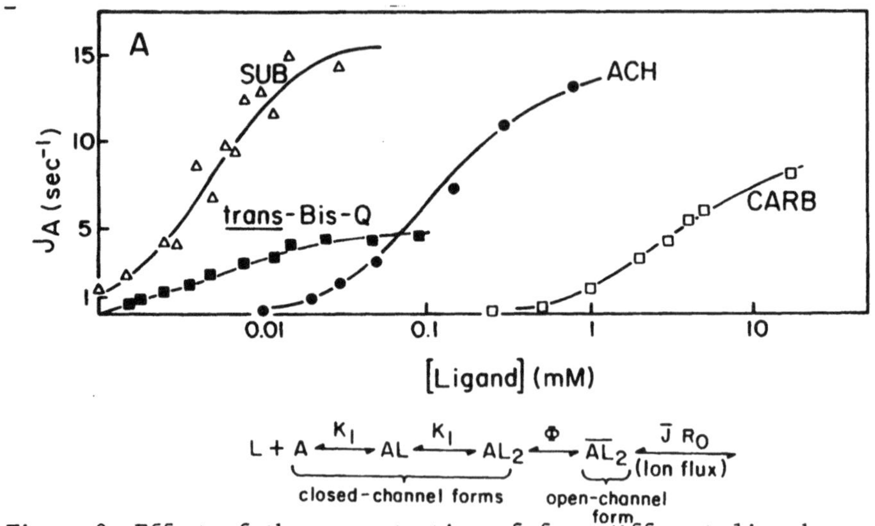

Figure 3: Effect of the concentration of four different ligands on the rate of ion translocation at pH 7.0 and 1° C.

The specific reaction rate, $\bar{J}$, of the receptor-controlled ion translocation process can be calculated from the $J_{A(max)}$ value and a knowledge of R_0, the moles of receptor sites per liter internal vesicle volume (44). Our ability to separate vesicles with functional receptors from those without (26) and to obtain vesicle preparations with uniform distributions of vesicle size and receptor concentration (28) allowed us to determine R_0 and therefore $\bar{J}$ (44), which has a value of $3 \times 10^7\ M^{-1}\ s^{-1}$ at pH 7.0 and 14° C.

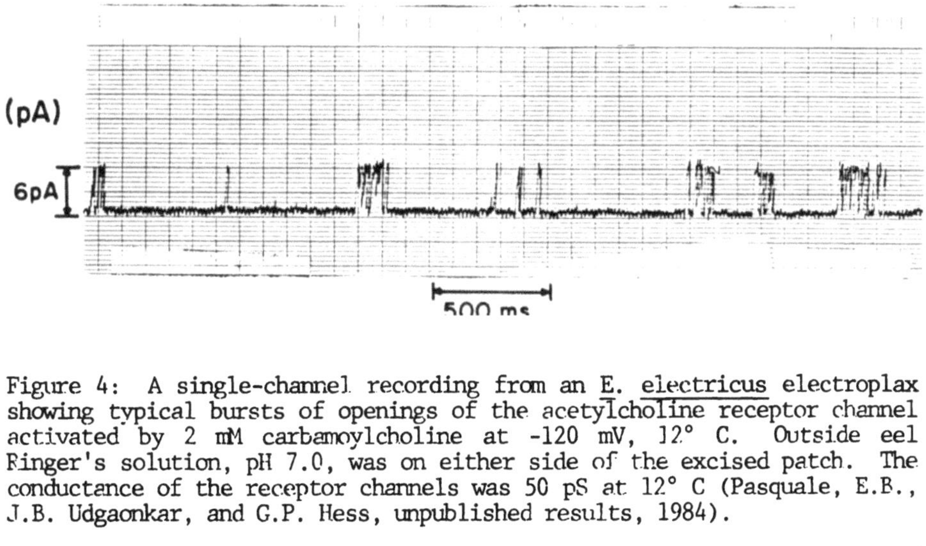

Figure 4: A single-channel recording from an E. electricus electroplax showing typical bursts of openings of the acetylcholine receptor channel activated by 2 mM carbamoylcholine at -120 mV, 12° C. Outside eel Ringer's solution, pH 7.0, was on either side of the excised patch. The conductance of the receptor channels was 50 pS at 12° C (Pasquale, E.B., J.B. Udgaonkar, and G.P. Hess, unpublished results, 1984).

In Figure 4 is a recording from acetylcholine receptor channel in an electroplax cell in eel Ringer's solution, pH 7.0. The acetylcholine receptor from this cell was also used in the chemical kinetic measurements with vesicles. The deviations from the baseline represent the currents passing through a single receptor-formed channel. Short bursts of channel activity are observed and those bursts are separated by much longer intervals of no channel activity. These periods of channel activity and the interval between them have been interpreted as representing the lifetimes of two states of the acetylcholine receptor: an active form that can give rise to transmembrane channels and an inactive state that can not (50). The relationship between the mean lifetime of the burst of channel activity, τ (burst), determined in single-channel current measurements and the rate coefficient for receptor inactivation, α, determined in chemical kinetic measurements with vesicles (40), has been derived (45), where τ^{-1}(burst) = α at high ligand concentrations. The frequency of occurrence distribution of the bursts of channel activity is given in Figure 5. τ^{-1} has a value of 0.1 s corresponding to an α value of 10 s^{-1}.

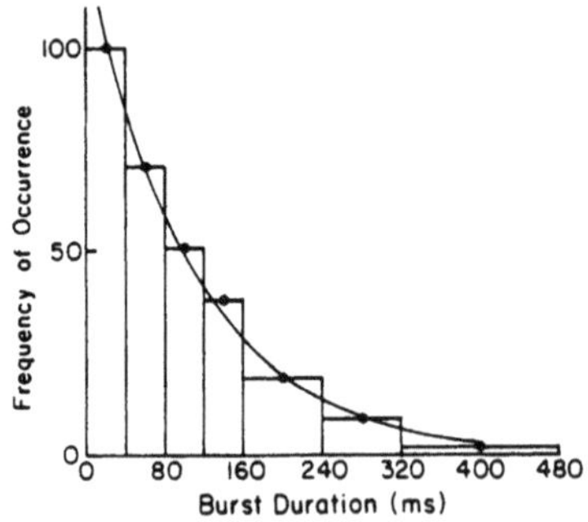

Figure 5: A distribution of burst durations of single-channel currents through the acetylcholine receptor of the _E. electricus_ electroplax activated by 2 mM carbamoylcholine at -80 mV, 12° C. The data was fitted with a single exponential and a mean burst duration of 0.1 s was obtained. Periods of activity were accepted as bursts if they were preceeded and followed by inactive periods whose durations (in this case 50 ms) were more than five times the mean gap duration measured between bursts. Periods of activity were not accepted for analysis if they had higher than unit conductance levels. The mean interval between bursts defined in this way was about 0.2 s (Pasquale, E.B., J.B. Udgaonkar and Hess, G.P., unpublished results, 1984).

A comparison of the values of $\bar{J}$ and α obtained from chemical kinetic measurements made with vesicles and from single-channel current measurements made with the electroplax cell from which the vesicles were prepared is given in Table I. The agreement between the two types of measurements is good. This agreement is important, not only because we are comparing measurements with vesicles and with the cells from which the vesicles were prepared, but also because the techniques are complementary and the assumptions made are quite different. The interpretation of the specific reaction rate, $\bar{J}$, and the rate coefficient for receptor inactivation, α, depend on the model (36). The assumptions made in evaluating the single-channel conductance and the lifetime of various receptor states, obtained from single-channel current measurements made with cells, are quite different (45); for instance,

Table I: Comparison of Chemical Kinetic Measurements and Single-channel Current Measurements, Eel Ringer's Solution, pH 7.0.

Ligands	Values calculated from: Chemical kinetics (Vesicles)	Values calculated from: Single-channel current (Cell)
Acetylcholine, carbamoylcholine, suberyldicholine	$\bar{J} = 3 \times 10^7\ M^{-1}s^{-1}$ (a)	$\bar{J} = 5 \times 10^7\ M^{-1}s^{-1}$ (b)
Carbamoylcholine	$\alpha = 14\ s^{-1}$ (b)	$\alpha = 10\ s^{-1}$ (b)

(a) 1°C (b) 12°C

the assumption that the lifetime of the bursts of channel activity represents the lifetime of the active state of the receptor, A (see Figure 2).

Chemical kinetic measurements of receptor-controlled ion translocation in cells: How can one make the chemical kinetic approach for investigating receptor function generally useful? Chemical kinetics allows one to obtain information about the effect of ligand concentration on receptor-controlled ion flux, receptor inactivation, and receptor inhibition, but at the moment the approach is restricted to receptor-rich vesicles prepared from the electroplax of fish. Single-channel current measurements allow one to obtain information about the single-channel conductance and the lifetimes of receptor states in a great variety of cells, but at the moment, because of receptor desensitization, are restricted to very low concentrations of ligands.

Recently Stühmer and Almers in Seattle (51) have developed a technique that allows one to measure the currents from many channels of a cell, with a time resolution of 50 μs and an excellent signal-to-noise ratio. So far these measurements can be made only with channels that are opened in response to a change in the transmembrane voltage of a cell, for instance, protein channels that are selective for sodium ions. A typical current trace, produced by opening such channels by a change in transmembrane voltage of 40 mV in frog muscle cells, is shown in Figure 6. The technique requires that a number of control measurements are made before the channel currents are measured. Exposing the receptor to acetylcholine during this time would inactivate the receptor and drastically reduce the signal. In order to overcome this problem, compounds are needed that can be converted rapidly from an inactive precursor to an active receptor ligand. In principle, the compound cis-Bis Q developed by Erlanger and his colleagues (48) is such a compound. (See the Chapter by Delcour and Hess in this volume). Table II indicates the wealth of information that can be obtained from integrated current measurements. All the constants, and the inactivation (desensitization) rate, that can be obtained from chemical kinetic measurements (including the apparent dissociation constants of receptor:inhibitor complexes) with vesicles can also be obtained from integrated current measurements with cells. One can also measure the channel-opening and -closing rates, the bimolecular rate constants, and the dissociation rate constant for receptor:ligand complexes. The method should eventually be useful for making measurements with most cells of the nervous system, with brain slices, and with living organisms.

In conclusion, a few experiments have been described to demonstrate that the fast reaction techniques developed for investigation of kinetic processes in solution can also be used to investigate the dynamic properties of membrane-bound proteins involved in translocation of ions and molecules across membranes. As in kinetic measurements in solution, it is possible to know and vary the solution compositions inside and outside

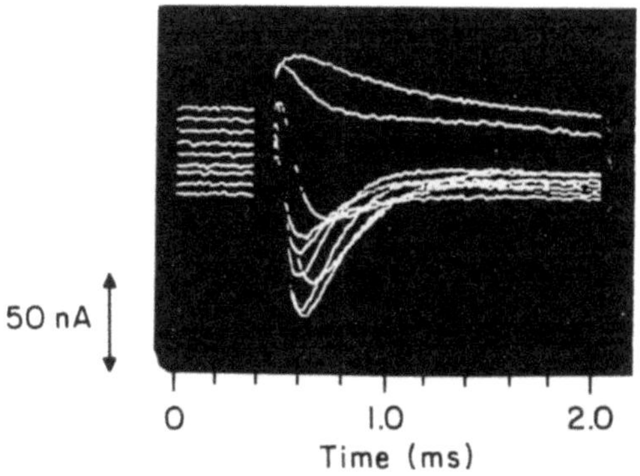

Figure 6: Integrated current measurements made with frog muscle cells, 22° C, pH 7.0 (51).

the vesicles. Similarly, it is possible to know and vary the concentrations of the reactants, namely the receptor and acetylcholine. The precision and time resolution are those obtained previously in investigations of the dynamic properties of proteins in solution. We have described (i) the method used to prepare membrane vesicles of uniform size and receptor content, (ii) the development of a pulsed quench-flow technique with a time resolution of 5 ms, and (iii) the development of a stopped-flow spectrophotometric technique in which we monitor the influx of cesium ions into vesicles. We were able to break down a complex ion translocation process, mediated by the acetylcholine receptor, into individual steps which could be investigated over a wide range of acetylcholine concentration. This has led to a self-consistent model in which the dissociation constants of the active and the inactive forms of the receptor, the channel-opening equilibrium constant, the specific reaction rate, and the rate coefficient for receptor inactivation and reactivation could be determined. We have calculated the effect of acetylcholine concentration on the three kinetic parameters, which we can measure. The calculated and experimental values agree over a 5000-fold concentration range of acetylcholine and over a wide concentration range of suberyldicholine and carbamoylcholine. The inactivation rate coefficient and the specific reaction rate of receptor-controlled ion translocation obtained with kinetic measurements made with vesicles are in excellent agreement with the corresponding parameters that were obtained from single-channel measurements made with the cells from which the vesicles were prepared.

Table II: Chemical Kinetic Measurements with Cells. The Relationship between Integrated Current Measurements and Chemical Kinetic Constants.

I. Time interval < 1 ms

$$A + L \underset{k_{21}}{\overset{k_{12}}{\rightleftharpoons}} AL \underset{k_{32}}{\overset{k_{23}}{\rightleftharpoons}} \overline{AL}_2 \text{(open channel)}$$

(i) Fast step, $AL_2 \rightleftharpoons \overline{AL}_2$: $k_{12} = k_{23},\ k_{21} = k_{32}$

$$k_{obs} = 2(k_{12}S_0 + k_{21})$$

(ii) Slow step, $AL_2 \rightleftharpoons \overline{AL}_2$: $k_{obs} = k_{closed} + k_{open}\left(\frac{L}{L + K_1}\right)^2$

II. Time interval 1 - 10 ms

(i) $$I_{obs} = I_{max} \frac{L^2}{L^2(1 + \Phi) + 2K_1L\Phi + K_1^2\Phi} \cdot \frac{K_R}{K_R + L_0}$$

(ii) $$\frac{I_{obs}}{I_{obs(I_0)}} = 1 + \frac{I_0}{K_I} \text{ (a)}$$

III. Time interval > 10 ms

$$(I_{obs})_t - (I_{obs})_{t=\infty} = [(I_{obs})_{t=0} - (I_{obs})_{t=\infty}]\, e^{-\alpha t}$$

---

a) I_Q represents the molar concentration of a non-competitive inhibitor and I_{obs} and $I_{obs}(K_Q)$ represent currents measured in the absence and presence of an inhibitor respectively.

The following are some of the physiological implications of the results obtained in chemical kinetic measurements. The affinity of acetylcholine for the active form of the receptor is 100 times lower than the value obtained from physiological experiments and is comparable to the amount of acetylcholine released between cells. Therefore, changes in the concentration of acetylcholine can have major effects on the number of receptor sites involved in channel opening and on the transmission of signals between cells. The chemical kinetic measurements indicate that inactivation of the receptor in the ms time region occurs almost 100 times faster than was deduced from electrophysiological experiments. Reactivation occurs in the 200 ms time region. The inactivation-reactivation of the receptor could have an important physiological function, namely the storage of information for short periods of time. Chemical kinetic investigations have led to the discovery of a voltage-dependent regulatory site for acetylcholine. Depending on the amount of acetylcholine released from the nerve terminal and the resting transmembrane potential of the cell, a slight change in transmembrane potential (10 mV) can have a dramatic effect on the receptor molecules that can form transmembrane channels.

One of the new approaches being developed involves the synthesis of compounds that can be photolysed to acetylcholine. We hope to use these compounds to measure integrated currents in cells and that this will allow us to extend chemical kinetic techniques to most types of cells in the nervous system.

References

1. Nachmansohn, D., A.L. Machado. 1943. J. Neurophys. 4, 348.

2. Wilson, I.B., L.T. Kremzner. 1964. Biochemistry 3, 1902.

3. Rosenberry, T.L 1975. Adv. Enzymol. 43, 103.

4. Wilson, I.B., D. Nachmansohn. 1954. In: Ion Transport across Membranes (H.T. Clarke and D. Nachmansohn, eds.). Academic Press. p. 35.

5. Wilson, I.B., S. Ginsburg. 1955. Biochim. Biophys. Acta 18, 168.

6. Nachmansohn, D. 1953. Harvey Lectures 1953-1954 69, 57.

7. Olsen, R., J. Meunier, J.-P. Changeux. 1972. FEBS Lett. 28, 96.

8. Wright, C.S., G.P. Hess, D.M. Blow. 1972. J. Mol. Biol. 63, 295.

9. Davis, L.A., G.P. Hess. 1974. J. Mol. Biol. 82, 27.

10. Blow, D.M. 1971. In: The Enzymes (P. Boyer, ed.) Academic Press. 3, 185.

11. Hess, G.P. 1971. In: The Enzymes (P. Boyer, ed.) Academic Press. 3, 213.

12. Roughton, F.J.W., B. Chance. 1953. In: Techniques of Organic Chemistry (S.L. Friess and A. Weissberger, eds) (1st. ed). Interscience. p. 669.

13. Eigen, M., L. DeMaeyer, 1963. In: Techniques of Organic Chemistry (A. Weissberger, ed.) John Wiley and Sons. 8, 895.

14. Chance, B., R.H. Eisenhardt, Q.H. Gibson, K.K. Longberg-Holm (eds.) 1964. Rapid Mixing and Sampling Techniques in Biochemistry. Academic Press.

15. Gibson, Q.H. 1966. Ann. Rev. Biochem. 35, 435.

16. Monod, J., J. Wyman, J.-P. Changeux. 1965. J. Mol. Biol. 12, 88.

17. Eigen, M. 1967. Nobel. Symp. 5, 333.

18. Koshland, D.E., G. Nemethy, D. Filmer. 1966. Biochemistry 5, 365.

19. Hammes, G.G., C.W. Wu. 1974. Ann. Rev. Biophys. Bioeng. 3, 1.

20. Hammes, G.G. 1982. Enzyme Catalysis and Regulation. Academic Press.

21. Nachmansohn, D., E. Neumann. 1975. Chemical and Molecular Basis of Nerve Activity. Academic Press.

22. Schoffeniels, E., D. Nachmansohn. 1957. Biochim. Biophys. Acta 26, 1.

23. Higman, H.B., T.R. Podleski, E. Bartels. 1963. Biochim. Biophys. Acta. 75, 187.

24. Changeux, J.-P., T.R. Podleski. 1968. Proc. Natl. Acad. Sci. USA 59, 944.

25. Kasai, M., J.-P. Changeux, 1971a,b,c. J. Membr. Biol. 6, 1, 24, 58.

26. Hess, G.P., J.P. Andrews. 1977. Proc. Natl. Acad. Sci. USA 74, 482.

27. Bernhard, J., E. Neumann. 1978. Proc. Natl. Acad. Sci. USA 75, 3756.

28. Sachs, A.B., B. Lenchitz, R.L. Noble, G.P. Hess. 1981. Analyt. Biochem. 115, 34.

29. Changeux, J.-P. 1981. Harvey Lectures 75, 85.

30. Fersht, A.R., R. Jakes. 1975. Biochemistry 14, 3350.

31. Hess, G.P., D.J. Cash, H. Aoshima. 1979. Nature 282, 329.

32. Cash, D.J., G.P. Hess. 1981. Analyt. Biochem. 112, 39.

33. Ramseyer, G.O., G.H. Morrison, H. Aoshima, G.P. Hess. 1981. Analyt. Biochem. 115, 34.

34. Karpen, J.W., A.B. Sachs, D.J. Cash, E.B. Pasquale, G.P. Hess. 1983. Analyt. Biochem. 135, 83.

35a. Katz, B., Thesleff, S. 1957 J. Physiol. 138, 63-80.

35b. Moore, H.-P.H., M.A. Raftery. 1980. Proc. Natl. Acad. Sci. USA 77, 4509.

36. Hess, G.P., D.J. Cash, H. Aoshima. 1983. Ann. Rev. Biochem. Bioeng. 12, 443.

37. Hess, G.P., S. Lipkowitz, G.E. Struve. 1978. Proc. Natl. Acad. Sci. USA 75, 1703.

38. Cash, D.J., G.P. Hess. 1980. Proc. Natl. Acad. Sci. USA 77, 842.

39. Changeux, J.-P., T.R. Podleski. 1968. Proc. Natl. Acad. Sci. USA 59, 944.

40. Aoshima, H., D.J. Cash, G.P. Hess. 1981. Biochemistry 20, 3467.

41. Lester, H.A., J.-P. Changeux, R.E. Sheridan. 1975. J. Gen. Physiol. 65, 797.

42. Del Castillo, J., G.D. Webb. 1977. J. Physiol. (London) 270, 271.

43. Takeyasu, K., J.B. Udgaonkar, G.P. Hess. 1983. Biochemistry 22, 5973.

44. Hess, G.P., H. Aoshima, D.J. Cash, B. Lenchitz. 1981. Proc. Natl. Acad. Sci. USA 78, 1361.

45. Hess, G.P., H.-A. Kolb, P. Läuger, E. Schoffeniels, W. Schwarze. 1984. Proc. Natl. Acad. Sci. USA 24, 5281.

46. Neher, E., B. Sakmann. 1976. Nature 260, 779.

47. Katz, B., R. Miledi. 1972. J. Physiol. 224, 665.

48. Deal, W.J., B.F. Erlanger, D. Nachmansohn. 1969. Proc. Natl. Acad. Sci. USA 64, 1230.

49. Gardner, P., D.C. Ogden, D. Colquhoun. 1984. Nature 309, 160.

50. Sakmann, B., J. Patlak, E. Neher. 1980. Nature 286, 71.

51. Stühmer, W., W.M. Roberts, W. Almers. 1983. In: Single-Channel Recording (ed. B. Sakmann and E. Neher). Plenum Press. p. 123.

PHOTOAFFINITY LABELING OF RECEPTOR STATES WITH MILLISECOND TIME RESOLUTION

Alfred Fahr[&], Leander Lauffer[+], Ferdinand Hucho[+]
+ Institut für Biochemie and & Fachbereich Physik, Freie Universität Berlin, Thielallee 63, 1ooo Berlin 33

Summary:

Photoaffinity labeling of membrane bound nicotinic acetylcholine receptor from Torpedo marmorata electric tissue with the ion channel blocker ^{3}H-TPMP$^+$ reveals various functional states of the receptor protein if labeling is performed with msec time resolution. In the resting and in the activated state most of the label is incorporated into the α-polypeptide chains of the receptor complex. Equilibrated with agonists and antagonists predominantly the δ-polypeptide chain (and to a lesser extent the ß-chain) reacts with the photolabel. Reactivity of the δ-chain increases after exposure to cholinergic effectors with a half-life slower than the kinetics of receptor activation or rapid desensitization.

Agonists and antagonists stimulate photolabeling of the δ-chain with different kinetics: For acetylcholine, carbamoylcholine and suberyldicholine half-life of the reactivity increase is 4oo - 5oo msec, for the antagonists hexamethonium, D-tubocurarine and flaxedil it is about 1o sec. The latter slow kinetics is also observed when the receptor is preequilibrated with agonists or antagonists prior to mixing with ^{3}H-TPMP$^+$ and starting the photoreaction.

We conclude that time resolved photoaffinity labeling can covalently mark protein structures involved in receptor functions.

Molecular Basis of Nerve Activity

Of special interest is the observation that antagonists too induce a conformational change in the receptor protein.

Introduction:

The nicotinic acetylcholine receptor (AChR) comprises two functions: Signal recognition (ligand binding) and ion channel gating. The latter function is thought to be related to conformational changes within the receptor protein. Such changes have been observed through a variety of methods and are summarized in an allosteric model of the AChR (1,2). One of the principal unsolved questions concerning the functional mechanism of the receptor is the following: What is the difference between agonists which bind to the AChR and cause opening of the ion channel, and antagonists, which bind competitively but have no effect on the gating mechanism? We believe the best way to tackle this question is to investigate the gating mechanism itself by means of noncompetitive receptor ligands which have been shown to interact with and to block the ion channel directly (3,4). The interaction of some of these noncompetitive blockers with the AChR ion channel can be a sensitive indicator monitoring structural changes within the channel or its gating device (5-7). Here we use the channel ligand $TPMP^+$ (8) and a rapid photolabeling technique (9,1o) for the covalent labeling of AChR polypeptide chains at different time lapses after mixing the membrane bound receptor with cholinergic effectors. We were especially interested in the kinetics of the antagonist-stimulated photolabeling which we had observed previously (11). The allosteric model (1,2) of the AChR does not include antagonist-induced conformational states; only agonists are considered as capable to activate (open) and to inactivate (desensitize) the receptor and its ion channel.

Results and Discussion:

Previously we showed that the four different polypeptide chains of Torpedo marmorata AChR α,β,γ and δ, react differently in different functional states of the receptor. While in the resting and in the activated states the α-polypeptide chains are the primary target of the photoreaction (9) the reaction site shifts to the δ-chain (and less pronounced to the β-chain) when the receptor is equilibrated with the agonist carbamoyl-choline (11). This shift is time dependent with a half-life of about o.4 sec (1o).

Fig.1 shows a scheme of the experimental setup for rapid photo-affinity labeling with msec time resolution. Basically the apparatus is a combination of a stopped-flow apparatus with a high energy pulse laser which can be triggered at various time intervals after mixing e.g. AChR-rich membranes and ^{3}H-$TPMP^{+}$ with agonist or antagonist. Details of the experimental design are published elsewhere (12).

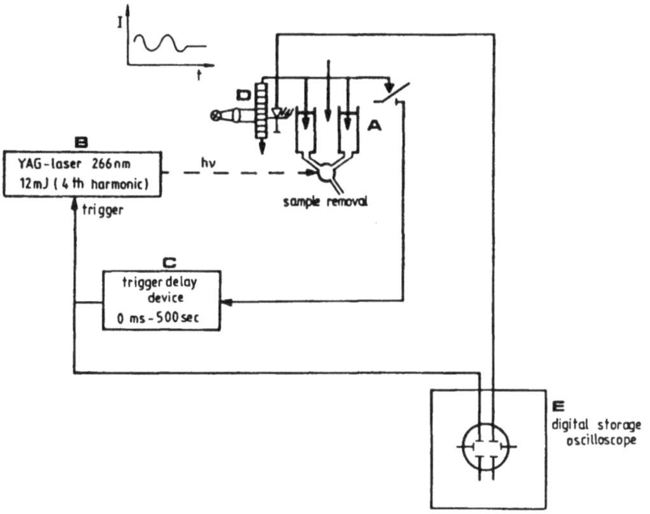

Fig.1: Scheme of the stopped-flow laser flash photoaffinity apparatus. (A) Stopped-flow apparatus; (B) high energy laser; (C) trigger delay device; (D) Optoelectronic device for monitoring the velocity of driving pistons and the time lapse between stop switch and actual stop; (E) digital storage oscilloscope.

Fig.2A shows that acetylcholine acts similar to carbamoylcholine (9) with respect to the stimulation of δ-labeling (1o). At about 5oo msec after rapid mixing of AChR-rich membranes with ^3H-$TPMP^+$ and acetylcholine δ-labeling is maximal. Antagonists too cause this change in the labeling pattern (11). Kinetics in the presence of the antagonist hexamethonium however are quite different (Fig.2 B): Here, the δ-polypeptide chain becomes labeled only several seconds after mixing. Note that photolabeling after prolonged mixing results in decreased amounts of radioactivity incorporated into the α-polypeptide chain (Fig. 2 A and B, trace 9).

The α-polypeptide chain becomes labeled immediately after mixing (dead time of the stopped-flow apparatus was 2.4 msec) in both cases. Interestingly in the presence of hexamethonium the peripheral receptor-associated α'-peptide is labeled as well. The α'-peptide is located on the inside of the AChR-rich vesicular membrane of the receptor complex (1). Its labeling therefore is not surprising since $TPMP^+$ is a lipophilic cation known to be able to cross membranes.

In order to evaluate the results in a more quantitative manner, in the following experiments we excised the bands of the coomassie-stained gels corresponding to the α- and δ-polypeptide chains and determined their radioactivity by liquid scintillation counting.

Carbamoylcholine if mixed simultaneously with AChR-rich membranes and ^3H-$TPMP^+$ (Fig.3 A,B) promotes the δ-reactive state in a similar manner as acetylcholine (completion of the reaction after about 1o sec, $t_{1/2}$= 45o msec). In contrast, after preincubation with carbamoylcholine, a different kinetic pattern emerges (completion after about 1oo sec, $t_{1/2}$ = 9.5 sec) (Fig 3 A,B). As observed before (cf.Fig.2 and ref.1o), the amount of label incorporated in the α-polypeptide chain re-

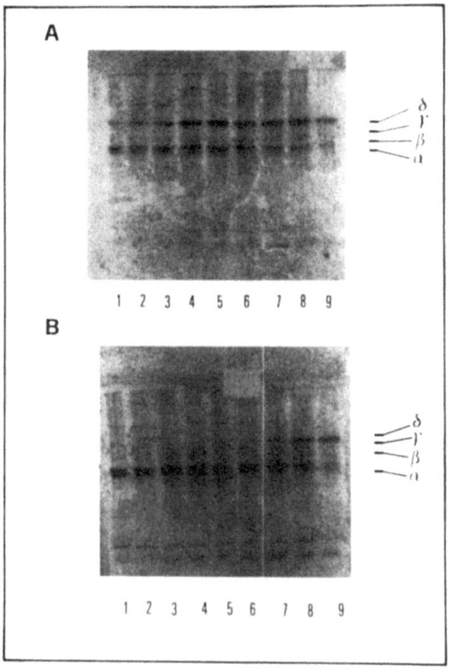

Fig.2: Time resolved photoaffinity labeling of AChR stimulated by an agonist (A) or a competitive antagonist(B). The autoradiograms of SDS-polyacrylamide gel electrophoresis show AChR-polypeptide chains photoaffinity-labeled with the noncompetitive antagonist ^{3}H-TPMP$^+$ at different time lapses after mixing AChR-rich membranes with a mixture of ^{3}H-TPMP$^+$ and acetylcholine or hexamethonium, respectively. After mixing in a stopped-flow apparatus the laser was triggered after (1) 8msec, (2) 23 msec, (3) 5o msec, (4) 2oo msec, (5) 5oo msec, (6) 2 x 1o^3 msec, (7) 5 x 1o^3 msec, (8) 5o x 1o^3 msec, (9) 33o x 1o^3 msec. In the experiment (B) the time lapses were identical as in (A). In the stopped-flow apparatus the one syringe contained receptor -rich membrane (1mg/ml) in Ringer solution; the other syringe contained ^{3}H-TPMP$^+$ (o.4 μM) and (A) acetylcholine and (B) hexamethonium, respectively, at o.1 mM. The irradiated volume was 22 μl, the dead time of the stopped-flow apparatus was 2.4 msec, irradiation time (duration of 1 pulse of the laser) was about 4nsec with a pulse energy of 15 mJ at 266 nm wavelength. The experiment was performed at room temperature. After irradiation the sample was removed from the mixing chamber and analysed by SDS-polyacrylamide gel electrophoresis as described in the Materials and Methods section.

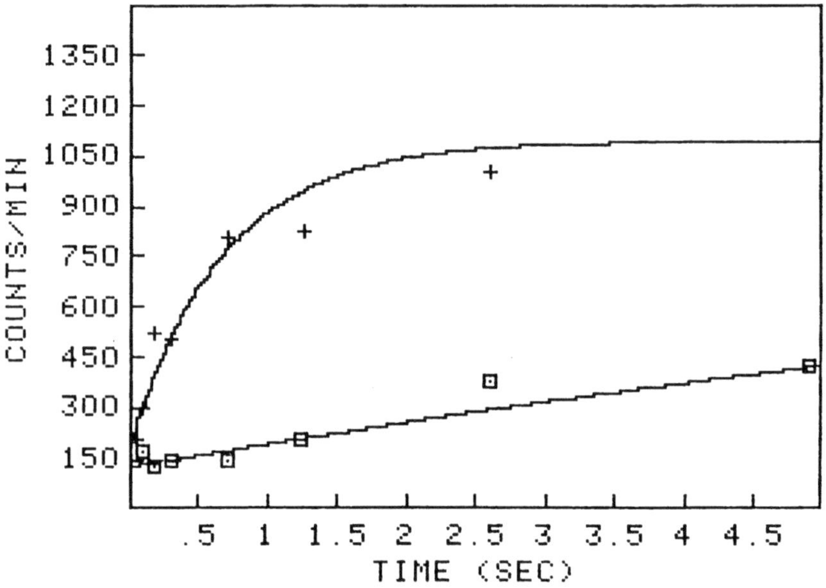
1350
1200
1050
900
750
600
450
300
150
COUNTS/MIN
.5
1
1.5
2
2.5
3
3.5
4
4.5
TIME (SEC)

3A

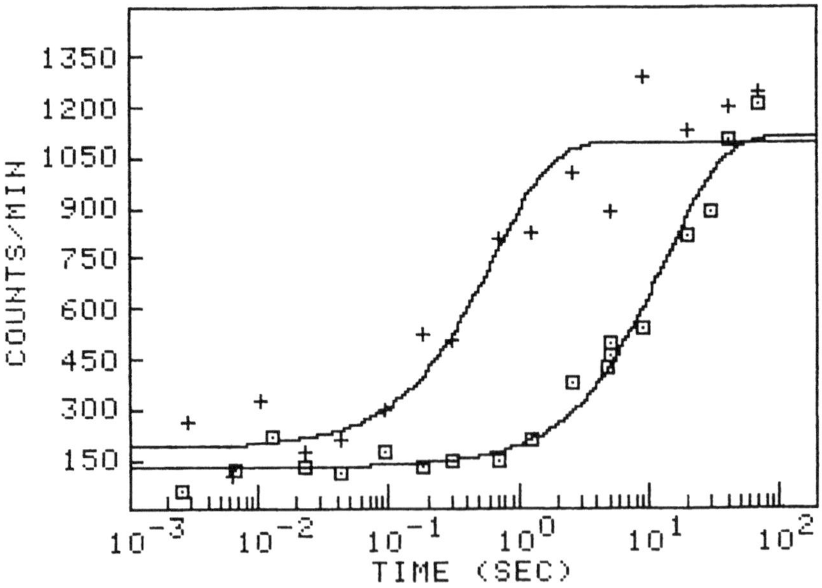
1350
1200
1050
900
750
600
450
300
150
COUNTS/MIN
10^{-3}
10^{-2}
10^{-1}
10^{0}
10^{1}
10^{2}
TIME (SEC)

3B

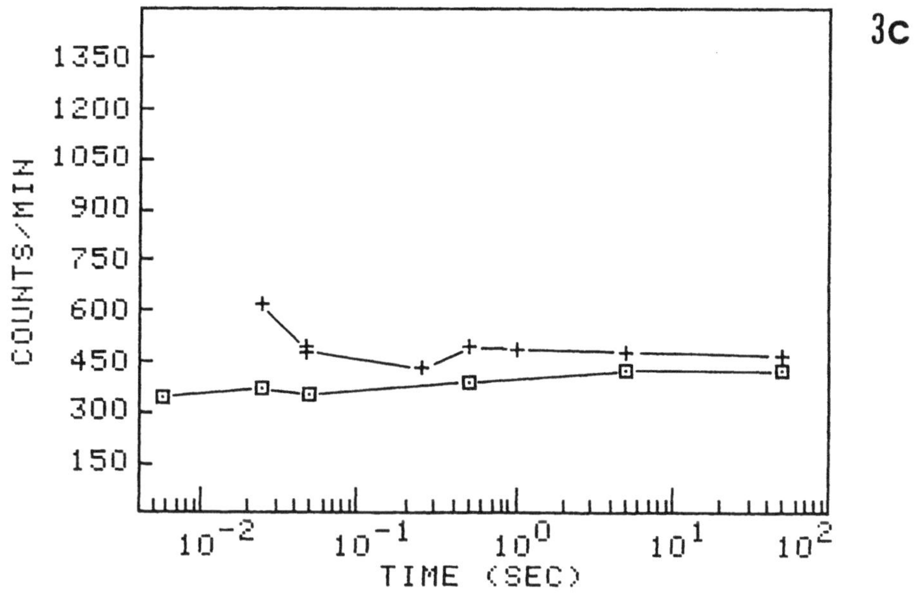

Fig.3: Time resolved photoaffinity labeling of AChR with ^3H-$TPMP^+$ stimulated by carbamoylcholine. (A) and (B) labeling of the δ-polypeptide chain, (C) labeling of the α-polypeptide chain. + - + - + o.1 mM carbamoylcholine added to receptor-rich membranes simultaneously with the photolabel; □ - □ - □ receptor-rich membranes preincubated with o.1 mM carbamoylcholine for about 1o min, then mixed in the stopped-flow apparatus with ^3H-$TPMP^+$/carbamoylcholine. (A) and (B) show the same experiment, plotted with linear and with logarithmic time scale, respectively, to visualize the kinetics in the different time ranges. For the other experimental details see legend to Fig.1.

mains more or less the same or decreases slightly (Fig.3 C).

Another agonist, suberyldicholine, has the same effect as carbamoylcholine: fast kinetics after simultaneous mixing, slow kinetics after preincubation (Fig.4).

Competitive antagonists also stimulated the photoreaction with the δ-polypeptide chain. With all antagonists tested, i.e. hexamethonium (Fig.2B), flaxedil (Fig.5) and D-tubocurarine (Fig.6) we found no fast stimulation of δ-labeling irrespectively if the membrane suspension was mixed simultaneously

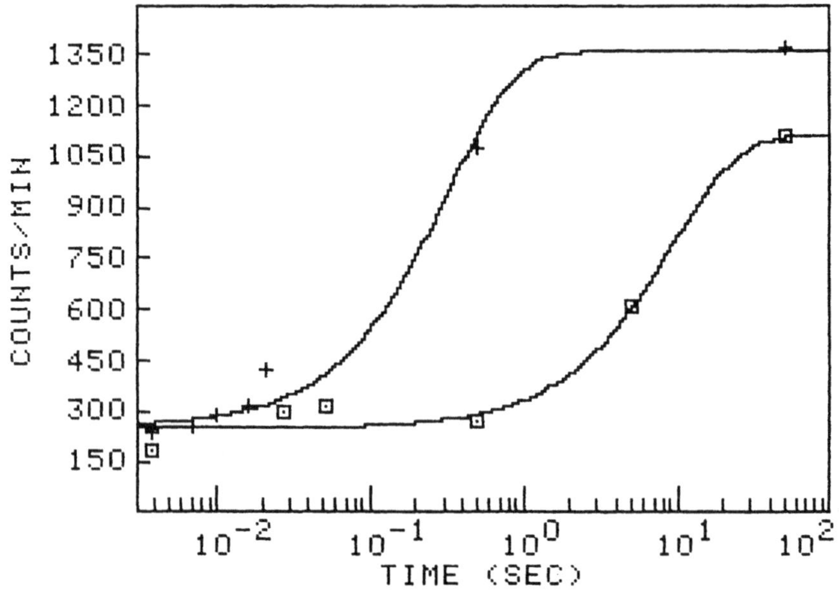

Fig.4: Time resolved photoaffinity labeling of AChR with ^{3}H-TPMP^{+} (o.75 μM) stimulated by suberyldicholine (o.1 mM), δ-polypeptide chain. + - + - + agonist and photolabel added simultaneously, □-□-□ receptor-rich membranes preincubated with the agonist. For the other experimental details see legend to Fig.1. No quantitative evalutation of the rate constants was attempted because of the scarcity of data points.

with antagonist and ^{3}H-TPMP^{+} or had been preincubated with antagonist before mixing. The kinetics in the presence of antagonists differs slightly from the one obtained after preincubation with agonists. The calculated time constant of the former is 11 sec, of the latter 14 sec (carbamoylcholine).

The experimental data were evaluated assuming a single exponential rate equation. Evaluation with two exponentials would fit the data only slightly better. Therefore with the number of data points available we cannot decide whether the reaction is (pseudo) first order or bimolecular. In our experiments concentration of bungarotoxin binding sites were chosen to be

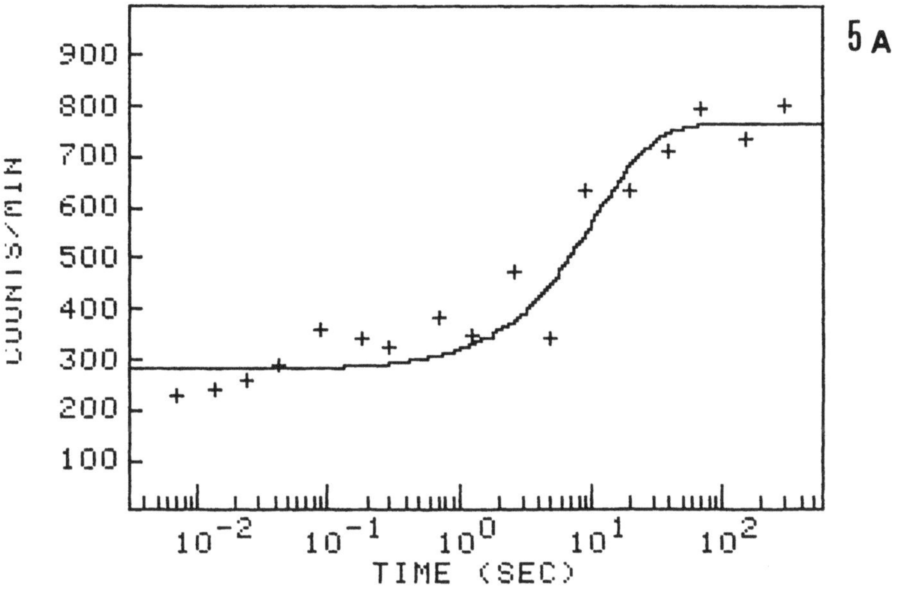
5 A
900
800
700
600
500
400
300
200
100
COUNTS/MIN
10^{-2}
10^{-1}
10^{0}
10^{1}
10^{2}
TIME (SEC)

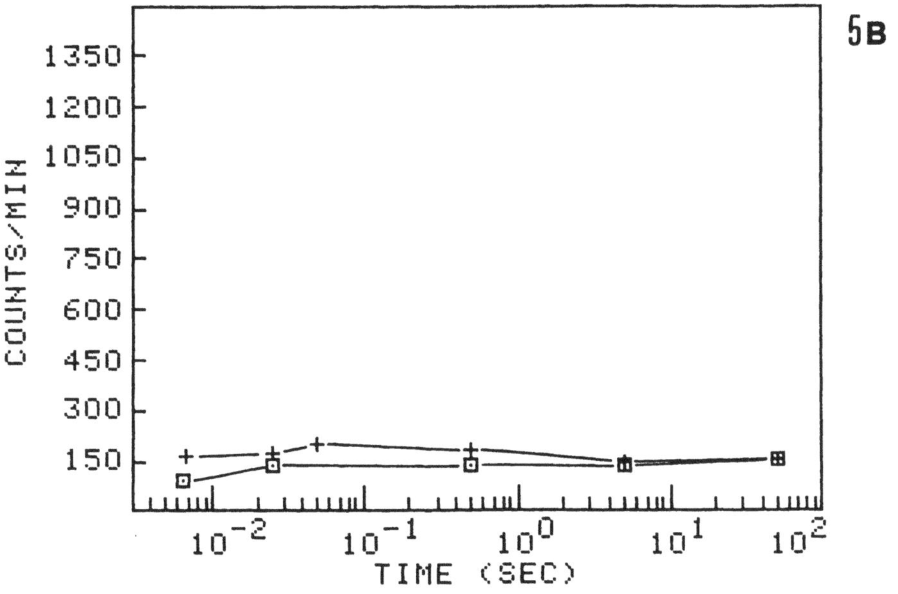
5B
1350
1200
1050
900
750
600
450
300
150
COUNTS/MIN
10^{-2}
10^{-1}
10^{0}
10^{1}
10^{2}
TIME (SEC)

Fig.5: Antagonist stimulated photoaffinity labeling of AChR. (A) δ-polypeptide chain (B) α-polypeptide chain; + - + - + flaxedil (o.1 mM) and ^{3}H-TPMP^{+} (o.7 μM) added simultaneously, ◻-◻-◻ receptor-rich membranes preincubated with o.1 mM flaxedil. For the other experimental conditions see legend to Fig.1.

twentyfold in excess over the ^{3}H-TPMP^{+} concentration to favour a pseudo first order reaction mechanism.

Rapid mixing of AChR-rich membranes with ^{3}H-TPMP^{+} alone and laser-flash photolabeling at different times after mixing resulted in no significant changes of the labeling pattern. Similarly preincubation of membranes with antagonist prior to rapid mixing with ^{3}H-TPMP^{+} and time dependent photolabeling yielded no unequivocal kinetics.

The allosteric model (1) postulated the following gating mechanism for the AChR-coupled ion channel: A preformed equilibrium between receptor molecules in the resting state (channel closed) and the active state (channel open) is shifted to the latter by preferential binding of agonists to the active state. Competitive antagonists according to this model stabilize the receptor in the resting state by preferential - though not necessarily exclusive - binding to protein molecules existing in this state. The model is completed by two desensitized states (channel closed, affinity high) which also bind preferentially agonists. This four state model (13) describes AChR activity as a series of conformational changes brought about by agonist molecules. Competitive antagonists stabilize the receptor protein in its resting state, i.e. they are not thought to induce conformational changes.

From our eyperiments (Fig 2 A, 3 A/B) it is evident that the channel blocker ^{3}H-TPMP^{+} applied as a photoaffinity label can be used to monitor structural changes within the receptor-protein brought about by agonists as e.g. acetylcholine, car-

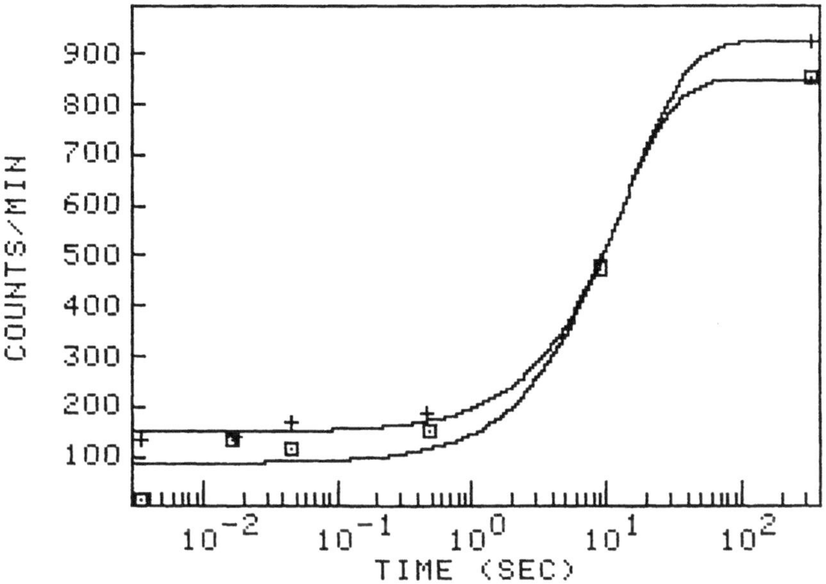

Fig.6: Antagonist stimulated photoaffinity labeling of AChR (δ-polypeptide chain) + - + - + D-tubocurarine (o.1 mM) and ^{3}H-TPMP$^+$ (o.75 μM) added simultaneously, □-□-□ receptor-rich membranes preincubated with o.1 mM D-tubocurarine. For the other experimental conditions see legend to Fig.1.

bamoylcholine and suberyldicholine. But figures 2 B, 5 A/B, 6, prove that competitive antagonists also can induce structural changes monitored by an increase in reactivity towards the photolabel. The rate of the reactivity increase is different for agonists ($t_{1/2}$ = o.5 sec) and competitive antagonists ($t_{1/2}$ = 1o sec); both velocities are much slower than gating of the channel (completed for the individual channel within 2o μsec) and channel open times (1-5 msec). In both the resting and activated state it is actually the α-polypeptide chain which incorporates the photolabel covalently. It has to be shown by finger printing whether or not the same site within the primary structure of the α-chain is labeled in the two states.

For interpreting the kinetics of the changes in photolabeling the following possibilities have to be considered: (i) The slow increase in reactivity of the δ-polypeptide chain reflects a slow conformational change, or (ii) it monitors slow reversible binding of $^{3}H\text{-}TPMP^{+}$ preceding the covalent photoreaction.

The first interpretation, assuming a slow conformational change as the reason for the reactivity increase is probably applicable for the effect of simultaneously adding agonist and $^{3}H\text{-}TPMP^{+}$ to the AChR-rich membranes. The kinetics of the effect ($t_{1/2}$ = 5oo msec) is slightly slower than rapid and faster than slow desensitization having a half-life of 1oo msec and seconds to minutes, respectively (1).

The second interpretation, slow reversible binding of the photolabel, is unlikely to be the cause for the $t_{1/2}$ = 5oo msec increase in photoreactivity of the δ-chain induced by agonists which have been mixed with AChR-rich membranes simultaneously with $^{3}H\text{-}TPMP^{+}$. In Li^{+}flux experiments we have shown previously (8) that $^{3}H\text{-}TPMP^{+}$ blocks the ion channel immediately when added to Li^{+}-loaded AChR-rich membrane vesicles together with the agonist. But the even slower ($t_{1/2}$ = 1o sec) reactivity increase observed after preincubating the receptor-rich membranes with agonists could well be caused by slow binding of $^{3}H\text{-}TPMP^{+}$ to a high affinity site (8) the access to which is sterically hindered. A similar slow binding has been observed with other noncompetitive antagonists (14). It was interpreted as a property of the receptor in its desensitized state. We consider this interpretation as unlikely because not only the desensitizing agonists but also non-desensitizing antagonists stimulate photolabeling with similar kinetics ($t_{1/2}$ = 1o sec).

The rate of the antagonist-stimulated photolabeling is similar to the rate of slow desensitization. It appears to be of the same order of magnitude as the rate of photolabeling obtained with AChR-rich membranes which have been equilibrated (i.e. desensitized) with agonist prior to photolabeling. But the time constants are not quite identical (11 sec vs. 14 sec). Furthermore competitive antagonists do not increase the receptor's affinity for either agonists or antagonists, an effect characteristic for desensitization, with the one exception of hexamethonium (15).

When AChR-rich membranes are preincubated with agonists or competitive antagonists the structural changes induced by these effectors are already completed when the photolabel ^3H-$TPMP^+$ is added in the stopped-flow mixing chamber. The slow increase of reactivity of the δ-chain observed in this experiment is best explained by a shift of the rapidly and reversibly bound ^3H-$TPMP^+$ to a site of higher affinity from where it undergoes the photoreaction. $TPMP^+$ exerts its channel-blocking effect when bound rapidly to the ion channel (or to a regulatory site involved in gating the channel). The function of the slowly accessible site to which it is directed by agonists and antagonists, is at present unknown, nor is it clear that this site is the same in the case of agonists or antagonists.

The allosteric four-state model is proposed as a minimum model describing AChR functions satisfactorily (1,13). Alternative models have been proposed (for a recent review see ref. 16), but they all may be too simple especially with respect to AChR inactivation and competitive antagonists. The latter quite obviously are not simply ligands of the resting AChR with no other effect than to stabilize the protein in this state. Not only has it been observed that certain competitive antagonists as e.g. D-tubocurarine can simultaneously directly block the ion channel (17) (D-tubocurarine can even open ion channels (18)), others termed metaphilic (19,2o) have been shown to bind to the desensitized state too (21). But also flaxedil, a pure competitive antagonist with no desensitizing (15) properties causes an albeit slow conformational change as monitored by various methods (22,23). Here we confirm this notion and provide a method for further investigating this "antagonist state" by labeling covalently a site on the receptor protein obviously characteristic for this state; the antagonist induced conformation is probably not identical with the desensitized state. At least if influenced by flaxedil it is lacking one of the two characteristics of the desensitized state, the high affinity for agonists (17,24) (the other characteristic being the closed channel). By localising the label in the primary structure it will be possible to compare the slowly desensitized and the slowly antagonist-induced AChR-conformation.

At present only the α-neurotoxins from snake venoms appear to comply to the definition of a pure competitive antagonist. All others obviously do not bind exclusively to AChR in its resting state but interact with the open-channel and desensitized states as well. α-bungarotoxin indeed was the only competitive antagonist tested so far causing no shift in the labeling pattern under our experimental conditions.

We tend to consider the α-neurotoxins the exception rather than the rule. The rule appears to be that competitive antagonists are not specific for the resting state of the receptor but exhibit complex binding and cause conformational changes of the receptor protein.

Agonists appear to bind to the receptor very rapidly, probably in a diffusion controlled process (16), and antagonists bind and act on the receptor with similar velocities. The slowly occuring antagonist state observed here undoubtly is too slow to be responsible for the blocking action of competitive antagonists. Its relationship to the physiological fast processes and the slow desensitization remains to be determined.

Conclusion:

Rapid photoaffinity labeling of AChR has been proposed recently by several laboratories (1o,26,27). Only the laser-flash method (12) provides a time resolution comparable to physiological events. The kinetic analysis presented here shows that at least three receptor states can be discriminated: Either the α-chain is the target of the photoreaction (i), or the δ-chain becomes accessible with a $t_{1/2}$ = 5oo msec (ii), or finally the δ-chain becomes accessible with a $t_{1/2}$ of the order of seconds (iii). (i) comprises the resting and the activated (channel open) state, (ii) represents an agonist-induced state, and (iii) is induced by competitive antagonists and by desensitizing conditions. By chosing the appropriate time scale and in combination with finger print analysis detailed insights into dynamic processes involving the ion channel of the receptor are to be expected.

Acknowledgement:

We thank P.Muhn for many helpful discussions and for his help in planning the experiments shown in Fig 2 and 4. Excellent technical assistance by G.Bandini was an essential part of our experiments. Financial support by the Land Berlin, FGS-Biomembranen, the Deutsche Forschungsgemeinschaft and the Fonds der Chemischen Industrie is gratefully acknowledged.

References:

1. Changeux,J.-P. (1981), Harvey Lect.75, 85-254

2. Changeux,J.-P., Bon,F.,Cartaud,J., Devillers-Thiery,A., Giraudet,G.S.H., Heidmann,Th., Holton,B., Nighiem,H.-O., Popot,J.-L., van Rapenbusch,R. & Tsartos,S. (1983), Cold Spring Harbor Symp. Quant.Biol.48, 35-52.

3. Steinbach,A.B. (1968), J.Gen.Physiol.52, 144-161

4. Cohen,J.B., Weber,M. & Changeux,J.P.(1974), Mol.Pharmacol. 1o, 9o4-932.

5. Eldefrawi,M.E., Aronstam,R.S., Bakry,N.M., Eldefrawi,A. & Albuquerque,E.X. (198o), Proc.Natl.Acad.Sci.USA 77, 23o9-2313.

6. Eldefrawi,M.E., Eldefrawi,A.T., Aronstam,R.S., Maleque, M.A., Warnick,J.E. & Albuquerque,E.X. (198o), Proc.Natl. Acad.Sci.USA 77, 7458-7462.

7. Oswald,R.E., Heidmann,Th.,& Changeux,J.P. (1983), Biochemistry 22, 3128-3136.

8. Lauffer,L. & Hucho,F. (1982), Proc.Natl.Acad.Sci.USA 79, 24o6-24o9

9. Muhn,P.,Fahr,A. & Hucho,F. (1984) Biochemistry 23, 2725-273o

1o. Muhn,P., Fahr,A., & Hucho,F.(1984) FEBS Letters 166, 146-15o

11. Muhn,P. & Hucho,F., (1983) Biochemistry 22, 421-425

12. Fahr,A. & Hucho,F. (1984), Anal. Biochem. submitted for publication.

13. Neubig,R. & Cohen,J.B. (198o), Biochemistry 19, 277o-2775.

14. Oswald,R.E., Heidmann,T.,& Changeux,J.-P. (1983), Biochemistry 22, 3128-3136.

15. Weber,M., David-Pfeuty,Th & Changeux,J.-P. (1975),Proc. Natl.Acad.Sci.USA 72, 3443-3447.

16. Maelicke,A. (1984), Angew.Chemie 96, 193-219.

17. Marty,A., Neild,T., & Asher,P. (1976), Nature 261, 5o1-5o3

18. Trautmann,A. (1982), Nature 298, 272-275.

19. Rang,H. & Ritter,J.M.(197o), Mol.Pharmacol.6, 357-382.

2o. Rang.H & Ritter,J.M. (197o), Mol.Pharmacol.6, 383-39o

21. Weiland,G.,& Taylor,P. (1979), Mol.Pharmacol.15, 197-212

22. Grünhagen,H. & Changeux,J.-P.(1976), J.Mol.Biol.1o6, 517-535.

23. Oswald,R.E.& Changeux,J.-P. (1981), Biochemistry 2o, 7166-7174

24. Quast,U., Schimerlik,M.,Lee,T., Witzemann,V., Blanchard, S. & Raftery,M.A. (1978), Biochemistry 17, 24o5-2414

25. Prinz,H.& Maelicke,A. (1983), J.Biol.Chem.258, 1o273-1o282

26. Cox,R.N.,Kaldany,R.-R., Brandt,P.W., Ferren,B.,Hudson,R.A. & Karlin,A. (1984), Anal.Biochem.136, 476-486.

27. Heidmann,T. & Changeux,J.-P. (1984), Proc.Natl.Acad.Sci. USA 81, 1897-19o1.

BINDING OF [^{3}H]ACETYLCHOLINE TO ACETYLCHOLINE RECEPTOR FROM TORPEDO CALIFORNICA UNDER EQUILIBRIUM CONDITIONS: STOICHIOMETRY AND EVIDENCE FOR LONG-LIVED METASTABLE STATES.

H.W. Chang, E. Bock
Department of Neurology, Columbia University, New York, NY, USA

E. Neumann
Department of Biophysical Chemistry, University of Bielefeld, Bielefeld 1, D-4800, F.R. Germany

Introduction

The binding of acetylcholine (AcCh) by the acetylcholine receptor (AcChR) *in vivo* is the primary event that leads to the increased membrane permeability since it triggers the opening of transmembrane ion channels. A number of recent studies (see a recent review 1) suggest that this channel opening is associated with a conformational change of the AcChR as proposed by Nachmansohn as early as 1953(2). In *in vitro* experiments, however, prolonged exposure of the AcChR to AcCh or other agonists causes a gradual loss of the permeability response to agonists. This phenomenon is referred to as pharmacological desensitization. Recent biochemical studies with *Torpedo* membrane fragments suggest that while receptor activation is transient in nature and associated with lower-affinity for AcCh (K_d>1 μM) (3), desensitization is correlated with an increase in agonist affinity insofar as both of these latter phenomena follow qualitatively the same time course (4-9). Furthermore, evidence for a putative "pre-desensitized" state (10-11) is concurrent with more recent electrophysiological evidence that desensitization develops in two phases: an ini-

Molecular Basis of Nerve Activity

tial fast phase (subsecond) followed by a slower phase which extends over tens of seconds to minutes (12-15). The recovery from desensitization was also found to display both a fast and a slow exponential phase with time constant of 12 sec and 4-5 min, regardless of the concentration or nature of the agonist. The fast and slow phases of desensitization onset and offset have been reported not to be due to independent causes but rather to be coupled (14). Thus the classical two-state cyclic scheme of desensitization scheme of Katz and Thesleff (15), would appear to require serious modifications. At present neither the mechanism of desensitization internally coupled to the complex receptor functional system nor its physiological significance in vivo is understood.

In binding studies under equilibrium conditions, where the receptor is exposed to the agonist for prolonged periods, the data represents the high-affinity binding properties of either a preexisting desensitized form or of AcChR "desensitized" through isomerization during the binding. In spite of recent advances (see recent reviews, 1, 17-20), there is still a great deal of disagreement about the equilibrium AcCh binding properties of isolated and even of membrane-bound AcChR. In direct binding studies of $[^3H]$ AcCh with receptor-rich membrane fragments, overall equilibrium constants, $\bar{K}$ from $\sim$10 to 40 nM, either with positive cooperativity (20-24) or without cooperativity (25-27) have been reported. While some laboratories report one high-affinity AcCh binding site per α-toxin site (20-23, 26), others have found a ratio of 0.5 AcCh sites per α-toxin site (25, 27). In this paper we present evidence for the involvement of long-lived metastable states in the high -affinity binding of AcCh to the AcChR and we describe the effect of nonionic detergents on the apparent number of α-bungarotoxin (α-Bgtx) sites which may explain many of the aforementioned discrepancies (28).

Methods

AcChR-rich membrane fragments. The method for preparing membrane fragments from electric organ of *Torpedo californica* was essentially that of Sobel et al (29) and has been described elsewhere (28). The number of [^{125}I]α-Bgtx binding sites was determined by the DE-81 filter disk method as previously described (28, 30). Residual acetylcholinesterase activity of samples for [^{3}H]AcCh binding studies was inhibited as described in (28).

Modes of introducing [^{3}H]AcCh in binding studies. The following three modes of introducing [^{3}H]AcCh to receptor samples were used: Dialysis mode. [^{3}H]AcCh is introduced to the sample contained in a dialysis bag by slow diffusion through the membrane from a large pool of almost constant [^{3}H]AcCh concentration. From the point of view of the AcChR, the dialysis mode implies an open-system with respect to [^{3}H]AcCh. The true free concentration of [^{3}H]AcCh inside the dialysis bag at the end of the dialysis time was determined by the ultracentrifugation method (28). For binding studies with detergent solubilized receptor, the free concentration of [^{3}H]AcCh was determined from the counts in the filtrate derived from the dialysis bag content after 10 min centrifugation at 2000 g using the Centrifree (Amicon) micropartition system; Pulse-mode. Aliquots of [^{3}H]AcCh solution are directly added to aliquots of membrane fragments and rapidly mixed by vortexing. The pulse-mode implies a closed-system with respect to [^{3}H]-AcCh. The free ligand concentration was determined by the ultracentrifugation method with appropriate corrections for radioimpurities as described previously (28); Combination-mode. [^{3}H]AcCh is added to the membrane fragments by pulse-mode and this mixture is dialyzed against either a buffer containing the same free [^{3}H]AcCh concentration found in each sample or against plain buffer to yield a much lower final free [^{3}H]AcCh concentration.

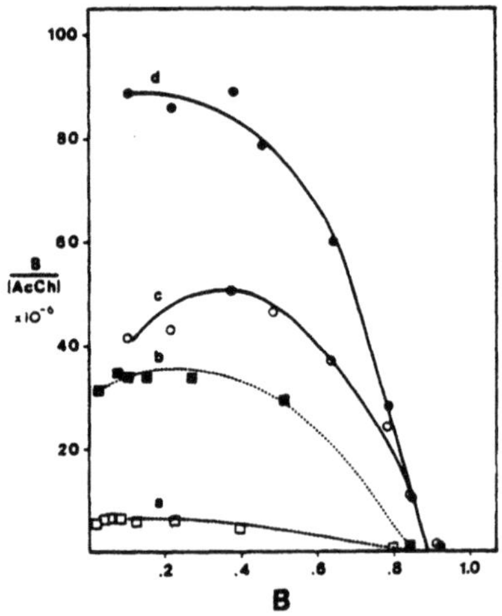

Figure 1. Scatchard representations demonstrating the time dependent increase in the affinity of membrane-bound receptor for AcCh and the lack of attainment of equilibrium in [^{3}H]AcCh equilibrium dialysis binding studies. Dialysis bags each containing receptor-rich membrane fragments (0.59 μM in α-Bgtx site) were subjected to equilibrium dialysis (two bags per flask) against varying [^{3}H]AcCh concentrations. One bag was removed after 3.5 hrs (---) and the other after 23 hrs (——) and AcCh, the free [^{3}H]AcCh concentration, and B, nmol of AcCh-bound per nmol of α-Bgtx binding sites, at these times was determined. a and c: the free ligand concentration was assumed to be equal to the concentration of [^{3}H]AcCh in the dialysis bath at the time of sampling. b and d: the free AcCh concentration was determined from the radioactivity in the supernatant after ultracentrifugation of the bag content.

Results and Discussion

Time dependence of [^{3}H]AcCh binding to AcChR in dialysis-mode.

When either plain buffer or α-Bgtx-blocked membrane bound or detergent solubilized AcChR solution is dialyzed against a [^{3}H]AcCh containing solution, >98% equilibration is achieved in less than 2 h, regardless of the initial [^{3}H]AcCh concentration. However, when [^{3}H]AcCh binding studies with the receptor-rich membrane fragments are carried out under the same equilibrium dialysis condition, the resulting binding curves vary with the dialysis time, as can be seen by comparing

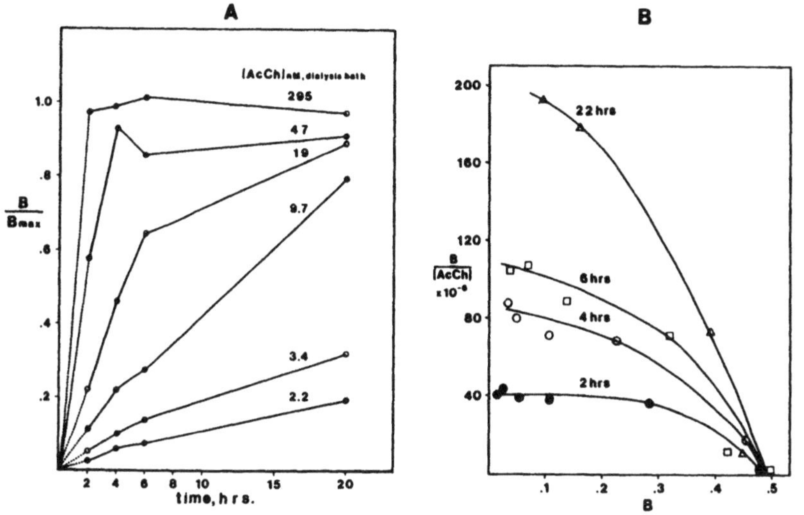

Figure 2. Time dependent AcCh affinity increase of detergent solubilized AcChR during equilibrium dialysis. Purified AcChR-rich membrane fragments (Band 3) were solubilized with 1% Lubrol WX and centrifuged. A. The fraction of AcCh binding sites occupied (B/B_{max}) at various AcCh concentrations (2.2 - 295 nM) in the dialysis bath, as a function of the dialysis time. The true free concentration of [^{3}H]AcCh inside the dialysis bag at the time of sampling was determined by centrifugation of the bag content at 2000 x g for 10 min using the Centrifree (Amicon) micropartition system. B. Scatchard representations generated from 2, 4,6 and 22 hr time points.

Figure 1b (3.5 h) and 1d (23 h). While the maximum value for AcCh bound, B, is the same for both dialysis times, there is a large increase in B values with time in the low [^{3}H]AcCh concentration range. It is apparent that in the low [^{3}H]AcCh concentration range equilibrium between the inside and outside of the dialysis bag is still not attained even after as long as 23 h. The two different Scatchard plots for both the 3.5 h (Figure 1a and b) and the 23 h (Figure 1c and d) time points

differ because the actual free concentration of AcCh, determined by the ultracentrifugation method, is much lower than the concentration in the dialysis bath which is, in the equilibrium dialysis method, assumed to be equal to the free ligand concentration. Furthermore, the concentration of bound AcCh has been underestimated in the dialysis method since it

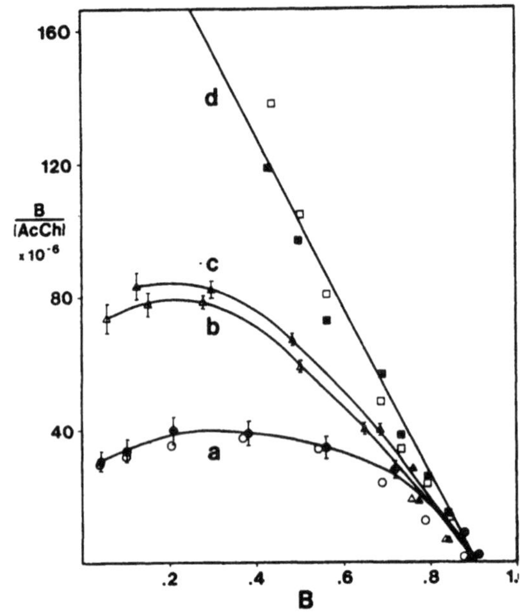

Figure 3. Scatchard representations of [^{3}H]AcCh binding to AcChR-rich membrane fragments at 4° C arising from different modes of AcCh addition. The same 0.2 mM DFP treated membrane preparation and the same final receptor concentration (0.48 μM in α-Bgtx binding sites assayed in the absence of detergent) was used in all of these experiments. The buffer composition was 10 mM PIPES (pH 7.0), 4 mM Ca^{++}, 0.5 mM EDTA, 100 mM NaCl, 0.8 mM Tetram. B, nmol of ligand bound per nmol of α-Bgtx binding sites. AcCh, molar concentration of free [^{3}H]AcCh at the time of sampling. a. ●——● A fixed amount of varying concentrations of [^{3}H]AcCh was added (pulse-mode) to aliquots of membrane fragment suspension preincubated with 0.1 mM Tetram, to give total [^{3}H]AcCh concentrations ranging from 15 nM - 1 μM. A portion of each sample was immediately centrifuged at 140,000 x g for 15 min and concentrations of bound and free [^{3}H]AcCh were determined from the radioactivity in 100 μL solution taken before and after ultracentrifugation with appropriate corrections for radiochemical impurities. Open circles (o), are data points from samples incubated for 17 hrs prior to ultracentrifugation. b. Δ——Δ An aliquot (550 μL) of each [^{3}H]AcCh-containing sample prepared in a was placed in a dialysis bag and dialyzed for 3 hrs against buffer containing the concentration of [^{3}H]AcCh determined in a to be the free [^{3}H]AcCh concentration. The content of each dialysis bag was then subjected to ultracentrifugation and the concentration of free and bound [^{3}H]AcCh was determined as in a. c. ▲——▲ Same as in b except that that the dialysis time was 17 hrs. d. ■——■ To aliquots of membrane suspension was added [^{3}H]AcCh to give concentrations ranging from 1 M to 20 μM. Each sample was dialyzed against plain buffer for 17 hrs to achieve final free concentrations in the 3 - 400 nM range. Free and bound [^{3}H]AcCh concentrations were determined from the counts before and after ultracentrifugation. This data yields $\bar{K}$ = 5 (± 1) x 10^{-9} M. □——□ Same as above except that the free [^{3}H]AcCh concentration was assumed to be equal to the [^{3}H]AcCh concentration of the dialysis buffer outside the bag after 17 hrs.

is inferred from the differences between the total and the free concentrations.

The time and concentration dependences observed in the binding of [^{3}H] AcCh with solubilized AcChR were qualitatively similar to that of membrane fragments and are presented in Figure 2. Although at high [^{3}H] AcCh concentrations maximum binding, ~0.5 AcCh per α-Bgtx site, has already been attained after 2 h, there is a large difference in binding between the 2 and 22 h points at lower concentrations. The fraction of bound AcCh at a given receptor concentration is a function of dialysis time and the rate of increase in $\underline{B}/\underline{B}_{max}$ to reach a time-independent value, presumably the equilibrium value, is faster at higher AcCh concentrations (Figure 2A). Below 10 nM AcCh concentration, the maximum value of $\underline{B}/\underline{B}_{max}$ has not been reached even after 22 h. Scatchard representations (Figure 2B) of the data clearly demonstrate that a slow progressive increase in the affinity of the solubilized AcChR for AcCh occurs with dialysis time and the curvature in the Scatchard plot suggests that true equilibrium binding data was not obtainable even after 22 h of dialysis.

Effects of different modes of introducing [^{3}H] AcCh on the binding of [^{3}H] AcCh to AcChR. Binding data of pulse-mode addition of [^{3}H] AcCh to freshly prepared receptor-rich *Torpedo* membrane fragments in the 2-200 nM free [^{3}H] AcCh concentration range, typically results in curved Scatchard plots, as shown in Figure 3a. Analysis of several similar experiments indicates that, if one assumes the binding to be in a state of equilibrium, the data may be characterized by $\bar{K}$ = 20 ± 10 nM and positive cooperativity (Hill coefficient, n_H = 1.7 ± 0.2), consistent with some of the earlier reports (see in 17, 24). Under these conditions no significant change in [^{3}H] AcCh binding was observed regardless of whether the ultracentrifugation of the incubation mixture was carried out within 30 min or after 24 h (Figure 3a, closed and open circles). The measured $\underline{B}$ values appear to be time-independent. However, when the same incubation mixture was placed in a dialysis bag and subjected to the open-system condition afforded by dialysis, against buffer containing the same free [^{3}H]AcCh concentration found by the initial ultracentrifugation method, a continuous increase in binding as a function of time occurred, particularly in the low free [^{3}H]AcCh concentration range ([A]<0.1 μM), while the maximum value of $\underline{B}$ remained constant, as indicated by the abscissa intercept of Figures 3b and 3c. These results suggest that the population of AcChR in a higher affinity state is increasing during dialysis.

In another experiment equilibrium dialysis was carried out with all of the [^{3}H]AcCh usually introduced into the 50 mL of outside buffer solutions added instead to the 550 μL of membrane fragment samples, thus exposing the AcChR to a much higher concentration of [^{3}H]AcCh (1-20 μM). These samples were then individually dialyzed against 50-200 mL of plain buffer for 17 h to attain the final free [^{3}H]AcCh concentrations in the 3-400 nM range. The Scatchard plot generated by this experiment (Figure 3d) yielded a straight line with $\bar{K}$ = 5 ± 1 nM and no indication of positive cooperativity. Furthermore, the free [^{3}H]AcCh concentration determined by ultracentrifugation of the dialysis bag content and counting the radioactivity of the supernatant (solid triangles) was equal to the [^{3}H]AcCh concentration found in the outside dialysis solution (open triangles), indicating that a state of equilibrium had been attained under these conditions. These results show that, in addition to the time element, the transition to the higher-affinity state is facilitated by exposure to higher AcCh concentrations. This effect was found to be reversible since after exhaustive dialysis, membrane fragments preincubated with high concentrations of AcCh yielded essentially the same Scatchard curve as that derived from membrane fragments not previously exposed to AcCh. Therefore the intermediate curved Scatchard plots (Figure 3) obtained either by pulse-mode or

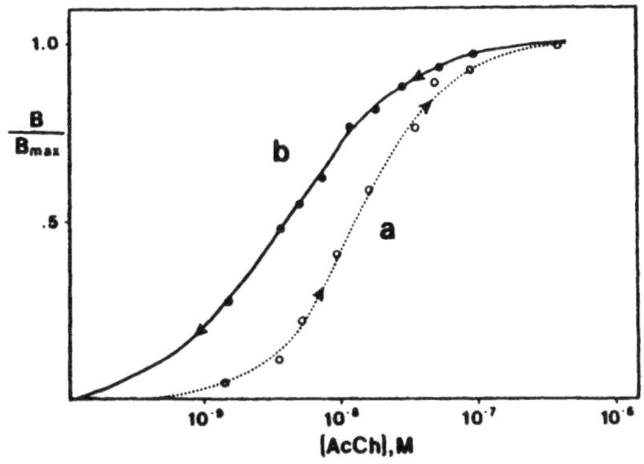

Figure 4. The degree of binding of [^{3}H]AcCh (B/B_{max}) at 4° C to Torpedo receptor-rich membrane fragments as a function of cyclic changes in the concentration of AcCh (logarithmic scale), showing hysteresis. a. o---o Binding curve obtained by "pulse mode addition" of [^{3}H]AcCh to the receptor, followed by ultracentrifugation to determine bound and free [^{3}H]AcCh concentrations (see Figure 3a); b. •——• Binding data obtained by incubation of AcChR (0.48μM in α-Bgtx sites) with high AcCh concentrations (1 - 20 μM) followed by dialysis against initially AcCh free buffer to reach final free AcCh concentrations of 3 - 400 nM (see Figure 3d). The curve is a calculated line for $\bar{K} = 5 \times 10^{-9}$ M.

---

dialysis-mode AcCh addition cannot be analyzed in terms of overall equilibrium cooperativity since they appear to represent nonequilibrium distributions of rather long-lived, metastable high- and lower-affinity conformers and do not reflect equilibrium positive cooperativity, as previously believed (20-24).

Examination of our binding data indicates that when AcChR is exposed to cyclic changes of increasing and decreasing AcCh concentrations (see Figure 3a and 3d), a very pronounced hysteresis loop results, as shown in Figure 4. Since the binding of ligands acting as receptor antagonists, such as decamethonium, Bis-Q and dimethyltubocurarine, does not result in the unusual binding characteristics exhibited by AcCh (unpublished work), the hysteresis phenomenon appears to be related to a complex mechanism involving desensitization of AcChR by an agonist. The hysteresis which results from the occurrence of long-lived metastable states (see a review 31) classifies the AcChR as a macromolecular memory system.

Stoichiometry of [^{125}I]α-Bgtx and [^{3}H]AcCh binding sites. During our investigation of the [^{3}H]AcCh and [^{125}I]α-Bgtx binding properties of acetylcholine-rich membrane fragments prepared from Torpedo electric organ in the presence of Ca^{++} and antiproteolytic agents, we made the following significant observations: preincubation of a given volume of these membrane fragments with 0.1% Triton X-100 or Lubrol WX results in a dramatic (up to 100%) increase in [^{125}I]α-Bgtx bound, relative to the same volume without detergent, and this increase is critically dependent on the particular density gradient fraction analyzed (Table I). In our recent experiments we employ a 42%, 37% and 30% sucrose step gradient for fractionating the crude membrane fragments. Toxin binding assays of these bands reveal that Bands 1 and 2 typically bind 50% to 70% more [^{125}I] α-Bgtx in the presence of detergent, and Band 3 consistently shows an even larger increase - often approaching 100% - under the same conditions, as shown in Table I.

There are at least two interpretations for this observation: contrary to other types of membrane preparation (32), a portion of these microsacs may be "outside-in" and therefore are not accessible to the toxin; or the detergent may loosen the membrane and expose a second sterically hindered α-Bgtx binding site or increase its affinity. Comparison of many [^{3}H]AcCh binding studies with membrane-bound versus detergent solubilized receptor and the fact that detergent treatment of membrane fragments does not increase the total number of AcCh binding sites, led us to favor the latter interpretation. The smaller detergent effect (∿50% increase) observed in membrane-bound receptor from the upper band (Band 1) taken from sucrose gradients may be attributable to loosening of the membrane through proteolysis, resulting in exposure of some of the second α-Bgtx bind-

Table I Characterization of a representative preparation of AcChR-rich membrane fragments isolated by discontinuous sucrose density gradient centrifugation.

Membrane fragments[a]	Band 1	Band 2	Band 3	Band 3 DFP treated
Location, % sucrose[a]	0/30	30/37	37/42	37/42
Volume, mL	20	16	16	10
α-Bgtx bound[b] no detergent, nmol/mL	1.36	1.02	1.52	2.30
α-Bgtx bound[b] in 0.1% Triton nmol/ml	2.08	1.70	2.96	4.60
% increase in α-Bgtx[b] bound with detergent	53	67	95	100
Specific activity[c] nmol/mg	0.8(1.2)	1.1(1.8)	1.3(2.3)	1.4(2.8)
AcChE activity[d] Units/nmol α-Bgtx site	33	14	5	0.03

a AcCh-rich membrane fragments were prepared from 90 g of electric organ of *Torpedo californica*. Three opaque bands were collected from the region of the interfaces of discontinuous sucrose density gradients.

b Nanomoles of [^{125}I]α-Bgtx bound, determined by DE-81 filter disc assays, in the absence and present of 0.1% Triton X-100.

c Specific activity: apparent moles of α-Bgtx binding sites determined in the absence or presence of 0.1% Triton X-100 (values in parentheses) per mg of protein.

d 1 enzyme activity unit is defined as 1 μmol of AcCh hydrolyzed per min. The activity in membrane fragments is expressed as units per mole of [^{125}I]α-Bgtx sites assayed in the presence of 0.1% Triton X-100.

ing sites. It should be recalled that while the kinetics of binding of α-Bgtx to Torpedo membrane-bound receptor is monophasic (20, 32-34), that for solubilized receptor is biphasic (34), suggesting that either the affinity of one of the two α-Bgtx sites on the receptor in membrane is strongly altered by solubilization (34), or that a sterically hindered second site becomes available to α-Bgtx upon detergent treatment.

Our [^{3}H]AcCh binding studies with membrane-bound receptor, using carefully standardized [^{3}H]AcCh (28), showed the stoichiometry of AcCh to α-Bgtx sites to be close to 1. This ratio becomes about 0.5 if the number of α-Bgtx sites determined in the presence of detergent is used, in agreement with the ratio obtained with our solubilized and purified AcChR. Therfore we conclude that the stoichiometry of the high-affinity AcCh binding site is 1 AcCh per receptor monomer (i.e., per M_r 290,000) or per two α-Bgtx sites assayed in the presence of detergent). It is pertinent to note that the investigators who have reported the stoichiometry of AcCh to α-Bgtx sites in membrane fragments to be 1 have assayed for toxin binding in the absence of detergent (20, 23, 26, 33), while those who have reported it to be 0.5 appear to have used 0.1% Triton X-100 (25, 35).

It may very well be that there are two widely different binding sites for AcCh and other cholinergic ligands per receptor monomer (i.e., per two α-Bgtx sites) (36, 37), and that only one of these can exhibit high-affinity for AcCh when binding studies are performed under the conditions described here. It should be recalled that the nonequivalent nature of the two α-toxin binding sites of the receptor has been demonstrated by showing that after disulfide reduction, only one of the α-toxin sites is affinity alkylatable by a low concentration of 4-(N-maleimido)benzyltrimethylammonium (MBTA) or bromoacetylcholine (36, 38-40).

Proposed reaction scheme. In Figure 5 we present a reaction scheme designed to rationalize the two main puzzling questions

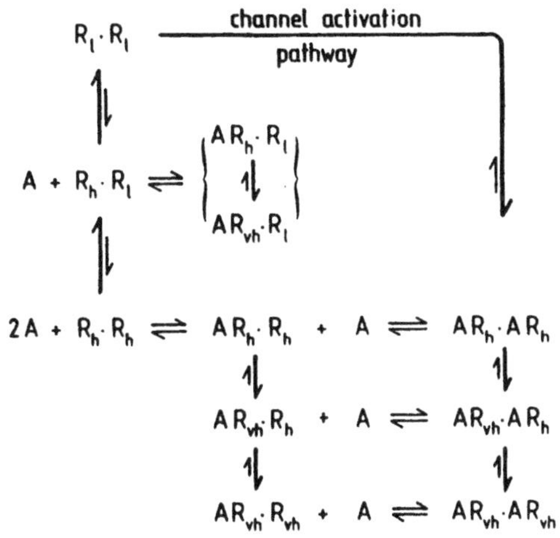

Figure 5. Proposed reaction scheme of AcCh binding to AcChR to account for the difference in AcCh-binding profile resulting from "single pulse mode addition" (closed-system with respect to AcCh) and "dialysis mode addition" (open-system). The simple bimolecular processes are represented by the horizontal sequences, whereas the vertical steps model the various slow structural isomerizations. R_l: low affinity conformer, (10^{-6} M<$\bar{K}$<10^{-4} M); R_h: high affinity conformer, (10^{-8} M<$\bar{K}$<10^{-6} M); R_{vh}: very high affinity conformer, ($\bar{K}\sim5$ (±1) x 10^{-9}M); $R_h \cdot R_l$: hybrid form of the two binding sites. The thick arrows indicate the preferential position of the isomerization equilibria. Large upper arrow indicates the channel activation pathway. A high AcCh concentrations (10^{-6} M), the low affinity conformer, R_l, is directly involved in the binding which results in channel opening and is subsequently transformed to its high affinity state. At low AcCh concentrations (10^{-6} M), the R_h and R_{vh} conformers are the dominant direct reaction partner for AcCh and thus constitute the direct route of the high affinity state pathway of the scheme. "Single pulse mode addition" of AcCh (closed system) favors the $AR_h \cdot R_l$ and $AR_{vh} \cdot R_l$ binding states, whereas "dialysis mode addition" (open system) ultimately leads to the $AR_{vh} \cdot AR_{vh}$ conformer.

---

raised by our AcCh binding experiments: 1) why AcChR is so slow in reaching an equilibrium state in AcCh binding during the dialysis-mode? 2) why pulse-mode AcCh binding results in less AcCh bound than the dialysis-mode and appears to be time -independent? In an attempt to describe the dialysis-mode binding data, a reaction model must account for the observed inequality of free [^{3}H]AcCh concentration, $[A_f]$, $[A_f]_{in} < [A_f]_{out}$, which develops between the inside and the outside of the dialysis bag (see Table I) even when the initial conditions are $[A_f]_{in} = [A_f]_{out}$. The inequality can arise from the additional binding of AcCh to higher-affinity states of the AcChR, generated by conversion of unoccupied low affinity sites to high affinity sites, or unoccupied high affinity to still higher affinity sites. Such newly created high affinity sites would now bind a part of the remaining AcCh such that the free AcCh concentration inside the dialysis bag decreases below the initial value. In this situation the AcCh flux through the dialysis membrane is rate-limiting for the overall rate of net AcCh binding to AcChR. When the most stable state R_{vh}, associated with the individual equilibrium constant $K_{vh} \simeq 10^{-9}$ M ($K_{vh} < \bar{K}$ = 5 nM), is the final very high affinity conformer, the conformers R_h must have K_h values which are larger than 10^{-9} M, perhaps $\sim 10^{-7}$ M (see 3). Hence at final equilibrium the concentrations of the intermediate AR_h state may be negligible compared to that of AR_{vh}. Since our data suggest that cooperative interactions between at least two AcCh binding sites exist and given the stoichiometry of one AcCh binding site per monomer (9S), the receptor dimer (13S), R·R, is required for the smallest cooperative element in the scheme.

When the AcCh concentration is increased more rapidly (pulse-mode) than in the dialysis-mode of AcCh binding, rather long-lived nonequilibrium distribution of AcChR states with fewer occupied AcCh binding sites are maintained. Because these metastable nonequilibrium distributions depend on the rate of AcCh increase they must result from rate processes. In any case, the pulse-mode data suggest that the overall transitions from $R_l \cdot R_l$ to more high-affinity conformers can be "interrupted". On an elementary scale the overall equilibrium $R_lR_l \rightleftharpoons R_h \cdot R_h$ very likely involves the intermediate form $R_h \cdot R_l$. This hybrid $R_h \cdot R_l$, as a part of the sequence, $R_l \cdot R_l \rightleftharpoons R_h \cdot R_l \rightleftharpoons R_h \cdot R_l$ represents a suitable branching point where different rate processes can occur according to the scheme:

$$
\begin{array}{ccccccc}
 & & R_l \cdot R_l & & & & \\
 & & \text{slow} \updownarrows & & & & \\
A & + & R_h \cdot R_l & \overset{\text{rapid}}{\rightleftharpoons} & AR_h \cdot R_l & \rightleftharpoons & AR_{vh} \cdot R_l \\
 & & \text{slow} \updownarrows & & & & \\
 & & R_h \cdot R_h & & & &
\end{array}
$$

In the above reaction scheme the binding of AcCh to the conformer R_h of the hybrid $R_h \cdot R_l$ conserves the adjacent low affinity state. The longevity of this low affinity conservation suggests that, in addition to simple AcCh binding, structural transitions to a complex with higher life-times, $AR_{vh} \cdot R_l$, may also be involved. In the pulse-mode, the binding of AcCh to the hybrid $A + R_h \cdot R_l \rightleftharpoons AR_h \cdot R_l$, initiated by a bimolecular step, appears to be faster than the intramolecular transition $R_h \cdot R_l \rightleftharpoons R_h \cdot R_h$. Furthermore, under closed system conditions with respect to AcCh the conversion to the final R_{vh} conformation may be extremely slow. Thus under such conditions of fixed total AcCh concentration, nonequilibrium distributions with less AcCh bound states may be maintained for a very long time. The low affinity conservation pathway can therefore be considered as a frozen-in or deadend pathway.

Subsequent transfer of these metastable nonequilibrium distributions to the open system condition afforded by dialysis apparently enhances the coupling of the hybrid to the intramolecular step $R_h \cdot R_l \leftrightarrows R_h \cdot R_h$ of the reaction scheme (see Figure 3b, c). The reaction steps $2A + R_h \cdot R_h \rightleftharpoons AR_{vh} \cdot R_{vh} + A \rightleftharpoons AR_{vh} \cdot AR_{vh}$, apparently give rise to larger driving forces to diverge the complexes from the low-affinity conservation pathway and induce a maximum shift to the side of the high-affinity complexes.

This research was supported by the National Institutes of Health (Grant RO1-NS-13744, the Muscular Dystrophy Association of America, and the Deutsche Forschungsgemeinschaft (Grant NE 227/3-5).

References

1. Changeux, J.-P. 1981. In: Harvey Lect. 75, Academic Press, New York, p. 81.

2. Nachmansohn, D. 1955. In: Harvey Lect. 1953-1954, 49, Academic Press, New York, p. 57.

3. Boyed, N.D., J.B. Cohen. 1980. Biochemistry 19, 5344.

4. Weber, M., T. David-Pfeuty, J.-P. Changeux. 1975. Proc. Natl. Acad. Sci. U.S.A. 72, 3443.

5. Sugiyama, H., J.-L. Popot, J.-P. Changeux. 1976. J. Mol. Biol. 106, 485.

6. Weiland, G., B. Georgia, S. Lappi, C.F. Chignell, P. Taylor. 1977. J. Biol. Chem. 252, 7648.

7. Bernhardt, J., E. Neumann. 1978. Proc. Natl. Acad. Sci. U.S.A 75, 3756.

8. Barrantes, F.J. 1978. J. Mol. Biol. 124, 1.

9. Quast, U., M. Schimerlik, T. Lee, V. Witzemann, S. Blankard, M.A. Raftery. 1978. Biochemistry 17, 2405.

10. Heidmann, T., J.-P. Changeux. 1979. Eur. J. Biochem. 94, 281.

11. Cohen, J.B., N.D. Boyd. 1979. In: Catalysis in chemistry and biochemistry: Theory and Experiments, B. Pullman, ed. D. Reidel Publ. Co., p. 293.

12. Sakmann, R., J. Patlak, E. Neher. 1980. Nature (London) 286, 71.

13. Clark, R.B., R.P. Adams. 1981. Biophys. J. 33, 16a.

14. Feltz, A., A. Trautmann. 1982. J. Physiol. (London) 322, 257.

15. Chesnut, T.J. 1983. J. Physiol. (London) 336, 229.

16. Katz, B., S. Thesleff. 1957. J. Physiol. (London) 138, 63.

17. Karlin, A. 1980. In: Cell surface reviews (Poste, G., G.L. Nicolson, C.E. Cotman, Eds.) Elsevier North Holland, p. 191.

18. Conti-Tronconi, B.M., M.A. Raftery. 1982. Annu. Rev. Biochem. 51, 499.

19. Taylor P., R.D. Brown, D.A. Johnson. 1983. Curr. Top. Membr. Transp. 18, 407.

20. Weber, M., J.-P. Changeux. 1974. Mol. Pharmacol. 10, 1. and 15.

21. Damle, V.N., S. Hamilton, R. Valderrama, A. Karlin. 1976. Pharmacologist 18, 146.

22. Schiebler, W., L. Lauffer, F. Hucho. 1977. FEBS Lett. 81, 39.

23. Neubig, R.R., J.B. Cohen. 1979. Biochemistry 18, 5464.

24. Fels, G., E.K. Wolff, A. Maelicke. 1982. Eur. J. Biochem. 127, 31.

25. Raftery, M.A., R.L. Vandlen, K.L. Reed, T. Lee. 1975. Cold Spring Harbor Symp. Quant. Biol. 40, 193.

26. Sugiyama, H., J.-P. Changeux. 1975. Eur. J. Biochem. 55, 505.

27. Blankard, S.G., S.M.J. Dunn, M.A. Raftery. 1982. Biochemistry 21, 6258.

28. Chang, H.W., E. Bock, E. Neumann. 1984. Biochemistry 23, 4546.

29. Sobel, A., M. Weber, J.-P. Changeux. 1977. Eur. J. Biochem. 80, 215.

30. Chang, H.W., E. Bock. 1977. Biochemistry 16, 4513.

31. Neumann, E. 1973. Angew. Chem. Intern. Ed. 42, 356.

32. Hartig, P.R., M.A. Raftery. 1979. Biochemistry 18, 1146.

33. Weiland, G., G. Georgia, R.T. Wee, C.F. Chignell, P. Taylor. 1976. Mol. Pharmacol. 12, 1091.

34. Blanchard, S.G., U. Quast, K. Reed, T. Lee, M.I. Schimerlik, R.L. Vandlen, T. Claudio, C.D. Strader, H.-P. H. Moor, M.A. Raftery. 1979. Biochemistry 18, 1875.

35. Schimerlick, M., U. Quast, M.A. Raftery. 1979. Biochemistry 18, 1884.

36. Delegeane, A.M., M. McNamee. 1980. Biochemistry 19, 890.

37. Dunn, S.M.J., B.M. Conti-Tronconi, M.A. Raftery. 1983. Biochemistry 22, 2512.

38. Damle, V., A. Karlin. 1978. Biochemistry 17, 2039.

39. Moore, H.P.H., M.A. Raftery. 1979. Biochemistry 18, 1862.

40. Wolosin, J.M., A. Lydiott, J.O. Dolly, E.A. Barnard. 1980. Eur. J. Biochem. 109, 495.

THE DIMER OF TORPEDO ACETYLCHOLINE RECEPTOR: SYNCHRONIZED DOUBLE CHANNEL

E. Neumann, F. Spillecke
Physical and Biophysical Chemistry, University of Bielefeld, P.O.Box 8640, D-4800 Bielefeld 1, F.R.Germany

H. Schindler
Biozentrum, University of Basel, CH-4056 Basel, Switzerland

Introduction

Recent biochemical and biophysical observations on the acetylcholine receptor channel protein may possibly lead to a re-evaluation of classical data on gated ion flow in bioelectricity (1).

It is well known that the nicotinic acetylcholine receptor (AcChR) of electric fish can be isolated in two macromolecular forms, depending on the experimental conditions of isolation (2-4). When sulfhydryl-alkylating agents are present in the first tissue homogenization step the dimer (D), consisting of two disulfid-linked monomers (M), is the predominant species. The monomeric form prevails when reducing agents such as dithiothreitol (DTT) are used in the homogenization medium. As outlined in various summaries (see, e.g., ref. 4) the monomer or light (L) form has a subunit stoichiometry of two α-chains, one β-, one γ- and one δ-chain ($\alpha_L\beta\gamma\delta$). The relative molar mass of the monomer is M_r 290,000; the dimer or heavy (H) form is twice as large.

It is still disputed which of the two isolation products, the D-species or the M-molecule is the native and functionally

Molecular Basis of Nerve Activity

operative unit in the biomembrane. Electronmicrographs of intact electrocytes (5) and of receptor rich postsynaptic membranes show the receptor proteins as rosette-doublets lined-up in rows (6). The functional significance of this topology: doublets assigned to receptor dimers and the dimer rows, is not yet understood.

In this context two specific questions of the receptor organization are discussed: (i) Are there functional differences between the D- and the M-species? More explicitely, does it matter whether the monomer is functioning as an isolated monomer or as a part of the dimer? (ii) Which of the two macromolecular forms is the in vivo functional unit?

Vesicle flux studies and planar lipid bilayer reconstitution experiments of the isolated and purified receptor proteins have so far been interpreted in terms of functional equivalence of the reconstituted monomers and dimers (4, 7-13). Recently a combined biochemical and biophysical approach to the questions raised above has provided new information for a possible resolution of this particular structure/function problem of the AcChR channel proteins. In the present account some key data of this biophysicochemical study on the AcChR monomer/dimer problem are summarized. The actual observations suggest an extension and re-interpretation of classical electrophysiological and biophysical data on the gating of ion transport by the nicotinic acetylcholine receptor, in terms of a concerted and synchronized opening and closing of the dimer double-channel unit.

The main conclusions are: (i) Measured single channel events must not necessarily reflect one single channel protein; rather, a single channel event may result from two or more cooperatively synchronized channel proteins. (ii) Positive cooperativity in dose-response curves must not necessarily be caused by the binding of two (or more) agonist or activator molecules to one channel protein; rather, a Hill coefficient of two can reflect a pair (or more) of channels like that of the AcChR dimer. (iii) Different conductance states may simply result from differences in the transient binding of different types of ions to the open channel conformers of the AcChR protein.

AcChR Oligomerization in Solution

In the course of studies to reconstitute the isolated and purified AcChR proteins in lipid bilayers it was found that the rereptor monomers exhibit an inherent tendency to specific noncovalent association in vesicular lipid bilayers as well as in solution (1,14,15). In particular the reversible formation of stable dimers according to $2M \rightleftharpoons M_2$, i.e. without reformation of the δ-δ sulfide bond, readily occurs in the presence of phospholipids (14,15). Noncovalent aggregations of monomeric receptors to dimers [and higher oligomers (16)] might be one of the reasons why, until recently (1), no characteristic functional differences between monomers (M) and dimers (D) have been observed.

Due to the intrinsic tendency to noncovalent dimer formation in the presence of phospholipids many monomer preparations may have actually dealt with dimers of the type M_2. Although phospholipids are not a necessary condition to form M_2 the stability of the noncovalent dimer strongly increases in the presence of lipids ($K_d \approx 10^{-10}$ M) compared to that in the absence of lipids ($K_d \approx 10^{-6}$ M); see ref. 14,15. Several lines of evidence support the specificity of this dimer formation and exclude artefactual crosslinking: (i) Gel electrophoresis and the reversibility of complex formation observed on the time scale of ultracentrifugation experiments ($\approx$ 24 h) confirms the noncovalent nature of the dimeric associates. (ii) The stoichiometry of the complex is 1:1; no aggregates of variable size occur under the given conditions. (iii) The dimerization is independent of inhibitors of free radical lipid oxidation, thus any oxidative crosslinking by detergent impurities was absent. (iiii) The dimerization is hardly affected by agents which modify sulfhydryl groups. However, chemical modification of carboxylate groups of the (acidic) receptor protein reduces the stability of the dimers (14,15).

Crosslinking (14,15). Under experimental conditions where the formation of dimers M_2 is favoured crosslinking with glutardialdehyde leads to a spectrum of oligomers that, after sucrose gradient centrifugation, is consistent with the presence of dimers and specific oligomers of the dimer (of lower stability): tetramers $(M_2)_2$, hexamers $(M_2)_3$, etc. Therefore, in addition to the specificity of the stable dimer formation there is some kind of 'long range specificity' for oligomers of the dimer excluding random association of monomers to species with uneven numbers of monomers.

The crosslinking data support the results from experiments at very low detergent concentrations (0.05% Lubrol PX) where the AcChR proteins occur as M_2 species and larger aggregates. By addition of an excess of phospholipid/cholate mixture the large aggregates dissolve and, concomitantly, the amount of receptors at the dimer position increases.

Single Channel Events of Dimers and Monomers

The functional reconstitution of the isolated and purified AcChR proteins from Torpedo californica electric tissue involved reincorporation of the channel proteins into lipid vesicles, formation of monolayers and finally of planar bilayers from these vesicles (17). For the protein reincorporation into vesicles a novel technique was applied (fast dilution method) because conventional cholate dialysis caused perturbed and irregular channels and receptor aggregations.

Provided the planar bilayers are adjusted to cohesive pressures of 30 and 32 mN/m, 16 mole per cent cholesterol of the total lipid, both the dimer and the monomer exhibit single channel events when agonists such as AcCh, carbamoylcholine or subероyldicholine are added (Fig. 1). As depicted in Figure 1

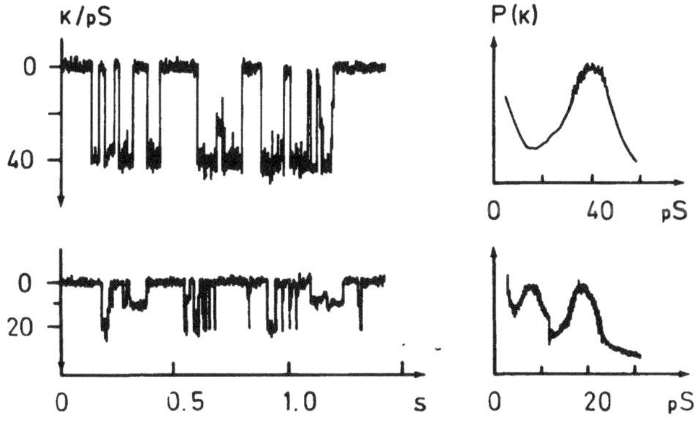

<u>Fig. 1</u> Conductance κ/time traces and channel histograms P(κ) (probability of occurrence) of reconstituted Torpedo californica AcChR in planar lipid bilayer (16 mole % cholesterol), 5 min after addition of suberoyldicholine (0.5 μM); 20°C, 0.5 M NaCl. Upper trace: the native dimer (D) at 100 mV; lower trace: monomer (M) at 150 mV.

the most striking difference between the channel events is that the maximum channel conductance (≈20 pS) of monomers (M) is about half as large as that (≈40 pS) of dimers (D). Furthermore, the amplitude and time characteristics of the channels formed by the isolated dimers (≈40 pS) in lipid bilayers are almost quantitatively the same (≈40±10 pS) as for unextracted receptors (microsacs) from a different Torpedo species: T. marmorata (18).

<u>Effect of reducing agents on the dimer (1).</u> When the disulfide reducing agent DTT was added to a planar bilayer containing

dimers (D), the initial dimer channel events started to change: at first conductance levels characteristic of the monomeric receptor occurred at the expense of dimer events; after about 10 min the channel events became undefined and within 1 h the channel activity disappeared. It thus seems that disulfide reduction first abolishes one of the two channels within a dimer and with progressing time the second one also disappears. These observations may relate to inhibitory effects of DTT on ligand binding and on receptor activity (19,20). In view of the dimer conductance level being twice as large as that of the monomer the dimer appears to cause a "double-channel" with apparent "single channel" characteristics.

AcChR Channel Oligomerization

The isolated and purified AcChR monomers (M) and dimers (D) reconstituted in planar bilayers may give rise to a synchronized opening and closing of up to several dimeric channel units caused by protein association (1,21). For instance, when the cohesive pressure of a bilayer with monomer channels was raised from 32 to 40 mN/m by the addition of $CaCl_2$ to a final concentration of 0.5 mM, large conductance peaks developed at the expense of the monomer levels; they are grouped in activity clusters of multiple dimer levels. The conductance histogram of such a cluster is shown in Figure 2. Most remarkably, only even multiples of the monomer level (20 pS) are populated corresponding well to the dimer level (40 pS) and multiples thereof (80,120 and 160 pS); peaks at uneven multiples do not occur.

The specific channel oligomerization in multiples of dimers corresponds well with the oligomerization of receptor proteins in solution where, after crosslinking, the dimer complexes $(M_2)_n$, n = 1,2,3... may be identified (14,15). The channel characteristics of the dimer M_2 is hardly distinguishable from that of the (native) dimer D.

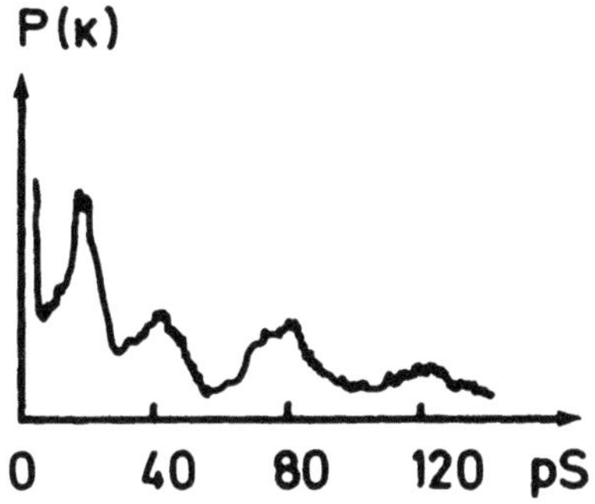

Fig. 2 Probability of occurrence P(κ) of channel conductances κ of AcChR monomers (M) and oligomers at 0.5 M NaCl, 15 min after increase in cohesive membrane pressure from 32 to 40 mN/m by 0.5 mM $CaCl_2$. Conductance oligomers are multiples of the dimers (M_2): $(M_2)_2$, $(M_2)_3$, $(M_2)_4$; no intermediate oligomers of M occur.

Cation Dependence of Open-Channel States

The AcChR monomer channels always exhibit two different conductance levels for Na^+ ions (Figure 1) and for K^+ ions: a maximum conductance of 20±2 pS and a substate of 9±2 pS. The relative population of the two levels is, however, dependent on the type of cation which is transported (1).

The dimers also show substates of lower conductance. For Na^+ ions the main level is 40±3 pS; a substate of 30 pS occurs to 20%. For K^+ ions there is only 6% of the 40 pS events, the larger fraction are 30 pS events. The data are summarized in Table I.

It may be concluded that the different conductance levels of

Table I

Channel Conductance Levels (and Relative Populations) of <u>Torpedo Californica</u> Acetylcholine Receptor (AcChR) Proteins Reconstituted in Planar Lipid Bilayers. (Cohesive membrane pressure of 32 mN/m, 0.5 μM suberoyldicholine, pH 7.4, 20°C.)

AcChR protein	Conductance levels (pS)			
	0.5 M KCl		0.5 M NaCl	
	low	high	low	high
Dimer (D; H-form):	31±2	40±3	30±1	40±1
	(12:1)		(1:5)	
Monomer (M; L-form):	9±2	19±2	9±2	20±2
	(7:1)		(1:2)	
Dimer (D):	31±2	40±3	30±1	40±1
+ 10 mM DTT (≈5 min):	9±2	19±2	9±2	20±2
Dimer (D) at 0.3M Na^+/0.2M K^+:	30±2	40±2		
Monomer at 40 mN/m :	9±2	19±2	9±2	20±2
after 10 min:	9±2	31±2		40±1
>> 10 min:	multiples of 31,		of	40±1
Torpedo marmorata microsacs (18):				40±10

the receptor monomers and dimers are due to the existence of two open-channel states. The transient associations of the Na^+ and K^+ ions with these channel states are different. K^+ usually favors the receptor state of lower conductance.

Multiple conductance states in single channel activity of AcChR have been observed in patch-clamp studies on cultured embryonic muscle cells, but unfortunately did not include K^+ ions (22). A direct comparison with the data of Torpedo AcChR (summarized in Table I) is thus limited to the conductance values for Na^+ ions.

Double Channel and Pair Cooperativity

Recalling the oligomerization tendency of the isolated AcChR monomers in solution and in the planar lipid bilayer it appears that contact formation between the monomers alone leads not only to stable complexes but also to pronounced changes in the channel characteristics, especially in the channel conductance. The effect of the monomer-monomer interactions within the dimer does apparently not depend on the intactness of the δ-δ disulfide bridge. The channel conductance of two noncovalently associated monomers in the dimer M_2 equals that of the isolated dimer D containing the (native) disulfide bridge.

The factor of 2 in the conductance difference between monomers and dimers and the occurrence of multiples of dimers in the protein complexes as well as in the channel conductances can be consistently interpreted in terms of a simple model: synchronous opening and closing of two monomeric channels within a dimer (D and M_2). Therefore, the operationally defined "dimer-channel" with apparent "single-channel" characteristics has to be visualized as two cooperative monomeric channels that act fully in concert.

The fact that the same channel conductances were observed with isolated dimers (D and M_2) and with unextracted microsacs (18, 21) suggests that the basic functional units of the observable AcChR channels in the biomembrane are, in reality, synchronized double-channels.

At present it is not known if the principle of cooperative double-channel gating applies to other cholinergic systems as well. Remarkably corresponding patch-clamp data for "single-channel" conductances (22,23) are closer correlated to the channel conductance of the reconstituted Torpedo dimer than to that of the monomers. However, the comparison is obscured by the different ionic conditions applied besides possible species-dependent differences.

Cooperativity. Two distinct cooperative interactions within AcChR complexes are apparent from biochemical studies in solution as well as from the channel oligomerization of reconstituted monomers and dimers: (i) Strong (short range) cooperativity between pairs of monomers and monomeric channels both in the absence and presence of the disulfide linkage between the δ-subunits of two monomers. (ii) Weak (long range) cooperativity between several pairs of dimers and dimer-channels.

The extent of this "pair cooperativity" is, however, smaller in monomer associates, on the average between two pairs of channels, compared to dimer(D) associates under otherwise the same membrane conditions, on the average between four pairs of channels (18,21). Thus, the S-S bridge in the (native) dimer species enhances long range cooperativity probably by allowing a more rigid arrangement of the dimers(D) compared to that in monomer associates (M_2).

Since the AcChR in the native postsynaptic membrane can be visualized in double rows of dimers (6), it appears likely that the oligomeric channel cooperativity (consistent with the receptor oligomerization in solution) is the functional counterpart of the native organization in double rows. This feature and the enhancement of long range cooperativity by the intact disulfide bridge between the monomers of the D-species together with the effect of reducing agents on the ratio of monomers and dimers in the protein isolation (1), suggest that the disulfide-bridged dimer D is the native and functionally operative channel unit.

Channel Gating Scheme of the AcChR Dimer

If the "dimer-channel" is indeed the functional conductance unit in the biomembrane, the Hill coefficient of about 2 for the channel activation (for review see ref. 4) may be interpreted in a different manner compared to previous analyses. The dimer channel activation may result from the successive association of two agonist molecules (A) with the two monomeric low affinity conformers R_ℓ of the dimer species $R_\ell \cdot R_\ell$; i.e. each monomer channel within the dimer binds only one agonist molecule. The functionally relevant step is then the concerted transition of the complex $AR_\ell \cdot AR_\ell$ to the conducting conformer $AR_\ell^* \cdot AR_\ell^*$.

$$2A + R_\ell \cdot R_\ell \rightleftharpoons AR_\ell \cdot R_\ell + A \rightleftharpoons \begin{array}{c} AR_\ell \cdot AR_\ell \\ \upharpoonleft\downharpoonright \\ AR_\ell^* \cdot AR_\ell^* \end{array} \qquad (1)$$

The Hill coefficient of 2 is thus related to the highly cooperative, practically synchronous transition of the receptor dimer, in its low affinity state for agonist binding. The occupation of only one of the α-subunits within a monomeric part of the dimer compares well with the channel activation by irreversibly binding agonists. It is sufficient for channel activation when only one "irreversible agonist" binds per monomer after reduction of disulfide bridges in the vicinity of a binding site for AcCh on one of the two α-subunits (24-26); see also ref. 27. The covalent agonist labeling of the second α-subunit requires much higher concentration of irreversible agonists (28).

There are still other arguments in favor of the activation scheme(1) in terms of two agonist molecules and the dimer. Recent dialysis studies have confirmed that the final equilibrium state for the binding of [^{3}H]AcCh to membrane-bound receptors (microsacs) and to purified dimers is associated with only one AcCh molecule bound per monomer (K ≈ 5 nM, 4°C, Torpedo cal.), independent of whether detergent is present or not; there is no positive cooperativity to this final high affinity state (29). If the channel activation would require two agonist molecules per receptor monomer, then the final slow phase of receptor inactivation (associated with a Hill coefficient of one) leading to one agonist bound per monomer would have to release one of the two previously bound agonist molecules. Although such a process could occur via allosteric structural transitions (including the two α-subunits of R) the sequence $2A + R_\ell = R_\ell A_2 = R_\ell^* A_2 = R_h A + A$ appears very unusual. Furtheron, according to Neubig et al. (1982) the functional unit of $^{23}Na^+$ efflux gating in Torpedo membranes (microsacs) comprises two α-bungarotoxin (α-Bgt) sites (30). Since in the absence of detergents there is only one α-Bgt molecule bound with high affinity per monomeric AcChR (29), the functional unit of ion transport gating must be the dimer (D).

Receptor inactivation procedes in at least two phases (see e.g. ref. 4), a fast cooperative one (31,32) and a slow noncooperative one (30,32). Recent dialysis data show that at low AcCh concentrations (≦ 50 nM) the slow transition to the final equilibrium state of very high affinity denoted by R_{vh} (K ≈ 5 nM) is induced by agonist binding (29). The preexisting high affinity receptor state R_h to which the agonists may directly bind is of smaller affinity ($K_h \approx 0.1$ μM for AcCh).

In order to facilitate a comparison with schemes developed previously, the following correspondences to ref. 4 hold:
$R_\ell \hat{=}$ R of ref. 4; activatable, low AcCh affinity ($K_\ell \approx 10^{-4}$ M).
$R_\ell^* \hat{=}$ A of ref. 4; activated, conducting ($K_\ell^* \approx 10^{-5}$ M).
$R_h \hat{=}$ I of ref. 4; inactivated, high AcCh affinity ($K_h \approx 10^{-7}$ M).
$R_{vh} \hat{=}$ D of ref. 4; fully desensitized, very high AcCh affinity ($K_{vh} = 5 \times 10^{-9}$ M).

Channel activation from desensitized AcChR. Channel activity is also observed under conditions where the AcChR protein is usually assumed to be inactivated or desensitized (23,1). Whereas the response immediately after agonist addition are frequently occurring short single channel events, longer exposure to agonists gives rise to occasional bursts of repetitive opening and closing events (23,12,13,1): "Nachschlag" or flickering (23). See Fig. 3. The occasional dimer-channel activation under desensitizing conditions requires an extension of the receptor activation scheme by the cooperative coupling (Hill coefficient two) of the open conformation $AR_\ell^* \cdot AR_\ell^*$ to inactivated states according to

$$AR_\ell^* \cdot AR_\ell^* \rightleftharpoons AR_h \cdot AR_h \qquad (2)$$

The reaction scheme depicted in Fig. 3 is an abbreviated form of the possibly more complicated reaction network of AcChR activation and inactivation processes. In detail, the inactive (intermediate) state AR_h may be populated either via the open receptor states or by direct binding of activator to the R_h conformers. The slow noncooperative phase of inactivation (30) to the final state AR_{vh} of very high AcCh affinity involves structural transitions of the type $AR_h \rightleftharpoons AR_{vh}$, where each of

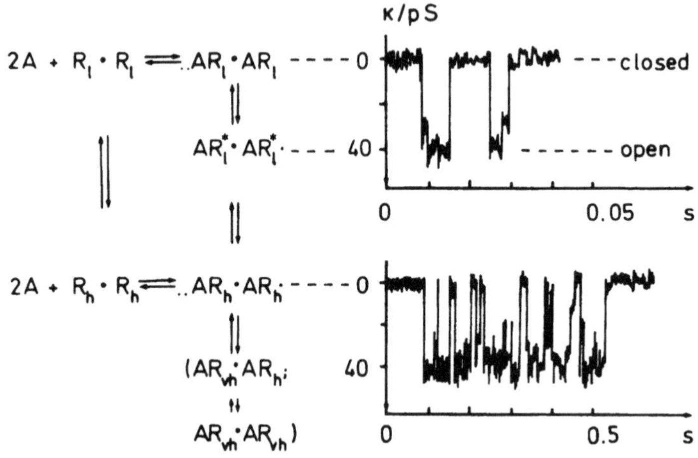

Fig. 3 Reaction scheme (short form) for the AcChR dimer channel activation. Upper part: normal low affinity pathway of activation leading to "single-channel" events according to $AR_{\ell} \cdot AR_{\ell}$ (closed) $\rightleftharpoons$ $AR^{*}_{\ell} \cdot AR^{*}_{\ell}$ (open). Lower part: burst of repetitive opening and closing events under desensitizing conditions, modelled by $AR_h \cdot AR_h$(closed) $\rightleftharpoons$ $AR^{*}_{\ell} \cdot AR^{*}_{\ell}$(open).

the two AR_h conformers within the dimer $AR_h \cdot AR_h$ can independently convert to the AR_{vh} state (Hill coefficient one, ref. 30,32).

Monomer. The isolated AcChR monomer (fragment) in detergent solution (27) as well as reconstituted in planar lipid bilayers (1) appears to have a more loose structure compared to the dimer. The high affinity AcCh binding to the M-species is a factor of about ten ($K \approx 5 \times 10^{-8}$ M) less stable (2,29) than to the dimer ($K \approx 5 \times 10^{-9}$ M). The hydrodynamic properties of the M-fragment suggest a more extended, perhaps partially unfolded structure (27). As seen from Table I the occurrence of the substates of lower conductance are much more frequent for the monomers than for the dimers. The monomer channels are thus more "flexible" than the dimer channels. The more loose structure of the monomer may also be the reason why membrane fragments in the presence of detergents (0.1% Triton X-100) bind two α-Bgtx molecules per monomer compared to only one high affinity bound α-Bgt per monomer in the absence of detergents (29).

In the absence of detergents high affinity toxin binding to the second α-subunit of the monomer part does not occur because of sterical hindrance (33); or the high affinity occupation of one α-subunit turns the second α-subunit allosterically into a state of very low affinity for the binding of agonists and inhibitors. Unusually high concentrations may then be necessary to also occupy the second α-subunit.

Acknowledgements

This study is dedicated to the memory of Professor David Nachmansohn (1899-1983), the pioneer of the molecular bioelectricity. We gratefully acknowledge the support by the Deutsche Forschungsgemeinschaft (Grant Ne 227/3-5 to E.N.) and the Swiss National Science Foundation (Grant 3.773.80 to H.S.).

References

1. Schindler, H., F. Spillecke, E. Neumann. 1984. Proc. Natl. Acad. Sci. USA 81, 6222-6226.

2. Chang, H.W., E. Bock. 1977. Biochemistry 16, 4513-4520.

3. Hamilton, S.L., M. McLaughlin, A. Karlin. 1977. Biochem. Biophys. Res. Commun. 79, 692-699.

4. Changeux, J.-P., A. Devillers-Thiéry, P. Chemouilli. 1984. Science 225, 1335-1345.

5. Heuser, J.E., S.R. Salpeter. 1979. J. Cell Biol. 82, 150-173.

6. Cartaud, J. 1980. In: Ontogenesis and Functional Mechanisms of Peripheral Synapses (J. Taxi, ed.) Elsevier, New York, pp. 199-210.

7. Miller, D.S., H.-P. Moore, P.R. Hartig, M.A. Raftery. 1978. Biochem. Biophys. Res. Commun. 85, 632-640.

8. Anholt, R., J. Lindstrom, M. Montal. 1980. Eur. J. Biochem. 109, 481-487.

9. Wu, W., M.A. Raftery. 1981. Biochem. Biophys. Res. Commun. 99, 463-444.

10. Epstein, M., E. Racker. 1978. J. Biol. Chem. 253, 6660-6662.

11. Lindstrom, J., R. Anholt, B. Einarson, A. Engel, M. Osame, M. Montal. 1980. J. Biol. Chem. 255, 8340-8350.

12. Boheim, G., W. Hanke, F.J. Barrantes, H. Eibl, B. Sakmann, G. Fels, A. Maelicke. 1981. Proc. Natl. Acad. Sci. USA 78, 3586-3590.

13. Montal, M., P. Labarca, D.F. Fredkin, B.A. Suarez-Isla, J. Lindstrom. 1984. Biophys. J. 45, 165-174.

14. Spillecke, F. 1984. Thesis, University of Konstanz. Martinsried.

15. Spillecke, F., E. Neumann. 1985. Eur. J. Biochem. (in press)

16. Rüchel, R., D. Watters, A. Maelicke. 1981. Eur. J. Biochem. 119, 215-223.

17. Schindler, H. 1980. FEBS Lett. 122, 77-79.

18. Schindler, H., U. Quast. 1980. Proc. Natl. Acad. Sci. USA 77, 3052-3056.

19. Karlin, A., E. Bartels. 1966. Biochim. Biophys. Acta 126, 525-535.

20. Walker, J.W., R.J. Lukas, M.G. McNamee. 1981. Biochemistry 20, 2191-2199.

21. Schindler, H. 1985. EMBO J. (in press)

22. Hamill, O.P., B. Sakmann. 1981. Nature (London) 294, 462-464.

23. Sakmann, B., J. Patlak, E. Neher. 1980. Nature (London) 286, 71-73.

24. Silman, I., A. Karlin. 1969. Science 164, 1420-1421.

25. Cox, R.N., M. Kawai, A. Karlin, P.W. Brandt. 1979. J. Membrane Biol. 51, 145-159

26. Lester, H.A., M.E. Krouse, M.M. Nass, N.H. Wassermann, B.F. Erlanger. 1980. J. Gen. Physiol. 75, 207-232.

27. Maelicke, A. 1984. Angew. Chem. 96, 193-219.

28. Wolosin, J.M., A. Lyddiatt, J.O. Dolly, E.A. Barnard. 1980. Eur. J. Biochem. 109, 495-505.

29. Chang, H.W., E. Bock, E. Neumann. 1984. Biochemistry 23, 4546-4556.

30. Neubig, R.R., N.D. Boyd, J.B. Cohen. 1982. Biochemistry 21, 3460-3467.

31. Pennefather, P., D.M. Quastel. 1982. Br. J. Pharmacol. 77, 395.

32. Heidmann, T., J. Bernhardt, E. Neumann, J.-P. Changeux. 1983. Biochemistry 22, 5452-5459.

33. Rehm, H., H. Betz. 1981. Biochem. Biophys. Res. Commun. 102, 1385-1392.

SYNCHRONIZATION OF ELECTRICALLY AND CHEMICALLY EXCITABLE ION CHANNELS

H. Schindler
Biozentrum, Dep. Biophysical Chemistry, University Basel, CH-4056 Basel

W. Schreibmayer
Inst. for Med. Physics and Biophysics, University Graz, A-8010 Graz

Introduction

In the preceding presentation Dr. Neumann introduced data on reconstitution of purified Acetylcholine Receptor (AChR) molecules in planar membranes. It was evidenced that the two channels of the AChR-dimer open and close fully synchronized upon chemical excitation by agonist (1). Such channel synchronization has been observed earlier for Matrix-Protein from Escherichia coli (2). Unfortunately, the physiological significance of channel cooperativity remained obscure in these two examples although intriguing possibilities have been proposed. This appears different for the latest example of this kind, I'll report on today. When studying the Hodgkin-Huxley type potassium channel present in heart plasma membranes, we found voltage induced cooperativity among these channels, which provides a mechanistic clue to electrical excitability. Such a connection has been theoretically considered earlier, see such as ref. 3.

Molecular Basis of Nerve Activity

Results

The procedure to generate planar membranes containing potassium channels was as follows. The heart of a rat was dissociated into cells by collagenase treatment. The cells, a considerable fraction of which were still beating, were homogenized and centrifuged twice which yielded vesicular membrane fragments predominantly originating from the cardiovascular membranes. These membrane vesicles were mixed with (soybean) lipid vesicles. The mixture was introduced into one compartment of a planar membrane cell (cis side) where the other compartment was filled with only (soybean) lipid vesicles (trans side). Upon spontaneous formation of monolayers at the two water-air interfaces from the vesicles in the subphases, the two monolayers were combined to a planar bilayer by raising the water levels above an aperture.

Application of positive voltage (positive at the cis-side) reproducibly elicited conductances as shown in Fig. 1 for 80 mV. These conductances could be abolished by TEA (tetraethylammonium) as it is expected for a potassium selective channel. The current voltage relation, Fig. 2, is asymmetrical and shows a steep onset, at 40 mV for the conditions prevailing in the planar bilayer (symmetrical lipid and asymmetrical salt concentrations: 120 mM KCl and 20 mM NaCl at the cis side, 20 mM KCl and 120 mM NaCl at the trans side). Such a relation is characteristic for a (delayed) rectifying potassium channel and matches almost quantitatively the original data of Hodgkin and Huxley (4) for the potassium channel of nerve, taking only a shift along the voltage axis by some 50 mV into account. Towards a better understanding of the physical nature of this voltage dependence we investigated the conductances at the single channel level. For this, we reduced the amount of native membranes in the mixture with lipid vesicles. At rather

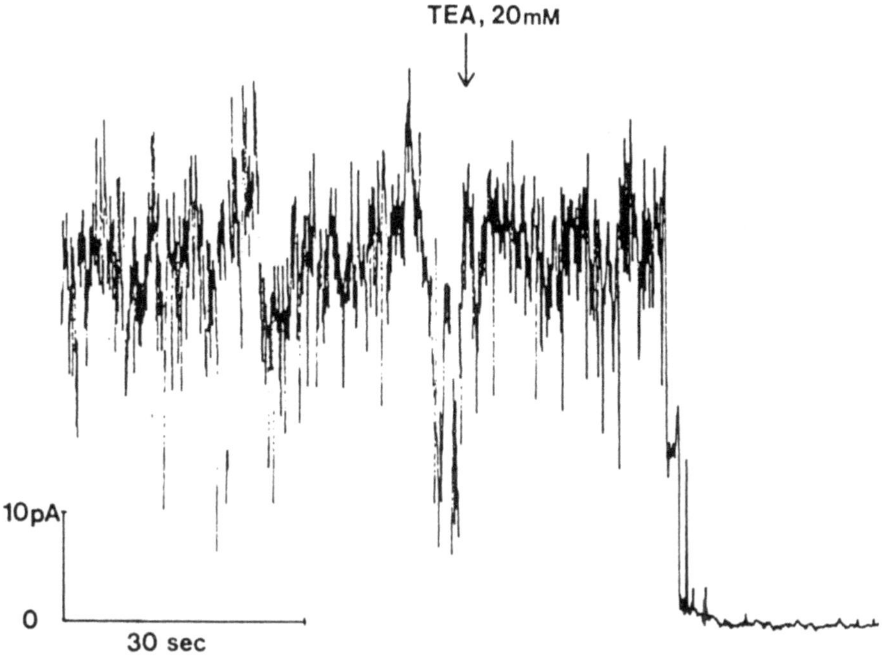

Fig. 1. TEA effect on membrane current.

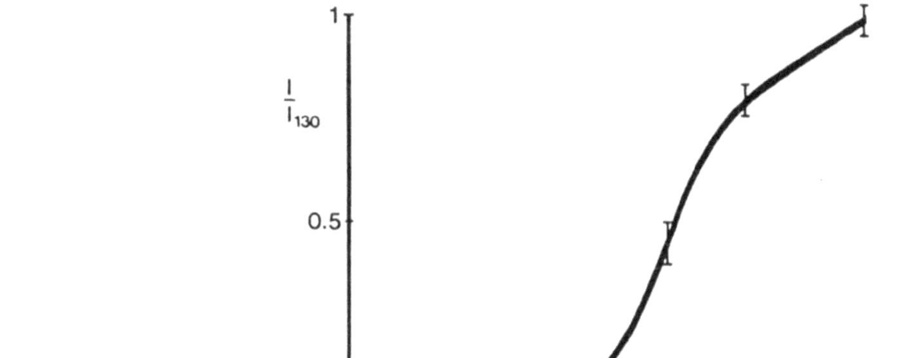

Fig. 2. Current voltage relation.

extreme dilutions (about 10^9 lipids per one channel, assuming a channel density in the native membrane of about 10 per μm^2, and 10^{10} lipids per bilayer) we observed single rectangular conductance changes. Their appearance changed dramatically with voltage. This to illustrate, we start from high negative voltage and proceed to the region of the steep conductance increase at positive voltage. Between -150 mV and about +30 mV single channels of 6 pS conductance were observed as exemplified in Fig. 3. The trace at -150 mV shows occasional chance overlappings of independent 6 pS events. The trace at +30 mV contains, in addition to the normal 6 pS steps, one considerably larger event of up to 4x6 pS. The question emerged whether such routinely observed large events were in any relation with the 6 pS events. That there is indeed an intimate relation became obvious at higher positive voltages. One example is shown in Fig. 4. A 4x6 pS event showed sublevels corresponding to 1 and 2x6 pS in a fashion which excluded the possibility of chance overlappings of independent 6 pS channel events. Other isolated events exhibited 3x6 pS levels as well.
The occurrence of time-correlated multiples of 6 pS levels became even more pronounced at still higher positive voltage. At 100 mV, Fig. 5, long lasting conductance levels were observed, exhibiting all even multiple of 6 pS up to 16 x 6 pS as sublevels, as indicated in the figure. At 150 mV repetitive closing is observed, Fig. 6, from predominantly 16 or 12x6 pS with intermediate restings at other even multiples of 6 pS.

The data clearly reflect cooperative transitions of a channel or a channel associate between zero and a voltage dependent conductance state, always being a multiple of 6 pS where lower multiples are only transiently populated. This conclusion does not require any assumption about structure or organization of

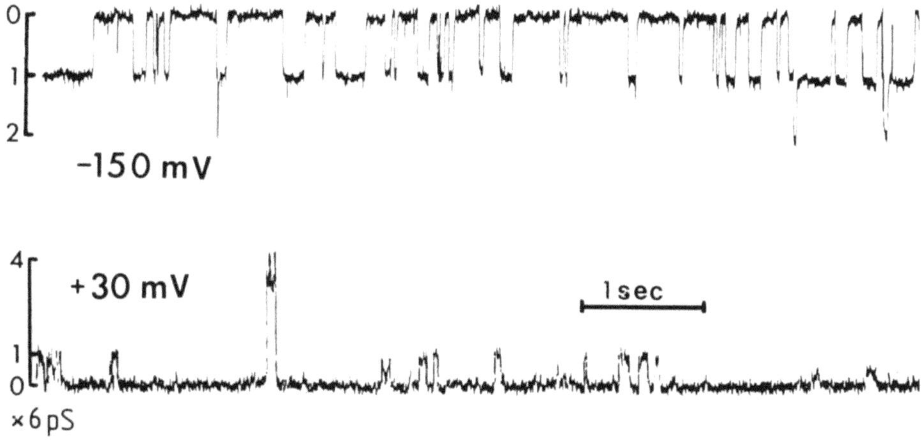

Fig. 3. Single channel traces representative for negative voltages to up to +30 mV.

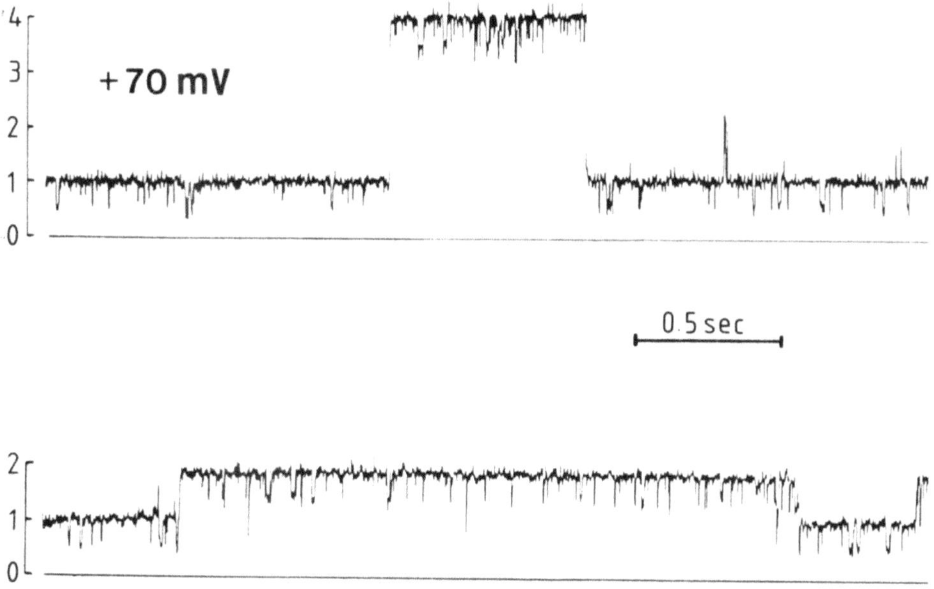

Fig. 4. Channel events typical for the voltage range between 40 and 80 mV.

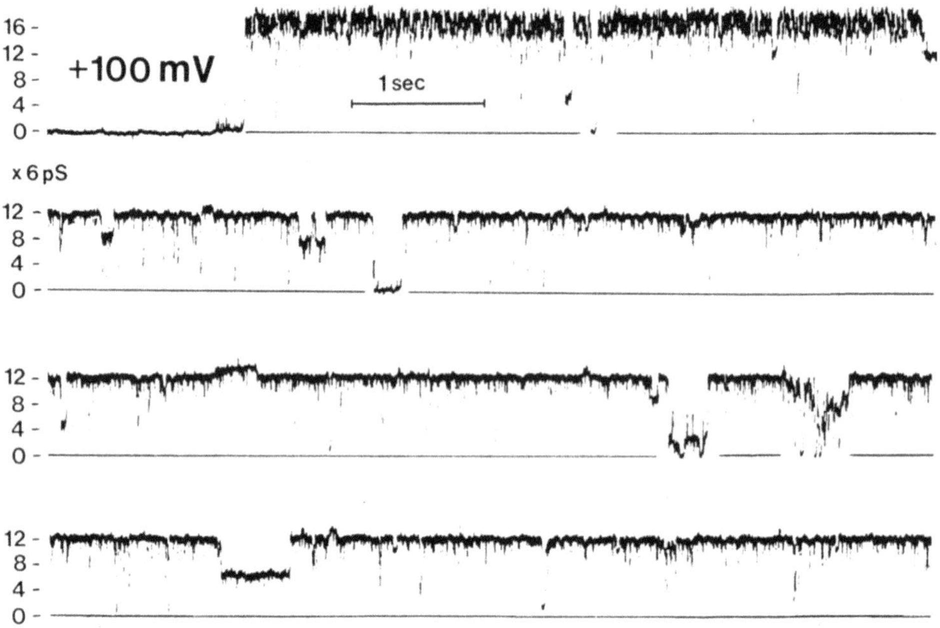

Fig. 5. Stabilized event at 100 mV showing multiples of 6 pS (indicated).

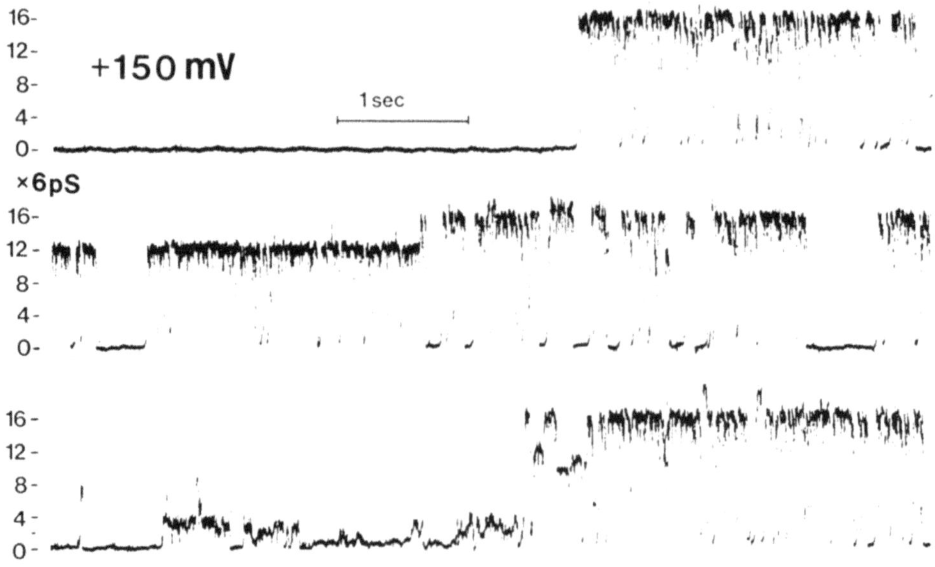

Fig. 6. Fast transient closing of multiples of 6 pS at 150 mV.

ion pathways. With respect to the latter, two possibilities may be considered. First, the time correlated equally spaced conductance levels may represent transitions between substates of one elementary channel, where the channel undergoes a cooperative transition from zero to its highest state when the voltage is increased from negative to positive values. Second, the equal spacing of time correlated conductance levels may reflect synchronized activity of an associate of topologically separate channels, where the channel associate undergoes a cooperative transition from all closed to all open when voltage is increased. In view of Fig. 5 and 6 the first possibility leads to rather unlikely constructions; it would require up to 16 structural changes of one channel each leading to the same conductance increment. A clear decision between the two possibilities is provided by the data in Fig. 7. All observed conductance levels (1x6 pS, 4x6 pS and 12x6 pS are included in the figure) showed the same potassium selectivity, which is evident from the same reversal potential of -26 mV (corresponding to a permeability ratio p(K)/p(Na) equal to about 6). It is not to see how one channel (first possibility) could increase its conductance by a factor of 16 without modification of its ion selectivity. We are therefore left with the second possibility, an associate of channels, which provides a natural explanation for the observed conservation of ion selectivity and the large multiple values of 6 pS.

If one accepts that the data reflect the synchronized action of associated channels there are again two main possibilities for the effect of voltage. One may consider an association-dissociation equilibrium of channel proteins with open channels only in oligomers where the equilibrium is shifted by voltage towards oligomers at more positive voltage. This to apply would require a current-voltage relation which is dependent on the concentration of channels in the membranes. However, the same I-V characteristic could be constructed from single

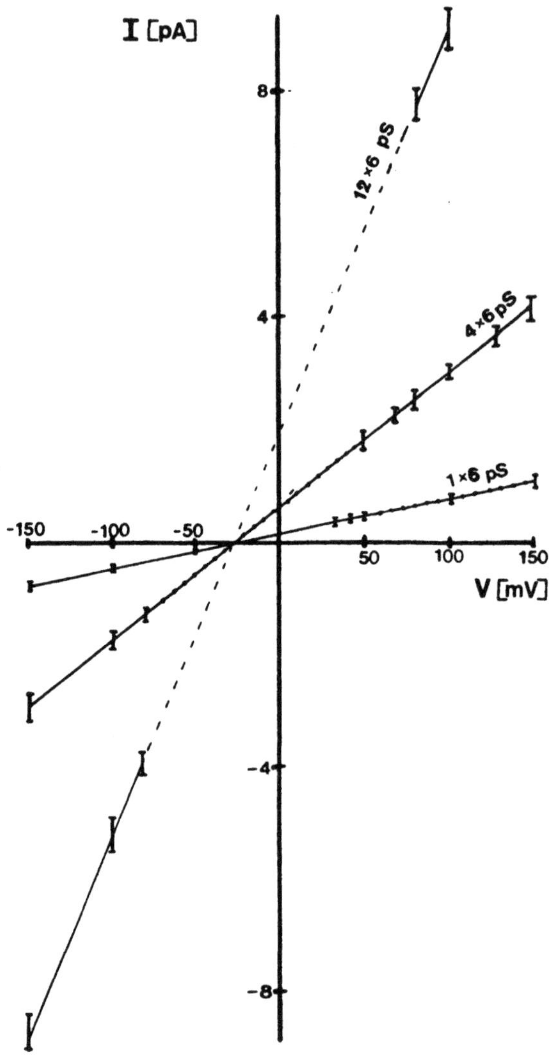

Fig. 7. Current-voltage relations of prominent current levels corresponding to multiples of 6 pS.

channel data as it was found in many-channel experiments (normalized current scales). From this we conclude that association dissociation is not involved but that there exist stable channel associates which undergo a voltage driven cooperative transition from all channels open to all channels closed. Each channel associate has the observed current-voltage relation as an intrinsic characteristic.

These conclusions are schematically summarized in Fig. 8.

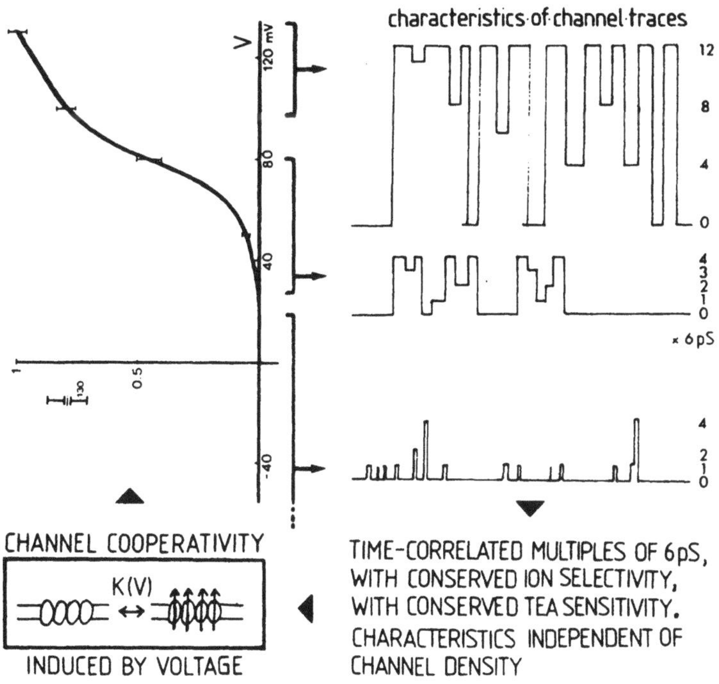

Fig. 8 Schematic representation of the main conclusion that the observed electrical excitability is due to channel cooperativity induced by voltage. The cartoon at the bottom-left exemplifies an all-or-none cooperative transition of 4 channels. At high positive voltage cooperativity extends over up to 16 channels.

How do these observations match with the channel traces observed on intact cells using patch clamp techniques (W. Schreibmayer et al., this issue)? The latter exhibit also a voltage dependence of the conductance level of isolated events, from 24 pS to up to about 100 pS. This indeed corresponds to main levels observed in the planar bilayer, of 4x6 pS up to 16x6 pS. Being aware of the reconstitution results, reinvestigation and particular design of patch clamp experiments now start to reveal also transiently populated multiples of 6 pS. It is of interest that a recently published patch clamp study on a potassium channel of heart (anomaleous rectifier) revealed equally spaced sublevels, which were multiples of 7 pS (5). In these and our patch clamp experiments, the sublevels are, however, too rarely populated in order to base any conclusion on their occurrence. In contrast to this, reconstitution provides experimental tools (parameters not specified here) to observe either ungraded (all-or-none) cooperativity as in patch clamp experiments or sufficiently graded cooperativity as in Fig. 4 to 6 which renders cooperativity apparent from direct inspection of the traces.
Formal description of such channel cooperativity leads to expressions which are similar to the Hodgkin-Huxley equations for electrical excitability of potassium channels (4). It will be interesting to investigate whether the electrically excitable sodium channel shows cooperativity as well.

Acknowledgement

This study was supported by a fund from the Roche Research Foundation to W.S. and by the Swiss National Fond, project 3.303.-0.82.

References

1. Schindler, H., F. Spillecke, E. Neumann. 1984. Proc. Natl. Acad. Sci. USA 81, 6222-6226.

2. Schindler, H., J.P. Rosenbusch. 1981. Proc. Natl. Acad. Sci. USA 78, 2302-2306.

3. Changeux, J.-P. 1969. In: Nobel Symposium 11 (A. Engström and B. Strandberg, eds.). Almquist & Wiksell, Stockholm, p. 235-256.

4. Hodgkin, A.L., A.F. Huxley. 1952. J. Physiol. 117, 500-544.

5. Sakmann, B., G. Trube. 1984. J. Physiol. 347, 641-657.

EFFECTS OF HISTRIONICOTOXIN ON THE SYNAPTIC AND CONDUCTING MEMBRANES OF THE ELECTROPLAX

E. Bartels
Universidad delle Valle, Cali, Columbia

The resting and action potentials of the excitable membrane of the electroplax of Electrophorus electricus were measured using conventional intracellular microelectrodes. Histrionicotoxin (HTX) at low concentrations (5-10 μM) blocks the postsynaptic potential. At 100 fold higher concentrations HTX blocks the directly evoked action potentials of the conducting membrane. The pH dependence of the blockade by HTX at synaptic channels is different from that at the conducting membrane; at the synapse HTX is more potent at acid pH, while at the conducting membrane it is more potent at basic pH. HTX at high concentration antagonizes the effects of batrachotoxin similar to local anesthetics, which indicates an effect on the BTX sensitive sodium channels involved in action potential generation. While the effects of HTX on synaptic channels are dependent on concentration, time and pH, the effects on the channels are, in addition, dependent on previous channel activity suggesting interaction of HTX with the activated forms of these channels similar to some local anesthetics.

Since the ionophore is part of the acetylcholine receptor it is possible that Nachmansohn's theory is supported by the results of the action of local anesthetics; there is no other alternative satisfactory explanation for the control of the ion channels in the conducting membrane than a bimolecular reaction to account for the very rapid change in permeability.

This work was supported by a grant from UNESCO.

Molecular Basis of Nerve Activity

THIAMINE TRIPHOSPHATE AS SPECIFIC OPERATING SUBSTANCE IN AXONAL CONDUCTION

E. Schoffeniels

Laboratory of general and comparative Biochemistry, University of Liège, 17, place Delcour, B-4020 Liège, Belgium

D.G. Margineanu

Laboratory of Biophysics, Faculty of Biologic Sciences, University of Bucarest, Spl. Independentei 91-95, Bucarest 76201, Romania

Introduction

Otto Meyerhof whose deep influence on the scientific thought of David Nachmansohn is well recognized, used to call ATP the specific operating substance in muscle contraction. In his well known theory of bioelectrogenesis, David Nachmansohn borrowed from Meyerhoff this expression and applied it to acetylcholine. Since this symposium is dedicated to the memory of David Nachmansohn,we find it appropriate to follow suit by taking up David's favorite expression to define the property of thiamine triphosphate which is certainly an important substrate in the biochemical cycle controlling the impedance variation in axonal membranes.

The study of the mechanisms by which various types of excitable cells respond to stimuli grew up in the last decades as an important interdisciplinary chapter of contemporary biological sciences, a topic to which not only innumerable research papers, but also many treatises, monographs (1,2,3) or symposia (4,5) are devoted (these references are mere examples). More specifically, the generation and conduction of the nerve impulse viewed as a well defined sequence of changes, in axolemmal ionic permeability stands by now as the example of a major biological process whose description in mechanistic terms is most detailed.

Molecular Basis of Nerve Activity

Surprisingly enough, within a flood of literature, the biochemistry of impulse conduction, i.e. the molecular bases of this phenomenon, is now almost a non-existing subject, while it seems to be a dogma of electrophysiology that the generation and propagation of the action potential does not involve any biochemical process, though the transmission of excitation between adjacent cells is acknowledged to be most often chemically mediated. As an illustration of this state of affairs, we notice that in quite recent presentations of excitable cells (6), the molecular basis of the nerve impulse are not even mentioned at all, in contrast to muscle contraction.
The main aim of this paper is to question - in the light of some new data - the completeness of the commonly held electrophysiological image of the nerve action potential, to reevaluate the almost forgoten possibility that electrogenesis in excitable cells could have molecular bases involving biochemical processes, and to plead for a less mechanistic description of this phenomenon. Upon concluding on the plausibility of biochemical conditioning of the nerve impulse, we propose a cycle involving phosphorylation-dephosphorylation of membrane proteins with thiamine triphosphate as phosphate donor thus defining a biochemical basis for the impedance variation in electrically excitable cells.

Electrophysiological description of the action potential and its failures at molecular level

The basic electrophysiological tool which allowed the analytical approach of the nerve impulse was the voltage clamping of the squid giant axon, i.e. the maintaining (for arbitrarily long time intervals) of the transmembrane potential E_m at a fixed value. This allows to eliminate the capacitive currents so that all the transmembrane current is carried by the ionic flows which cross the membrane down the electrochemical gradients. If one thinks in microscopical (atomo-molecular) terms, the constancy of the transmembrane electric field implies that the electric charges belonging to the membrane will no longer receive energy input from the electric field once the steady distribution was attained.
Based on this possibility, A.L. Hodgkin and A.F. Huxley (7,8) discovered the quite different time evolution of the separate currents carried by

potassium and by sodium ions. In voltage clamp conditions, the sodium current (E_{Na}), and thus the sodium partial conductance of the membrane (g_{Na}), rapidly raises and then falls down up to the resting values, while the potassium current (E_K) and its corresponding conductance (g_K) increase up to steady values, which are stable as long as E_m is kept constant.
Hodgkin and Huxley proposed empirical equations describing g_{Na} and g_K as functions of the membrane potential, the parameters appearing in those equations being chosen so as to fit the experimental data. Later on, this description was interpreted as implying that the excitable membrane possess some structural units (the ionic conducting sites) which can be either in a high or low conducting state, the transition among these states being controlled by the electric field. Their equations not only accomodated the data on the ionic currents but also predicted the existence of distinct conducting sites through which Na^+ and K^+ cross the membrane, and of intramembrane charge displacements when the conducting sites change their configuration. Much later, the surface density of the ionic sites was estimated on the basis of pharmacological and refined electrophysiological studies (9,10) and the intramembranous currents of both sodium (11,12) and potassium (13) were readily measured.
The success of H-H approach led to similar descriptions for myelinated nerve fibres as well as for cardiac and muscle fibres (for comprehensive reviews see 1, 2) and shaped the conceptual frame of the cellular electrophysiology. Contemporary ionic theory of the action potential epitomizes Bernstein's view of bioelectrogenesis, describing it in purely physical terms with the postulate that no molecular transformations in the chemical sense (i.e. involving redistributions of chemical bonds) are involved.
The generation and conduction of the action potential are considered solely as a specific sequence of ionic flows driven by the electrochemical gradients which are maintained by the activity of transport ATPases ($Na^+ + K^+$-ATPase and Ca^{2+}-ATPase). The axolemma is viewed as an assembly of conducting sites specific for either Na^+, K^+ or Ca^{2+}, distributed at surface densities of the order of $10-10^{-3}$ μm^{-2} in the lipid bilayer which also accomodates a denser distribution of pumps. The conducting sites would act as mere dissipators of the metabolic free energy stored only in the form of electrochemical gradients by the ionic pumps.

Though such an image might seem at a first glance self-consistent, every molecular biologist would feel uneasy by its mechanistic appearance. The functional components of the excitable membrane appear as granted once for ever, without any other connection with the metabolic dynamics of the cell, except for the fuelling of the pumps, In terms of this description, the propagation of the action potential appears as an unique example of a major biological event not being directly coupled with the chemical dynamics of the cell. Thus, it is no surprise that this description fails to account for the molecular aspects of bioelectrogenesis - some of which we shall discuss here - and it has to be completed with biochemical approaches.

It is a general agreement - reinforced by noise analysis (14,15) - that both potassium and sodium conducting sites assume two conductance states : a resting, low conductance state (LCS) and an active, high conductance one (HCS). But the clear distinction revealed by voltage clamp experiments, namely that at maintained depolarizing voltages the high conductance state of potassium sites is stable while for sodium sites it is transitory, shows that the sodium conductance of the excitable membrane cannot be simply field-dependent. In our view, this very fact is a strong argument for another form of energy involved in the operation of excitable membranes, along with the changes in electric field.

With the assumption that the time course of the transmembrane flows of sodium in the squid giant axon simply represents the time course of a change in permeability, Hodgkin and Huxley (2) predicted that at lower temperatures the extra influx of sodium associated with a nerve impulse would be higher (because the specific sites remain in a HCS for a longer time) with a Q_{10} factor 1/3. But undisputed experimental data (16) revealed a smaller extra influx of sodium at low temperature in opposition with a simple physical operation of the gateways but in accord with the assumption of chemical changes controlling their conductance.

It is commonly known that, in the study of all biological phenomena, the energetic aspects are the key step on the way from macroscopic appearances to the understanding of the intimate molecular mechanisms. As for the particular case of the nerve impulse,though everybody agrees that it is a dissipative and entropy producing process, its energetics is by now poorly understood.

Very accurate calorimetric measurements (for review see 17) showed that the extra production of heat in the active nerves directly related to impulse propagation (the so-called initial heat), is always biphasic, comprising a heat production during the rising phase of the action potential (AP), followed by the reabsorption of about 80 % of it during the falling phase. All the attempts to explain the initial heat of activity, and particularly the heat absorption in terms of only the ionic theory failed so far to reach any quantitative agreement with experimental data (18). Previously, we analyzed at length (19) the possible nature of the endergonic processes which occur during the action potential and concluded that the membrane molecular machinery underlying the nerve impulse has to be considered a thermodynamically open system which proceeds not only energy by means of the electric field, but also matter. The consideration of the structural transitions in the ionic gateways as a result of specific biochemical processes (along with charge redistributions and dissipative ionic flows) actually led to a fair quantitative agreement between the computed values and the microcalorimetric measurements and fully accounted for the timing of heat production and reabsorption (20).

Another failure of the electrophysiological description of bioelectrogenesis is its newly revealed incompatibility with classical thermodynamics (21). The voltage-dependent rate constants of the H-H first order rate equations for the ionic currents are incompatible with Boltzmann distributions of the charged components supposed to control membrane ionic conductances. The conclusion of this analysis is that the only possibility to save the first order rate equations is to assume that the components involved in conductance variation during the impulse have variable valencies, i.e. their electric charges undergo cyclic variations. Obviously, the most plausible mechanism for such variation is the cyclic chemical modification of the membrane components.

To conclude this discussion, it must be observed that all the above mentioned failures of the electrophysiological description of the action potential do not refer to minor or marginal aspects subject to revision following technical improvements but to some of the most basic features. Before presenting, the proposed biochemical changes controlling the action potential, we think it is worth evaluating how plausible appears now such an approach.

Plausibility of biochemical changes connected with bioelectrogenesis

The conception that the nerve impulse would be simply a bioelectric event strictly located within the excitable membrane and not involving any biochemical processes was probably best favoured by the discovery that internally perfused giant axons (apparently being mere membranous tubes devoid of axoplasm) had almost unchanged excitability and conduction properties when perfused with adequate saline solutions (for a detailed presentation of this technique see 22).

However, it was proved that in internally perfused giant axons energy metabolism (such as ATP synthesis) does continue and an enzyme system which specifically reacts with glutamate and aspartate is located directly on the inner face of the axolemma (23). More recently, a network of interwoven filaments firmly attached to the inner face of the axolemma was visualized (24), and several lines of evidences (such as the fact that electrical stimulation and the application of TTX and STX cause the release of proteins from this network (25) accumulated supporting the idea that this cytoskeletal network is involved in excitability. Even without accepting the interpretations offered by these authors, their findings confirm that the nerve membranes, along with every plasma membrane, are under cytoplasmic metabolic control (26).

Conversely, it becomes also evident that the changes in transmembrane electrical potentials do not represent solely the way by which signals are propagated in excitable membranes as in electric cables with batteries but they occur in many other cells in response to various chemical agents, apart the well known chemical synaptic transmitters. Thus, catecholamines and a number of peptide hormones either hyperpolarize or depolarize the membranes of target cells (27),the cells of endocrine (such as pancreatic beta cells) and exocrine (such as the pancreas and the parotid) glands show changes in electrical potential when the secretion is stimulated (28) and the development of a wide variety of biological systems (among which insect follicles, frog embryos, amputated amphibian limbs and even mammalian skin) is directly associated and probably controlled by ionic currents (29,30). Thus, it clearly appears that the changes in transmembrane potential do not occur only in nerve and muscle cells by mechanisms supposed to be unrelated to the cellular biochemistry, but in many other cells in connection with various functions.

Over several decades, Nachmansohn (31,32) provided experimental evidences that most nerves contain the enzymatic equipment needed for the synthesis and hydrolysis of acetylcholine (ACh) and forcefully argued that electric signal propagation along nerve fibres is an epiphenomenon generated by the opening and closing of axonal ionic gateways operated by an ACh regulated system similar to that existing in synaptic membranes. However two types of arguments were invoked to discard this conception.

First is the presumed slowness of any sequence of biochemical changes which are supposed - mostly on sensorial and intuitive basis (!) - not to meet the rapidity of nerve impulse propagation. However, the time course of the transition of the ionic conducting sites, of the order of 1 ms, is fully compatible with a cyclic sequence of reactions taking place within a highly ordered structure such as the membrane (for more details see 33). It should be observed that the acceptance of chemical reactions as a basis of conductance variation in nerve membrane does not necessarily imply that the ionic passage through the membrane is in itself a chemical process. Anyhow, it is unanimously accepted that the single file diffusion of ions through the conducting sites is "formally identical to simple enzyme kinetics" (34) and we want to stress that the resemblance could well be more than formal.

The absence of acetylcholinesterase from some nerves, such as the sensitive fibres of mammalian dorsal roots, was also invoked against the universality of the ACh role in nerve conduction. But, on the one hand, in most cases when the components of the ACh system were seriously checked in conducting tissues, they were indeed found (32 pp. 20-44), and on the other hand, even in some axons lacking the receptor for ACh, there is a component having great affinity for nicotine, which ressembles the ACh receptor from many viewpoints (35) so that the occurrence of a biochemical cycle of impedance variation (though possibly different from that of ACh) is once again supported.

The importance of thiamine derivatives

Thiamine compounds are largely distributed in pratically all types of cellular differentiation where, as the pyrophosphate derivative, thiamine acts as a coenzyme in transketolase reactions and in the oxidative decarboxylation of pyruvate and α-ketoglutarate. It has generally been assumed that the symptomatology of thiamine-deficient animals should be explained in terms of an impairment of the metabolic pathways requiring thiamine pyrophosphate as coenzyme (36). Others have proposed a direct role for thiamine in nerve conduction but without reference to a precise model (37,38,39,40). Our results suggest for these compounds a specific role related to the generation of the action potential.
The presence of thiamine-phosphorylated derivatives in the nervous system, central or peripheral, is a cell documented fact (41,42). It is also known that the electrical activity of nerves or the application of neurotropic compounds such as tetrodotoxin (TTX), local anaesthetics, acetylcholine, diisopropylfluorophosphonate (DFP), 5-hydroxytryptamine (5-HT), etc. are accompanied by a contemporary release of thiamine in the extracellular fluid (40,43). Since then, much work has been devoted to the study of thiamine and its derivatives. However no clear cut role has been assigned to this compound in the functioning of the nervous system : is it acting in the metabolism and release of neurotransmitters in relation to its property of coenzyme or are we dealing with a more specific property ?
Cooper and his colleagues (44,45) have rejected the idea that thiamine could be involved in the permeability change of the sodium conducting sites following their observation that thiamine antimetabolites or thiamine itself had no appreciable effect on either the magnitude or time course of ionic currents. This does not seem to us a conclusive argument since we do emphasize the idea that it is the triphosphate derivative of thiamine which is the specific operating substance.

The endogenous source of phosphate

While looking for a substrate that could be implicated in the bimolecular reaction steps to explain the energetics or the sodium conductance change in axonal membranes, we have found that specific proteins were phosphorylated and that the level of phosphorylation is affected by electrical stimulation and various neurotropic compounds (46,47,48,49,50). The protein forming the sodium conducting site, the Na^+-conductin, is also phosphorylated (51), a result that has been confirmed by others (52).
That the protein fraction eluted from the lectin column (see Dandrifosse et al., this volume) contains the specific protein is demonstrated by a specific binding of labelled TTX solely noticeable in this fraction.
As indicated by the results of Table 1, the addition of TTX during the

Table 1. Thiamine-derivatives (pmol/mg prot.) associated with the Protein Fraction containing the Na^+-conductin isolated from the main electric Organ of *Electrophorus electricus* L.

Fraction	ThTP	ThDP	ThMP	Th
DEAE wash				
control	-	2.6 - 4.6	-	-
TTX	-	1.1 - 6.6	14.0 -15.1	0.92 - 4.3
Veratridine	0.3 - 0.9	2.6	0.3 - 1.1	-
Na^+-conductin				
control	7.4 -11.6	10.6 -15.7	33.3 -98	-
TTX	0.7 - 2.2	2.2 - 3.7	5.1	1.5
Veratridine	7.4 -17.2	12.3 -38	34.4 -67.5	-

TTX (3×10^{-7}M) and veratridine (10^{-4}M) are added during the purification procedure before the DEAE step. Thiamine derivatives are determined as fluorescent compounds in HPLC. The figures quoted are extreme values obtained by measuring several aliquots of the same sample (after 53,54).

purification procedure displaces the thiamine derivatives as revealed by the amounts found in the DEAE wash and also seems to shift the equilibrium toward the less phosphorylated forms. The lack of effect of veratridine on thiamine-phosphorylated compounds binding is not at all surprising since it fits with previous results showing that there at least four different bindings sites in addition to the Na^+-binding sites (for a short review, see 55). Thus veratridine together with BTX and dihydrograyanotoxin II bind to a site category different from that of TTX (56).
Notice that at the concentrations used, veratridine does not affect the lipid structure as indicated by results of spin labelling showing no effect of veratridine on the ESR spectrum (M. Delmelle, unpublished results).

A biochemical model of impedance variation cycle

Our results as well as those of others (52) being so far in agreement with the idea that, in axonal membrane, the specific protein forming the sodium site is phosphorylated, it can reasonably be suggested that the net state of phosphorylation directly controls the conductance of the membrane (46,47).

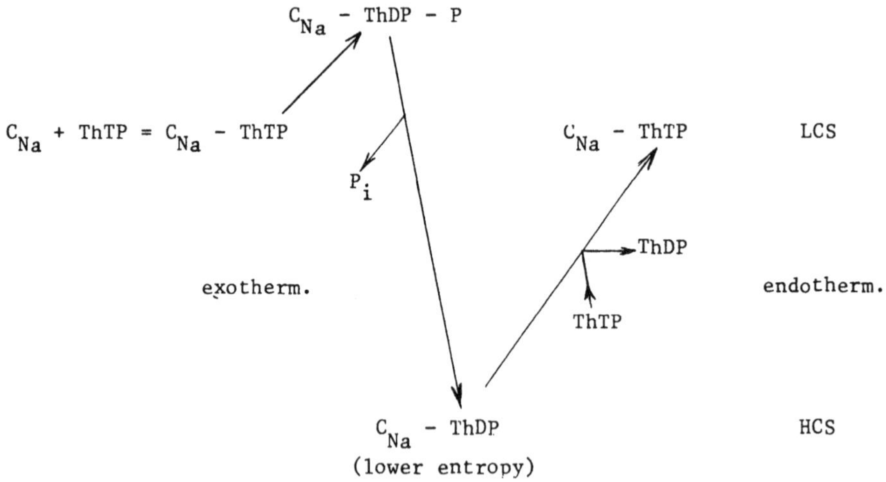

Fig. 1. Role of thiamine triphosphate (ThTP) in the control of the conductance to sodium ions (after 53).

The use of pathological models will necessarily be the very next step of our inquiries as soon as other experimental evidences will permit to confirm the following working hypothesis. Our data can be integrated to fit in a biochemical cycle involving phosphorylation-dephosphorylation processes explaining the impedance variation cycle.

This is done in Fig. 1 where on the right hand side the conducting state of the sodium-conductin is indicated. LCS means low conductance state while HCS is the high conductance state. In this model the sodium-conductin, C_{Na}, can assume a so-called closed configuration or LCS and the open configuration or HCS.

In resting conditions, the LCS is made of a population of C_{Na} that is partially associated with thiamine triphosphate (C_{Na}-ThTP). Upon electrical stimulation i.e. an adequate change in the electric field across the membrane, the terminal phosphate is liberated cooperatively in a large population of C_{Na}-ThTP, but remains fixed on the protein. This is the activated complex C_{Na}-ThDP-P that undergoes a dephosphorylation leading to the HCS or open configuration of the sodium-conductin : C_{Na}-ThDP. This gives rise to the ascending phase of the action potential. It is an exothermic process characterized by a decrease in entropy (20). The change in electric field that follows favours the displacement of ThDP which is replaced by ThTP, an endothermic process corresponding to the descending phase of the AP because it brings back the sodium-conductin to its resting state (LCS). Notice that C_{Na}, as shown in patch clamp studies of reconstituted liposomes (57) can be in two configurations, closed or open, in an electric field, but it does not exhibit the cooperativity or the threshold behaviour. However, what gives an action potential is a synchronization of the opening of many Na-conductin molecules under the influence of a threshold depolarization. In our model this cooperativity can only be obtained if C_{Na} is associated to ThTP. The form C_{Na}-ThTP does not change configuration in the electric field as is the case with C_{Na} and should be electrically silent in reconstituted liposomes. To activate the complex C_{Na}-ThTP one needs to apply a threshold depolariation.

Finally, it is worth recalling as mentioned above, that it is already known that TTX, as well as electrical stimulation or various neurotropic compounds, induces a release of thiamine from the axonal membrane (40). We therefore like to suggest that thiamine triphosphate could play a fundamental role in controlling the conductance to sodium ions of axonal membranes by providing the adequate means of bringing among Na-conductin molecules the cooperativity needed to explain the all-or-none response as well as the threshold behaviour. The above cycle of impedance variation in the axonal membrane appears fully compatible with the observed heat changes associated with impulse propagation. Though by now reliable calorimetric studies about the release of the terminal inorganic phosphate from the thiamine triphosphate and the subsequent rephosphorylation are lacking, we can use as a merely orientative value for the heat evolved when P_i is released from C_{Na}-ThTP that one accurately measured by Woledge (58,59) in the case of P_i release from the complex myosin-ADP-P_i. He found that this is an exothermic reaction producing a heat of 150 KJ/mol. Such a figure would imply that the passage of each conducting site from the LCS to the HCS produces $\pm$ 8 x 10^{-20} cal, i.e. a nerve whose axons have diameters of $\pm$ 100 μm and $\pm$ 100 conducting sites per μm^2 will produce $\pm$ 1 μcal/g fresh tissue. The agreement of this estimation with the heat production of the nerves (17,18) is so encouraging that we think it cannot be simply coincidental but strongly favors at least the plausibility of the proposed cycle.

The overall reaction of ATP splitting in the muscle was found to liberate only 65 KJ/mol (39) which implies that from the 150 KJ/mol evolved in the exothermic step, 85 KJ/mol are reabsorbed in the endothermic ones, when the diphosphate is replaced by the triphosphate. Thus a reabsorption of about 58 % of the heat produced is immediately accounted for when refering to the dephosphorylation-rephosphorylation cycle of impedance variation in the nerve membrane. Once again, even a rough estimation based only on related thermochemical data, agrees quite satisfactorily with the measurements on the nerve so that what we (19,20) previously termed as heat associated with the structural changes in the excitable membrane, appears now as largely connected with the dephosphorylation-rephosphorylation cycle of sodium conductin.

Conclusion

The description of action potential propagation in excitable membranes solely in terms of physical permeability changes without any biochemical correlates, though impressing by details and mechanistic coherence, is based on a static image of the membrane and on the assumption that the only dependence on the cellular metabolism is by means of the ionic pumps which maintain the transmembrane electrochemical gradients. Thus it lacks almost completely molecular basis, so that it is no surprise that it fails to accomodate many data at subcellular and molecular level, and is beset by thermodynamic inconsistency.
Without claiming at all to discard the ionic theory of excitation we argue that it is not simply justified but inescapable to consider the biochemical metabolic processes which underlie the permeability changes in excitable membranes. As a further step toward the unraveling of the molecular basis of the nerve impulse, a cycle of sodium conductance variation involving membrane proteins phosphorylation and thiamine triphosphate as specific operating substance, is proposed.

Acknowledgement

This work was supported by grant nr 2.4518.80 from the Fonds de la Recherche Fondamentale Collective (Belgium).

References

1. Cole, K.S. 1968. Membranes, Ions and Impulses. University of California Press, Berkeley.

2. Luttgau, H.C., H. Glitsch. 1976. Membrane Physiology of Nerve and Muscle Fibres. Gustav Fischer Verlag, Stuttgart.

3. Vasilescu, V., D.G. Margineanu. 1982. Introduction to Neurobiophysics. Tunbridge Wells.

4. Moore, J.W. (ed.). 1975. Fed. Proc. 34, 1317-1407.

5. Schoffeniels, E., E. Neumann (eds.). 1980. Neurochem. Internat. 2, 1-354.

6. Keynes, R.D., D.J. Aidley. (1981). Nerve and Muscle. Cambridge University Press, Cambridge.

7. Hodgkin, A.L., A.F. Huxley. 1952. J. Physiol. Lond. 116, 449-472.

8. Hodgkin, A.L., A.F. Huxley. 1952. J. Physiol. Lond. 117, 500-544.

9. Keynes, R.D., J.M. Ritchie, E. Rojas. 1971. J. Physiol. Lond. 213, 235-254.

10. Ritchie, J.M., R.B. Rogart. 1977. Proc. Natl. Acad. Sci. USA 74, 211-215.

11. Armstrong, C.M., F. Bezanilla. 1973. Nature 242, 459-461.

12. Keynes, R.D., E. Rojas. 1974. J. Physiol. Lond. 239, 393-434.

13. Bezanilla, F., M.M. White, R.E. Taylor. 1982. Nature 296, 657-658.

14. Conti, F., E. Neher. 1980. Nature 285, 140-143.

15. Sigworth, F.J. 1977. Nature 270, 265-267.

16. Landowne, D. 1973. Nature 242, 457-459.

17. Abbott, B.C., J.V. Howarth. 1973. Physiol. Rev. 53, 120-158.

18. Howarth, J.V. 1975. Philos. Trans. Roy. Soc. Lond. B 270, 425.

19. Schoffeniels, E., D.G. Margineanu. 1981. J. Theor. Biol. 92, 1-13.

20. Margineanu, D.G., E. Schoffeniels. 1977. Proc. Natl. Acad. Sci. USA 74, 3810-3813.

21. Chapman, J.B. 1980. J. Theor. Biol. 85, 487-495.

22. Tasaki, I. 1968. Nervous Excitation : A Macromolecular Approach. C.C. Thomas, Springfield, Illinois.

23. Martin, K., T.I. Shaw. 1970. J. Physiol. Lond. 208, 171-185.

24. Metuzals, J., I. Tasaki. 1978. J. Cell Biol. 78, 597-621.

25. Baumgold, J., P. Gallant, S. Terakawa, H. Pant. 1981. Biochem. Biophys. Res. Comm. 103, 653-658.

26. Singer, S.J. 1974. Ann. Rev. Biochem. 43, 805-833.

27. Zierler, K., E.M. Rogus. 1981. Fed. Proc. 40, 121-124.

28. Williams, J.A. 1981. Fed. Proc. 40, 128-134.

29. Jaffe, L.F., R. Nuccitelli. 1977. Ann. Rev. Biophys. Bioeng. 6, 445-476.

30. Jaffe, L.F. 1981. Fed. Proc. 40, 125-127.

31. Nachmansohn, D. 1953. Metabolism and Function of the Nerve Cell (Harvey Lectures, appeared 1955). Academic Press, New York.

32. Nachmansohn, D., E. Neumann. 1975. Chemical and Molecular Basis of Nerve Activity. Academic Press, New York.

33. Neumann, E., J. Bernhardt. 1977. Ann. Rev. Biochem. 46, 117-141.

34. Hille, B. 1979. In : Membrane Transport Processes (C. Stevens and R.W. Tsien, eds.). Vol. 3. Raven Press, New York, pp. 5-16.

35. Denburg, J.L., M.E. Eldefrawi, R.D. O'Brien. 1972. Proc. Natl. Acad. Sci. USA 69, 177-181.

36. Peters, R.A. 1963. Biochemical Lesions and Lethal Synthesis. Pergamon Press, Oxford.

37. Minz, B. 1938. C.R. Soc. Biol. Filiales 127, 1251-1253.

38. Von Muralt, A. 1947. Vitam. Horm. 5, 93-118.

39. Itokawa, Y., J.R. Cooper. 1969. Science 166, 759-760.

40. Itokawa, Y., J.R. Cooper. 1970. Biochem. Pharmacol. 19, 985-992.

41. Gubler, C.J., Fujiwara, M., P.M. Dreyfus (eds.). 1976. Thiamine. Wiley, New York.

42. Sable, H.Z., C.J. Gubler (eds.). 1982. Thiamin, Twenty Years of Progress. Ann. N.Y. Acad. Sci. 378.

43. Itokawa, Y., R.A. Schulz, J.R. Cooper. 1972. Biochim. Biophys. Acta 266, 293-299.

44. Goldberg, D.J., J.R. Cooper. 1975. J. Neurobiol. 6, 435-452.

45. Goldberg, D.J., T.B. Begenisich, J.R. Cooper. 1975. J. Neurobiol. 6, 453-462.

46. Schoffeniels, E. 1980. Adv. Chem. Series, Amer. Chem. Soc. 188, 285-297.

47. Schoffeniels, E. 1980. Neurochem. Internat. 2, 81-93.

48. Dandrifosse, G., E. Schoffeniels. 1980. Neurochem. Internat. 2, 95-100.

49. Bontemps, J., G. Dandrifosse, E. Schoffeniels. 1980. Neurochem. Internat. 2, 101-110.

50. Schoffeniels, E., G. Dandrifosse. 1980. Proc. Natl. Acad. Sci. USA 77, 812-816.

51. Bontemps, J., G. Dandrifosse, E. Schoffeniels. 1981. IRCS Med. Sci. 9, 70.

52. Costa, M.R.C., J.E. Casnellie, W.A. Catterall. 1982. J. Biol. Chem. 257, 7918-7921.

53. Schoffeniels, E. 1983. Arch. Internat. Physiol. Biochim. 91, 233-242.

54. Schoffeniels, E., G. Dandrifosse, L. Bettendorff. 1984. J. Neurochem. 43, 269-271.

55. Schoffeniels, E. 1982. In : Membranes and Transport (A.N. Martonosi, ed.). Plenum Press, New York, pp. 379-383.

56. Catterall, W.A. 1975. Proc. Natl. Acad. Sci. USA 72, 1782-1786.

57. Barchi, R.L. 1983. 9th Meet. Int. Soc. Neurochemistry, Vancouver, B.C. July 10-15.

58. Woledge, R.C. 1977. In : Applications of Calorimetry in Life Sciences. W. de Gruyter, Berlin, pp. 183-197.

59. Curtin, N.A., R.C. Woledge. 1978. Physiol. Rev. 58, 690-761.

DOSE RESPONSE CURVES AT THE NEUROMUSCULAR JUNCTION

R. Sterz, K. Peper
II. Physiologisches Institut, University of Saarland,
D-6650 Homburg/Saar

Introduction

Experimental dose response curves are a valuable tool in the investigation of receptor function. Experiments were carried out on nerve muscle preparations of frog (M. cutaneus pectoris) and rat or mouse (M.omohyoideus, diaphragm, M.soleus and M.extensor digitorum longus). It was necessary to work on monolayer preparations of muscles permitting visual control under a microscope fitted with Normarski optics (500x) (1). Muscle fibres were voltage clamped at the neuromuscular junction site, and current changes were measured which result from the interaction of drugs (agonists) with the acetylcholine--receptor (AChR) and associated ionic channels. The amount of channels opened is a nonlinear function of the agonist concentration and is, among others, dependent on the properties of the particular agonist.

For quantitative measurements the agonist concentration can be altered by adding different amounts of agonist to the bath fluid in which the nerve muscle preparation is kept (Ringer's solution for frog, Tyrode solution for mammalian preparations). Bath application, however, has two major disadvantages. Firstly, many receptors are activated at the same time resulting in the opening of a large number of ionic channel at higher agonist concentrations. Thus, endplate currents might be too high to be handled by a conventional voltage clamp circuit. Secondly, receptors can desensitize. This means that the endplate current decreases even though the agonist concentration

Molecular Basis of Nerve Activity

is still the same. Both problems can be somehow circumvented by quantitative application of the agonist as described below.

Results

A) Equilibrium dose response curves using quantitative ionophoresis.

The tip of a glass micropipette filled with a saturated solution of the agonist is placed under visual control at a given distance above a geometrically prominent point of the endplate. In case of frog, this point is the end of a single straight running nerve terminal. On mammalian endplates, the tip of the pipette is placed above the center of the visible endplate (see fig. 1).

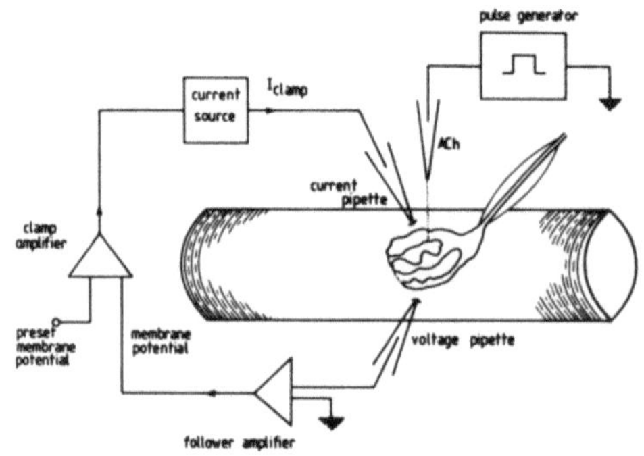

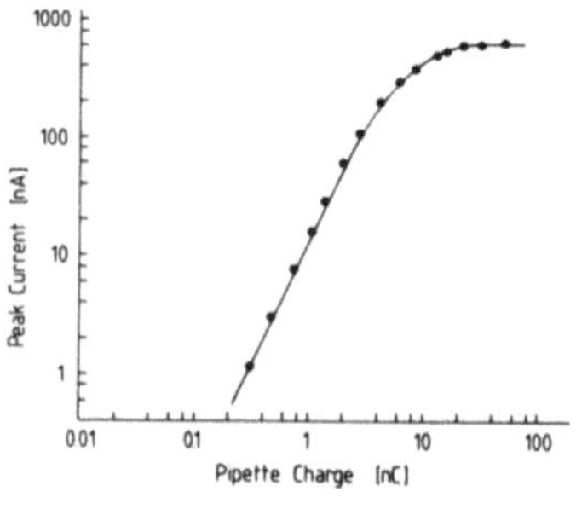

Fig. 2: Dose response curve, ACh, mouse endplate.

Fig. 1: Experimental arrangement for measuring dose response curves by quantitative ionophoresis.
Two glass micropipettes (voltage pipette, current pipette) are inserted on opposite sides of the muscle fibre underneath the endplate. One electrode measures the membrane potential, the other is used as a current pipette. Both electrodes are connected via a clamp amplifier. Acetylcholine is applied via a glass micropipette the tip of which is placed 25 µm above the center of the endplate under visual control. The amount of acetylcholine can be controlled by applying a short current pulse of known amplitude and duration to the micropipette.

After release of a given number of agonist molecules from the tip of the pipette, the agonist diffuses to the AChR in the postsynaptic membrane. The concentration change as a function of time and location on the endplate can be calculated using the diffusion equation. The temporary presence of agonist at the receptor site leads to receptor activation and opening of ionic channels. The shape of the ionophoretically elicited endplate current is characterized by its peak value I_p, and the time, t_p, to reach this peak. I_p is a function of the agonist concentration, t_p is a function of the distance of the tip of the pipette from the endplate.

By increasing the amount of agonist molecules released from the ionophoretic pipette (which is a linear function of the electric charge passed through the pipette), I_p is also increased. If one plots I_p (response) vs. the charge through the pipette (dose), one obtains characteristic dose response curves. Under the assumption that the concentration change at the receptor site is slow compared to the interaction between agonist and receptor (equilibrium reaction), and by using a stochiometrical model for the agonist receptor interaction, the empirical dose response curves can be fitted with computer calculated dose response curves. We found that the equilibrium condition can be met by placing the tip of the ionophoretic pipette at least 15 µm or higher above the endplate. Our experimental dose response curves can best be described using a cooperative model for agonist receptor interaction (2).

A characteristic dose response curve obtained on a mammalian endplate is shown in fig. 2.

Three parameters can be obtained from mathematical analysis of the experimental dose response curves: the Hill coefficient, n_H, from the slope of the dose response curve for low agonist concentrations, the average maximum agonist induced current per unit surface of the endplate, I_{max}, and K_{app}, the apparent dissociation constant for the agonist receptor interaction (K_{app}= agonist concentration for yielding $I_{max}/2$). The parameter I_{max} can be converted into the average maximum conductance change per unit endplate surface g_{max} (nS/μm^2) if the driving force is known (normally endplates were voltage clamped at -80 mV, the equilibrium potential for ionic endplate channels was determined at -5 mV). Dividing g_{max} by the average single channel conductance γ, obtained from spectral analysis of agonist induced current fluctuations, yields N, the maximum number of ionic channels per unit endplate surface which can be opened by saturating agonist doses. Table 1 summarizes average values for dose response parameters as well as average single channel conductances and mean channel life times from noise spectra.

	Rat			Frog
	Omohyoideus	E.d.l.	Soleus	cut.pectoris
Parameter				
g_{max} [nS/μm^2]	24.2±6.1 (116)	21.8±5.5 (38)	8.2±3.9 (20)	169±67 (132)
K_{app} [μM]	58.5±26.7 (116)	65.9±26.2 (38)	43.5±14.4 (20)	28±12 (132)
n_H	2.3±0.2	2.3±0.1	2.2±0.1	2.7±0.3
single channel conductance [pS]	25.3±4.6 (30)	24.6±4.5 (14)	23.9±5.8 (14)	21.4±4.2 (30)
mean channel life time [ms]	1.32±0.19 (30)	0.8±0.18 (14)	0.71±0.14 (14)	1.06±0.2 (30)

Table 1: Characteristic parameters of dose response curves and fluctuation analysis in rat and frog endplate.
Mean values ± s.d. (number of endplates investigated). Experimental conditions: T, 20°-23°; membrane potential, -60 mV - -80 mV; agonist, acetylcholine. Notice that g_{max} for frog is measured in terms of unit length of endplate [nS/μm].

It can be calculated from the parameters in table 1, that in rat omohyoideus and Edl about 1000 ionic channels/μm^2 can be opened, but only about 350 channels/μm^2 in soleus. In frog about 8000 channels per μm of length of endplate can be opened. In order to interpret these results one should bear in mind that for the analysis of experimental dose response curves the mammalian endplate is treated as a circular surface over which all receptors are homogenously distributed, and the frog endplate as a straight line of infinite length.

The receptor distribution in mammalian endplates, however, is clearly not homogenous as can be seen in photographs of immunofluorescently labeled endplates (3,4). In fact, morphometry of these stained endplates show that only about half or one third of the total endplate area is densely stained (5). Correcting for the densely stained areas in mammalian endplate and assuming a value of 2-3 μm for the width of frog endplate average values of about 3000 channels/μm^2 can be estimated in the receptor dense areas of frog M.cutaneus pectoris and rat M.omohyoideus.

B) The effect of anticholinesterase

If the cholinesterase in the synaptic cleft is inhibited by Tensilon or prostigmine, the dose-response curve is left shifted, i.e., g_{max} and n_H are unchanged. The only measured effect is a reduction in K_{app} by a factor of 3. This means that the presence of the esterase reduces, in equilibrium, the effective ACh-concentration at the receptor site by this factor. It should be pointed out, however, that the release of transmitter from a terminal is not an equilibrium condition, here it can be calculated that all esterase molecules at the release site are easily saturated, leaving enough ACh-molecules for reaction with the receptors (6).

C) Properties of some different agonists

In addition to acetylcholine, we measured dose response curves for several different agonists. As reported earlier (2) carbamylcholine (CCh) yields similar dose response curves as acetylcholine, however, the dose response parameters are somewhat different. In particular, carbamylcholine is not as effective as acetylcholine in opening ionic channels in frog endplate at saturating doses.

In experiments in which dose response curves were measured on the same endplate under similar conditions for acetylcholine and carbamylcholine, g_{max} values were consistently smaller (about 75-85%) for carbamylcholine.

This indicates that the "relative efficacy" (7,8) of carbamylcholine is smaller than that of acetylcholine. However, from experimental dose response curves it is not clear whether a saturation below the ACh-level is due to "efficacy" or whether receptor desensitization/open channel block have occurred. We have shown earlier (9) that the addition of ethanol (3% at frog, 0.5% at mouse) to the bathing fluid increases the affinity of the receptor to the agonist (K_{app} reduced). Thus saturation can be achieved with lower agonist concentrations. We found for ACh and carbamylcholine the parameters g_{max}, n_H and relative efficacy unchanged. Therefore, it appears likely that we have determined true efficacy.

D) Choline seems to be a strong agonist

The application of choline by ionophoresis proved to be problematic since even at the largest usable currents through the ionophoretic pipette only very small responses could be elicited in the endplate. We therefore applied 3% ethanol to the bathing solution in an attempt to increase the sensitivity of receptors to the agonist. Indeed, the current responses for a given dose of ionophoretically applied choline increased several fold in the presence of ethanol as can be seen in fig. 3.

This finding for the effect of ethanol on the choline dose response curve is quite different from the observations made for acetylcholine and carbamylcholine: ethanol changes the shape of the choline dose response curve drastically. It should be mentioned that ethanol is much more effective at mammalian endplates (10). Infact, we have found relative efficacies as high as 62% for choline in rat endplate in the presence of 0.5% ethanol.

In experiments in which 500 µM - 1mM choline was added to the bath solution, we found reductions in miniature endplate current (mepc) amplitudes by about 50% as well as shortening of the decay time constants.

In an attempt to interpret these results one might assume that choline cannot only specifically open but also unspecifically block open ionic channels (11).

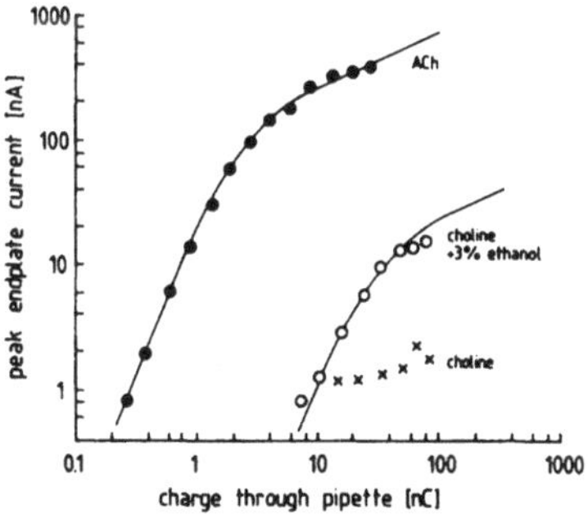

Fig. 3: Experimental dose response curves for acetylcholine and choline.
All dose response curves measured on the same frog endplate (●-acetylcholine, X-choline in the absence and O-choline in the presence of 3% ethanol in the bath). Dose response parameters for acetylcholine (choline +3%, ethanol in bath): g_{max}, 150 nS/µm (8nS/µm^2); K_{app}, 25 µM (850 µM) and n_H, 2.6 (2.0).

Since choline has a comparatively high K_{app} compared to acetylcholine (see Fig. 3), a large number of choline molecules are needed in order to open a given number of ionic channels. At the same time, the amount of channel blocking increases with the concentration of the blocking agent. In the presence of ethanol the sensitivity of the receptors to any agonist is increased (K_{app} is decreased). With ethanol, fewer agonist molecules are needed in order to open a given number of ionic channels. The degree of blockade of ionic channels is then much smaller due to a lower concentration of the blocking agent. Thus the overall number of open channels is higher.

Normally choline is classified a "week agonist", because even saturating doses elicit only a week response. Our results suggest that this is not due to an ·enherent low efficacy, rather than to open channel block. Thus choline (and, similarly, succinylcholine and nicotine) is to be classified a strong agonist (efficacy > 60%) and, in addition, strong open channel blocker.

E) <u>Effect of anti-receptor antibodies an receptor function</u>

The animal model of human myasthenia gravis, experimental autoimmune myasthenia gravis (EAMG), shows many similar symptoms: measurable serum levels of anti-AChR antibodies, decremental responses in stimulated electromyograms, increased muscular exhaustion and reduced α-bungarotoxin binding capacity at diseased endplates. EAMG can be elicited by injection of foreign AChR into experimental animals. The functional lesion at the endplate appears to be a reduction in the number of acetylcholine receptors. The characteristic properties of ionic channels which remain functional (single channel conductance, channel life time), do not seem to be very much different (12,13).

Correspondingly, we found a 50% reduction in g_{max} in rats with EAMG as reported earlier (13). However, we also found increased (by about 50%) dissociation constants in dose response curves on rats with EAMG. This decreased affinity of the acetylcholine receptor to acetylcholine presents a further impairment to neuromuscular transmission.

Similar symptoms as in EAMG can be found after passive transfer of human mysthenic serum to mice (14). In contrast to EAMG, which takes several weeks to develop and results from an active production of anti-AChR antibodies, passive transfer myasthenia gravis (PTMG) develops a few days after injection of the IgG-fraction of human myasthenic serum. We have measured dose response curves on mice with PTMG. The results are summarized in fig. 4.

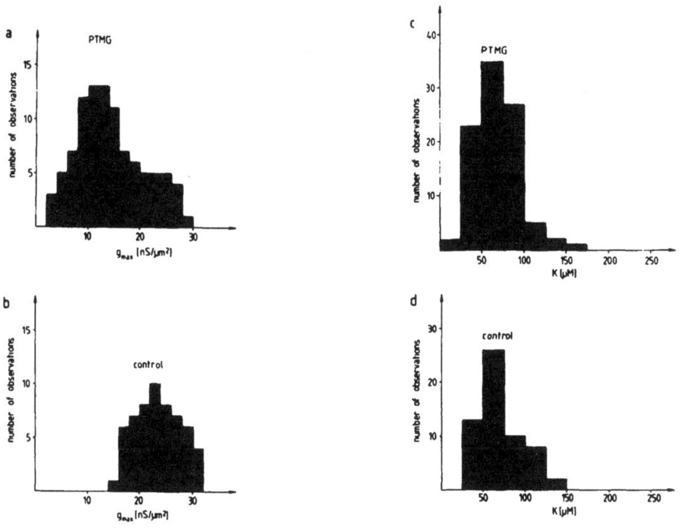

Fig. 4: Histograms of maximum acetylcholine induced conductance changes (g_{max}) in control mice and mice with PTMG (a and b) and histograms of the corresponding apparent dissociation constant (c and d).

control: 56 endplates (17 mice)
PTMG : 95 endplates (20 mice)

Fig. 4a shows the reduction in endplate conductance (g_{max}) in mice with PTMG compared to control (4b). This finding is similar to our earlier observations in EAMG. It can also been seen that the distribution of g_{max} values in mice with PTMG is skewed to the left whereas the control distribution resembles a normal distribution. This was also found in EAMG. In fact we observed that even in severely afflicted animals with strongly reduced receptor densities some endplates could be found with normal receptor densities.

This observation is remarkable. The endplate conductance can also be reduced by adding the receptor specific poison α-bungarotoxin. We have found that a mean reduction of g_{max} by α-bungarotoxin to the same degree as in EAMG or PTMG does not result in a skewed distribution, but rather in a shift of the control distribution to the left.

Thus it is not quite clear how this phenomenon in EAMG and PTMG can be interpreted. Perhaps some endplates are more resistant to the attack of the antibodies than others. One might speculate whether a trophic influence of nerve may play a role in this "protection" of some endplates.

Surprisingly, in contrast to the observations in EAMG, mice with PTMG show no increase in apparent dissociation constants (fig. 4c) in comparison to control (fig. 4d).

Again, the physiological explenation for this finding is not quite clear. Possibly the pathological changes in the morphology of the synapitc cleft observed in EAMG could be responsible for this difference. This pathological change could lead to an increased diffusional barrier. It should be noted that normally the synaptic cleft does not present a diffusional barrier for the measurement of dose response curves as experimental evidence shows (2,6).

Another tentative explanation could be the synthesis of a different receptor population with a reduced affinity to acetylcholine in the course of the development of EAMG, which takes several weeks in comparison a few days for PTMG as pointed out earlier.

However, a final interpretation cannot be given by this type of experiment.

Finally one observation deserves comment. The effect of α-bungarotoxin an nicotinic acetylcholine receptor has often been compared with the effect acetylcholine receptor antibodies. Our results show that this is not quite correct, at least not for PTMG. Bungarotoxin leads to an increased dissociation constant in experimental dose response curves, as reported earlier (4,6). This is not the case in PTMG.

Acknowledgement

This work was supported by the Deutsche Forschungsgemeinschaft, SFB 38,N.

References

1. McMahan, U.J., N.C. Spitzer and K. Peper. 1972. Proc.R.Soc. London Ser.B 181, pp. 421

2. Dreyer, F., K. Peper and R. Sterz. 1978. J.Physiol.London 281, pp. 395

3. Anderson, M.J.and M.W. Cohen. 1974. J.Physiol.London 237, pp. 385

4. Peper, K., R. Sterz and R.J. Bradley. 1981. Proc.NY.Acad. Sci 377, pp. 519

5. Sterz, R., M. Kaul and V. Moll: unpublished observations

6. Peper, K., R.J. Bradley and R. Sterz. 1982. Physiol.Rev. 62, pp. 1271

7. Stephenson, R.P. 1956. Br.J.Pharmacol. 11, pp. 379

8. Werman, R. 1969. Comp.Biochem.Physiol. 30, pp. 997

9. Bradley, R.J., K. Peper and R. Sterz. 1980. Nature London, 284, pp.60

10. Sterz, R.: unpublished observations

11. Adams, P.R. 1976. J.Physiol.London 260, pp. 531

12. Cull-Candy, S.G., R. Miledi and A. Trautmann. 1979. J.Physiol. 287, pp. 247

13. Hohlfeld, R., R. Sterz and K. Peper. 1981. Pflügers Arch. 390, pp. 156

14. Toyka, K.V., D.B. Drachman, D.E. Griffin, A. Pestronk, J.A. Winkelstein, K.H. Fischbeck and I.Kao. 1977. N.Engl.J.Med. 196, pp. 125

NBD-5-ACYLCHOLINE IS A FLUORESCENT AGONIST OF THE MEMBRANE-BOUND ACETYLCHOLINE RECEPTOR FROM *TORPEDO MARMORATA*

M. Covarrubias, A. Maelicke
Max-Planck-Institut für Ernährungsphysiologie,
Rheinlanddamm 201, D-4600 Dortmund, FRG

Abstract

We have studied the pharmacology and the binding properties of NBD-5-acylcholine using AChR-rich membrane vesicles from *Torpedo marmorata*. We report the following results:
(i) Micromolar concentrations of NBD-5-acylcholine stimulate the Tl^+-influx into these vesicles in the millisecond time range. The stimulated Tl^+-influx is prevented by preincubation of the vesicles with α-toxin. Thus, NBD-5-acylcholine is a cholinergic agonist also at the *Torpedo* receptor.
(ii) In contrast to the complete quenching of fluorescence observed upon binding to the purified receptor protein from *Electrophorus* (1), interaction with the membrane-bound *Torpedo* receptor results only in partial quenching of the ligand's fluorescence. Furthermore, quenching is different for the monoliganded and the diliganded receptor. This indicates structural differences in the two binding sites for NBD-5-acylcholine per receptor monomer (2). To make NBD-5-acylcholine a suitable marker for binding and kinetic studies of the *Torpedo* receptor, we have quantitatively compared the results of direct binding assays with those employing the change in fluorescence as the monitoring signal. In this way, the relative quantum yields of the receptor-ligand complexes were determined.

Taken together, the results presented here form the basis for the application of NBD-5-acylcholine as a fluorescent marker of the *Torpedo* receptor.

Molecular Basis of Nerve Activity

Introduction

Quantitative studies of the mechanism of ligand recognition, channel activation and inactivation, and response modulation of the nicotinic acetylcholine receptor all require the availability of assay systems of millisecond time resolution. Fluorescence stopped-flow systems have proven to be of particular advantage in this respect because (i) they can be employed for measurements of both ligand binding (3,4) and agonist-induced ion-flux (5), and (ii) by keeping receptor and ligand separated prior to the experiment, the problems of predesensitized receptor preparations can be avoided (6). The full potential of fluorescence stopped-flow studies, however, is only obtained if fluorescent probes of simple pharmacology are employed. The NBD-n-acylcholines first synthesised in our laboratory (1) and employed in detail to the purified Electrophorus receptor (4, 7) have proven to be of particular advantage because (i) they are pure agonists of acetylcholine (8) and (ii) their fluorescence effects upon interaction with the eel receptor are large and simple. In order to expand these earlier studies to a receptor preparation that provides a rich source of well defined membrane-bound receptor, we have performed the basic studies necessary to establish NBD-5-acylcholine as an agonist and fluorescent marker of the acetylcholine receptor from Torpedo.

Materials and Methods

AChR-rich membrane vesicles from Torpedo marmorata or Torpedo californica were prepared as described previously (9). Rapid ion-flux measurements were performed according to Moore and Raftery (5) employing the fluorescence of 1,3,6,8-pyrene tetrasulfonate (PTS) as monitoring signal for the rapid influx of Tl^{+}, which acts on PTS as a heavy-metal quencher.The excitation light coming from a xenon lamp was passed through two filters:

Schott UV-filter (λ_{max} = 379 nm) and a Carl-Zeiss UG-5 filter. The scattered light was blocked from entering the photomultiplier with a Schott 408 nm cut-on filter. The binding assays and fluorescence titrations will be described in detail elsewhere (2). The basic experimental set up and the methods of computer analysis were similar to those previously described (7).

Results and Discussion

NBD-5-acylcholine stimulates Tl^+-influx into *Torpedo* membrane vesicles. - The influx of Tl^+ into *Torpedo* vesicles consists of two components, a non-stimulated and an agonist-stimulated one. The non-stimulated component (Fig. 1, upper trace) is observed in the absence of stimulating ligands and, after blockade of the agonist binding sites by antagonists including α-toxins, also in the presence of agonists. The stimulated Tl^+-influx (Fig. 1, lower trace) is a specific result of agonist interaction with the receptor and can be employed to determine the potency of a ligand as cholinergic agonist. As shown in Fig. 1, NBD-5-acylcholine stimulates Tl^+-influx into *Torpedo* vesicles, and the effect can be blocked by α-cobratoxin. The initial rate of stimulated Tl^+-influx obtained in the presence of 36 μM NBD-5-acylcholine was 22 s^{-1}; the same size of response was also obtained by 300 μM carbamoylcholine. We conclude that NBD-5-acylcholine acts as a powerful cholinergic agonist at the *Torpedo* receptor. It differs in this respect from decamethonium (10) and Bis-Q (11) which act as competitive antagonists at the postsynaptic membranes isolated from *Torpedo* but as agonists at the frog motor endplate and electroplaques. Similarly a fluorescent derivative of suberyldicholine behaves as a competitive antagonist (12).

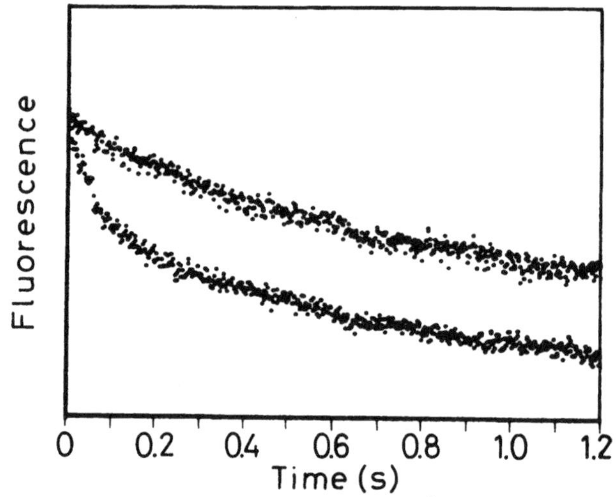

Fig. 1: NBD-5-acylcholine stimulated Tl^+-influx into AChR-rich vesicles from Torpedo californica.
Upper trace: PTS-loaded AChR-rich membrane vesicles from Torpedo californica suspended in 160 mM $NaNO_3$ and 10 mM HEPES, pH 7.4, were pretreated with α-cobratoxin (~ 10 fold excess over receptor sites) over a period of 6 hrs and then rapidly mixed with a buffer solution containing 36 µM NBD-5-acylcholine in 130 mM $NaNO_3$, 30 mM $TlNO_3$ and 10 mM HEPES, pH 7.4. The change in fluorescence was recorded as a function of time. Lower trace: Same experimental conditions as above except that toxin was not employed. Each trace represents the average of 6 and 4 experiments, respectively.

Binding of NBD-5-acylcholine to Torpedo receptor. - As previously observed with the receptor protein from Electrophorus (7), the fluorescence of NBD-5-acylcholine is quenched upon interaction with Torpedo membranes. This quenching is without spectral shift, and it is specifically cholinergic in that it can be completely inhibited or reversed by preincubation or addition of non-fluorescent agonists and antagonists. Contrary to the fluorescence effects observed with the eel receptor, however, the quenching of ligand fluorescence is not complete with the Torpedo receptor. The quenching of fluorescence accompanying binding to the initially occupied sites can be measured directly: At low concentrations of ligand (0.2-3 nM) and large excess of receptor (200 nM), the observed quenching was approximately 50%.

To assess the size of the quenching effect by a second independent set of experiments and also for the other NBD-5-acylcholine-receptor complexes, binding experiments by two different methods were performed in parallel: In the first, the incubation mixture of membrane fragments and fluorescent agonist were centrifuged with the airfuge (9), and the concentration of bound NBD-5-acylcholine was determined as the difference in fluorescence of total ligand (determined in the presence of a suitable blocking agent prior to centrifugation) and free ligand (supernatant after centrifugation). In this way, the quenching effect did not enter the calculations. These data were compared with those obtained by fluorescence titration experiments in homogenous solution (2). Since all other experimental conditions were kept identical, the two sets of experiments must lead to identical results if expressed in molar terms. The first set of experiments can therefore be employed to fit the titration experiments to molarities. In this way, the relative quantum yields of the various receptor-ligand complexes were estimated.

Fig. 2 shows the related data. As is obvious from the left panel of the figure, the plots of $\Delta F/F$ versus F (of titration experiments) differ dramatically from the true Scatchard presentations of the centrifugation experiments. To fit the quenching to the centrifugation data, the following reaction scheme was employed (7,9):

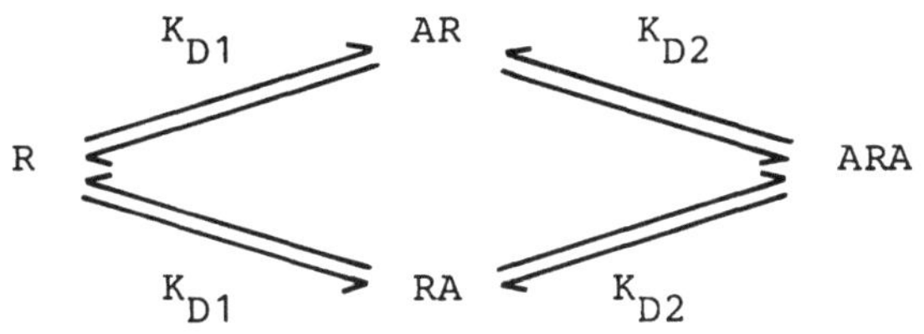

(I)

Because quenching accompanying binding is only partial, the fluorescence of the reaction mixture is composed of the following components:

$$F = C\ (A + 2\ Q_1\ \ AR + 2\ Q_2\ \ ARA)$$

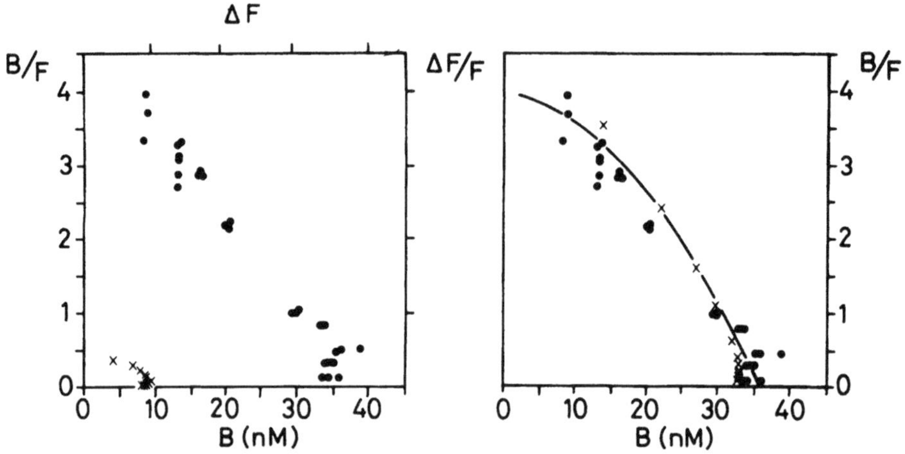

Fig. 2: Equilibrium binding of N-5-C to AChR-rich membrane vesicles from Torpedo marmorata.
The membrane vesicles were diluted in standard assay buffer (100 mM NaCl, 4 mM $CaCl_2$, 2 mM $MgCl_2$, 0.05 mM Tetram and 10 mM PIPES, pH 6.8). The mixture was then used to perform two parallel experiments: In the first, 2 ml of the membrane suspension were titrated stepwise with small aliquots of a stock solution of NBD-5-acylcholine, and after 15 min the fluorescence change was recorded (x). The change in fluorescence was calibrated in concentration units (nM). In the second, (●), increasing amounts of NBD-5-acylcholine were added to 10 different samples of the same membrane suspension (one concentration for each sample) and filled up to the same final volume (2.5 ml). After 15 min of incubation, 3 aliquots of each sample were centrifuged and the free concentration of ligand was determined. The concentration of bound ligand was indirectly determined by displacement. Each point represents an independent determination. The data were fitted simultaneously employing scheme I.

with C being a calibration factor, A, AR and ARA being the concentration of free ligand, monoliganded and diliganded receptor, respectively, and Q_1 and Q_2 being the relative quantum yields of the ligand bound as monoliganded and diliganded complex, respectively.

Fig. 2 (right panel) shows the best simultaneous fit of the binding data (solid line) and the fluorescence titration (x) corrected accordingly. Using $K_{D2} = 0.5\ K_{D1}$ (a ratio obtained from the analysis of several binding curves from different membrane preparations) and Q_1 = 0.5 (obtained independly as described previously) the best fit parameters obtained were K_{D1} = 9.0 $\pm$ 0.64 x 10^{-9}M and Q_2 = 0.8 $\pm$ 0.02. The last value is in good agreement with the experimentally estimated relative quantum yield for the fully occupied receptor (2).

Based on the employed scheme, the monoliganded receptor (AR and RA) quenches the ligand's fluorescence to an extent of 50%, while the fully occupied receptor (ARA) quenches even less (25-30%). These are overall effects, however, that at the present level of analysis do not consider the properties of each complex separately (see scheme).

Conclusions

Although smaller and more complicated fluorescence effects accompany the binding of NBD-5-acylcholine to the membrane-bound Torpedo receptor as compared to the solubilized Electrophorus receptor, the ligand appears to be suitable for in-depth analysis also of this receptor system. In particular, its pharmacological properties identify NBD-5-acylcholine as a true analog of acetylcholine (agonist).

Acknowledgement

This work was supported in part by grants of the Alexander-von-Humboldt foundation and the Max-Planck society (M.C.) and by the Deutsche Forschungsgemeinschaft, SFB 168.

References

1) Jürss, R., H. Prinz, A. Maelicke. 1979. Proc.Natl.Acad.Sci. U.S.A. 76, 1064-1068.

2) Covarrubias, M., H. Prinz, A. Maelicke. 1984. submitted.

3) Heidmann, T., J.-P. Changeux. 1979. Eur.J.Biochem. 94, 255-279.

4) Prinz, H., A. Maelicke. 1983b. J.Biol.Chem. 258, 10273-10282.

5) Moore, H.-P.H., M.A. Raftery. 1980. Proc.Natl.Acad.Sci. U.S.A. 77, 4509-4513.

6) Maelicke, A., H. Prinz. 1983. In: Modern Cell Biology (B.H. Satir, ed.). Alan R. Liss, Inc., New York, 1, pp. 171-197.

7) Prinz, H., A. Maelicke. 1983a. J.Biol.Chem. 258 10263-10272.

8) Meyers, H.-W., R. Jürss, H.R. Brenner, G. Fels, H. Prinz, H. Watzke, A. Maelicke. 1983. Eur. J. Biochem. 137, 399-404.

9) Fels, G., E.K. Wolff, A. Maelicke. 1982. Eur.J.Biochem. 127, 31-38.

10) Popot, J.-L., H. Sugiyama, J.-P. Changeux. 1976. J.Mol.Biol. 106, 469-483.

11) Hess, G.P. (personal Communication)

12) Bolger, M.B., V. Dionne, J. Chrivia, D.A. Johnson, P. Taylor. 1984. Mol.Pharmacol. 26, 57-69.

DO SIGMOID DOSE-RESPONSE CURVES REQUIRE THE ASSUMPTION OF ALLOSTERICITY?

H. Prinz, A. Maelicke
Max-Planck-Institut für Ernährungsphysiologie,
Rheinlanddamm 2o1, D-46OO Dortmund, West-Germany

Introduction

Dose-response curves correlate concentrations and physiological effects of drugs. Since the physiological effect of a drug may be separated by any number of coupled reactions from the initial interaction of drug and specific receptor site, response is not by necessity directly proportional to receptor saturation (binding). Notwithstanding this fact, sigmoid (S-shaped) dose-response curves are nowadays mainly discussed in terms of the allosteric model (1) implying cooperative binding of drug and receptor. It is worthwile to remember, however, that the allosteric model (1) is only concerned with binding and that additional assumptions (2,3) are required to interpret on its basis dose-response measurements.

Although the existence of positive cooperative interactions of agonist binding sites at the membrane-bound acetylcholine receptor from *Torpedo* is not disputed (4-6), it should be pointed out that much simpler assumptions than allostericity already suffice to explain sigmoid dose-response curves. Indeed even hyperbolic binding curves agree with sigmoid dose-response curves if binding is to more than one site at the receptor and if response is assumed to be proportional to the formation of higher liganded receptor only (7-9).

Molecular Basis of Nerve Activity

Theory and Model Calculations

Sigmoid dose-response curves can be explained by several models. The earliest model (equation 1) assumed simultaneous binding of several (n) ligand molecules (A) to the receptor (R) with the response being a consequence of complete saturation (equation 2):

$$nA + R \rightleftharpoons A_nR \tag{1}$$

$$\text{Resp} \sim [A_nR] = \frac{[R]_o \, ([A]/K_D)^n}{1 + ([A]/K_D)^n} \tag{2}$$

Equation 2 (10) is linear when log (Resp/max.Resp)/(1-Resp/max. Resp) is plotted versus log [A] (Hill plot (11)). The slope of this plot is called the Hill coefficient n_H and, according to equation 2, is identical to the number of binding sites n. Experiments, however, rarely show integer Hill coefficients. This fact has led to the proposal of two-state models with or without allostericity.

In the allosteric two-state models, two conformational states (one active, one inactive) are assumed for the receptor. The affinity of the ligand for the active conformation is assumed to be higher and, hence, the active conformation is stabilized in the presence of ligand. Allosteric interactions between receptor subunits (2) or receptor protomers (3) then cause the binding curve and, by inference the dose-response curve to be sigmoid. As a consequence, the related Hill plot also is sigmoid in shape with limiting slopes of 1 at very low and very high ligand concentrations. The maximal slope at the point of inflection of the Hill plot is equal to or smaller than the number of interacting units.

An equally valid assumption to explain sigmoid dose-response curves and Hill coefficients larger than 1 is the response being initiated only when more than one ligand molecule is bound to the receptor. With this assumption and the now generally accepted scheme of consecutive (12) rather then parallel (10) binding steps

$$R + nA \underset{}{\overset{K_D1}{\rightleftharpoons}} AR \rightleftharpoons \ldots\ldots \rightleftharpoons A_{n-1}R \underset{}{\overset{K_Dn}{\rightleftharpoons}} A_nR \qquad (3)$$

the following can be concluded:

(i) As shown in Fig. 1 for the case n = 3 and identical K_D-values for each reaction step, the concentration of total bound ligand is the expected hyperbolic function of the free ligand concentration. (ii) In contrast, however, the concentrations of each individual complex are related to the free ligand concentration by more complicated functions and, for the higher complexes (A_2R and A_3R in Fig. 1) are sigmoid in shape. Accordingly, if response is proportional to the concentration of one or both of these complexes, the dose-response curve also is sigmoid in shape. The Hill plots for total bound ligand and the two complexes A_2R and A_3R are shown in Fig. 1b: The Hill plot is a straight line of slope 1 for total bound ligand but is curved for each individual complex: For complexes with more than one ligand bound, the initial slope is equal to the number of bound ligand molecules. This slope decreases monotonously with increasing ligand concentration approaching the slope of one for the fully saturated complex (12).

As is evident from Fig. 1, the patterns of curves 2,3 closely resemble in shape the dose-response curves obtained experimentally by electrophysiological techniques. Not only are curve 2,3 in Fig. 1a sigmoid but they result in Hill plots with the maximal slope at low ligand concentrations as is observed experimentally. This slope is equal to the number of bound ligand molecules per individual complexes. The decrease of the Hill

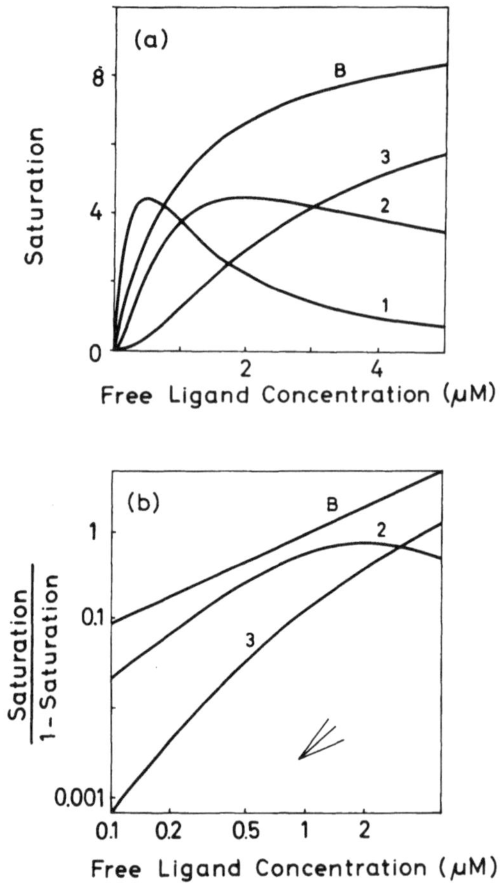

Fig. 1: Saturation of receptor as a function of free ligand concentration. Three binding sites with identical $K_D = 1 \times 10^{-6}$M were assumed. Concentration of complexes AR (1), A_2R (2) and A_3R (3) are given in fractions of receptor concentrations. Concentration of bound ligand relates to the concentration of complexes by the following equation:

$$B = [AR] + 2[A_2R] + 3[A_3R]$$

(a) Concentration of complexes versus free ligand concentration.
(b) Hill plot of (a).

coefficient at high ligand concentrations is another feature frequently observed experimentally (13) while the sigmoid Hill plot predicted by the allosteric model has not yet been observed in the related experiments.

Two more properties of dose-response curves are better explained by the multiple-sites-linear model than by the allosteric model: (i) In the presence of competing agonists but not antagonist, dose-response curves of the nicotinic acetylcholine receptor are changed from sigmoid to hyperbolic (14). (ii) In experiments with iontophoretically applied agonists (13), the Hill coefficients tend to vary from experiment to experiment. The first effect can be rationalized as follows: As seen in Fig. 1a, a sigmoid dose-response curve requires that the formation of the active complex is preceeded by an intermediate. If this intermediate is formed by the competing agonist, formation of the final (active) complex then follows a hyperbolic function. To understand the second effect, it should be remembered that in physiological experiments the response is plotted versus the total ligand concentration instead of the free ligand concentration. The resulting error is sufficiently small only as long as the receptor concentration is below the equilibrium dissociation constant (K_D) of this interaction (15). This can be appreciated by comparing Fig. 2a and 2b in which for the simplest case of only two binding sites the concentration of complexes is plotted versus free (Fig. 2a) and versus total (Fig. 2b) ligand concentration. To expose the effect under discussion, a receptor concentration of 10 times the K_D value was assumed. Under these conditions the half-maximal response obtained from plots versus total concentration differs drastically from the real dissociation constant and is of the order of the receptor concentration (Fig. 2b). Moreover, the Hill plot (Fig. 2c) now is sigmoid. In the shown case of two binding sites, the initial slope of the Hill plot is two and the final slope is one. The maximal slope depends entirely on the concentration of receptor relative to the K_D value.

It is a central limitation of most physiological experiments, that the receptor concentration cannot be determined quantitatively. As a consequence, if the area under investigation contains only densely packed receptors such as at a single endplate, quantitative errors can easily occur. Thus, the closer the iontophoretic pipette is applied to the endplate and the richer in receptor these endplates are, the larger is the local

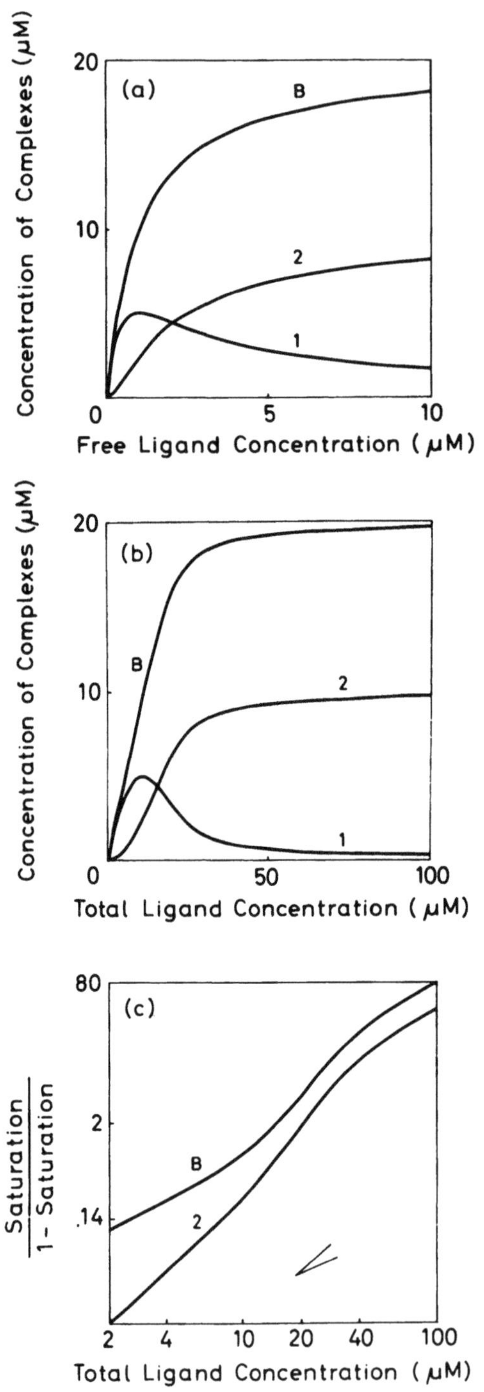

Fig. 2: Concentrations of receptor-ligand complexes as functions of free and total ligand concentration.
Two binding sites with identical $K_D = 10^{-6}M$ and a receptor concentration $R_o = 1 \times 10^{-5}M$ were assumed. Concentration of complexes AR (1) and A_2R (2) and of total bound ligand (B) are given in molar concentrations.
(a) Concentration of complexes versus free ligand concentration.
(b) Concentration of complexes versus total ligand concentration.
(c) Hill plot of (b).

receptor concentration and the greater is the difference between free and total concentration of ligand. Consequently, the Hill coefficient will depend on the experimental conditions as is indeed observed (13). This effect is independent of the proposed reaction scheme and can only be controlled by independently monitoring the free ligand concentration in iontophoretic experiments.

Conclusions

We have described here a simple means of explaining sigmoid dose-response curves on the basis of agonist binding to several non-interacting sites at a receptor. Exploring in this way the simple scheme of consecutive binding reactions (12) to a receptor with several ligand binding sites, sigmoid dose-response curves, Hill coefficients larger than 1, the often observed discrepancies between K_D-values and concentrations of half-maximal response, the competition dose-response curves observed with the nicotinic receptor and the dependence on experimental conditions of electrophysiologically obtained Hill coefficients, among other experimental results, can be explained without the assumption of allostericity. Thus, the above listed experimental effects cannot be used as supportive arguments for an allosteric mechanism of ligand-receptor interaction.

Acknowledgements

This work was supported in part by grants from the Deutsche Forschungsgemeinschaft (SFB 168) and the Fond der Chemischen Industrie.

References

1. Monod, J., J. Wyman, J.-P. Changeux. 1965. J.Mol.Biol. 12, 88.

2. Karlin, A.. 1967. J.Theoret.Biol. 16, 306.

3. Changeux, J.-P., J. Thiery, Y. Tung, C. Kittel. 1967. Proc.Natl.Acad.Sci.USA 57, 335.

4. Weber, M., J.-P. Changeux. 1974. Mol.Pharmacol. 10, 15-34.

5. Neubig, R.R., J.B. Cohen. 1979. Biochemistry 18, 5464-5475.

6. Fels, G., E.K. Wolff, A. Maelicke. 1982. Eur.J.Biochem. 127, 31-38.

7. Prinz, H., A. Maelicke. 1983. J.Biol.Chem. 258, 10263-10271.

8. Maelicke, A., H. Prinz. 1983. In: Modern Cell Biology, Vol. 1 (B.H. Satir, ed.). Alan R. Liss, Inc., New York. pp. 171-917.

9. Prinz, H.. 1983. J.Rec.Res. 3, 239-248.

10. Clark, A.J.. 1937. In: Handbuch der experimentellen Pharmakologie, Ergänzungswerk 4 (W. Heubner and J. Schüller, eds.). Springer Verlag, Berlin.

11. Hill, A.V.. 1909. J.Physiol., Lond. 39, 361.

12. Werman, R.. 1969. Comp.Biochem.Physiol. 30, 997-1017.

13. Peper, K., F. Dreyer, K.-D. Müller. 1976. Cold Spring Harbor Symp. Quant. Biol. XL, 187.

14. Changeux, J.-P., T.R. Podleski. 1968. Proc.Natl.Acad.Sci. USA, 59, 944.

15. Straus, O.H., A.J. Goldstein. 1943. J.Gen.Physiol. 26, 559.

FOUNDATIONS OF THE ION FLUX METHOD

J. Bernhardt
Max-Planck-Institut für Biochemie, D-8033 Martinsried, FRG

E. Neumann
Physikalische und Biophys. Chemie, Universität Bielefeld, D-4800 Bielefeld 1, FRG

Introduction

The control of passive transmembrane ion transport by special gating systems plays a fundamental role in many cellular signal-transfer processes. Examples for such control systems, that regulate the opening and closing of ion channels, are divers neuroreceptors, and the sodium and potassium channel gating molecules connected with the generation of action potentials in nerves and muscles. For several decades the investigation of the molecular foundations of gating processes was dependent on crude pharmacological methods of measurement. The analysis of the resulting data often led to the determination of empirical parameters whose significance in relation to the fundamental phenomena usually remained obscure. Through progressive improvement of experimental techniques it has recently become possible to carry out much more detailed investigations.

A promising new method for studying ion gating processes is the ion flux technique originally introduced by Kasai and Changeux (1971). The measurement of the gating process dependent exchange of ions between a large bath and suspended

Molecular Basis of Nerve Activity

microscopic closed membrane structures (CMS) permits determination of detailed information about gating reactions. The technique entails measurement of the total amount of a certain ionic species contained in the CMS. Any parameter that directly reflects the ion content can be used as a variable of measurement (e.g., CPM of a radioactive ion species). A change of the total ion content can be determined under two separate experimental conditions: (a) when there is a net efflux of ions from the CMS into the bath, (b) when there is a net influx of ions from the bath into the CMS. The resulting data will reflect either the time course of the respective increase or decrease of ion content, or when flux is very rapid on the time scale of measurement, the time independent overall change of ion content (i.e. the so-called flux amplitude). Examples for CMS that have been used in ion flux studies are ion channel containing membrane fragments, isolated cells and reconstituted lipid vesicles.

A number of investigations have verified that the ion flux method constitutes a potentially broadly applicable procedure for investigating the elementary reaction events underlying channel gating processes. Here, unlike with more indirect measurement techniques, that rely on the detection of a fluorescence signal connected with a specific ligand or label, the ion flux process itself serves as a signal source. The system to be investigated is therefore practically unperturbed by the process of measurement. In comparison with techniques based on measuring electrical signals, that also fulfil this criterion (e.g., voltage clamp and current fluctuation methods), the ion flux method offers several advantages. For example the amounts of the chemical species that participate in gating processes (e.g., neuroactivator ligands, and gating molecules) are more accessible to accurate experimental control, when suspensions of small uniform membrane structures are used. Further, with such suspensions, it is possible to employ many experimental manipulations conventionally used in the study of chemical reactions in homogeneous solution (e.g., effective mixing through stirring).

Three fundamental aspects that had to be delt with in the progressive refinement of quantitatively accurate procedures for studying ion flux were: (a) the development of reliable experimental of techniques for rapidly initiating gating processes, and for measuring suitable signals reflecting the subsequent ion transport events (b) the development of procedures for preparing suspensions of CMS particularly suited for flux studies (c) the development of analytic schemes for extracting information about channel gating processes from flux data.

Techniques of Flux Measurement

In excitable biomembranes, ion transport phenomena controlled by a channel gating system are activated by special physical processes (e.g., release of a transmitter substance, or change of the membrane potential). In ion flux experiments, corresponding transport phenomena must be initiated by experimental simulation of these activation events. Up to now only experimental procedures for the investigation of ligand induced gating processes have been developed. These entail initiation of ion fluxes by rapid mixing of a suspension of CMS with a solution of a suitable neuroactivator substance. The course of the flux process is most commonly measured by rapid filtration using synthetic membrane filters. The individual filtrations are carried out at set intervals after the initial mixing step. Corresponding to the experimental conditions chosen, this allows determination of the increase or decrease of the ion content of the CMS retained on the filters. Information about the gating processes that control flux can then be obtained through an analysis of the flux time course and flux amplitudes.

The channel gating system that has been most extensively investigated is the nicotinic acetylcholine receptor - an intrinsic membrane protein that regulates nerve-nerve and nerve-muscle signal transmission processes. At neuromuscular and electromotor synapses the action of acetylcholine initiates receptor processes, that lead to a change in the postsynaptic membrane potential. This involves two types of reaction events. Short pulses (millisecond) of the presynaptically released transmitter at submillimolar concentrations, lead to receptor activation connected with a net opening of ion channels in the postsynaptic membrane. If acetylcholine is present for longer periods (fractions of seconds to minutes) at sufficiently high concentrations (micromolar), the initial channel opening phase is followed by a complex reduction of the number of open channels - the so-called desensitization of the acetylcholine receptor.

Most fundamental ion flux studies have been performed using acetylcholine receptor containing membrane fragments derived from the electric organs of Torpedo and Electrophorus electricus. In early investigations attempts were made to measure the ligand-induced efflux of $^{22}Na^{+}$ on the time scale of minutes (Kasai & Changeux, 1971; Hess et al., 1975; Popot et al., 1976) using a filter assay technique. Bernhardt & Neumann (1978) found that such methods permit the determination of ligand concentration dependent flux amplitudes, but not the actual time course of receptor controlled ion flux. With the introduction of more complex measurement techniques, it is now possible to measure flux processes on the millisecond to minute time range. These refined methods require the use of rapid mixing instruments, that were originally designed for the kinetics of fast chemical reactions. It is thereby possible to measure the time course of the change in ion content of a suspension of CMS using quench-flow (Hess et al., 1979; Neubig & Cohen, 1980) or stopped-flow (Moore & Raftery, 1980) techniques.

Investigations of ion flux regulated by the acetylcholine receptor permit several conclusions about the gating events connected with receptor processes. The time course of the ion flux initiated by neuroactivator ligand binding to the receptor consists of at least two separate, apparently exponential phases that occur on a progressively more rapid time scale, with increasing neuroactivator concentration (Neubig & Cohen, 1980; Hess et al., 1982; Heidmann et al., 1983). This flux behavior is compatible with the assumption that, following a rapid initial channel opening step (receptor activation), the receptor process induced by neuroactivators leads to at least two successively slower channel closing steps (receptor inactivation). Such a finding can also be deduced from measurements with other techniques (Heidmann & Changeux, 1979).

Further it could be shown that at low neuroactivator concentrations the slow receptor inactivation step, also known as desensitization, leads to the appearance of concentration dependent flux amplitudes (Bernhardt & Neumann, 1978; Walker et al., 1982). The amplitude was found to increase steadily with increasing neuroactivator concentration, until a limiting maximum amplitude is reached. A reverse dependence was found in inhibition studies, where acetylcholine receptor containing membrane fragments were inhibited by preincubation with an irreversibly binding snake toxin. Inhibition by sucessively greater amounts of toxin leads to a stepwise reduction of the flux amplitudes induced by neuroactivator ligands (Moore et al., 1979; Neubig & Cohen, 1980; Bernhardt & Neumann, 1982b).

Preparative Techniques

Not all CMS suspensions are equally suitable for flux studies. It is to be expected that sizeable admixtures of open and nonfunctional membrane structures may, on the one hand, complicate data analysis, and, on the other hand, lead to artifacts.

Special preparative techniques are therefore required to separate a suspension of highly homogeneous vesicular structures from crude homogenates of diverse membrane fragments and other cell components.

Generally applicable methods for producing such suspensions were first developed for acetylcholine receptor containing membrane fragments. Homogenization of whole electric organs from Torpedo or Electrophorus electricus can be used to generate a crude cell fraction. In initial studies (Kasai & Changeux, 1971; Duguid et al., 1973) sedimentation on sucrose gradients was used to separate fractions from such homogenates, that are rich in acetylcholine receptor content, but poor in acetylcholinesterase content. Due to progressive refinement of such techniques, it is now possible to produce uniform suspensions of membrane fragments with a high surface density of receptors (Sobel et al., 1977; Elliot et al., 1980). Unfortunately, such suspensions are not optimally suited for flux studies (a) because using gradient fractionation under these conditions does not necessarily lead to a separation of open and closed membrane structures, and (b) because non-functional vesicles, which contain proteolytically cleaved receptors, are not separated.

To minimize the contribution of open and non-functional membrane fragments in flux measurements, special methods were developed. Thus, through addition of the neuroactivator carbamoylcholine in the presence of $^{22}Na^{+}$ to a suspension obtained through gradient separation, a selective filling with tracer of closed, functionally intact vesicles was achieved (Hess et al., 1975). In tracer flux measurements with such preincubated suspensions, only the selectively filled vesicles contribute to the flux signal. Unfortunately, the open and non-functional membrane fragments still present as a background, nevertheless constitute an undesirable admixture. Due to the receptor binding sites they contain, they lead to a reduction of the neuroactivator concentration which is difficult to estimate. A more efficient gradient separation was achieved by preincubating the crude homogenate with CsCl. In the subsequent sucrose gradient sedimentation there is an enrichment of receptor containing closed membrane vesicles (Hess & Andrews, 1977).

A consideration of the detailed factors that control gated ion flux Bernhardt & Neumann, 1980 a,b) led to the proposal of more stringent criteria CMS suspensions optimally suited for quantitative flux experiments must fulfill: (a) there should be as little variation as possible in the ion channel content and volumes of the CMS (b) these parameters should lie in a range for which the flux time course is slow enough to be measured, and is on the same time scale as the gating process to be investigated. Bernhardt et al. (1981) developed a procedure for generating suspensions of acetylcholine receptor containing membrane fragments, that comply with these criteria. A crude homogenate of electric organs is first incubated in 0.2M CsCl. A selective exchange of Li^+ for Cs^+ in solely the functionally intact closed membrane structures is then effected by the stepwise addition of acetylcholine in the presence of 0.2M LiCl. These can then be separated from the heavier non-functional fragments, filled with CsCl, and other cell components, by sedimentation on a percoll gradient. Uniform suspensions of vesicular fragments having roughly the same volume and receptor content, can be obtained from fractions of the gradient derived from different density regions. With minor modifications this procedure should be generally applicable for arbitrary CMS suspensions.

Methods of Analysis

The aim of the analysis of flux data is to derive quantitative information about channel gating events from the time course and amplitude of the flux process. In initial studies, empirically defined parameters were employed to analyse curves (Kasai & Changeux, 1971). By explicitely considering the physical foundations of flux and channel gating processes, it was possible to develop theoretically well-founded, generally applicable methods of analysis (Bernhardt & Neumann, 1978, 1980 a,b, 1981). A main postulate in the derivation was, that the overall flux process, measured on a collection of CMS, is composed of a superposition of independent contributions of the individual CMS. The quantity actually measured in ion flux experiments - the total amount X(t) of a certain ion species present in the CMS at time t - is thus given as the sum of the ion content of the separate CMS. To derive practically applicable expressions for X(t) it was first necessary to examine the time course of ion flux from or into a CMS. It was possible to formulate a kinetic equation that explicitly encompasses all factors that determine flux. The rate of flux was found to be directly proportional to the rate constant for flux through a channel k', to the number of channels n_i, the fraction of these $\alpha_i(t)$ that are open at time t, and inversely proportional to the CMS volume v_i

$$\frac{dx_i}{dt} = - (n_i/v_i) \cdot k' \cdot \alpha_i(t) \cdot x_i(t) \qquad (1)$$

where $x_i(t)$ is the CMS ion content at time t, and the index i implies that the corresponding quantities may vary from CMS to CMS.

In principle the explicit evaluation of this equation should lead to an expression for X(t). Unfortunately such a direct mode of solution is complicated by the following circumstances (Bernhardt & Neumann, 1978, 1980 a,b): (a) In every actual suspension of CMS a variation of the values n_i and v_i is to be expected. Since the flux rate depends on these magnitudes, this entails that the time course of X(t) will reflect a complex superposition of the time course of ion flux from or into the individual CMS. (b) Even in the ideal case of a suspension of perfectly uniform CMS a complex mathematical espression for X(t) is expected to result, due to an always present variation of the fractions $\alpha_i(t)$. This follows from the fact that the fraction of open channels $\alpha_i(t)$ on a given CMS is not necessarily identical to the total fraction of open channels on all CMS $\alpha(t)$. Rather, on the basis of probability laws, a statistical variation of the $\alpha_i(t)$ values is to be expected.

A quantitatively exact evaluation of X(t) is only possible, if the variations of n_i, v_i and $\alpha_i(t)$ are explicitly incorporated into the framework of the mathematical treatment. Despite these difficulties it was nevertheless possible to derive simple expressions for X(t) by introducing several approximations (Bernhardt & Neumann, 1978, 1980 a,b). If in equation (1) one substitutes the mean values $\bar{n}$ and $\bar{v}$ for the corresponding parameters n_i and v_i, and $\alpha(t)$ for $\alpha_i(t)$, one obtains an easily solvable equation, the solution of which is a simple exponential expression for X(t) - the approximate flux equation, which, for efflux conditions, is given by

$$X(t) = X(0) \cdot \exp\left[-\bar{n} \cdot \bar{k}' \cdot \int_0^t \alpha(\tau)d\tau \right]. \qquad (2)$$

This expression, or variants thereof having the same functional form, have been frequently employed to analyze flux data. However, due to the implicit approximations, the range of applicability remained uncertain. Heidmann et al. (1983) recently compared the results of an ion flux analysis based on equation (2), with corresponding results of a separate stopped flow kinetic investigation carried out under identical conditions. They were able to verify that this approximate expression leads to quantitatively correct rate constants and dissociation constants for a ligand induced gating process.

That the applicability of equation (2) is limited, follows from a number of inhibition studies. Thus, Moore et al. (1979), and Neubig & Cohen (1980) were forced to postulate the presence of "spare" or "excess" receptors, in order to explain the stepwise reduction of flux amplitude accompanying inhibition of receptors by snake toxin. Bernhardt & Neumann (1982b) were able to show that such ad hoc assumptions are superfluous, and result from an incorrect modelling of the underlying flux phenomena by expression (2).

Recently Bernhardt & Neumann (1981) were able to derive generally applicable flux equations, that constitute exact solutions of the kinetic equation (1):

$$X(t) = \begin{cases} X(0) \sum_n P_n \left[\langle e^{-kt} \rangle^n \right]_v & \text{(efflux)} \quad (3) \\ X(\infty) \sum_n P_n \left\{ 1 - \left[\langle e^{-kt} \rangle^n \right]_v \right\} & \text{(influx)} \quad (4) \end{cases}$$

where the summation is over subpopulations of CMS having n gated channels per CMS. The quantity actually measured in ion flux experiments - the time dependent overall ion content of a collection of CMS - is thus seen to depend on: (a) the mean single channel flux contribution $\langle e^{-kt} \rangle$, which is a function solely of the kinetic parameters for ion flux and channel gating, and (b) the weight factors P_n and an average over a volume distribution function (expressed by the square brakets $[\ldots]_v$), which explicitly incorporate the variations in CMS ion channel content and volume. Employing equation (3) Bernhardt & Neumann (1982 a,b) were able to analyze snake toxin inhibition of acetylcholine receptor controlled ion flux, without recourse to additional assumptions.

References

1. Bernhardt, J., E. Neumann (1978). Proc. Natl. Acad. Sci. USA 75, 3756.

2. Bernhardt, J., E. Neumann (1980a). Theor. Biol. 86, 649.

3. Bernhardt, J., E. Neumann (1980b). Neurochem. Int. 2, 243.

4. Bernhardt, J., K.M. Moss, R.M. Luckinger, E. Neumann (1981). FEBS Letters 134, 245.

5. Bernhardt, J., E. Neumann (1981). Biophys. Chem. 14, 303.

6. Bernhardt, J., E. Neumann (1982a). Biophys. Chem. 15, 317.

7. Bernhardt, J., E. Neumann (1982b). Biophys. Chem. 15, 327.

8. Duguid, J.R., M.A. Raftery (1973), Biochemistry 12, 3593.

9. Elliot, J., S.G. Blanchard, W. Wu, J. Miller, J. Strader, P. Hartig, H.-P. Moore, J. Racs, M.A. Raftery (1980). Biochem. J. 185, 667.

10. Heidmann T., J.P. Changeux (1979). Eur. J. Biochem. 94, 255.

11. Heidmann, T., J. Bernhardt, E. Neumann, J.P. Changeux (1983). Biochemistry 22, 5452.

12. Hess, G.P., J.P. Andrews, G.E. Struve, S.E. Coombs (1975). Proc. Natl. Acad. Sci. USA 72, 4371.

13. Hess, G.P., J.P. Andrews (1977). Proc. Natl. Acad. Sci. USA 74, 482.

14. Hess, G.P., D.J. Cash, H. Aoshima (1979). Nature (London) 282, 329.

15. Hess, G.P., E.B. Pasquale, J.W. Walker, M.G. Mc Namee (1982). Proc. Natl. Acad. Sci. USA 79, 963.

16. Kasai, M., J.P. Changeux (1971). J. Membr. Biol. 6, 1.

17. Moore, H.P., P.R. Hartig, M.A. Raftery (1979). Proc. Natl. Acad. Sci. USA 76, 6265.

18. Moore, H.P., M.A. Raftery (1980). Proc. Natl. Acad. Sci. USA 77, 4509.

19. Neubig, R.R., J.B. Cohen (1980). Biochemistry 19, 2770.

20. Popot, J.L., H. Sugiyama, J.P. Changeux (1976). J. Mol. Biol. 106, 469.

21. Sobel, A., M. Weber, J.P. Changeux (1977). Eur. J. Biochem. 80, 215.

22. Walker, J.W., M.G. Mc Namee, E. Pasquale, D.J. Cash, G.P. Hess (1981). Biochem. Biophys. Res. Commun. 100, 86.

SURFACE PROCESSES IN ION TRANSPORT AND EXCITATION

Martin Blank

Biological Sciences Division, Office of Naval Research
Arlington, Virginia 22217

Foreward

I dedicate this paper to the late David Nachmansohn. Over the years at Columbia University my biophysical perspective was frequently at odds with his biochemical approach. However, I shared his view of science and interest in bioelectric phenomena, and I found his search for unifying principles and enthusiasm for research particularly stimulating. Early on he convinced me of the need for different approaches to elucidate the mechanism of bioelectricity, even though it appeared to be essentially a biophysical problem. This review of my theoretical work on excitable membranes, while quite different in detail from the work of David Nachmansohn, is very much in the spirit of his approach to science.

Abstract

Surface processes are critical in ion transport, and we have included them in a description of ion fluxes during a voltage clamp. By treating the surface regions of membranes as compartments, we can write standard physical chemical equations to describe the effects of surface concentrations, charges, potentials and capacitances. In the presence of a non-selective voltage gated channel, we obtain the currents normally observed during a voltage clamp (i.e. early inward currents and later outward currents that are functions of the applied voltage). A model of the ion channel, based on the effect of surface charge on oligomer association equilibria, enables us to determine the physical properties (e.g. gating current conductance, surface capacitance) that can contribute to ion

Molecular Basis of Nerve Activity

specificity. A ligand gated channel may function by the same mechanism, since ligand binding can lead to changes in surface charge.

The Surface Compartment Model

Physiologists have generally formulated ion transport processes using membrane equations that involve the intra- and extra-cellular solutions only. This has frequently led to elaborate schemes that make little or no sense in terms of physical chemistry. This theoretical framework is obviously oversimplified, and a more realistic model could remove some of the confusing results. In the last few years we have been able to include ionic processes at membrane interfaces in the simple model, and this has helped to simplify the description of the unusual ion transport processes during excitation and to account for them in terms of physical chemistry(1).

Based on our research on ion transport through interfaces and monolayers, we have developed a way of approximating the ionic processes in the electric double layer region at charged surfaces (2) using the Surface Compartment Model (SCM). The SCM treats the electrical double layers at the surfaces of natural membranes as compartments, and includes two solutions with different ionic concentrations, two surface charges and potentials as well as two surface capacitances. It also considers ion concentration changes and cation binding and release in the surface layers. The equations for this system are based on: 1) conservation of mass for ions, 2) conservation of surface charges, 3) conservation of charges during current flow across the different interfaces, and 4) the reaction kinetics of cation binding and release at the membrane surfaces. In the equations, the individual ion flows are proportional to the electrochemical potential differences. The equations for the system have been solved under several sets of conditions, using numerical values from studies of the squid axon and relevant physical models (1-5).

Using the SCM system of differential equations together with an equation for a channel, where the cation permeability increases upon depolarization, we have obtained the membrane currents normally observed in voltage clamp experiments on squid axons, i.e., an initial inward current, followed by an outward current (figure 1). The dependence of the current on clamp voltage, generally characterized in terms of the peak inward current and the steady state outward current (figure 2), is similar to the curves published for squid axon. Changes in the values of the parameters can cause large changes in the curves. This has enabled us to obtain information about the relevant ranges of physical properties that affect excitation, and specifically about the channel proteins.

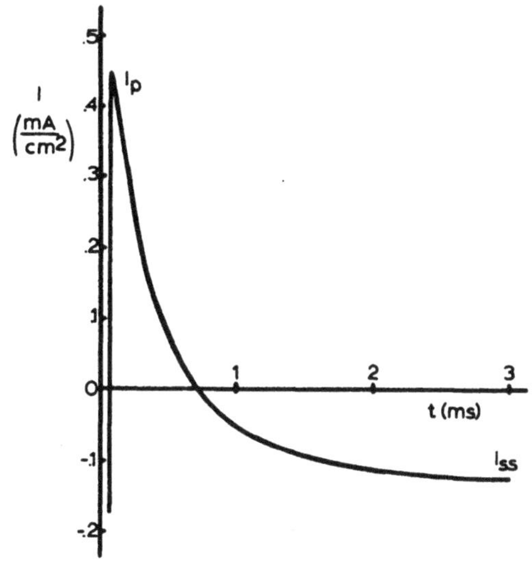

Figure 1. The membrane current, I (in mA/cm^2), versus the time, t (in ms) at a clamp voltage of +20mV. The early current is inward (positive) with a peak, Ip, at about 0.1ms, and the steady state current, I_{ss}, is outward (negative). Reproduced in modified form, with permission, from Bioelectrochem. Bioenerg. 10:451, 1983.

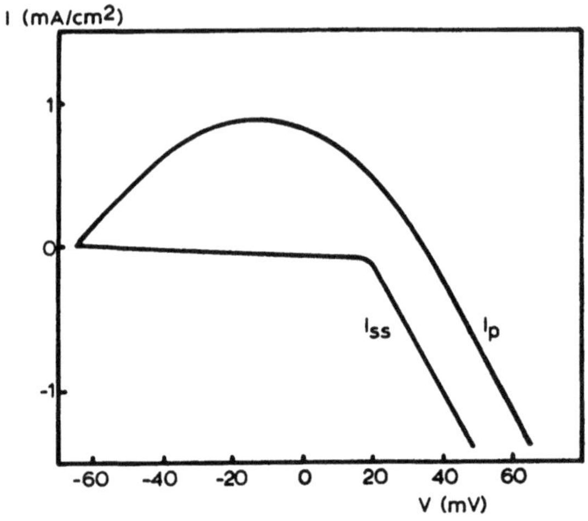

Figure 2. The peak current, Ip, and the steady state current, I_{ss}, both in units of mA/cm^2, versus the clamp voltage, V (in mV). Reproduced in modified form, with permission, from Bioelectrochem. Bioenerg. 10:451, 1983.

The voltage clamp current shown in figure 1 can be separated into an early inward component ascribed to sodium and a later outward component ascribed to potassium, by setting the permeability to either of the ions equal to zero. However, in the SCM both ion fluxes occur simultaneously through the same channel during a normal voltage clamp. We have shown that even though there is an increase in permeability to both cations, the ionic current appears specific because of the asymmetry of the resting concentration gradients. Greater apparent selectivity of a channel is possible and depends upon physical properties of the channel proteins, such as the conductance during the "gating" of the channel and the magnitude of the surface capacitances. Decreases in either of these properties can cause an apparent sodium selectivity to shift to a potassium selectivity.

A Voltage Dependent Ion Channel Based on Oligomeric Equilibria

Although any mechanism having a voltage dependent ion channel will give the above results with the SCM, we have developed a rationale for a voltage gated channel based on the dissociation of oligomeric proteins with increases in charge (3). Figure 3 shows how the small and rapid shifts of charge called "gating currents" could cause the permeability changes if the channel is oligomeric and dissociation is the molecular process that leads to opening. Because the charge distribution in the resting state is known to be asymmetric, the channel is not uniformly associated. It is dissociated (i.e., open) on the outer surface, where the charge is high, and associated (i.e., closed) on the inner surface, where the charge is low. Upon depolarization, the shift of negative charge from the outer to the inner surface causes the charge density to increase to the point where dissociation is energetically favored. The dissociation of hemoglobin tetramers into dimers occurs when the surface charge density of the tetramer, calculated as the total area divided by the titratable net positive or net negative charge, is approximately equal to 1 charge per 10 nm^2. The surface charge density on the inner face of the resting squid axon is about the same, so the channel proteins may be on the point

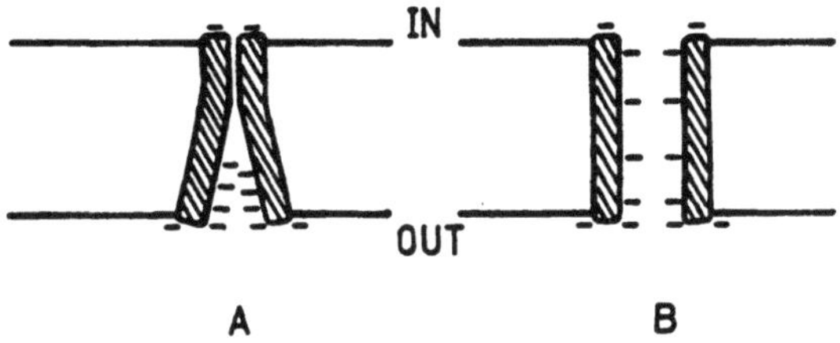

Figure 3. A diagram of a voltage-sensitive oligomeric pore within the membrane matrix. (A) In the resting state, the charge density causes the outer region to be dissociated while the inner region is associated. (B) Upon depolarization, the charge density shifts causing the inner region to dissociate and the pore to open. Reproduced with permission from Bioelectrochem. Bioenerg. 9:615, 1982.

dissociating due to the depolarizing current. Knowing the oligomer dissociation constant as a function of charge for a molecule like the hemoglobin tetramer, together with the measured number of channels in the axon membrane, we have related the fraction of open channels to the surface charge density (3) and shown that the gating currents give rise to voltage dependent changes in ion permeability. The properties of this voltage gated ion channel are compatible with the observed direction and magnitude of the shift of charge during depolarization, the range of surface charge where opening occurs, the cation selectivity of the channel, and the cation binding.

In principle, there can be different kinds of channels in a membrane due to chemical differences in the oligomers, but the simplest assumption is that an open channel should be available to all ions that can fit through the orifice. (In a charged channel there are also restrictions on the basis of charge.) This assumption, made in the SCM approach, is in line with the observed virtual interchangeability of Na^+ and Li^+ ions in one channel, and of K^+ and Tl^+ ions in another. It is also consistent with the similar time courses of different channels, e.g. K^+ and Ca^{++} in the cell bodies of excitable cells. The many experimental procedures that are used to define separate distinct channels, i.e. the passing of conditioning (hyperpolarizing or depolarizing) currents or the addition of pharmacologic agents (e.g. pronase), all cause significant changes in the structure of the channels, and it may be possible to explain many of the observations in line with the SCM. For example, the expected actions of pronase, rather than selecting for a particular channel, could lead to changes in the SCM membrane current that are consistent with the observations (1).

The Ligand Gated Channel

Another structure in excitable membranes that can be considered in terms of the same mechanism is the ligand gated acetylcholine receptor, which binds the neurotransmitter acetylcholine. In the electric organ of the electric eel, the receptor has a molecular weight of about 270.000 daltons, an isoelectric point below 5, and is composed of five protein chains ($\alpha_2\beta\gamma\delta$) that form a 10 nm long channel through the membrane. When two acetylcholines bind to the outer surface of the two α chains, the channel opens.

At first sight the reaction of ligands with receptors causing a channel to open appears to be quite different from the mechanism of voltage gating illustrated in figure 3. Two acetylcholine molecules interacting with groups on the two α chains of the receptor is quite specific in comparison with the gating currents caused by depolarization of the sodium channel. However, certain aspects of the reaction suggest parallels between the molecular mechanisms of the two types of gating.

Under normal conditions, the acetylcholine receptor is negatively charged and binds about 60 calcium ions per 300,000 molecular weight (6). When acetylcholine binds to the receptor, about 4-6 of those calcium ions are released. The displacement of calcium ions by acetylcholine ions leads to an increase in the total negative charge on the protein, which could trigger greater dissociation. It has been suggested (7) that the opening of a channel is due to this displacement reaction, and the partial disaggregation of the receptor protein could be the molecular mechanism. Biopolymer aggregation-disaggregation phenomena appear to be charge mediated (8) and could be the basis for opening and closing of oligomeric channels. Our recent unpublished results on the pH dependence of disaggregation in the biopolymers phycocyanin and hemocyanin support this general view. Our calculations of the Bohr effect in the oxygen binding of hemoglobin (9) and on the concentration dependence of the viscosity of hemoglobin solutions (10), which are based on the same model of the energetics of biopolymers, are also in line with this view of ion channel operation.

Acknowledgement

This work was done at Columbia University and supported by contract N00014-83-K-0043 from the Office of Naval Research.

References

1. Blank, M. 1983. The Surface Compartment Model with a Voltage Sensitive Channel. Bioelectrochem. Bioenerg. 10, 451-465.

2. Blank, M. and J. S. Britten. 1978. The Surface Compartment Model of the Steady State Excitable Membrane. Bioelectrochem. Bioenerg. 5, 528-540.

3. Blank, M. 1982. The Surface Compartment Model-Role of Surface Charge in Membrane Permeability Changes. Bioelectrochem. Bioenerg. 9, 615-624.

4. Blank, M. and W. P. Kavanaugh. 1982. The Surface Compartment Model During Transients. Bioelectrochem. Bioenerg. 9, 427-438.

5. Blank, M., W. P. Kavanaugh and G. Cerf. 1982. The Surface Compartment Model-Voltage Clamp. Bioelectrochem. Bioenerg. 9, 439-458.

6. Chang, H. W. and E. Neumann. 1976. Dynamic Properties of Isolated Acetylcholine Receptor Proteins: Release of Calcium Ions Caused by Acetylcholine Binding. Proc. Nat. Acad. Sci. USA 73, 3364-3368.

7. Nachmansohn, D. and E. Neumann. 1975. "Chemical and Molecular Basis of Nerve Activity". Academic Press, New York

8. Blank, M. 1980. A Surface Free Energy Model for Protein Structure in Solution: Hemoglobin Equilibria. Colloids and Surfaces 1, 139-149.

9. Blank, M. 1975. A Model for Calculating the Bohr Effect in Hemoglobin Equilibria. J. Theoret. Biol. 51, 127-134.

10. Blank, M. 1984. Molecular Association and the Viscosity of Hemoglobin Solutions. J. Theoret. Biol. 108, 55-64.

CHEMICAL KINETIC INVESTIGATIONS OF THE EFFECTS OF CIS AND TRANS BIS-Q ON THE ACETYLCHOLINE RECEPTOR IN ELECTROPHORUS ELECTRICUS VESICLES

Anne H. Delcour, George P. Hess
Section of Biochemistry, Molecular and Cell Biology, Division of Biological Sciences, 270 Clark Hall, Cornell University, Ithaca, New York 14853 U.S.A.

Abstract

A quenched flow technique was used to study the effects of *cis* and *trans* Bis-Q on the acetylcholine receptor-controlled ion translocation in vesicles prepared from the electric organ of *E. electricus*. The response elicited by the agonist *trans* Bis-Q was analyzed in terms of a minimum mechanism which accounts for the properties of activation, inactivation and regulation of the receptor. Two molecules of *trans* Bis-Q must be bound in order for channel opening to occur, but, at concentrations greater than 50 μM, the population of open channels decreases with ligand concentration because of the additional binding of one molecule of *trans* Bis-Q to a regulatory site that is independent of the activating sites. Although not an agonist, *cis* Bis-Q was found to inhibit the response of the receptor to acetylcholine and to induce inactivation (desensitization).

Introduction

The design of photoactivatable substrates allows the study of chemical reactions in the micro- to millisecond time region by the rapid conversion of an inactive compound to an active

Molecular Basis of Nerve Activity

one. Cis-trans isomerizations of azobenzenes are particularly suitable because they occur within microseconds with a high quantum yield (1). Initially these compounds were used as inhibitors of chymotrypsin (2), of acetylcholinesterase (3), and of the acetylcholine receptor (4). More recently, several photoisomerizable cholinergic ligands with different effects on the nicotinic acetylcholine receptor, were designed and synthesized by M.H. Wassermann and B.F. Erlanger (5). Because of the different geometrical arrangements of the *cis* and *trans* isomers, the two configurations differ in their physical and pharmacological properties and can therefore be used as inactive precursors of one another, depending on the reaction investigated.

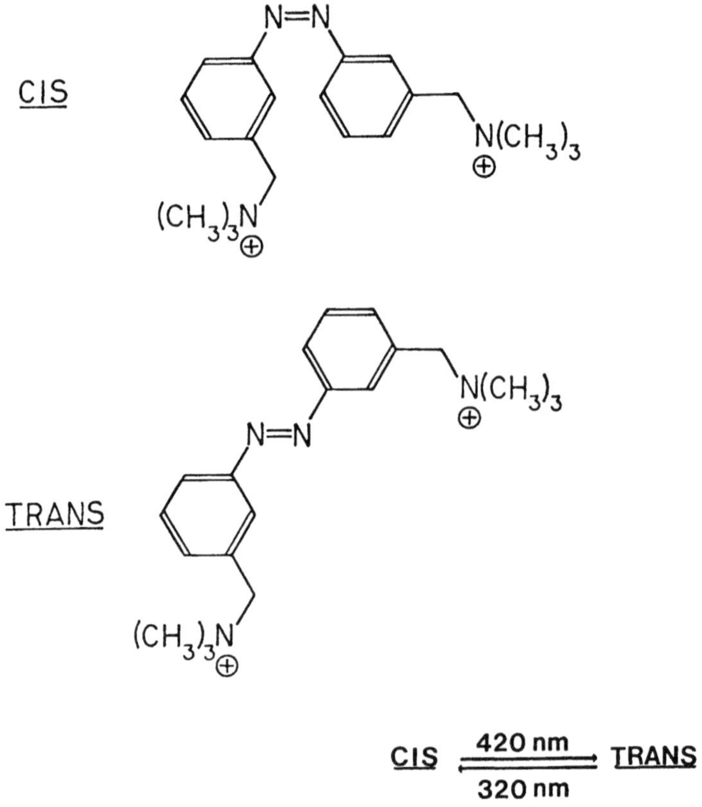

Figure 1: Structure of the photoisomers of Bis-Q

For example, the *cis* form of Bis Q (3,3' bis (trimethyl-ammoniomethyl)-azobenzene bromide) (Fig. 1) has been used by H.A. Lester and coworkers in experiments involving the light-induced generation of an agonist (*trans* Bis-Q) in the neighborhood of acetylcholine receptors of *E. electricus* electroplaques, with the aim of studying the activation of the receptor and the channel-opening process (6). In less than 1 µs, production of 350 nM *trans* Bis Q could be obtained from 600 nM *cis* Bis Q, and the lag before detection of an electrical response, defined as the channel opening latency, was 10 µs. *Trans*-*cis* isomerizations of bound *trans* molecules, after the voltage is jumped from +50 mV to -150 mV, produce a transient conductance decrease, which has been associated with channel closing (7). The authors estimated that the rate of channel closing is increased 100-fold when bound *trans* molecules are isomerized to *cis* Bis-Q. This effect occurs in less than 80 µs. This result implies that *cis* Bis-Q is a much poorer agonist than *trans* Bis-Q at the *E. electricus* electroplaques.

However, since these results were obtained in the presence of a mixture of the two isomers, an imperative in the analysis of these experiments is the availability of a rigorous and quantitative characterization of the effects of both isomers on the acetylcholine receptor in a well-defined system. Kinetic measurements of ion flux in vesicles, prepared from *E. electricus* electroplaques, using fast reaction techniques have provided direct information on the rates of activation and inactivation (desensitization) of the receptor and have led to the proposal of a minimum mechanism that accounts for the properties of ligand binding, activation, inactivation, regulation of the receptor, and for single channel current measurements with *E. electricus* electroplaques from which the vesicles were prepared (8,9,10,11). Our ability to separate *cis* from *trans* Bis-Q (12) and the use of a quenched flow technique (13) allowed the determination of the kinetic parameters characterizing the effects of cis and trans Bis-Q on the acetylcholine receptor of *E. electricus*. From the experiments reported here, the action of the agonist trans Bis-Q follows the model derived for the response to suberyldicholine (9), where high concentrations of the ligand inhibit the flux activity of the receptor. cis Bis-Q is not an agonist, but appears to compete with acetylcholine for the binding to the receptor, and induces receptor inactivation (desensitization).

Materials and Methods

Receptor-rich vesicles from the electroplaques of *E. electricus* were prepared as described (14,15). The vesicles were equilibrated in eel Ringer's solution (169 mM Nacl, 5 mM KCl, 3 mM $CaCl_2$, 1.5 mM $MgCl_2$, 1.5 mM sodium phosphate pH 7.0). trans Bis-Q was purchased from Molecular Probes (Junction City, OR). Pure cis Bis-Q was prepared by high performance liquid chromotography from the photoequilibrium mixture obtained by exposure to a N_2 laser beam at 337 nm, according to published procedures (12). The experiments were done at 1°C and in a dark room using Sylvania red striplight F40R Lifeline as a safelight.

A quenched flow technique (13) was used to follow the acetylcholine receptor-mediated $^{86}Rb^+$ flux in the millisecond-to-second time region in the presence of solutions of the pure isomers.

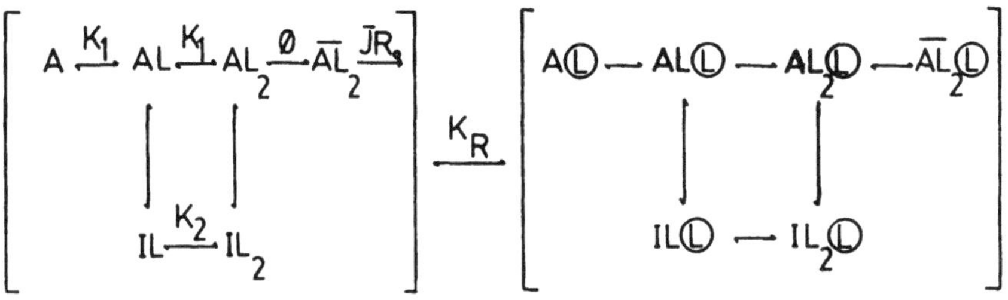

$$\frac{M_t}{M_\infty} = 1 - \exp -[J_A(\frac{1 - \exp(-\alpha t)}{\alpha}) + J_I t] \quad (1)$$

$$J_A = \frac{\bar{J} R_0 L^2}{[L^2(\emptyset + 1) + K_1^2\emptyset + 2K_1\emptyset L][1 + L/K_R]} \quad (2)$$

Figure 2: Minimum mechanism relating ligand binding to the acetylcholine receptor and receptor-controlled cation flux. Three states of the receptor are represented: an active closed form (A), an active open form ($\bar{A}$) and an inactive (desensitized) form (I). K_1 and K_2 represent the intrinsic dissociation constants of the active and inactive forms respectively. Φ is the channel-closing equilibrium constant, and K_R is the dissociation constant of the inhibitory site. The ligand molecules binding to that site are shown as Ⓛ. Equation (1) describes the change in the ratio of the radioactivity contained in the vesicles at time t to that after complete filling of the vesicles (∞) as a function of time. J_A is the rate constant for ion flux before the onset of inactivation. J_I is the rate constant for ion flux mediated by the population of receptors remaining after inactivation (desensitization) has been completed and α is the rate constant for inactivation. Equation (2) describes the dependence of J_A on the ligand concentration (L). $\bar{J}$ is the specific reaction rate for the ion translocation process and R_0 is the number of receptors per liter internal vesicle volume.

The rate constant for ion flux, J_A, was obtained by fitting the experimental points to equation (1) (Fig. 2). The rate constant for inactivation (desensitization), α, was obtained either from the fits of the ion flux curves or from direct measurements of the remaining activity of the receptor population after exposure to the ligand for various periods of time. The procedure for and the analysis of the desensitization experiments have been described (16).

Results

Ion flux curves obtained with three concentrations of trans Bis-Q are shown in Figure 3. The points are the mean of three determinations and the lines were fitted to equation (1) (Fig. 2) using a non-linear least-squares computer program.

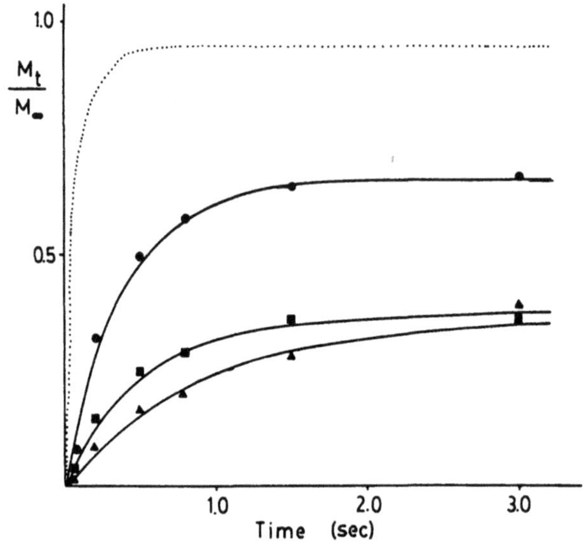

Figure 3: trans Bis-Q induced $^{86}Rb^+$ flux. (▲) 1.5 μM, (●) 50 μM, (■) 300 μM. The dotted line represents the theoretical influx curve obtained with a saturating concentration of acetylcholine (1 mM). The lines were fitted to equation (1).

The final $^{86}Rb^+$ content of the vesicles reached after 3 seconds (Fig. 3) increases with increasing trans Bis-Q concentration up to 50 μM, but then decreases at higher concentrations. The final amplitude of the flux response is determined by the ratio J_A/α, as can be seen from equation (1) (Fig. 2). The concentration dependence of the 3-second amplitude measurement can, therefore, be explained by a decrease in J_A, an increase in α, or both. Figure 4 shows that only J_A decreases at high concentrations of agonist while α, the rate constant for inactivation, remains constant.

A similar pattern was reported for suberyldicholine when no voltage was established across the membrane (9) and for acetylcholine at a transmembrane voltage, V_m, of -45 mV (10). The results can be accounted for by a model which assumes the binding of two molecules of agonist to the activating sites and of an additional molecule to a separate regulatory site present on all forms of the receptor (Fig. 2) (9).

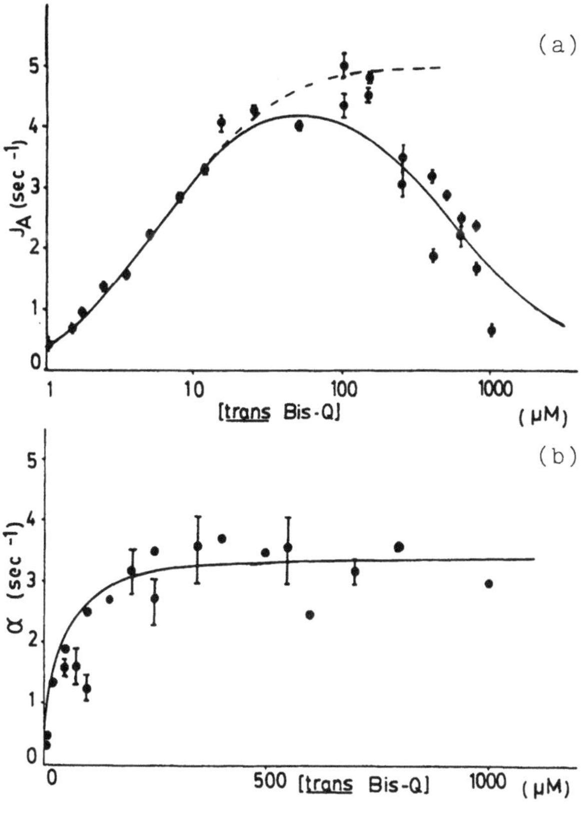

Figure 4: (a) Concentration-dependence of J_A. The dotted line corresponds to an equation similar to equation (2) where non-competitive inhibition by the ligand does not occur; it shows what J_A (max) would be in that case. The solid line was calculated according to equation (2) using the following values for the constants: K_1 = 3 μM, K_R = 530 μM, Φ = 6.3. (b) Concentration dependence of α. The points without error bars were obtained by fitting the influx curves to equation (1) and the points with error bars were obtained from direct measurements of α (16).

The solid line in Figure 4a was fitted to the experimental points using equation (2) (Fig. 2) and the following parameters were obtained: K_1, dissociation constant of the activating sites, is 3 μM; K_R, dissociation constant of the regulatory site, is 530 μM; Φ, channel-closing equilibrium constant, is 6.3.

The response of the acetylcholine receptor to concentrations of pure cis Bis-Q ranging from 1 μM to 1 mM was studied by 1-second and 20-second flux amplitude measurements. No significant flux activity was detected. However, cis Bis-Q can inhibit acetylcholine-induced flux. The pattern of the cis Bis-Q inhibition in the presence of three concentrations of acetylcholine is shown in Figure 5, and suggests a competitive interaction between acetylcholine and cis Bis-Q for binding to the receptor in the micromolar range.

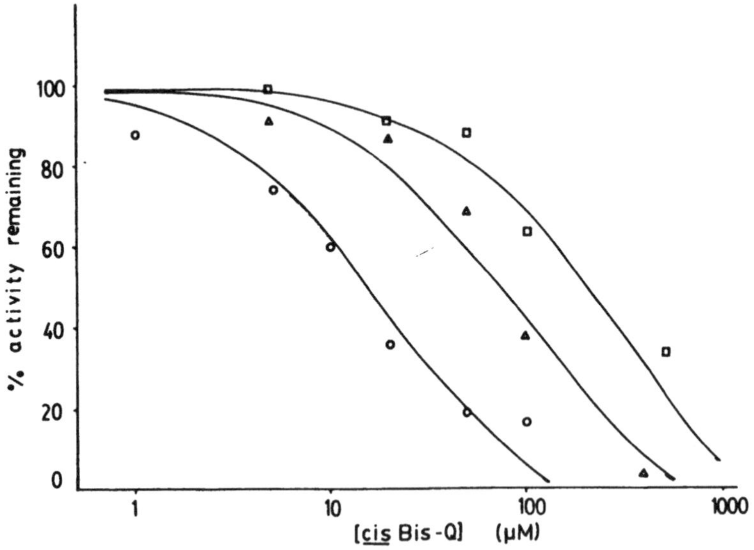

Figure 5: Inhibition of acetylcholine-induced flux by cis Bis-Q. 3 concentrations of acetylcholine were used: (o) 0.1 mM, (Δ) 0.8 mM, (□) 4 mM.

When vesicles are preincubated with a solution of pure cis Bis-Q (100 μM) for times up to 25 seconds, a time-dependent decrease in the flux amplitude subsequently elicited with 5 mM acetylcholine is observed, which corresponds to the inactivation of the receptor in the presence of cis Bis-Q (Fig. 6). The half-time of the process was found to be 11 seconds ($\alpha = 0.06\ sec^{-1}$).

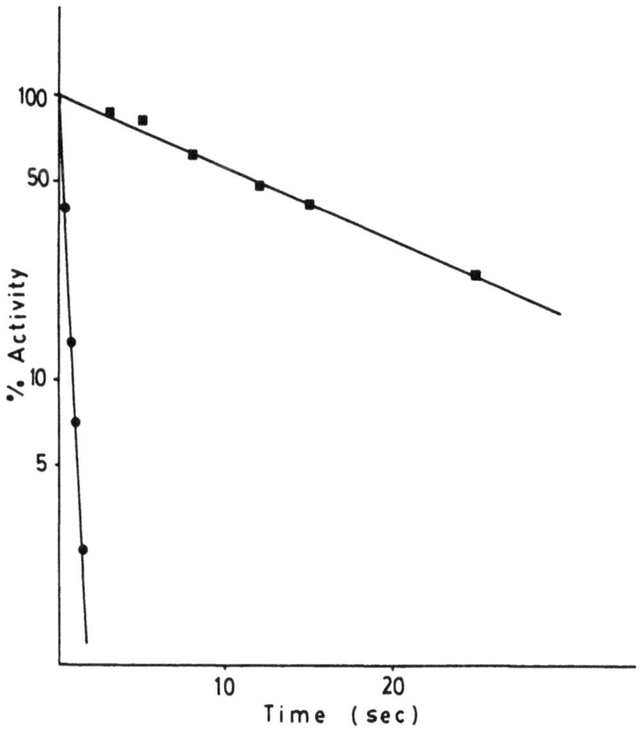

Figure 6: Inactivation (desensitization) induced by 100 μM cis Bis-Q. % activity is measured by a 800 msec flux amplitude in the presence of 5 mM acetylcholine with and without preincubation with cis Bis-Q. The preincubation times are given on the abscissa. For cis Bis-Q (■), $\alpha = 0.06\ sec^{-1}$. A control was run with 5 mM acetylcholine but without cis Bis-Q in the preincubation mixture (●), $\alpha = 2.8\ sec^{-1}$.

Discussion

Chemical kinetic studies of the response of the acetylcholine receptor to three ligands (acetylcholine, carbamoylcholine, and suberyldicholine) in the absence and the presence of a transmembrane voltage have led to the proposal of a minimum mechanism relating ligand binding to ion flux (8,9,10). They allowed the determinations of kinetic parameters that are intrinsic to the receptor and the ligand: the intrinsic dissociation constants of the activating sites, K_1, and of the regulatory site, K_R, the channel-opening equilibrium constant, Φ^{-1} and the specific reaction rate for the ion translocation process, $\bar{J}$. $\bar{J}$ is directly related to the single-channel conductance, γ, determined in electrophysiological experiments with cells (11). The values of $\bar{J}$ and γ are in good agreement with each other (11), and have been found to be independent of the ligand used. The maximum J_A value reached with <u>trans</u> Bis-Q is 5.1 sec^{-1}, which is relatively low compared to acetylcholine (14.8 sec^{-1}) or suberyldicholine (18.5 sec^{-1}). J_A(max) depends on $\bar{J}$ and Φ, the channel-closing equilibrium constant. The low value of J_A(max) for Bis-Q is related to the high value of Φ (6.3) compared to acetylcholine (1.5) and suberyldicholine (1.0). Electrophysiological measurements indicate that the channel closing rate depends on the activating ligand used, while the channel opening rate does not (22,23). The difference in Φ values obtained for the different activating ligands is therefore accounted for by the agonists exerting a specific effect mainly on the rate constant for channel closing, while the rate constant for channel opening is not affected (9). In this view, the high value for Φ obtained with <u>trans</u> Bis-Q would be associated with a fairly large rate constant for channel closing. The fact that <u>cis</u> Bis-Q, like <u>trans</u> Bis-Q, can bind to the receptor and elicit inactivation, but unlike <u>trans</u> Bis-Q does not elicit flux response, suggests that the channel closing rate constant, and therefore Φ, is much larger for *cis* than for *trans* Bis-Q. This is in agreement with measurements which indicate that the lifetime of *cis* Bis-Q channels is much smaller than that of *trans* Bis-Q (17,18).

The isomers of Bis-Q have been used in electrophysiological experiments on eel electroplaques to investigate the channel opening process (6). Our purpose was to characterize these compounds first before using them in studies where *cis-trans* isomerizations of Bis-Q would rapidly produce *trans* Bis-Q in the neighborhood of acetylcholine receptors, and the agonist-induced currents due to channel opening could be followed using the loose patch clamp technique (24). Chemical kinetic measurements of receptor function in cells, which are in principle possible with photoactivable receptor ligands, require that measurements can be made over a wide range of ligand concentration and that the "inactive" ligand precursor does not inactivate receptor channel. The experiments in Figure 6 indicate that the inactive precursor of *trans* Bis-Q desensitizes the receptor and the experiments in Figure 4 indicate that a wide concentration range of *trans* Bis-Q cannot be used without inhibition of the receptor. Finally, the large value of Φ obtained with *trans* Bis-Q would represent a disadvantage in attempts to obtain the channel opening rate constant since the channel closing rate constant is expected to dominate the kinetic measurements. For these reasons, we feel that we would not be able to obtain quantitative information about acetylcholine receptor function in cells with Bis-Q, and that another photoactivatable ligand should be used (see Hess, G.P. *et al.*, this volume).

The binding of agonists to a voltage-dependent regulatory site has been observed with suberyldicholine (9) and acetylcholine (10). The results obtained with *trans* Bis-Q add to the generalization of the model for the nicotinic agonists. A blockade by agonists has also been observed in electrophysiological experiments with suberyldicholine (19), with decamethonium (20), and with both isomers of Bis-Q (18), and in kinetic studies with dansylcholine (21). The scheme presented in Figure 1 is the simplest one that accounts for all the data, and it assumes that the regulatory site is present on all forms of the receptor.

In conclusion, <u>trans</u> Bis-Q displays the typical bell-shaped response curve already observed with other agonists, which can be accounted for by the binding of two molecules to the activating sites and an additional one to the independent regulatory site which causes inhibition of the receptor. The use of the two isomers of Bis-Q is, therefore, limited to very low concentrations where agonist inhibition is not observed and where <u>cis</u> Bis-Q can be considered as an inactive precursor (<1 μM in the absence of transmembrane voltage).

Acknowledgements

A.H.D. is a recipient of a fellowship awarded by the Association of American University Women. This work was supported by a grant (NS 08527) awarded to G.P.H. by the National Institutes of Health.

References

1. Ross, R.L., Blanc, J. in G.H. Brown (Editor), Techniques of Chemistry, Vol. III, Photochromism, Wiley-Interscience, New York, 1971, p. 471.

2. Kaufmann, H., Erlanger, B.F. (1967) Biochemistry, 6, 1597.

3. Bieth, J., Vratsanos, S.M., Wassermann, N., Erlanger, B.F. (1969) Proc. Natl. Acad. Sci. USA, 64, 1103.

4. Deal, W.J., Erlanger, B.F., Nachmansohn, D. (1969) Proc. Natl. Acad. Sci. USA, 64 1230.

5. Lester, H.A., Nerbonne, J. (1982) Ann. Rev. Bioph. Bioeng., 11, 151.

6. Krouse, M.E., Nass, M.M., Nerbonne, J.M., Lester, H.A., Wassermann, N.H., Erlanger, B.F., in Neurotransmitters and Hormone Receptors in Insects, ed. D.B. Satelle, L.M. Hall, J.G. Hildebrand, p. 17, Amsterdam: Elsevier-North Holland.

7. Nass, M.M., Lester, H.A., Krouse, M.E. (1978) Biophys. J., 24, 135.

8. Hess, G.P., Cash, D.J.,Aoshima, H. (1983) Ann. Rev. Biophys. Bioeng., 12, 443.

9. Pasquale, E.B., Takeyasu, K., Udgaonkar, J.B., Cash, D.J., Severski, M.C., Hess, G.P. (1983) Biochemistry, 22, 5967.

10. Takeyasu, K., Udgaonkar, J.B.,Hess, G.P. (1983) Biochemistry, 22, 5973.

11. Hess, G.P., Kolb, H.-A., Läuger, P., Schoffeniels, E., Schwarze, W. (1984) Proc. Natl. Acad. Sci. 81, 5281.

12. Delcour, A., Cash, D.J., Erlanger, B.F., Hess, G.P. (1982) J. Chromato., 248, 461.

13a. Hess, G.P., Cash, D.J., Aoshima, H. (1979), Nature, 282, 329-331.

13b. Cash, D.J., Hess, G.P. (1981) Analyt. Biochem., 112, 39.

14. Kasai, M., Changeux, J.P. (1971) J. Membr. Biol., 6, 1.

15. Fu, J.-j. L., Donner, D.B., Moore, D.E., Hess, G.P. (1977) Biochemistry, 16, 678.

16. Aoshima, H., Cash, D.J., Hess, G.P. (1981) Biochemistry, 20, 3467.

17. Lester, H.A., Krouse, M.E., Nass, M.M., Wasserman, N.H., Erlanger, B.F. (1980) J. Gen. Physiol., 75, 207.

18. Chabala, L.D., Gurney, A.M., Lester, H.A. (1984) Biophys. J., 45, 387a.

19. Neher, E., Steinbach, J.H. (1978) J. Physiol. (London), 277, 153.

20. Adams, P.R., Sakmann, B. (1978) Proc. Natl. Acad. Sci., 75, 2994.

21. Cohen, J.B., Changeux, J.P. (1973) Biochemistry, 12, 4855.

22. Neher, E., Sakmann, B. (1976) Nature (London) 260, 799.

23. Katz, B., Miledi, R. (1973) J. Physiol. (London) 230, 707.

24. Stühmer, W., Roberts, W.M., Almers, W. (1983) in Single-Channel Recording, ed. B. Sakmann and E. Neher, p. 123, New York:Plenum.

DETERMINATION OF THIAMINE AND ITS PHOSPHATE ESTERS BY REVERSED-PHASE HPLC AT FEMTOMOLE LEVELS. APPLICATION TO BIOELECTROGENESIS

L. Bettendorff , Ch. Grandfils, E. Schoffeniels, J. Bontemps, D. Schmartz and G. Dandrifosse.

Laboratoire de Biochimie générale et comparée, Université de Liège, place Delcour, 17, B-4020 Liège, Belgique

Introduction

Nearly 30 years ago, Nachmansohn made the first attempt to propose a biochemical theory of nerve excitation (1, 2). According to this model, changes of permeability may be induced by the release of acetylcholine (AcCh) from a storage site and its binding to a receptor protein, thereby inducing a conformational change. Hydrolysis of AcCh by AcCh-esterase would permit the return of the receptor protein to its resting state.
More recently, a complete discussion of the thermodynamic aspects of the action potential is provided by Margineanu and Schoffeniels (4). From their considerations, it springs that the conformational transitions between the low conductance and the high conductance states of the sodium channel should be controlled by a chemical process with bimolecular reaction rates (5,6).
Now, the question arises as to the nature of this chemical input. Schoffeniels (7) considered the idea of a phosphorylation cycle controlling the impedance variation in excitable membranes. He and his colleagues (8,9) brought evidence for the phosphorylation of specific high molecular weight membrane proteins extracted from crab nerves. Independently this observation was confirmed by other authors (10).
This phosphorylation process could be influenced by some neurotropic compounds such as tetrodotoxin (TTX) and veratridine which are specific sodium conductin markers, as well as by AcCh and various other compounds (8,11).

Molecular Basis of Nerve Activity

Here, we present evidence, that phosphorylated esters of thiamine (thiamine triphosphate : ThTP, thiamine pyrophosphate : ThPP, thiamine monophosphate : ThMP, thiamine : Th) copurify with the TTX receptor protein (12,13). However due to the fact that these compounds are found in small quantities, it was necessary to improve an available method of thiamine identification by reversed phase HPLC.

Determination of thiamine and its phosphate esters by reversed phase high performance liquid chromatography

High performance liquid chromatography (HPLC) is the method of choice for the determination of thiamine and its phosphate esters because of its rapidity and simplicity. High sensitivity was obtained through the chemical oxydation of thiamine compounds in fluorescent derivatives prior to injection outo the column. This precolumn derivation procedure with potassium ferricyanide ($1.54.10^{-4}$M) and NaOH (15 %) was the same as described previously (14).

Instrumentation

The HPLC instrument was an Altex model (model 334-50) equipped with an LS-4 fluorescence spectrometer from Perkin-Elmer. The wavelengths were set at 365 nm for the excitation and at 433 nm for the emission. We used a PRP-1 analytical column (250 x 4.1 nm, 10 µm) from Hamilton (Reno, USA) protected by a precolumn (32 x 4.2 mm) dry-packed with Vydac-201 RP (30-44 µm) from Macherey-Nagel (Düren, F.R.G). The mobile phase was composed of 90 % phosphate buffer (25 mM, pH 8.5) and 10 % methanol for the separation of ThTP, ThPP and ThMP. The instrument was run at 40 % phosphate buffer and 60 % methanol for the analysis of thiamine. The solvent flow rate was 0.5 ml/min and the pressure around 700 psi. The injection volume was 20 µl.

Method

We studied the chromatographic behaviour of thiamine compounds by means of capacity factors, plate numbers and resolution (Fig. 1).

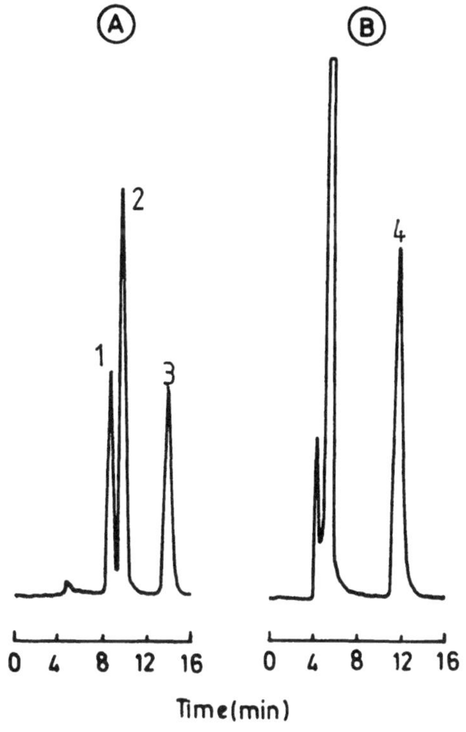

Fig. 1 : Chromatograms of thiamine phosphate derivatives (A) and thiamine (B). (ThTP : 1, ThPP : 2, ThMP : 3 and Th : 4).

The capacity factors (k') for ThTP, ThPP and ThMP are 0.82, 1.04 and 1.88 respectively, when the mobile phase was composed of 90 % phosphate buffer and 10 % methanol. T, which is eluted with a proportion of 60 % methanol in the mobile phase, has a k' of 1.4.

This method is characterized by relatively high plate numbers (N) as compared to the literature, N = 2500 for the phosphate esters and N = 1500 for Th(15).
The resolution between ThTP and ThPP is 1.28 and the one between ThPP and ThMP is 2.5. This was enough for all the applications we had to deal with.
The detection limit of this method is 5 femtomoles for each thiamine compound injected. A linear relationship between peak area and the quantity of each thiamine phosphate was tested from 0.01 to 20 pmoles injected. These considerations show that our method is about 5 times more sensitive than the most sensitive methods published until now (14,15,16).

Quantitative analysis

Once the characteristics of this method were settled, we applied it to the analysis of thiamine compounds in rat tissues (table 1). The sample preparation was the same as described by Ishii et al (17) except for the following points. Sciatic nerves were homogenized in 10 volumes of 5 % trichloracetic acid (TCA) instead of 3 volumes. After centrifugation (20 min, 5000 g, 4°C) the supernatant of all the samples was extracted 3 times with 4 volumes of water-satured diethyl ether. All the operations, except the homogeinization, took place in polypropylene vials.
Our results were compared to those obtained by other authors either by a HPLC technique (17) or by a low pressure method (18). They are in good agreement with those of these authors. Ishii et al, for example, indicate the total content of thiamine in nmol/g of wet weight as follows : brain, 6.39, heart, 15.7 and kidney, 14.9.

Table 1 : Contents of Thiamine and its Phosphate Esters in Rat Tissues (nmol/g of wet weight)

	Sciatic nerve	Brain	Heart	Kidney
Thiamine	0.18±0.01*	0.42±0.09	0.17 ±0.05	1.3 ±0.1
ThMP	0.52±0.20	0.87±0.07	1.4 ±0.3	2.2 ±0.42
ThPP	2.0 ±0.2	5.8 ±0.5	16.4 ±2.3	11.4 ±2.1
ThTP	0.03±0.01	0.07±0.01	0.10 ±0.03	0.11±0.01
Total thiamine	2.7 ±0.3	7.2 ±0.5	18.1 ±2.3	15 ±2.1

*Values represent the mean ± S.D. for 3 rats.

Analytical patterns of thiamine phosphate esters and thiamine in rat brain and heart are shown in fig. 2.
In all the samples we assayed, the detection of ThTP didn't provide any difficulty. This represents a considerable improvement over our previous method (14), where we have been unable to detect ThTP in several rat tissues as brain and kidney, because of an interfering front peak. The fluorescence spectrometer we used here offers a much better selectivity, than conventional filter fluorometers with broader passbands. When working under isocratic conditions on a reversed phase column, it is impossible to separate the four thiamine compounds in a single chromatography, because of the much greater retention of Th compared to its phosphate esters (19). We preferred two different isocratic conditions for the elution of thiamine phosphoesters (10 % methanol) and thiamine (60 % methanol) rather than a gradient system (14). Isocratic conditions offer a better reproductibility than gradient elution and avoid large baseline fluctuations, which are particularly bothersome for the determination of small quantities (< 1 pmol) of thiamine compounds.
The loss of peak height during isocratic conditions compared to gradient elution could be more than equaled by the much more powerful fluorometer.

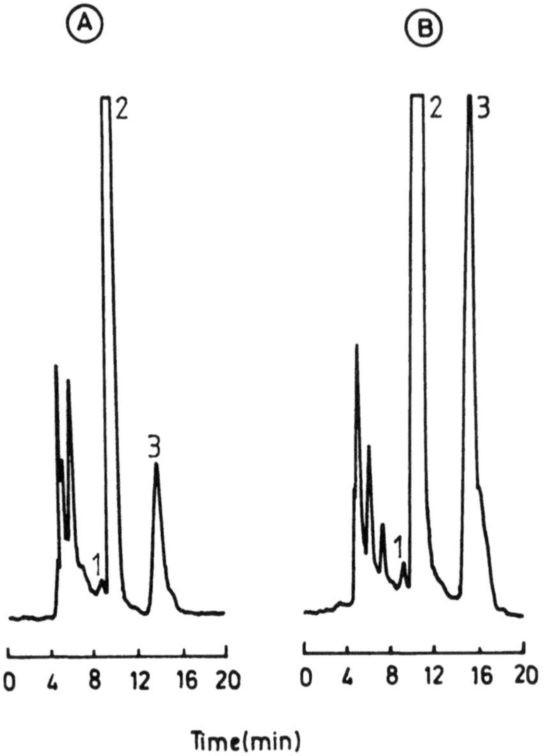

Fig. 2 : Chromatograms of thiamine phosphate derivatives (ThTP : 1, ThPP : 2, ThMP : 3) in rat brain (A) and heart (B).

Thiamine phosphorylated derivatives and sodium channel isolated from P. electricus excitable membranes

After having set up a new and powerful method for the determination of thiamine derivatives in biological samples, we wanted, in a first approach, to see whether phosphorylated derivatives of thiamine co-purify with the sodium conducting site protein, called the TTX receptor.

Purification of the TTX receptor

The TTX receptor of E. electricus conducting membranes was purified as described by Dandrifosse et al (20).
Briefly, membranes from the main electric organ of E. electricus were prepared and stored at - 70°C until use. The membranes were solubilized in 1 % Lubrol-PX and centrifuged at 150 000 g for 30 min. The membrane extract thus obtained is mixed with a suspension of DEAE-Sephadex A-50. The anion exchanger is washed several times (DEAE wash) at low ionic strength and then proteins fixed on the gel are eluted at high ionic strength. The fractions containing TTX binding activity were pooled (DEAE Pool) and chromatographed on a WGL-Sepharose 6MB column. Glycoproteins were eluted with a linear gradient of N-acetylglucosamine (lectine eluate). The degree of purification was measured through the binding of ED-diTTX(21) performed by rapid gel filtration on small Sephadex G-50 column as described by Bontemps et al (22).

Thiamine phosphorylated derivatives and sodium conducting site

During the whole purification procedure of the sodium conductin, the concentration of thiamine derivatives was estimated by the HPLC method described above.
The results obtained are quoted in table 2.
There is a clear enrichment of thiamine phosphate esters during the purification of the TTX receptor. The final specific activity in the lectin eluate is about 135 pmol of total thiamine per mg of protein. The same fractions contain about 400-600 picomoles of TTX bound/mg of proteins. This corresponds to a minimum degree of purification around 10 to 15 %.
Thus, there is a non-parallel increase in both TTX receptor and thiamine compounds. In the initial fraction (detergent extract), the total amount of thiamine is of the same order of magnitude as the number of TTX binding sites (6 pmol /mg of protein). Rapid gel filtration of the detergent extract on small Sephadex-G-25 columns shows that all of the phosphoryla-

ted thiamine derivatives found in this fraction are fixed to proteins. The non parallel increase of both thiamin derivatives and TTX receptor may thus result from a continuous shift in the equilibrium towards the formation of free species.

Table 2. : Concentration and distribution of thiamine derivatives purified with the TTX receptor isolated from the main electric organ of E. electricus

Fraction (n)	ThTP	ThPP	ThMP	Th	Total thiamine
		(% of total thiamine)			(pmol/mg prot.)
Detergent extract (3)	3.7±1.2*	56 ±13	31± 4	14.5±0.7	11.3± 1.8
DEAE pool (2)	14.5±2.1	24.5± 6.4	55±11	6.0±2.8	27 ± 5
Lectine eluate (6)	12 ±4	13 ± 6	74± 9	0 ±0	135 ±32

*mean ± SD, n = number of measurements.

Effect of TTX and veratridine

In order to test the influence of neurotropic compounds on the concentration and distribution of thiamine compounds, TTX (3.10^{-7}M) and veratridine (10^{-4}M) were added during the purification procedure before the DEAE Sephadex step (13).

The addition of TTX (3.10^{-7}M) to the detergent extract just before the DEAE-step causes a shift in the equilibrium towards less phosphorylated derivatives and a partial liberation of these forms during the DEAE wash (Table 3).

This observation fits well with those made by Ytokawa et al who revealed this same effect of TTX on intact nerves (23) and nerve membranes (24). We could not demonstrate any significant effect of veratridine on protein bound thiamine compounds. This observation is in fact not surprising, because TTX and veratridine share different binding sites on the sodium conducting site (25).

Table 3 : Thiamine derivatives (pmol/mg protein) associated with the Na^+ channel isolated from the main electric organ of Electrophorus electricus

Fraction	ThTP	ThPP	ThMP	Th
DEAE wash				
Control	-	2.6-4.6	-	-
TTX	-	1.1-6.6	14.0-15.1	0.9-4.3
Veratridine	0.3-0.9	2.6	0.3-1.1	-
Lectin eluate				
Control	7.4-11.6	10.6-15.7	33.3-98	-
TTX	0.7-2.2	2.2-3.7	5.1	1.5
Veratridine	7.4-17.2	12.3-38	34.4-67.5	-

Conclusion

The role of thiamine and its phosphorylated derivatives in the central and peripheral nervous system has been questionned for many years. It is well known that thiamine derivatives are localized throughout the nervous system and preferentially located in membrane structures (26). Minz (27), von Muralt (28) and later Itokawa and Cooper (23) noted that the stimulation of nerve fibers results in a liberation of Th and ThMP into the incubation medium and a concomitant reduction of the distribution in high phosphate esters of thiamine in the nerve fiber. The same result is produced, when instead of an electrical stimulation, TTX, acetylcholine, or a local anesthetic is added to the incubation medium (24).
Some pathological cases deserve special attention. Patients suffering from polyneuritis exhibit normal ThPP-dependent enzyme activities in the brain (29). Therefore the interpretation that Peters (30) gave of this biochemical lesion should be revised. Cooper et al (31) attribute a role to ThTP in subacute necrotizing encephalomyelopathy. In these patients, blood contains a factor inhibiting ThTP-ATP phospho-transferase of rat brain. Also, the content of ThTP in the brain is significantly reduced.

In view of these observations, we like to consider that phosphorylated esters of thiamine, and in particular ThTP, may be the neurophysiological active form of vitamine B1 in the nervous system, independantly of the role of ThPP as cofactor in dehydrogenase and transketolase reactions. However, until now, it has not been possible to define this function more exactly.

Considering our results and those of the cited authors, we think that phosphorylated compounds of thiamine could play a fundamental role in controlling sodium conductance in excitable membranes, possibly by a phosphorylation-dephosphorylation cycle (13). However, much more work is needed to support this model.

Acknowledgments

Our thanks are due to L. Roland for efficient technical assistance. This work was supported by grants n° 2.4518.80 to E. Schoffeniels from the Fonds de la Recherche Fondamentale Collective (Belgium) and 1.5.612.83F to G. Dandrifosse from the Fonds National de la Recherche Scientifique (Belgium). L. Bettendorff is "Aspirant du Fonds National de la Recherche Scientifique".

References

1. Nachmansohn, D. 1955. Harvey Lect. 49, 57-99.

2. Nachmansohn, D. 1970. Science 168, 1059-1066.

3. Abbott, B.C., A.V. Hill, J.V. Howarth. 1958. Proc. Roy. Soc. Ser. B 148, 149-187.

4. Margineanu, D.G., E. Schoffeniels. 1977, Proc. Natl. Acad. Sci. USA 74, 3810-3813.

5. Dorogi, P.L., E. Neumann. 1980. Neurochem. Internat. 2, 45-51.

6. Schoffeniels, E., D.G. Margineanu. 1981. J. Theor. Biol. 92, 1-13.

7. Schoffeniels, E. 1976. 2nd International Meeting on Torpedo, Arcachon, France, September 1976.

8. Schoffeniels, E., G. Dandrifosse. 1980. Proc. Natl. Acad. Sci. USA 77, 812-816.

9. Bontemps, J., G. Dandrifosse, E. Schoffeniels. 1980. Neurochem. Internat. 2, 101-110.

10. Costa, M.R.C., J.E. Casnellie, N.A. Catterall. 1982. J. Biol. Chem. 257, 7918-7921.

11. Dandrifosse, G., E. Schoffeniels, 1980. Neurochem. Internat. 2, 95-100.

12. Bettendorff, L., J. Bontemps, G. Dandrifosse, E. Schoffeniels. 1983. IRCS Med. Sci. 11, 1098-1099.

13. Schoffeniels, E., G. Dandrifosse, L. Bettendorff, 1984. J. of Neurochem. 43, 1984.

14. Bontemps, J., P. Philippe, L. Bettendorff, J. Lombet, G. Dandrifosse, E. Schoffeniels, J. Crommen. 1984. J. Chromatogr. 307, 283-294.

15. Sancmori, H., H. Ueki, T. Kawasaki. 1980. Anal. Biochem. 107, 451-455.

16. Kimura, M., Bh. Panijpan, Y. Itokawa. 1982. J. Chromatogr. 245, 141-143.

17. Ishii, K., K. Sarai, H. Sanemori, T. Kawasaki. 1979. J. Nutr. Sci. Vitaminol. 25, 517-523.

18. Rindi, G., L. De Guiseppe. 1961. Biochem. J. 78, 602-606.

19. Bontemps, J., L. Bettendorff, J. Lombet, Ch. Grandfils, G. Dandrifosse, E. Schoffeniels, F. Nevejans, J. Crommen. 1984. J. Chromatogr. 295, 486-491.

20. Dandrifosse, G., Ch. Grandfils, L. Bettendorff, E. Schoffeniels, J. Bontemps. (This volume).

21. Grandfils, Ch., L. Bettendorff, E. Schoffeniels, J. Bontemps, G. Dandrifosse. (This volume).

22. Bontemps, J., Ch. Grandfils, G. Dandrifosse, E. Schoffeniels. 1984. Arch. Int. Physiol. Biochim. 92, 39-45.

23. Itokawa, Y., J.R. Cooper. 1970. Biochem. Pharmacol. 19, 985-992.

24. Itokawa, Y., J.R. Cooper. 1970. Biochim. Biophys. Acta. 196, 274-284.

25. Catterall, W.A. 1975. Proc. Natl. Acad. Sci. USA 72, 1782-1786.

26. Sable, H.Z., C.J. Gubler (eds). 1982. Thiamine, Twenty years of Progress. Ann. New York Acad. Sci. 378.

27. Minz, B. 1938. C.R. Soc. Biol. (Paris) 127, 1251-1253.

28. Muralt, A.V. 1942. Pflügers Arch. 245, 604-632.

29. Cooper, J.R., J.H. Pincus. 1979. Neurochem. Res. 4, 223-233.

30. Peters, R.A. 1936. Lancet 1, 1161-1164.

31. Cooper, J.R., Y. Itokawa, J.H. Pincus. 1969. Science 164, 74-75.

IV. Immunology of the Acetylcholine Receptor: Myasthenia Gravis and Synaptic Plasticity

REGULATION OF THE IMMUNE RESPONSE TO THE ACETYLCHOLINE RECEPTOR AND OF EXPERIMENTAL MYASTHENIA

Sara Fuchs, Miry C. Souroujon and Steve Bogen

Department of Chemical Immunology, The Weizmann Institute of Science, Rehovot 76100, Israel

Introduction

Special interest in the immunological properties of the nicotinic acetylcholine receptor (AChR) arises from the involvement of an autoimmune response against this receptor in myasthenia gravis and its animal model, experimental autoimmune myasthenia gravis (EAMG) (for reviews see 1-4). In our laboratory we are studying various topics related to the immunology and immunochemistry of AChR, with an emphasis on the involvement of this receptor in myasthenia gravis (MG). Our studies include structure and function analysis of AChR, preparation of specific reagents (monoclonal antibodies) for muscle and brain receptor, and developing means for attempting specific regulation of experimental myasthenia.

Three approaches for specific regulation or immunosuppression of myasthenia are being employed in our lab. These include the use of a) non myasthenic derivatives of AChR, b) specific anti-idiotypes and c) AChR-specific T-cell lines and T-cell suppressor factors. In the following we will summarize briefly earlier studies related to structure and function analysis of AChR and report on some recent findings from studies on anti-idiotypes against anti-AChR antibodies, murine AChR-specific suppressor T cells and their secreted factors and on an AChR-specific helper factor from plasma of myasthenic patients.

Molecular Basis of Nerve Activity

Results and Discussion

Structure and function analysis of the acetylcholine receptor. A molecular dissection of AChR, followed by pharmacological and immunological analysis may lead to the isolation and identification of fragments or sites in the molecule which are responsible for its specific cholinergic and/or pathologic myasthenic activity. By following such an analysis it was possible to design derivatives of AChR with a potential to regulate specifically the immune response to AChR, and to immunosuppress EAMG, and to identify small fragments of AChR carrying the biological functions of the intact molecule.

The first derivative of AChR that we have studied in detail was a denatured preparation of the receptor (5). AChR was irreversibly denatured by reduction and carboxymethylation. The resulting reduced-carboxymethylated AChR (RCM-AChR) did not bind cholinergic ligands and did not induce EAMG in rabbits. Nevertheless, RCM-AChR was shown to be immunogenic in rabbits and elicited antibodies which cross-react with intact AChR. RCM-AChR was shown to be an appropriate derivative for regulation of EAMG (6). Thus, RCM-AChR, which by itself is devoid of any myasthenic activity was shown to have an immunosuppressive effect on EAMG, either by preventing the onset of the disease or by reversing the clinical symptoms in myasthenic rabbits. Now that the sequence of the entire AChR molecule is available, synthetic peptides and their antibodies can be employed both for regulation experiments as well as for localizing and identifying biologically important sites in the receptor molecule (see Neumann et al., this proceedings).

Proteolytic digestion of AChR with trypsin (7) and papain (7, 8) was employed for attempting the isolation and analysis of fragments of the receptor molecule which retain the original specific cholinergic binding site, and/or the pathologic myasthenic activity. It was demonstrated that the pharmacological and myasthenic activities of AChR reside within a conformation dependent structure in the molecule which is sensitive to denaturation and resistant to mild proteolytic digestion (7,8).

Monoclonal antibodies (mcAbs) against Torpedo AChR were prepared and characterized (9-11). Some monoclonal antibodies were directed against conformation-dependent antigenic determinants, and others were directed against determinants present also on denatured AChR.

Several of the anti-Torpedo AChR mcAbs cross react with muscle and brain AChR. These mcAbs were tested for their effects on various receptor properties and functions such as ligand binding, receptor degradation, sodium transport, single channel properties as well as for their ability to transfer EAMG (10-12).

Anti-idiotypes. One of the possible specific therapeutic approaches for myasthenia may be the application of anti-idiotypic antibodies. In previous studies using anti-idiotypic antibodies against polyclonal anti-AChR antibodies we have demonstrated a broad crossreactivity between idiotypes of anti-AChR antibodies from different mouse strains, as well as with anti-AChR antibodies from other species (13). In another study we have shown that anti-idiotypes raised against a purified polyclonal preparation of anti-AChR antibodies from rabbits with EAMG were successful in preventing the initial development of EAMG (14).

A useful way to analyze the fine specificity and repertoire of idiotypes which comprise the polyclonal anti-AChR reactivity, is to study the idiotypic specificities of monoclonal anti-AChR antibodies. Anti-idiotypes against three monoclonal anti-AChR antibodies which crossreact with muscle AChR were produced in rabbits and mice. The three monoclonal antibodies employed were mcAb 5.5, mcAb 5.14 and mcAb 5.34 (10,11). These mcAbs represent three different antigenic specificities which might play a role in the pathogenesis of EAMG. The specificity of the various rabbit or mouse anti-idiotypes was shown to be restricted each to the immunizing idiotype, as demonstrated in binding and inhibition experiments. There was no significant binding to other anti-AChR idiotypes or to non-related mcAbs. Inhibition experiments also demonstrated restricted specificity of the anti-idiotypes (Fig. 1).

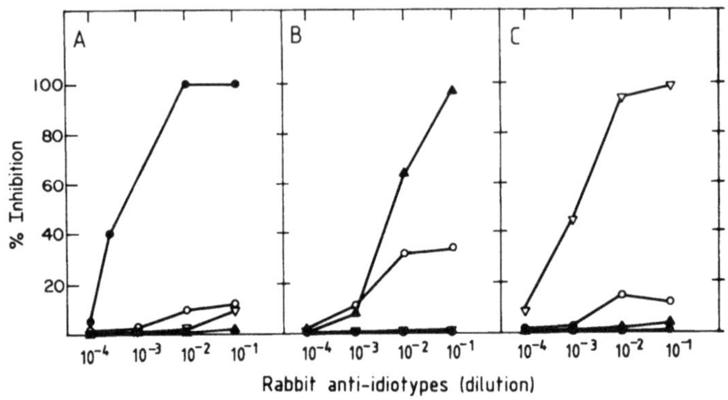

Figure 1. Inhibition of AChR binding to different idiotypes by rabbit anti-idiotypes. Inhibition by: A - anti-mcAb 5.5, B - anti-mcAb 5.14, and C - anti-mcAb 5.34, of the binding of AChR to: mcAb 5.5 (●——●), mcAb 5.14 (▲——▲), mcAb 5.34 (▽——▽) and polyclonal mouse anti-AChR antibodies (o——o).

It was of interest to test whether the specificity of the anti-idiotypes observed in binding and inhibition experiments is expressed also in vivo. For that we have first attempted to elicit a neuromuscular block in chickens by passive transfer of various anti-AChR antibodies. Passive transfer into chicken hatchings with mcAbs 5.5, 5.14 and 5.34, as well as polyclonal mouse anti-AChR antibodies was performed. Of these, only mcAb 5.5, which is directed to the cholinergic binding site (10), affected neuromuscular transmission to a clinically detectable degree and produced acute paralysis (Fig. 2).

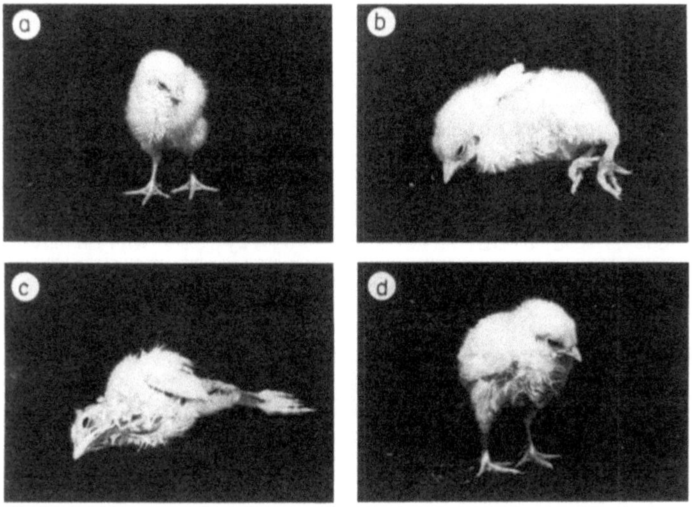

Figure 2. Induction of a neuromuscular block by transfer of mcAb 5.5 and its reversal by anti-idiotypes. One mg of mcAb 5.5 immunoglobulin was transferred to chickens b,c, and d. a) Control chicken, b) 24 hr after transfer of mcAb 5.5, c) 48 hr after transfer of mcAb 5.5, d) 48 hr after transfer of mcAb 5.5 and treatment twice with anti-5.5 anti-idiotypic antibodies 12 and 18 hr after transfer of mcAb 5.5

In order to test the regulatory effect of anti-idiotypes on the myasthenic symptoms in chickens, rabbit anti-idiotypes elicited against mcAb 5.5, 5.14 or 5.34 were administered into chickens before or after the passive transfer of the neuromuscular block by mcAb 5.5. Pretreatment of chickens with anti-idiotypes against mcAb 5.5 prevented the induction of a neuromuscular block by a subsequent injection of mcAb 5.5. Also, administration of anti-idiotypes against mcAb 5.5 to chickens in which a neuromuscular block has been induced by mcAb 5.5, led to a recovery from the neuromuscular block (Fig. 2). Rabbit anti-idiotypes prepared against other anti-AChR mcAbs or rabbit anti-normal mouse immunoglobulin had no such effects.

The recent demonstration of auto-anti-idiotypes in sera of myasthenic patients (15) suggests that anti-idiotypes may have a regulatory role in myasthenia gravis. Our results demonstrate that passive transfer of anti-idiotypes of the right specificity can reverse the pathogenic effect of anti-AChR antibodies. Since the anti-AChR immune response in both experimental and human myasthenia is heterogeneous, the question still remains which idiotypes should be best used for the elicitation of the appropriate anti-idiotypes, which might, in turn, be used in the treatment of myasthenia.

The *in vivo* effect of anti-idiotypes on the overall anti-AChR immune response was tested in mice preimmunized with different monoclonal anti-AChR antibodies and producing the respective anti-idiotypes. It was found that preimmunization with a certain idiotype leads to the preferential suppression of this particular idiotype in the polyclonal response to AChR. Moreover, preimmunization with either polyclonal or monoclonal anti-AChR antibodies resulted in a reduction of the overall anti-Torpedo AChR and anti-muscle AChR titers (Table 1). This reduction was greater than would be expected from the representation of each of the respective idiotypes in the polyclonal anti-AChR serum. As can be seen, preimmunization of mice with polyclonal anti-AChR antibodies resulted in a 98% reduction in the anti-AChR titer as compared to mice preimmunized with normal mouse immunoglobulins. Interestingly, pretreatment with either mcAbs 5.5, 5.14 or 5.34 resulted in a higher reduction of the anti-AChR titer than would be expected from the representation of each of the respective idiotypes in mouse anti-AChR sera. Thus, a decrease of 95%, 70% and 60% in the anti-AChR titer was obtained following immunization with mcAb 5.34, 5.5 and 5.14, respectively. This may imply that in addition to the immunizing idiotype other anti-AChR idiotypes are also suppressed.

Our results provide evidence that the humoral response to AChR can be manipulated by the anti-idiotypic approach, in spite of the complexity and heterogeneity of this immune response.

Table 1. Anti-AChR Titer in Mice Preimmunized with Idiotypes

	Anti - AChR titer[a]	
Idiotype injected	Anti-Torpedo AChR (titer x 10^7)	Anti-muscle AChR (titer x 10^9)
Normal mouse Ig	100.0 ± 18.0	8.6 ± 0.4
Mouse anti-AChR antibodies	1.8 ± 0.5[b]	0.2 ± 0.1[b]
mcAb 5.5 Ig	30.0 ± 4.8[c]	2.3 ± 0.2[c]
mcAb 5.14 Ig	40.0 ± 6.1[c]	2.0 ± 0.4[c]
mcAb 5.34 Ig	4.6 ± 0.6[b]	0.5 ± 0.2[b]

a) Titer is given in moles AChR precipitated by 1 liter serum.
b) Significantly lower than the control group $p < 0.001$
c) Significantly lower than the control group $p < 0.05$

Moreover, the use of the appropriate anti-idiotypes may have a potential in the regulation of the autoimmune response directed against the receptor in myasthenia.

Induction of AChR-specific suppressor T-cells and T-cell factors. It has been suggested that the underlying defect in myasthenia gravis (as well as other autoimmune diseases) might be a deficiency of suppressor T cells (16). Therefore, specific treatment of myasthenia gravis might be achieved if a suitable source of antigen (AChR)-specific suppressor T cells and their secreted factors could be developed. In light of its potential importance to the human disease, we have established the conditions to induce AChR-specific suppressor T cells capable of suppressing the *in vitro* anti-AChR response. Moreover, we have demonstrated that this suppressive effect can be mediated by a secreted factor found in the cell-free supernatant (17).

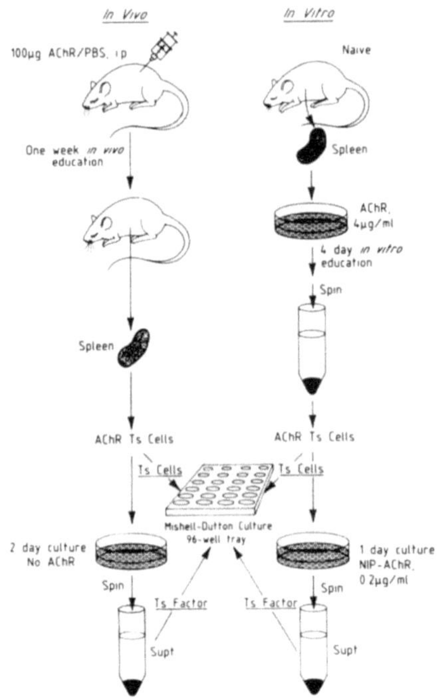

Figure 3. Education procedures for the preparation of AChR-specific suppressor T cells (Ts) and T cell factors.

We have induced both _in vivo_ and _in vitro_, murine AChR-specific suppressor T cells (Ts) (Fig. 3). _In vivo_ induction of Ts cells was achieved by injection of soluble AChR. A potent AChR-specific suppression was found in the spleen one week later. _In vitro_ induction of Ts cells involved the primary education of naive splenocytes by culturing them with high concentrations of AChR. Both the _in vivo_ and _in vitro_ induced Ts cells were shown to secrete AChR-specific factors which mediate their suppressive effects. The T cell factors were shown to adsorb to and elute from Sepharose-AChR columns (17). It was also shown that the Ts factors are not genetically restricted.

One of our goals in this research is to set the foundation for future attempts to develop large-scale sources of AChR-specific suppressor T cell factors. We have attempted to produce an

AChR-specific suppressor T cell hybridoma. Several T cell fusions of AChR-specific Ts cells with the BW147 thymoma were performed. One of these fusion products, termed 1.4E10, secreted an AChR-specific Ts factor which was capable of specifically suppressing the AChR-induced lymphoproliferative assay as well as the plaque forming cell response. The suppressive activity of this factor was adsorbed to and eluted from an AChR-Sepharose column. Kinetic studies with the 1.4E10 supernatant demonstrated that it was most effective when added on day 0. We have encountered a problem of stability with this hybridoma and are trying now to establish more stable hybridomas.

AChR-specific helper factor in plasma of myasthenic patients. We have recently initiated some studies with human sera in order to investigate possible regulatory factors involved in the anti-AChR autoimmune response in MG (Bogen, Safran and Fuchs, in preparation). We have devised a method for the isolation of AChR-specific T cell derived factors from the serum

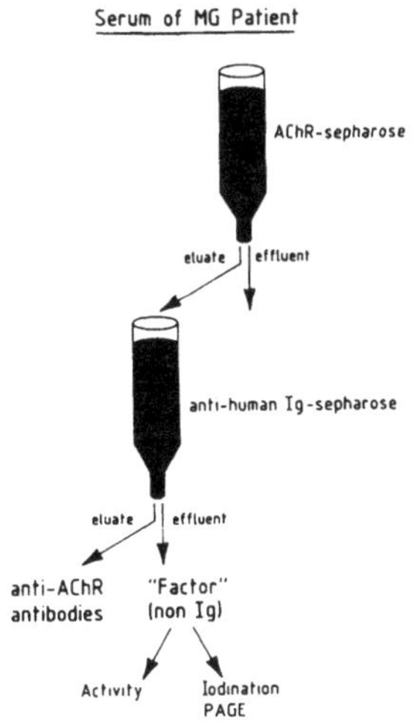

Figure 4. Isolation of helper factor from MG plasma

of MG patients (Fig. 4). Plasma from MG patients are passed first through a Sepharose-AChR column and absorbed material, following elution, is then passed through a Sepharose-anti-human Ig column and the effluent is collected. This second column is intended to remove anti-AChR antibodies. The effluent, containing the putative AChR-specific human factor is being checked in an anti-AChR Mishell-Dutton assay or radio-labeled for analysis. Our experiments have indicated that such factors can replace AChR-specific T helper cells in the *in vitro* anti-AChR response.

Examination of the radiolabeled putative human helper factor from eight MG patients on SDS-PAGE, consistently demonstrated one single non-immunoglobulin band at a molecular weight of 39-43 K. Control normal sera did not yield any such band.

In order to further study the nature of the human factor, an antiserum was developed by repeated injections into mice. This antiserum thus represents a mouse anti-human T cell factor which, although raised by an AChR-specific factor, may react with constant regions as well. Indeed, we have shown that this antiserum recognizes 15-20% of the total plastic non-adherent phytohemagglutinin (PHA) activated human T cell blasts (Fig. 5). Immunofluorescence microscopy demonstrated patchy membrane fluorescence. These results suggest that certain proteins present on the membrane of PHA-activated peripheral blood lymphocytes are related to the factor found in plasma. To our knowledge this is the first isolation of an antigen-specific immunoregulatory factor (a helper factor) from a human auto-immune disease. This factor is probably secreted by AChR-specific helper T cells.

It is especially interesting that the isolated factor has helper function. In MG, a defect in AChR-specific T cell suppression is believed to be responsible for development of anti-AChR autoantibodies (16). By this model, a deficiency of

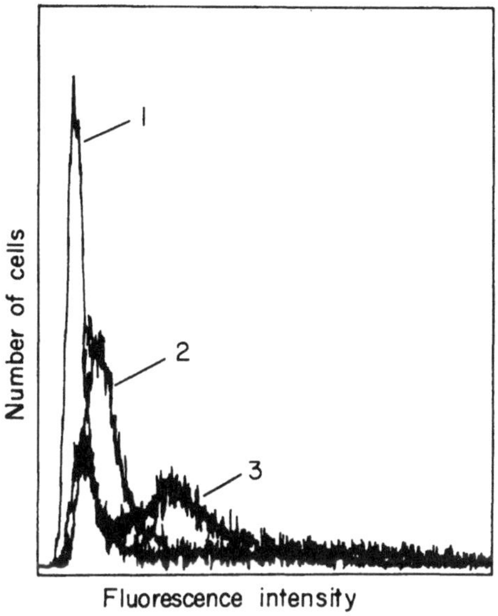

Figure 5. FACS II immunofluorescence staining pattern of PHA-activated peripheral blood lymphocytes from normal individuals using: 1) irrelevant immune C57BL/6 sera, 2) C57BL/6 anti-human T cell factor, or 3) monoclonal Leu 3 antibody followed by goat anti-mouse Ig - FITC.

AChR-specific suppressor T cells results in an exaggerated development of AChR-specific helper T cells and T cell factors. These helper factors stimulate B cells to proliferate, differentiate, and secrete anti-AChR autoantibodies. It is this AChR-specific helper T cell factor that we believe we have isolated. We believe that the use of antisera against immunoregulatory products present in plasma raises the possibility of new avenues of research and potential immunotherapy in diseases of abnormal immunoregulation.

Acknowledgements

This research was supported by grants from the Muscular Dystrophy Association of America, the Los Angeles Chapter of the Myasthenia Gravis Foundation and the United States-Israel Binational Science Foundation (BSF).

References

1. Drachman, D.B. 1978. New Engl.J.Med. 298, 136;186.

2. Fuchs, S. 1979. Current Topics in Microbiology and Immunology 85, 1.

3. Lindstrom, J. 1979. Adv.Immunol. 27, 1.

4. Vincent, A. 1980. Physiol.Rev. 60, 756.

5. Bartfeld, D., S. Fuchs. 1977. FEBS Lett. 77, 214.

6. Bartfeld, D., S. Fuchs. 1978. Proc.Natl.Acad.Sci.USA 75, 4006.

7. Bartfeld, D., S. Fuchs. 1979. Biochem.Biophys.Res.Commun. 89, 512.

8. Fuchs, S., D. Bartfeld. 1983. In: Cell Surface Receptors (P.G. Strange, ed.). Ellis Horwood, p. 126.

9. Mochly-Rosen, D., S. Fuchs, Z. Eshhar. 1979. FEBS Lett. 106, 389.

10. Mochly-Rosen, D., S. Fuchs. 1981. Biochemistry 20, 5920.

11. Souroujon, M.C., D. Mochly-Rosen, A.S. Gordon, S. Fuchs. 1983. Muscle and Nerve 6, 303.

12. Goldberg, G., D. Mochly-Rosen, S. Fuchs, Y. Lass. 1983. J. Membrane Biol. 76, in press.

13. Schwartz, M., D. Novick, D. Givol, S. Fuchs. 1978. Nature 273, 543.

14. Fuchs, S., D. Bartfeld, D. Mochly-Rosen, M. Souroujon, C. Feingold. 1981. Ann.N.Y.Acad.Sci. 377, 110.

15. Dwyer, D.S., R.J. Bradley, C. Kendrick-Urquhar, J.F. Kearney. 1983. Nature 301, 611.

16. Berrih, S., C. Gand, M.-A. Bach, H. Le Brigand, J.P. Binet, J.F. Bach. 1981. Clin.Exp.Immunol. 45, 1.

17. Bogen, S., E. Mozes, S. Fuchs. 1984. J.Exp.Med. 159, 292.

CONGENITAL AND AUTOIMMUNE DISEASES OF THE NEUROMUSCULAR JUNCTION

A. Vincent, P.J. Whiting and J. Newsom-Davis
Department of Neurological Science, Royal Free Hospital School of Medicine, London NW3 2PF UK.

Introduction

In recent years the sequence of events underlying the transmission of impulses from nerve to muscle has been investigated in great detail, and many of the techniques have been applied to human muscle biopsies allowing disorders of neuromuscular transmission to be classified as presynaptic or postsynaptic (see Table 1). In addition, it has become increasingly clear that two of these disorders have an autoimmune basis, whereas others are congenital and probably due to genetic defects.

This chapter concentrates on our current understanding of the role and nature of antibodies against the acetylcholine receptor in myasthenia gravis, and of the involvement of antibodies against a presynaptic determinant in the Lambert-Eaton syndrome. The experimental findings and probable site of defects in congenital forms will be summarised briefly.

These diseases not only compliment the physiological and biochemical studies described elsewhere in this volume, but provide us with potential probes for analysing the function at the molecular level.

Table 1. Disorders of the Neuromuscular Junction

Presynaptic		Postsynaptic
	Autoimmune	
Lambert-Eaton syndrome		Myasthenia Gravis
	Congenital	
Familial infantile myasthenia		Acetylcholinesterase deficiency
		Slow channel syndrome
		Acetylcholine receptor abnormalities
		Acetylcholine receptor deficiency

Myasthenia Gravis

Myasthenia gravis is a disease characterised by weakness and susceptibility to fatigue, which can be generalised or localised to certain muscle groups, particularly the extraocular or bulbar muscles. Anti-cholinesterase treatment often produces symptomatic improvement. The underlying defect is a reduction in the number of postsynaptic AChRs (1,2) which can be measured by binding of the snake toxin a-bungarotoxin (a-BuTx). In MG the reduction in endplate AChRs results in reduced amplitude of the miniature endplate potentials (m.e.p.p.s) (3). The endplate potentials (e.p.p.s) are accordingly either subthreshold on initial nerve stimulation, or become subthreshold during repetitive stimulation, resulting in neuromuscular block.

Anti-AChR in MG

Using ^{125}I-a-BuTx labelled solubilised human AChR it is possible to detect anti-AChR antibodies in 80-85% of MG patients (4,5). The results however, show a rather poor correlation between antibody levels and the severity of the disease between individuals, except that they are significantly lower in patients with weakness restricted to ocular muscles. Very few normal control sera have anti-AChR although a recent Japanese report demonstrates positive values in elderly normal subjects (6). We have not been able to confirm this finding in Caucasians (7).

The pathogenic significance of anti-AChR was demonstrated by passive transfer of MG immunoglobulins to mice which resulted in weakness and reduced m.e.p.p. amplitude (8). Moreover, during plasma exchange, which reduces circulating immunoglobulins, there was a striking inverse relationship between clinical improvement and the reduction in anti-AChR in each individual (9,10).

Anti-AChR antibodies are thought to mediate loss of endplate AChRs by several mechanisms. 1. An 'immunopharmacological' block of AChR by antibodies binding to the ACh binding site, the ion channel or some other functional determinant may be important in some patients, but antibodies against the a-BuTx binding site, for instance, are infrequent (11), except in a few patients (12). 2. Anti-AChR can be shown to increase the rate of receptor degradation both in vitro (13,14) and in mice with passively transferred MG (15). This process depends on the ability of divalent antibodies to cross-link AChRs (16) and indicates that multiple determinants are present on the AChR molecule. A correlation between the clinical state of the patient and the ability of their serum to increase degradation has been shown (17) but there is evidence that in mice with passively transferred MG AChR synthesis can increase to compensate for increased degradation (18).
3. The most important mechanism is probably complement-mediated lysis of the postsynaptic membrane. Engel and his associates (19,20) have demonstrated the presence of IgG and C3 at the endplates of MG intercostal muscles, and their distribution is similar to the distribution of AChRs. On the other hand the final lytic component of complement, C9, was

located principally on degenerated folds and synaptic debris (20). These observations suggest that the severity of the disease in individuals partly reflects the efficacy of the later complement components in producing postsynaptic lysis.

Heterogeneity of anti-AChR in indivuduals

Another explanation for the poor correlation between antibody levels and severity of the disease might lie in the heterogeneity of the anti-AChR. Several studies have shown that the antibody is usually both polyclonal and multi-subclass within an individual, and between individuals varies in its antigen binding characteristics (11, 21,22) and in its idiotype (the antigen combining site) (23,24). Recently Tzartos, Seybold and Lindstrom (25) have attempted to define the subunit localisation of anti-AChR binding sites. They used monoclonal antibodies against isolated AChR subunits to characterise other monoclonal antibodies, some of which bound to native human AChR. These were than used in competition experiments with the MG anti-AChR. The results, disappointingly, showed no correlation between clinical state and the proportion of antibodies binding to different antigenic regions of the AChR.

Clinical subgroups of MG and antibody heterogeneity

There is increasing evidence for an HLA-B8 association with MG (26,27) and in a study of 68 patients we correlated the presence of different HLA antigens with three main subgroups of patients (5). These were defined by their thymic pathology, age of onset, sex incidence and HLA associations (see Table 1). In addition, a group of patients with purely ocular symptoms can be distinguished by their low anti-AChR levels and high male incidence.

It was of interest to see whether any particular anti-AChR characteristics were associated with the different clinical subgroups. Although in initial studies the characteristics of anti-AChR did not differ significantly between the three main groups of patients with generalised MG (28), we have recently found some differences in the extent to which binding of their anti-AChR is inhibited by monoclonal antibodies against human AChR (29) (see Table 2).

Table 2. Clinical Subgroups of MG and Antibody Characteristics

Subgroup	Thymoma		Onset $<$40 years		Onset $>$40 years
Sex	Male=Female		Female Male		Male Female
Typical HLA	-		HLA B8		HLA B7
Inhibition of anti-AChR binding by monoclonal antibodies against human AChR					
Monoclonal antibody Group	% Inhibition				
1	18±15		25±33		31±27
2	40±31		57±36		40±32
3	29±29	$p<0.001$	70±25	$p<0.05$	35±31
4	42±11		56±39	$p<0.05$	81±17
5	38±26		38±38	$p=0.05$	65±25

Interestingly, monoclonal antibodies from group 1, which tended to compete less effectively with MG anti-AChR, bound strongly to denervated muscle AChR, but only with very low affinity to endplate AChR (30). These results suggest that monoclonal antibodies raised against human AChR may provide powerful tools for dissecting the antigenic regions.

There are also significant differences in anti-AChR from patients with pure ocular MG compared to patients with generalised disease, particularly in their reactivity with ocular AChR (28). We found, however, similar characteristics in sera of other patients with particularly low anti-AChR levels (31). Thus is appears, that a proportion of MG patients respond in a restricted manner to AChR; these tend to be male and have ocular or mild generalised disease. Competition experiments with monoclonal antibodies have not yet been performed using sera from these low titre patients.

The role of the thymus

The aetiology of myasthenia is unknown but there is ample evidence that the thymus gland is involved in at least some forms of the disease. About 60% of patients have a hyperplastic thymus with multiple germinal centres and many of these will improve or go into complete remission after thymectomy (32). There is a significant correlation between improvement and fall in anti-AChR but in most cases the changes occur relatively slowly (33). A further 15% of patients have a thymic tumour and often these do not do so well after removal of the thymus. The remaining patients, who tend to be those of the older age onset, or those with mild or localised disease, sometimes have an atrophied or involuted thymus.

Thymic lymphocytes from hyperplastic glands can be shown to synthesise anti-AChR in culture (34). This synthesis is spontaneous and correlates with the duration of the patients' disease suggesting that anti-AChR synthesis by thymic cells is not necessarily a primary event. Moreoever since anti-AChR falls only slowly after thymectomy, clearly the thymus is not the only sourse of antibody production. On the other hand, thymic cells can enhance antibody synthesis by autologous peripheral blood lymphocytes. This enhancement may depend on the presence of the follicular dendritic antigen-presenting cells which are in close relationship with germinal centres in the thymus (35,36,37).

The presence of germinal centres in the myasthenic thymus suggests that antigen is localised there. Muscle-like, or myoid, cells have been recognised in normal and MG thymus for many years, initially on the basis of their staining pattern with MG serum positive for anti-striated muscle antibodies. Although recent reports have indicated the presence of a-BuTx binding sites in extracts of normal mouse, rat and rabbit thymus (38,39,40), and thymic non-lymphoid cultures sometimes develop into myoblast-like cells with a-BuTx binding (41,41), the localisation of AChRs on the myoid cells has not yet been demonstrated convincingly.

The Lambert-Eaton Syndrome

The Lambert-Eaton syndrome (L-ES) is a disorder characterised by weakness which, in contrast to that found in MG, typically improves during repetitive effort. It occurs in a small proportion of patients with oat cell carcinoma of the lung but is also found in patients with no detectable carcinoma, many of whom have other autoimmune diseases or other autoantibodies (43).

The main physiological defect is a reduction in the quantum content which increases towards normal on repetitive stimulation at high frequency (44). The response to bath applied Ca^{++} is reduced (45).

The possible role of a circulating factor released from the carcinoma, or of an autoantibody was investigated by plasma exchange which produced a small but significant improvement in the muscle action potential. In addition, immunosuppression with prenisolone and azathioprine produced a long-lasting improvement in muscle strength in many patients (46).

The involvement of a plasma IgG in the pathogenesis of the

disease was investigated by passive transfer of IgG from L-ES patients to mice. Animals given daily injections of the IgG had a significant reduction in the quantum content of the e.p.p. when compared to animals receiving control human IgG. Thus there is evidence that the Lambert-Eaton Syndrome is an autoantibody mediated condition (46,47). The nature of the proposed autoantigen is poorly understood. Recently, Fukanaga and Engel (48) have shown that the active zone particles, which are thought to be close to the sites of quantal release, are disorganised and reduced in number in muscle biopsies from patients with this disease. Similar morphological changes have been found in mice with passively transferred L-ES (49), and recently, an alteration in the sensitivity of quantum release to external Ca^{++} has also been shown in these animals[50].

Congenital Disorders of Neuromuscular Transmission

Congenital disorders can be distinguished from the autoimmune conditions by their onset in early life or strong familial incidence. Anti-AChR and other autoantibodies are absent and the patients do not respond to plasma exchange. Within this group of patients, however, several different conditions can now be recognised, reviewed in (51).

Familial infantile myasthenia

Several families have been described in which weakness during the immediate postnatal period causes regurgitation and breathing difficulties. Frequently several siblings have been involved, and death in early infancy has occurred. The response to anticholinesterase treatment may be dramatic and life-saving. Intercostal muscle biopsies from two members of one family showed normal amplitude m.e.p.p. which became reduced on repetitive stimulation at low frequency.

The AChRs were normal in number and distribution. These findings suggested that the synthesis or packaging of ACh was abnormal (52).

We have also studied a family in whom the eldest child was weak from birth, and had very small m.e.p.p.s with apparently normal AChRs.(53). This 16 year old boy had not experienced the life-threatening feeding problems usually associated with the syndrome described above. His sister had normal m.e.p.p. amplitude at rest but reduced K^+ evoked ACh release (Molenaar, Vincent and Newsom-Davis unpublished observations).

Congenital absence of acetylcholinesterase

This has only been described in one Hindu boy who had generalised weakness and hyporeflexia since infancy (54). There was no response to anticholinesterases. Electromyography showed a decremental response to repetitive stimulation, and a repetitive response to a single nerve impulse. The most striking abnormality was a complete absence of the 16 S endplate-specific species of acetylcholinesterase (AChE). This resulted in prolongation of e.p.p.s with repetitive firing. Slightly reduced m.e.p.p. amplitude, reduced m.e.p.p. frequency and morphological changes of the nerve terminals were presumed to be secondary to the main defect. AChRs were reduced at some endplates but the changes were not substantial.

Abnormalities of the Acetylcholine receptor
Slow Channel Syndrome

This condition differs from those described above in that there is clear evidence of dominant inheritance. Symptoms present either in infancy or adolescence and weakness and wasting is most marked in cervical, scapula and dorsal forearm muscles. There was repetitive muscle action potentials in response to a single stimulus, associated with marked prolongation of the e.p.p.s. AChRs were normal or slightly reduced, probably due to degeneration of the junctional folds (55). The evidence points to a prolongation in the open time of each ACh-induced ion channel but there are other possible explanations, and until direct analysis of the time course of individual channels is performed, the precise nature of the defect cannot be confirmed.

We have examined a muscle biopsy from a woman with similar symptoms in which the m.e.p.p.s and e.p.p.s were also grossly prolonged (see fig 1). We hope to conduct noise analysis on further biopsies from this patient and her relatives (Wray, Prior, Vincent, Newsom-Davis and Oosterhuis unpublished observations).

It is more than likely that other abnormalities of the AChR exist which result in abnormal function. A case has been described in which the number of AChRs was essentially normal but there was a tenfold increase in the apparent affinity for d-tubocurarine (56).

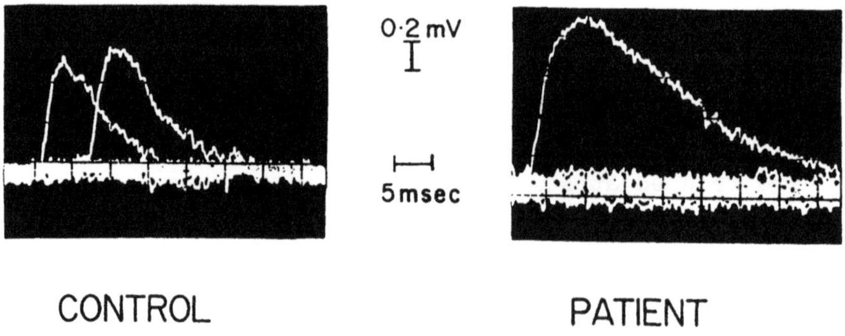

Fig 1. Representative m.e.p.p.s from a patients (kindly referred by Dr H Oosterhuis) with congenital weakness who showed some wasting of muscles and repetitive firing in response to single stimuli (courtesy of D Wray and C Prior).

Acetylcholine receptor deficiency

This form of congenital myasthenia is probably the most difficult to distinguish from MG of acquired aetiology. The distribution of weakness is similar to that found in MG but the symptoms tend to appear at birth or shortly afterwards. Anti-AChR is completely absent and there is no response to plasma exchange (10,53). Results from muscle biopsies are similar to MG with low amplitude m.e.p.p.s and marked reduction in the number of AChRs associated with elongated endplate regions. The characteristics of ACh noise were normal (53). A similar condition is found in dogs (57) as a recessively inherited syndrome. The cause of the AChR deficiency is not known but possibilities include an abnormality of AChR synthesis or insertion into the membrane.

Conclusions

The heterogeneity of diseasesinvolving the neuromuscular junction reflects the complexity of the processess involved in neuromuscular transmission. The autoantibody mediated diseases (MG and L-ES) provide us respectively with potential probes for AChRs and for the putative presynaptic antigen involved in quantal release. The congenital disorders lend themselves to genetic analysis, particularly when DNA probes for the human AChR genome become available. A combination of mapping antigenic regions and analysing the precise nature of genetic defects should be most productive in defining the relationship between primary structure and function.

Acknowledgement

I thank the Medical Research Council of Great Britain for support.

References

1. Famborough D, Drachman DB, Satyamurti S. 1973. Neuromuscular junction in myasthenia gravis. Decreased acetylcholine receptors. Science, 182, 293-295.

2. Ito Y, Miledi R, Vincent A, Newsom-Davis J. 1978. Acetylcholine receptors and end-plate electrophysiologyin myasthenia gravis. Brain, 101, 345-368.

3. Elmqvist D, Hofman WM, Kugelberg J, Quastel DMJ. 1964. An electrophysiological investigation of neuromuscular transmission in myasthenia gravis. J Physiol (Lond), 174, 417-434.

4. Lindstrom JM, Seybold ME, Lennon VA, Whittingham S, Duane DD. 1976a. Antibody to acetylcholine receptor in myasthenia gravis. Prevalence, clinical correlates, and diagnostic value. Neurology (Ny), 26, 1054-1059.

5. Compston DAS, Vincent A, Newsom-Davis J, Batchelor JR. 1980. Clinical, pathological, HLA antigen and immunological evidence for disease heterogeneity in myasthenia gravis. Brain, 103, 579-601.

6. Tanaka M, Miyatake T. 1983. Anti-acetylcholine receptor antibody in aged individuals J.Neuroimmunol. 4, 17-24

7. Robb SA, Vincent A, Newsom-Davis. Acetylcholine receptor antibodies in the elderly and in Down's Syndrome. In preparation.

8. Toyka KV, Drachman DB, Pestronk A, Kao I. 1975. Myasthenia gravis: passive transfer from man to mouse. Science, 190, 397-399.

9. Pinching J, Peters DK, Newsom-Davis J. 1976. Remission of myasthenia gravis following plasma exchange. Lancet, ii, 1373-1376.

10. Newsom-Davis J, Pinching AJ, Vincent A, Wilson SG. 1978. Function of circulating antibody to acetylcholine receptor in myasthenia gravis investigated by plasma exchange. Neurology (Minneapolis), 28, 266-272.

11. Vincent A, Newsom-Davis J. 1979. A-bungarotoxin and anti-acetylcholine receptor antibody binding to the human acetylcholine receptor. Adv Cytopharmacol, 3, 269-278.

12. Whiting PJ, Vincent A, Newsom-Davis J. 1983. Acetylcholine receptor antibody characteristics in myasthenia gravis: fractionation of a-bungarotoxin binding site antibodies and their relationship to IgG subclass. J Neuroimmunol, 5, 1-9.

13. Anwyl R, Appel SH, Narahashi T. 1977. Myasthenia gravis serum reduces acetylcholine sensitivity in cultured rat myotubes. Nature (Lond), 267, 262-263.

14. Bevan S, Kullberg RW, Heinemann S. 1977. Human myasthenic sera reduce acetylcholine sensitivity of human muscle cells in tissue culture. Nature, 267, 263-265.

15. Stanley EF, Drachman DB. 1978. Effect of myasthenic immunoglobulin on acetylcholine receptors of intact neuromuscular junctions. Science, 200, 1285-1286.

16. Drachman DB, Angus CW, Adams RN, Michelson JD, Hoffman GJ. 1978. Myasthenic antibodies cross-link acetylcholine receptors to accelerate degradation. N Engl J Med, 298, 1116-1122.

17. Drachman DB, Adams RN, Joseifek LF, Pestronk A, Stanley EF. Antibody mediated mechanisms of acetylcholine receptor loss in myasthenia gravis. Ann NY Acad Sci, 377, 175-188.

18. Wilson S, Vincent A, Newsom-Davis J. 1983. Acetylcholine receptor turnover in mice with passively transferred myasthenia gravis. II. Receptor synthesis. J Neurol Neurosurg Psychiat, 46, 383-387.

19. Engel AG, Lambert EH, Howard FM. 1977. Immune complexes (IgG and C3) at the motor endplate in myasthenia gravis. Ultrastructural and light microscopic localisation and electrophysiologic correlations. Mayo Clin Proc, 52, 267-280.

20. Sahashi K, Engel AG, Lambert EH, Howard FM. 1980. Ultrastructural localisation of the terminal and lytic 9th complement component (C9) at the motor endplate in myasthenia gravis. Neuropath Exp Neurol, 39, 160-172.

21. Lindstrom J, Campbell M, Nave B. 1978. Specificities of antibodies to acetylcholine receptors. Muscle & Nerve, 1:140-145.

22. Vincent A, Newsom-Davis J. 1980. Anti-acetylcholine receptor antibodies. J Neurol Neurosurg Psychiatr, 43, 590-600.

23. Lefvert AK. Anti-idiotypic antibodies against the receptor antibodies in myasthenia gravis. Scand J Immunol, 13, 493-497.

24. Lang B, Roberts A, Vincent A, Newsom-Davis J. Acetylcholine receptor antibody idiotypes in myasthenia gravis. Submitted for publication.

25. Tzartos SJ, Seybold ME, Lindstrom JM. 1981. Specificities of antibodies to acetylcholine receptors in sera from myasthenia gravis patients measured by monoclonal antibodies. Proc Natl Acad Sci, 79, 188-192.

26. Pirskanen R, Tiilikainen A, Hokkanen E. 1972. Myasthenia gravis and HLA antigens. Ann Clin Res, 4, 304-306.

27. Feltkamp TEW, van den Berg-Loonen PM, Nijenhuis LE, Engelfriet CP, van Rossum AL, van Loghem JJ, Oosterhuis HJGH. 1974. Myasthenia gravis, autoantibodies and HLA antigens. Br Med J, 1, 131-133.

28. Vincent A, Newsom-Davis J. 1982. Acetylcholine receptor antibody characteristics in myasthenia gravis. 1. Patients with generalised myasthenia or disease restricted to ocular muscles. Clin Exp Immunol, 49, 257-265.

29. Whiting PJ, Vincent A, Newsom-Davis J. Monoclonal antibodies to the human acetylcholine receptor. 1984 Biochem Soc Trans (in press).

30. Whiting PJ, Vincent A, Newsom-Davis J. Monoclonal antibodies can differentiate between normal and denervated human acetylcholine receptor. In preparation.

31. Vincent A, Newsom-Davis J. Anti-acetylcholine receptor antibody in myashtenia gravis. III. Patients with low antibody levels. Submitted for publication.

32. Buckingham JM, Howard FM, Bernatz PE, Spencer-Payne W, Harrison EG, O'Brien PC, Weiland LH. 1976. The value of thymectomy in myasthenia gravis. Ann Surg, 184, 435-458.

33. Vincent A, Newsom-Davis J, Newton P, Beck N. 1983. Acetylcholine receptor antibody and clinical response to thymectomy in myasthenia gravis. Neurology (Cleveland), 33, 1276-82.

34. Scadding GK, Vincent A, Newsom-Davis J, Henry K. 1981. Acetylcholine receptor antibody synthesis by thymic lymphocytes in vitro. Neurology (Minneap), 31, 935-943.

35. Newsom-Davis J, Willcox N, Scadding G, Calder L, Vincent A. 1981. Anti-acetylcholine receptor antibody synthesis by cultured lymphocytes in myasthenia gravis: thymic peripheral blood cells interactions. In: Grob D ed, Myasthenia Gravis: Pathophysiology and Management, Ann NY Acad Sci, 333, 393-402.

36. Thomas JA, Willcox N, Newsom-Davis J. 1982. Immunohistological studies of the thymus in myasthenia gravis. Correlation with clinical state and thymocyte culture response. J Neuroimmunol, 3, 319-335.

37. Willcox HNA, Newsom-Davis J, Calder L. 1983. Greatly increased auto-antibody production in myasthenia gravis by thymocyte suspensions prepared with proteolytic enzymes. Clin Exp Immunol, 54: 378-386.

38. Lindstrom JM, Lennon VA, Seybold ME, Whittingham S. 1976. Experimental autoimmune myasthenia gravis and myasthenia gravis: biochemical and immunochemical aspects. Ann NY Acad Sci, 274, 254-274.

39. Raimond F, Morel E, Bach JF. 1984. Evidence for the presence of immunoreactive acetylcholine receptors on human thymus cells. J Neuroimmunol, 6, 31-40.

40. Ueno S, Wada K, Kang J, Takahashi M, Tarui S. 1982. Acetylcholine receptor and thymus in experimental autoimmune myasthenia gravis and experimental myositis. Clin exp Immunol, 50, 563-571.

41. Wekerle H, Ketelsen U-P. 1977. Intrathymic pathogenesis dual genetic control of myasthenia gravis. Lancet, i, 678-680.

42. Kao I, Drachman DB. 1977. Thymic muscle cells bear acetylcholine receptors: possible relation to myasthenia gravis. Science, 195, 74-75.

43. Guttmann L, Crosby TW, Takamori M, Martin JD. 1972. The Eaton-Lambert syndrome and autoimmune disorders. Am J Med, 53, 354-356.

44. Eaton G, Lambert EH. 1975. Electromyography and electrical stimulation of nerves in diseases of the motor unit: observations on a myasthenic syndrome associated with malignant tumours. JAMA, 161, 1117-24.

45. Cull-Candy SG, Miledi R, Trautmann A, Uchitel OD. 1980. On the release of transmitter at normal, myasthenia gravis and myasthenic syndrome affected human endplates. J Physiol, 299, 621-638.

46. Lang B, Newsom-Davis J, Prior C, Wray D. 1983. Antibodies to motor nerve terminals: an electrophysiological study of a human myasthenic syndrome transferred to mouse. J Physiol (Lond), 34, 335-345.

47. Newsom-Davis J, Murray N, Wray D, Lang B, Prior C, Gwilt M, Vincent A. 1982. Lambert-Eaton myasthenic syndrome: electrophysiological evidence for a humoral factor. Muscle and Nerve, 5, S17-S20.

48. Fukunaga H, Engel AG, Osame M, Lambert EH. 1982. Paucity of presynaptic membrane active zones in the Lambert-Eaton myasthenic syndrome revealed by freeze fracture electron microscopy. Neurol (Ny) 32, A222.

49. Fukunaga H, Engel AG, Lang B, Newsom-Davis J, Vincent A. 1983. Passive transfer of Lambert-Eaton myasthenic syndrome with IgG from man to mouse depletes the presynaptic membrane active zones. Proc Nat Acad Sci, USA, 80, 7636-7640.

50. Lang B, Newsom-Davis J, Prior C & Wray D. Effect of passively transferred Lambert-Eaton myasthenic syndrome antibodies on the calcium sensitivity of transmitter release in the mouse. J Physiol, (Lond), in press.

51. Fenichel GM. 1978. Clinical syndromes of myasthenia in infancy and childhood. Arch Neurol, 39, 97-103.

52. Hart Z, Sahashi K, Lambert EH, Engel AG, Lindstrom J. 1979. A congenital, familial myasthenic syndrome caused by a presynaptic defect of transmiter resynthesis or mobilization. Neurology (Minneap), 29, 559.

53. Vincent A, Cull-Candy S, Newsom-Davis J, Molenaar PC, Polak RL, Trautmann A. 1981. Congenital myasthenia: endplate acetylcholine receptors and electrophysiology in fivew cases. Muscle and Nerve, 4, 306-318.

54. Engel AG, Lambert EH, Gomez MR. 1976. A new myasthenia syndrome with endplate acetylcholinesterase deficiency, small nerve terminals, and reduced acetylcholine release. Ann Neurol, 1, 315-330.

55. Engel AG, Lambert EH, Mulder DM, Torres CF, Sahashi K, Bertorini TE, Whitaker. 1982. A newly recognised congenital myasthenic syndrome attributed to a prolonged open time of the acetylcholine-induced ion channel. Ann Neurol, 11, 553-569.

56. Morgan-Hughes JA, Lecky BRF, Landon DN, Murray NMF. 1981. Alterations in the number and affinity of junctional acetylcholine receptors in a myopathy with tubular aggregates. A newly recognised receptor defect. Brain, 104, 279-295.

57. Lennon VA, Palmer AC, Pflugfelden C, Indreri RJ. 1978. Myasthenia gravis in dogs: acetylcholine receptor deficiency with and without anti-receptor antibodies. In: Genetic Control of Autoimmune Disease (Rose NR, Bigazzi PE, Warren NL eds.). Elsevier, North Holland.

ANTI-IDIOTYPIC ANTIBODY TO THE ACETYLCHOLINE AND ADENOSINE RECEPTORS

B.F. Erlanger*, W.L. Cleveland*, N.H. Wassermann*, B.L. Hill*, A.S. Penn+, H.H. Ku*, and R. Sarangarajan*

Departments of Microbiology* and Neurology+ and the Cancer Center/Institute of Cancer Research, Columbia University, New York, N.Y. 10032

One of us (BFE) has known David Nachmansohn throughout his (BFE's) professional life, becoming acquainted with David while a student in the Department of Biochemistry of Columbia University, in which David was a Professor. After BFE subsequently accepted a faculty position in the Department of Microbiology, there were many opportunities to discuss science with David, in particular in the university cafeteria, and to participate in collaborative projects with David's colleague, I.B. Wilson (1). The result of one of the conversations was the development of a compound called "BisQ" (2,3). The chemical name of BisQ is trans-3,3'-bis[α-(trimethylammonio)-methyl]azobenzene. It is this compound that led to the research to be described in this paper.

At the time of these discussions, we were engaged in the development of model photoresponsive systems that utilized photochromic low molecular weight effector molecules to photoregulate enzyme activity. Photochromic compounds are molecules that can assume two distinct configurations with different absorption spectra, and whose relative concentrations can be controlled by light of selected wavelengths (4). All of the compounds with which we worked were azobenzene derivatives. These compounds exist as cis or trans isomers, the relative concentrations of which can be regulated by exposure to long ultraviolet or to visible (blue) light. The azo

Molecular Basis of Nerve Activity

compounds we synthesized were designed to function as enzyme substrates (5), inactivators (6,7,8) or inhibitors (7,8), with the _cis_ and _trans_ stereoisomers differing markedly in activity. This can be easily understood if models of _cis_ and _trans_ azobenzene are examined. The _trans_ isomer is a planar molecule; in the _cis_ isomer the benzene rings, rather than being in the same plane, are at an approximately 90° angle to each other. It is unlikely that two molecules differing so markedly in structure would bind with equal affinity to enzymes and affect activity equally. Thus, photoregulation of their relative concentrations would, in effect, photoregulate the activities of enzymes for which they were specific. Among the enzymes whose activities were made responsive to light were trypsin, chymotrypsin (6) and acetylcholinesterase (7-9). Reviews of this research have appeared (10,11).

On discussing our experiments with David, especially those with acetylcholinesterase, it occurred to both of us to collaborate on the photoregulation of the acetylcholine receptor (AChR), in particular as it functions in electroplax preparations derived from _Electrophorus electricus_. We, therefore, began collaborative experiments to convert the electric fish into a photoelectric fish!

The result was the design and synthesis of BisQ. Its _trans_ isomer was found to be one of the most potent agonists of AChR in the electroplax system of _Electrophorus electricus_, having an apparent dissociation constant of 0.6×10^{-7}M (2). Surprisingly, the _cis_ isomer lacked measureable activity.

Molecular models of _trans_-BisQ revealed that it is a highly constrained, planar molecule (12, 3). Unlike flexible molecules, which can assume many conformations in solution, a structurally constrained molecule has few degrees of freedom. Hence, if its biological activity depends upon interaction with the binding site of a macromolecular receptor, it can either be highly active or of little activity, depending on whether its conformation (or small number of possible conformations) allows complementary binding to the target receptor. The extremely high activity of trans-BisQ as an agonist suggested to us that its structure was complementary to the combining site of AChR when the latter was in the activated state (13).

It had been shown by others (14,15) that macromolecules that bind the same ligand can show binding site homologies despite having different biological functions. It occurred to us, therefore, that the combining sites of antibodies raised to trans-BisQ would show homology with the binding site of AChR, when the latter is in the activated state. If so, these antibodies should be able to bind agonists of AChR in accordance with their activities in the AChR system, and should bind antagonists poorly. Put simplistically

Anti-BisQ ≡ AChR

A BisQ derivative was prepared (16) which could be covalently linked to bovine serum albumin (BSA). Its coupling to BSA was via the mixed anhydride reaction (17) and a conjugate containing 8 molecules of BisQ per BSA was synthesized and used to immunize New Zealand White rabbits. Binding specificities of the antisera were determined by a competitive radioimmunoassay using [^{3}H]-trans-BisQ as the tracer.

The results can be summarized as follows: First of all, the order of binding activities of the various agonists was similar to that seen with AChR. Exact comparisons were not possible because many of the AChR experiments were made with electroplax or membrane preparations as well as with soluble AChR (18,19). There were other striking findings:

(1) The IC_{50} value for trans-BisQ (0.15uM) was close to the 50% point on the dose-response curve of experiments carried out with Electrophorus electricus electroplax preparations (0.06uM) (2).

(2) Decamethonium, a potent agonist, was bound with an apparent affinity that was 500X that of hexamethonium, a potent antagonist. A priori, it would not be expected that an antibody would distinguish so markedly between two molecules

of such similar structure.

(3) d-Tubocurarine, a potent antagonist was bound poorly.

The experiments indicated that we had succeeded in showing binding homologies in AChR and anti-BisQ (i.e. that anti-BisQ ≡ AChR).

The findings suggested that anti-idiotypic antibodies (20) directed at determinants of the combining sites of anti-BisQ antibodies might share characteristics of anti-AChR, i.e. that the combining sites of anti-BisQ and anti-AChR might contain crossreactive epitopes. Put another way, if anti-BisQ ≡ AChR, then anti-anti-BisQ ≡ anti-AChR.

When rabbits are immunized with purified AChR, they make specific anti-AChR antibodies and simultaneously show signs of experimental myasthenia gravis (21). Therefore, if we were to succeed in raising anti-idiotypic antibodies directed against the combining sites of anti-BisQ (i.e. anti-anti-BisQ) and, if, indeed, these antibodies crossreacted with AChR, the rabbits in which these antibodies were raised should show signs of myasthenia gravis.

The anti-BisQ antibodies, therefore, were purified on an affinity column (16) and then used to immunize 3 New Zealand White rabbits. The following observations were made:

1. Of three rabbits immunized, two developed muscle weakness after the first booster injections.

2. After the second and third booster injections, all animals improved, i.e. the signs of muscle weakness appeared to be transient.

3. One of the rabbits that recovered was re-immunized 4 months later with anti-BisQ antibody, administered intradermally in saline. Shortly thereafter, it showed extreme muscle weakness which could be temporarily reversed by administration of neostigmine.

4. Assay of the rabbit sera for reaction with AChR by ELISA showed the transient appearance of antibody with the peak titers coinciding with the period during which the rabbits

were most ill; a fall in titer occurred during the animal's recovery.

The transient, cyclical nature of the immune response to rabbit anti-BisQ is typical of an anti-idiotypic response (cf 22). We, therefore, examined the sera of two of the rabbits for characteristics consistent with the presence of anti-idiotypic antibody. The reaction of anti-idiotypic antibodies with the combining site of the target antibody will usually inhibit binding of a ligand by the ligand-specific (target) antibody. We found this to be true for two sera taken from each of two of the rabbits, i.e. they inhibited the binding of [^{3}H]-BisQ by anti-BisQ. However, one serum taken two days after an inhibiting serum caused enhancement of binding of [^{3}H]-BisQ by anti-BisQ and, in fact, could itself be shown to bind [^{3}H]-BisQ. Presumably serum taken at this time contained a predominance of the next generation of antibody in the anti-idiotypic network, directed at the combining site of anti-anti-BisQ.

We now had direct evidence for participation of the anti-idiotypic network (23) in the response to an antigen (anti-BisQ). In other words, immunization with an antigen (anti-BisQ) not only led to an anti-antigen response (anti-anti-BisQ) but to the rise of auto-anti-idiotypic antibodies specific for BisQ. With this in mind, we decided to devise a more direct route to monoclonal anti-receptor antibodies. In this approach, we chose to immunize mice with BisQ-BSA, fuse their spleen cells with an appropriate myeloma line and screen hybridomas for cells producing anti-anti-BisQ immunoglobulins capable of crossreacting with AChR.

BALB/cCr mice were immunized intraperitoneally (i.p.) with 0.1 ml of a 1 mg/ml solution of BisQ-BSA conjugate in complete Freund's adjuvant. Twenty-three days later, the mice were boosted i.p. with the same preparation. After 5 days, the spleen cells were collected and fused with a non-secreting myeloma line (P3x63-Ag 8.653) (24) essentially by the protocol of Kohler and Milstein (25) as modified by

Sharon *et al*. (26). Supernatants from hybridoma clones were obtained by a replica transfer technique and screened for activity against BisQ-RSA, rabbit anti-BisQ and AChR (*Torpedo*) by enzyme immunoassay using peroxidase-labelled goat anti-mouse immunoglobulin as the second antibody.

About 14% of the wells were found to be positive for the antigen. Another set of wells, about 7.4% of the total, were positive for purified rabbit anti-BisQ; of these, about a third were also positive for AChR. Thus, not all anti-BisQ reactive supernatants also bound receptor. Although a significant number of the wells might have contained more than a single clone, and, certainly, the same clones could have been distributed in more than one well, we can conclude that immunization with BisQ-BSA induced a significant anti-anti-BisQ response.

Positive wells were then subcloned by the procedure of limiting dilution and several of the clones that produced antibody reactive with both anti-BisQ and *Torpedo* receptor were examined. One clone, F8-D5, was chosen for detailed study. The antibody produced by this clone was able to bind purified preparations of AChR of *Torpedo* and *Electrophorus* electricus, a rat muscle preparation and purified rabbit anti-BisQ. These binding properties were retained after two subsequent subclonings by limiting dilution.

To confirm that the same monoclonal antibody bound to receptor and to anti-BisQ, reciprocal inhibition experiments were carried out. In an experiment in which the binding of F8-D5 to anti-BisQ was inhibited by *Torpedo* receptor, as high as 67% inhibition was observed, with 50% inhibition occurring with about 8 ug of receptor. In experiments in which the amount of antibody was decreased fivefold to 2 ug per well, 50% inhibition occurred in the presence of 0.9 ug of *Torpedo* receptor. In a similarly designed series of reciprocal experiments, 50% inhibition of the binding of F8-D5 to *Torpedo* receptor was accomplished with 3 ug of purified rabbit anti-BisQ.

The binding of F8-D5 to *Torpedo* receptor and to anti-BisQ could be inhibited by BisQ and other agonists, as well as by the AChR inhibitors, hexamethonium and α-bungarotoxin. These results reinforce the conclusion that the specificity of F8-D5 is for determinants intimately associated with the combining sites of both the *Torpedo* receptor and rabbit anti-BisQ. With respect to the latter, this is consistent with a specificity for an idiotypic determinant. It is important to note that BisQ is a small molecule (molecular weight 486) and not likely to block reaction with non-idiotypic determinants near the combining site as a result of steric interference.

The strategy described here provides a powerful route to anti-receptor antibodies that are likely to be directed at determinants associated with the combining sites of the receptor. As in our previous findings with rabbits, purified receptor is not required for immunization; in fact, these anti-idiotypic antibodies, presumably, can be used to isolate receptor. We have preliminary evidence for the feasibility of this approach. F8-D5 was immobilized on diaminodiproplyamine agarose (Pierce Chemical Co.) using glutaraldehyde as the linking agent (27). A crude preparation of receptor obtained from *Torpedo californica* was incubated with the immobilized antibody. After subsequent washing, the receptor was eluted with hexamethonium bromide. Assay for the receptor was by binding to $[^{125}I]$-α-bungarotoxin. Essentially all of the receptor bound to the column and could be eluted by hexamethonium. We have not yet succeeded in obtaining pure receptor, as judged by SDS-PAGE electrophoresis but the experiments are continuing.

The utility of antibodies produced in this way is also illustrated by the demonstration, by immunofluorescence techniques, that F8-D5 binds to *Torpedo* tissue, giving a pattern that is similar to that given by rabbit antisera raised by immunization with purified AChR.

In collaboration with J. Lindstrom of the Salk Institute,

F8-D5 was found to block function in a vesicle system containing reconstituted *Torpedo* AChR (28). At an antibody to receptor ratio of 2:1, in the presence of 1 mM carbamylcholine, ^{134}Cs influx was inhibited by about 80%. This result also supports our contention that F8-D5 reacts with the binding site(s) of AChR.

Anti-idiotypic Antibodies to Adenosine Receptors

We have also succeeded in extending the monoclonal auto-anti-idiotypic procedure to the adenosine receptor (for reviews on the adenosine receptor, see refs. 29,30). First, antibodies were prepared in rabbits to N-6-adenosinecaproyl-BSA. The antisera were then specifically purified on an aminohexylsepharose column to which N-6-adenosinecaproic acid was linked. The purified antibodies thus obtained were then used to screen for mouse monoclonal anti-idiotypic antibodies (anti-Ad) prepared by immunizing Balb/c mice with the BSA conjugate of N-6-adenosinecaproic acid.

The products of two clones ($A^2$21 and $A^2$18) have been followed in detail. By radioimmunoassay, both inhibited the binding of $[^3H]$-adenosine to the purified rabbit anti-Ad described in the previous paragraph. Both antibodies bound to a partially purified preparation of adenosine receptor (31). Preliminary results indicate that binding to the receptor by $A^2$21 and $A^2$18 can be inhibited by N^6-cyclohexyladenosine,2-chloroadenosine and theophylline.

Discussion

Our experimental results lend support to a functioning idiotypic network as foreseen by Jerne (23). Our earlier studies in rabbits (16), in particular the finding of a transient anti-BisQ response when immunizing with anti-BisQ, supported the network theory, as did the report of Schechter *et al.* (32) that immunization with insulin caused the spontaneous appearance of insulin receptor-specific antibodies. Similarly, auto-anti-idiotypic antibody appeared during a normal human immune response to tetanus toxoid (33). Our present studies add to these findings in that in the initial screening of the hybridomas, the number of wells scoring positive for anti-idiotypic antibodies was as much as one-half the number positive for anti-antigen antibodies. Thus, not only is the idiotypic network functioning during a normal immunization process (34), it is functioning very actively, as early as 5 days after a first booster injection.

It should be appreciated that the ease with which we have detected auto-anti-idiotypic antibodies depends on the use of hybridoma technology. By immortalizing the clones active at the time of cell harvest, the cellular events are "frozen" in time, making it possible to study clones that may have only a transient existence *in vivo*. Moreover, as the immortalized clones are separated from each other, this approach avoids the technical problems that arise from the formation of immune complexes and which are an important limitation of serological studies.

Another biological implication of our findings relates to the origins of autoimmune diseases in which anti-receptor antibodies are implicated. At least some of these diseases (16, 32) might originate from an anti-idiotypic response to antibodies formed against biologically active ligands normally present *in vivo*, such as insulin or thyrotropin. The likelihood of an anti-idiotypic etiology increases if the patients' antibodies are found to be directed at the combining site of a receptor. Although not always the case (35), this has been found to be true in many patients with myasthenia gravis (36), particularly in those who are severely ill. Moreover, we have shown that experimental myasthenia gravis can be induced in rabbits *via* the anti-idiotypic route. Thus, it could have a role in at least some cases of myasthenia gravis in humans.

The nature of the primary immunogen is, of course, not known. On the other hand, in Graves disease, the specificity of circulating anti-thyroid receptor antibodies is almost always directed at the combining site of the thyrotropin receptor (37). These antibodies, therefore, have an excellent chance of being anti-idiotypic, directed at idiotypes of anti-thyrotropin antibody. Kohn and his colleagues (38) give positive evidence for this as do the results of Islam *et al.*(39). In the earlier studies (16,32), only low titers of circulating anti-receptor antibodies were detected and, in our case, the anti-idiotypic response appeared to be transient. Our present findings show that the anti-idiotypic response is, on the contrary, a very important aspect of the response to an antigen. In normal conditions, it must be modulated in some way. Under circumstances in which regulation is defective, pathological consequences might follow.

Acknowledgments

The research described in this article was supported by the NIH (NS-15581, NS-17904 and AI-17949) and the Muscular Dystrophy Association. We wish to acknowledge the skillful assistance of Mrs. F. Schneider in the preparation of the manuscript.

References

1. Wilson, I.B., B.F. Erlanger. 1960. An explanation of an anomalous antipodal specificity of chymotrypsin. J. Amer. Chem. Soc. 82, 6422.

2. Bartels, E., N.H. Wassermann, B.F. Erlanger. 1971. Photochromic activators of the acetylcholine receptor. Proc. Nat. Acad. Sci. USA 68, 1820-1823.

3. Wassermann, N.H., B.F. Erlanger. 1981. Agents related to a potent activator of the acetylcholine receptor of Electrophorus electricus. Chem. Biol. Inter. 36, 251-258.

4. Brown, G.H. 1971. Photochromism, Vol. III. Techniques in Chemistry (A. Weissberger, ed.). Wiley-Interscience, New York. pp. 1-11.

5. Wainberg, M.A., B.F. Erlanger. 1971. Investigation of the active center of trypsin using photochromic substrates. Biochemistry 10, 3816-3819.

6. Kaufman, H., S.M. Vratsanos, B.F. Erlanger. 1968. Photoregulation of an enzymic process by means of a light-sensitive ligand. Science 162, 1487-1488.

7. Bieth, J., S.M. Vratsanos, N.H. Wassermann, B.F. Erlanger. 1969. Photoregulation of biological activity by photochromic reagents. II. Inhibition of acetylcholinesterase. Proc. Nat. Acad. Sci. USA 64, 1103-1106.

8. Bieth, J., S.M. Vratsanos, N.H. Wassermann, B.F. Erlanger. 1970. Photoregulation of biological activity by photochromic reagents. IV. A model for diurnal variation of enzymic activity. Proc. Nat. Acad. Sci. USA 66, 850.854.

9. Bieth, J., S.M. Vratsanos, N.H. Wassermann, A.G. Cooper, B.F. Erlanger. 1973. Photoregulation of biological activity by photochromic reagents. Inactivation of acetylcholinesterase. Biochemistry 12, 3023-3027.

10. Erlanger, B.F. 1976. Photoregulation of biologically active macromolecules. Annu. Rev. Biochem. 45, 267-283.

11. Erlanger, B.F., N.H. Wassermann. 1982. Models of photoregulation. In: Trends in Photobiology. (C. Helene, M. Charlier, Th. Montenay-Garestier, G. Laustriat, eds.). Plenum Publishing Corp., New York. pp. 81-92.

12. Wassermann, N.H., E. Bartels, B.F. Erlanger, 1979. Conformational properties of the acetylcholine receptor as revealed by studies with constrained depolarizing ligands. Proc. Nat. Acad. Sci. USA 76, 256-259.

13. Heidmann, T., J.-P. Changeux. 1978. Structural and functional properties of the acetylcholine receptor protein in its purified and membrane-bound states. Annu. Rev. Biochem. 47, 317-357.

14. Beiser, S.M., S.W. Tanenbaum. 1963. Binding site topology of enzymes and antibodies induced by the same determinants. Ann. N.Y. Acad. Sci. 103, 595-609.

15. Hough, C.A.M., J.A. Edwardson. 1978. Antibodies to Thaumatin as a model of the sweet taste receptor. Nature 271, 381-383.

16. Wassermann, N.H., A.S. Penn, P.I. Freimuth, N. Treptow, S. Wentzel, W.L. Cleveland, B.F. Erlanger. 1982. Anti-idiotypic route to anti-acetylcholine receptor antibodies and experimental myasthenia gravis. Proc. Nat. Acad. Sci. USA 79, 4810-4814.

17. Erlanger, B.F. 1973. Principles and methods for the preparation of drug protein conjugates for immunological studies. Pharmacol. Rev. 25, 271-280.

18. Karlin, A. 1974. The acetylcholine receptor: Progress report. Life Sci. 14, 1385-1415.

19. Meunier, J.C., J.-P. Changeux. 1973. Comparison between the affinities for reversible cholinergic ligands of a purified and membrane bound state of the acetylcholine receptor protein from *Electrophorus electricus*. FEBS Lett. 32, 143-148.

20. Oudin, Y., M. Michael. 1963. Une nouvelle forme d'allotypie des globulines du serum de lapin apparement liée a la fonction et la specificité anticorps. C.R. Acad. Sci. (Paris) 257, 805-808.

21. Patrick, J., J.M. Lindstrom. 1973. Autoimmune response to acetylcholine receptor. Science 180, 871-872.

22. Couraud,P.-O., B.-Z. Lu, A.D. Strosberg. 1983. Cyclical anti-idiotypic response to anti-hormone antibodies due to neutralization by autologous anti-anti-idiotype antibodies that bind hormone. J. Exp. Med. 157, 1369-1378.

23. Jerne, N.K. 1974. Towards a network theory of the immune system. Ann. Immunol. (Inst. Pasteur) Paris 125C, 373-389.

24. Kearney, J.F., A. Radbruch, B. Liesegang, K. Rajewsky. 1979. A new mouse myeloma cell line that has lost immunoglobulin expression but permits the construction of antibody-secreting hybrid cell lines. J. Immunol. 123, 1548-1550.

25. Kohler, G., C. Milstein. 1975. Continuous cultures of fused cells secreting antibodies of predefined specificity. Nature 256, 495-497.

26. Sharon, J., S.L. Morrison, E.A. Kabat. 1979. Detection of specific hybridoma clones by replica immunoadsorption of their secreted antibodies. Proc. Nat. Acad. Sci. 76,

1420-1424.

27. Reichlin, M. 1980. Use of glutaraldehyde as a coupling agent for proteins and peptides. Meth. Enzymol. 70A, 159-165.

28. Suarez-Isla, B.A., K. Wan, J. Lindstrom, M. Montal. 1983. Single-channel recordings from purified acetylcholine receptors reconstituted in bilayers formed at the tip of patch pipets. Biochem. 22, 2319-2323.

29. Daly, J.W., R.F. Bruns, S.H. Snyder. 1981. Adenosine receptors in the central nervous system: Relationship to the central actions of methylxanthines. Life Sci. 28, 2083-2097.

30. Wolf, J., C. Londos, D.M.F. Cooper. 1981. Adenosine receptors and the regulation of adenylate cyclase. Adv. Cyc. Nuc. Res. 14, 199-210.

31. Daly, J.W., P. Butts-Lamb, W. Padgett. 1983. Subclasses of adenosine receptors in the central nervous system: Interaction with caffeine and related methylxanthines. Cell. Molec. Neurobiol. 3, 69-80.

32. Shechter, Y., R. Maron, D. Elias, I.R. Cohen. 1982. Auto-antibodies to insulin receptor spontaneously develop as anti-idiotypes in mice immunized with insulin. Science 216, 542-544.

33. Geha, R.S. 1982. Presence of auto-anti-idiotypic antibody during the normal human immune response to tetanus toxoid antigen. J. Immunol. 129, 139-144.

34. Binion, S., L.S. Rodkey. 1982. Naturally induced auto-anti-idiotypic antibodies. Induction by identical idiotypes in some members of an outbred rabbit family. J. Exp. Med. 156, 860-872.

35. Tzartos, S.J., M.E. Seybold, J.M. Lindstrom. 1982. Specificities of antibodies to acetylcholine receptors in sera from myasthenia gravis patients measured by monoclonal antibodies. Proc. Nat. Acad. Sci. USA 79, 188-192.

36. Fulpius, B.W., A.K. Lefvert, S. Cuenoud, A. Mourey.1981. Properties and serum levels of specific populations of anti-acetylcholine receptor antibodies in myasthenia gravis. Ann. N.Y. Acad. Sci. 377, 307-315.

37. Rees-Smith, B., R. Hall. 1974. Thyroid-stimulating immunoglobulins in Graves disease. Lancet ii, 427.

38. Kohn, L., W.A. Valente, P. Laccetti, C. Marcocci, M. DeLuca, P.A. Ealey, N.J. Marshall, E.F. Grollman. 1984. Monclonal antibodies as probes of thyrotropin receptor structure. In: Receptor Biochem. and Method. 4 (in press).

39. Islam, M.N., B. Pepper, R. Briones-Urbina, N.R. Farid. 1983. Biological activity of anti-thyrotropin anti-idiotypic antibody. Eur. J. Immunol. 13, 57-63.

ANALYSIS OF THE PROPHYLACTIC AND THERAPEUTIC ACTION OF ELECTROLECTIN IN EXPERIMENTAL AUTOIMMUNE MYASTHENIA GRAVIS

VIVIAN I. TEICHBERG AND GIOVANNI LEVI

Neurobiology Department. The Weizmann Institute of Science, Rehovot, Israel

Introduction

The existence in many vertebrate tissues of an endogenous lectin activity with a β-D-galactoside specificity was demonstrated in 1974 and attributed to a family of proteins named electrolectins (1). The name electrolectin originates from the fact that the first animal lectin described was isolated from the electric organ of the electric eel Electrophorus electricus (1,2,3). To date, electrolectins have been isolated from chick embryo skeletal muscle (4,5) kidney (6), heart (7) and intestine (8), from calf heart, lung (9) and spleen (7), from rat lung (10), monkey skeletal muscle (11), human lung and skeletal muscle (10,11). Remarkedly, the various electrolectins share properties very similar to those described for the electrolectin from Electrophorus electricus. All of them agglutinate trypsin-treated rabbit erythrocytes, have primarily an intracellular cytoplasmic localization, have molecular weight between 25,000-33,000, are composed of 2-4 subunits, possess the same saccharide specificity β-D-galactosyl-β thiogalactopyranoside (TDG) > lactose > galactose, and strictly require reducing agents for the maintenance of their agglutinating activity. Moreover,

Molecular Basis of Nerve Activity

antigenic cross-reactivity has been shown between the electrolectins from various tissues of the same animal (12,13) and between the electrolectins of homologous tissues in distinct mammalian species (11).

Biological functions of electrolectins: an open question

Their wide distribution in the animal kingdom and their common properties strongly suggest that the electrolectins possess biological function(s) of high evolutionary importance but no agreement has yet been reached on the nature of this function. A striking property of electrolectins is that their concentration in tissues is often developmentally regulated. This is well documented for the electrolectins from chick embryo pectoral muscle (4), brain (15), liver (15), spinal cord (16), skin (17), cartilage (18) and from rat embryo brain (14) and lung (19). Another possibly relevant observation is the association of an electrolectin-like activity with neoplastic cells. We observed this first in neuroblastoma cells (1) and it was later confirmed for five human melanomas, a human cervical adenocarcinoma and a murine fibrosarcoma (20).

It was proposed that the presence of high concentrations of electrolectin during development may be related to intercellular recognition processess and in particular to myogenesis (21,22,23,24) and synaptogenesis (14,19). However, the role of electrolectin in myogenesis has been challenged (25,26).

Other physiological functions have been proposed: electrolectin would be a blood group antigen receptor (27), be involved in mucus secretion (13), receptor-mediated internalization (11), adhesion of endoderm cells during embryogenesis (28), dermal condensation in the chick embryo skin (17), chondrogenesis (29) and cell rearrangement during epiboly (30).

Studies of the interactions between electrolectin and the acetylcholine receptor protein.

In the framework of our own efforts to elucidate the biological functions of electrolectins, we investigated in detail the possible interactions taking place in embryonic skeletal muscle between the acetylcholine receptor protein (AChR) and electrolectin (EL). The rationale of this study was based on the fact that the concentrations of EL and of AChR in muscle tissue are both developmentally regulated (31,32) and reach a maximum concomitantly with the process of synaptogenesis (32).

No direct interactions could be demonstrated between EL and the AChR either when the latter was present in Triton X-100 solutions or in the myotube membrane (33). More specifically, no coprecipitation of radioactivity was observed when Triton X-100 solubilized AChR was incubated with anti-AChR antibodies in the presence of ^{125}I-labeled EL or when EL was precipitated by anti-EL antibodies in the presence of ^{125}I-labeled AChR. Furthermore, EL and its specific inhibitor TDG were found not to interfere with the binding between AChR and either α-bungarotoxin or anti-AChR antibodies. EL also did not affect the rate of degradation of AChR present in rat myotubes kept in tissue culture.

Prophylactic effect of eel electrolectin on experimental autoimmune myasthenia gravis

In the course of the experiments described in the previous paragraph, we observed that eel electrolectin, although not interacting directly with the AChR protein, could nevertheless modulate the *in vivo* immune response to AChR. Eel electrolectin was found to have a prophylactic and therapeutic effect on the experimental autoimmune *Myasthenis gravis* (EAMG) in rabbits (33). EAMG is an experimental autoimmune disease (34) that can be induced by immunization with the purified acetylcholine receptor protein and is characterized by the occurrence of symptoms of hypoactivity, uncoordinated movements, muscle weakness, sinking of the head and paralysis of the limbs and trunk. These symptoms are alleviated temporarily upon injection of edrophonium chloride. In most cases, the disease manifests itself, after a single injection of AChR as a progressive flaccid paralysis with a life expectancy of two-three weeks. When boosting injections of AChR are performed, the evolution of EAMG is extremely rapid. The animals are lying flat with extended limbs and breathe only with difficulty. The life expectancy of these rabbits is never more than 6-7 days.

Table I summarizes the results of experiments designed to analyze in detail the effects, on the onset of EAMG, of eel electrolectin administered before of concomitantly with the AChR immunization. Group 1 consists of rabbits that were injected with AChR alone. All these rabbits developed the typical clinical symptoms of EAMG that led ultimately to their death following either a primary or a single boosting injection of AChR. Group 2 and 3 are groups of rabbits that received simultaneous injections of AChR and EL. None of these animals showed over a period of 4 months any of the clinical symptoms of EAMG either following boosting injections with AChR and eel electrolectin (group 2) or AChR alone (group 3). Groups 4 and 5 are control rabbits that received primary and boosting injections of either AChR and denatured EL (group 4) or AChR and bovine serum albumin (group 5). All the rabbits in these groups developed EAMG and ultimately died.

The animals (group 6) that were first exposed to EL for 5 days and then immunized and later boosted once with AChR, did not develop EAMG. However, when these animals were injected for a third time with AChR, they did develop the symptoms of EAMG.

Table 1: Effects of eel electrolectin on the onset of EAMG

Group	Treat-ment(a)	Nature of Booster	No. of Rabbits	Incidence (%)	Lethality (%)
1	AChR	AChR	14(c)	100 (14/14)b	100 (14/14)
2	AChR+EL	AChR+EL	12	0 (0/12)	0 (0/12)
3	AChR+EL	AChR	4	0 (0/ 4)	0 (0/ 4)
4	AChR+Den EL	AChR+Den EL	4	100 (4/ 4)	100 (4/ 4)
5	AChR+BSA	-	2(d)	100 (2/ 2)	100 (2/ 2)
6	AChR(5 days after EL presentation)	AChR	4	0 (0/ 4)	0 (0/ 4)

a) 100 μg of each antigen were injected intradermally
b) Numbers in parentheses represent the number of animals
c) Seven out of 14 animals died following the first immunization
d) The 2 animals died following the first immunization

Therapeutic effect of EL on EAMG

The administration of EL to myasthenic rabbits led, in most cases, to a complete recovery from the disease (Table 2). The therapeutic effects of EL were found to depend mainly on the severity of the EAMG symptoms displayed by the rabbits before the EL treatment. The EAMG symptoms were mild following a single priming injection of AChR (50% of the cases) and severe following boosting injections of AChR (100% of the cases). The best results were obtained when mild EAMG was treated with EL (group 1). However, 40% of the animals suffering from severe EAMG were also treated successfully with EL (group 3). The administration of BSA to animals suffering from either mild or severe symptoms of EAMG did not have any therapeutic effects (groups 2,4).

When AChR was administered to EL-treated and "EAMG-cured" animals, the symptoms of EAMG reappeared within five days indicating that their immunological memory towards AChR was maintained. When these myasthenic animals were reinjected with EL, the symptoms of EAMG disappeared within 2-4 days (Group 5). If BSA was administered, the rabbits died in 4-6 days (Group 6).

Table 2: Therapeutic effects of eel electrolectin on EAMG

Group	Symptoms of EAMG	Treat- ment	Previous Presentation	Recovery days	Recovery Incidence	Survival (days)
1	Mild	EL	-	10±2(b)	5/5(a)	>120
2	Mild	BSA	-	.	0/2	18±3(b)
3	Severe	EL	-	21±5	4/10	>120
4	Severe	BSA	-	.	0/2	4±2
5	Severe	EL	+	3±1	4/4	>120
6	Severe	BSA	+	.	0/2	4±2

a) Numbers represent the number of animals
b) Means ± standard deviation

Effect of electrolectin of the titer of anti-AChR antibodies

Since EAMG is induced upon immunization with Torpedo AChR which produces circulating antibodies cross reacting with self AChR (34,35), we have attempted to correlate the prophylactic or therapeutic effects of EL on EAMG, to the level of circulating anti-AChR antibodies. A solid phase radioimmunoassay with immobilized AChR was used.

As can be seen in Fig. 1, the sera of animals injected with EL and AChR (group 2, Table 1) have a lower capacity to bind to immobilized AChR, than the sera of animals injected with AChR only (group 1, Table 1). Indeed, the sera of animals treated with EL and AChR recognize only 60% of the antigenic determinants of the AChR that are recognized by the sera of animals immunized only with AChR.

However, since a saturation of 50% of the immobilized AChR binding sites available to each group of sera is reached at the same sera dilution (indicated by the arrows in Fig. 1), one infers that the concentrations of anti-AChR antibodies in the sera of both groups of animals are not significantly different, but that their spectrum of specificity is different.

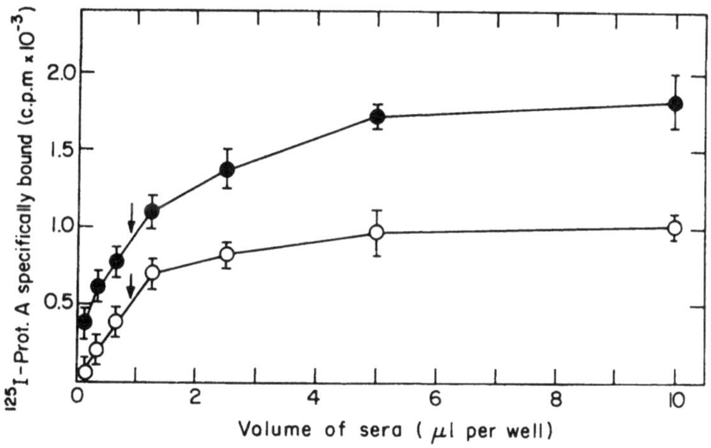

Figure 1: Anti-AChR antibody determination by solid phase radioimmunoassay with immobilized AChR. The anti-AChR antibodies bound to the AChR were reacted with radioactive protein A. Sera from rabbits immunized with: AChR (●—●) (Mean + SD, N=8), AChR + EL (○—○) (Mean + SD, N=8). The arrows indicate the half saturation point of binding. Taken from ref (33).

The effect of EL appears to be specific to AChR since EL does not affect the levels of anti-BSA antibodies in rabbits injected with both EL and BSA or BSA alone. Moreover, denatured EL does not modify the levels of anti-AChR antibodies in rabbits injected with both AChR and denatured EL or AChR alone (groups 1 and 4, Table 1). It is worth pointing out that these results indicate that the mechanism of action of EL in the prevention and therapy of EAMG is different from that of alpha-fetoprotein, a glycoprotein with immunosuppressive and therapeutic properties towards EAMG, which was reported to act by inhibiting the binding of anti-AChR antibodies to AChR (36).

Analysis of the cellular immune response after EL administration

Inguinal lymph nodes were harvested from rabbits immunized with AChR (group 1, Table 1) or with AChR and EL (group 2, Table 1). Lymphocytes were cultured *in vitro* and stimulated with AChR at different concentrations. The incorporation of radioactive thymidine into DNA was measured and taken as an index of the proliferation of lymphocytes.

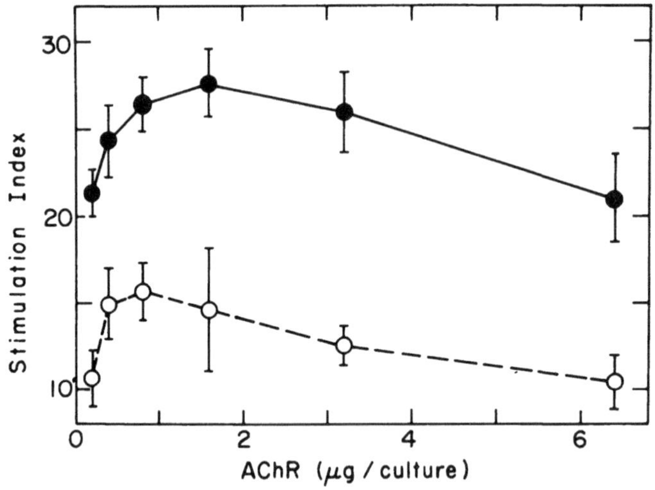

Figure 2: Lymphocyte proliferation. The response of lymphocytes from rabbits immunized with AChR (●) is compared to that of lymphocytes from rabbits immunized with AChR plus EL (O). Stimulation indexes were calculated from the ratio of the radioactivity accumulated in the presence and in the absence of AChR. Taken from ref (33).

It is clear from Fig. 2 that the lymphocytes from animals immunized simultaneously with AChR and EL gave a cellular response to AChR much lower than the lymphocytes harvested from animals immunized with AChR alone. This observation implies that the number of responding cells in the rabbits from group 2 is smaller than in group 1 or that they do not recognize the same determinants on the AChR protein.

These results, as those presented in figure 1, demonstrate that EL regulates the immune response to AChR.

The molecular mechanisms underlying the effect of EL on EAMG

An analysis of the possible mechanisms by which EL exerts its prophylactic and therapeutic effects on EAMG must necessarily take into account the following:

1. EAMG is an autoimmune disease involving the AChR or a major antigen (34,35).
2. Circulating antibodies directed against the AChR are of critical importance in the pathogeny of EAMG (37).
3. Thymus abnormalities are often associated with myasthenia gravis (38).
4. EL does not appear to interact with the AChR at the neuromuscular junction but is capable of modulating the immune response to AChR.

Accordingly, it is reasonable to assume that AChR and EL trigger in the immune system a series of events which at some point converge, overlap and partially antagonize each other. A place of possible convergence is the thymocyte surface. The thymocytes are indeed capable of binding anti-AChR antibodies via an "AChR-like" antigen present on their surface (39,40) and possess also binding sites for electrolectin (41). The thymocytes appear to play a crucial role in EAMG and in myasthenia gravis (MG) since the humoral immune response against AChR is a T-cell dependent process (42) and thymectomy is often followed by major improvements in the myasthenic symptoms (43).

On this basis, one may propose a mechanism of pathogenicity of EAMG in which the appearance of antibodies towards self-AChR is initially triggered by an impairment in the function of the thymus resulting from the binding of anti-foreign AChR antibodies to the "AChR-like" antigen on the thymocytes. If the latter phenomenon produces either an hyperactivity of the T helper cells or an hypoactivity of the T suppressor cells, a large and uncontrolled increase in B cell function and antibody production will take place including the production of anti-self AChR antibodies. Should EL counterbalance the deleterious effects of the anti-foreign AChR antibodies on the thymocyte functions, one might expect EL to prevent the occurrence of EAMG.

<u>Interaction of electrolectins from eel electric organ and mouse thymus with mouse immature thymocytes</u>

To establish the validity of the mechanism of action of EL described in the previous paragraph and investigate the possible natural involvement of electrolectins in the regulation of the immune response, we have studied (41) the interactions of eel electrolectin and mouse thymic electrolectin (MTL) with well defined cell populations derived from the mouse thymus. Eel EL was found to agglutinate selectively a subpopulation of mouse thymocytes. This cell population (EL^+) could be separated in the proportion of 70/30 from non-agglutinating cells (EL^-) by 1g sedimentation over fetal calf serum. Since the thymus is known to contain two classes of lymphatic cells, the immunocompetent medullary thymocytes and the immature cortical thymocytes which differ from each other by their separate cell-mediated reactivities and distinct surface antigens (44,45), we investigated whether the EL^- and EL^+ cells correspond respectively to these cell populations (44). The EL^+ cells were unambiguously identified as immature cortical thymocytes on the basis of the density of θ antigen they bear on their surfaces, their intrinsic mitotic activity and the absence of response to the mitogenic action of phytohemagglutinin. The immature thymocytes (i.e. EL^+ cells) were also found to bind the endogenous MTL, a thymic protein that displays the same saccharide specificity as the eel EL and with which it cross-reacts immunologically (46).

Can the immature thymocytes be the normal cellular targets of MTL? Since the thymic epithilium is known to contain and secrete various hormones involved in the differentiation and maturation of immature T lymphocytes, we investigated whether MTL was also localized in the epithelial cells and whether the latter could secrete it. We found that the specific hemagglutinating activity of MTL markedly increases after depleting the thymus of its thymocytes indicating that MTL is probably localized in the thymic epithelium. Furthermore, we obtained evidence of a secretion of MTL from thymic epithelial reticulocytes in tissue culture (41). These data suggest that MTL could behave as a thymic hormone and act on immature thymocytes.

Interactions of electrolectin and anti-AChR antibodies with mouse thymocytes

As stated above, the thymocytes possess at their surface an "AChR-like" antigen that bind anti-AChR antibodies. Since about 90% of the thymocyte population are immature lymphocytes (44) and 80% of the thymocytes bind anti-AChR antibodies (39), one can possibly deduce that the latter interact preferentially with immature lymphocytes i.e. with the same cell population that binds EL. We investigated therefore the question whether EL and the anti-AChR antibodies do interact with each other at the surface of the immature lymphocytes. Figure 3 shows the results of experiments in which thymocytes were incubated with rabbit anti-AChR antibodies after being exposed or not to EL. The anti-AChR antibodies bound to the thymocytes were then reacted with radiolabeled goat anti-rabbit IgG's. It can be seen that the exposure to EL, markedly reduced the ability of thymocytes to bind anti-AChR antibodies. Whether this EL effect is due to an overlap between the binding sites of EL and anti-AChR antibodies on the surface of the thymocytes or whether EL induces, for instance, the capping and internalization of the "AChR-like" antigen remain now to be investigated.

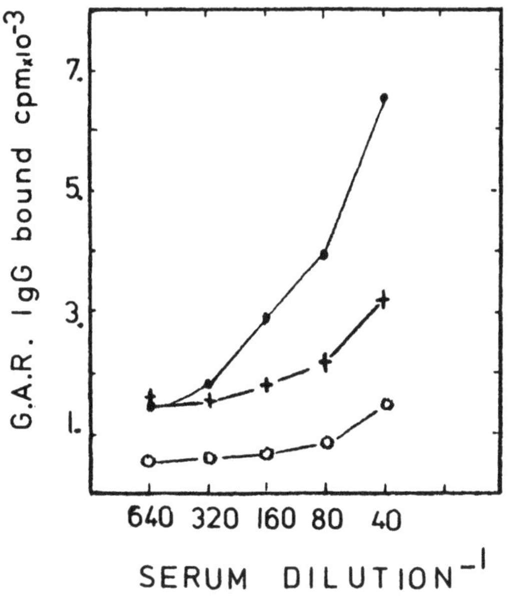

Figure 3: Inhibition of anti-AChR antibody binding to thymocytes by EL. Thymocytes (100000 cells) were cultured for 12 hours on polylysine coated plates and then exposed for 2 hours to eel EL (4 ug/ml in a phosphate buffered saline containing 1 mg/ml bovine serum albumin (PBS-BSA)). The cells were washed 3 times with PBS-BSA and exposed to various dilutions of sera raised in rabbits against AChR or DNP-BSA (open circle). After washing, the bound antibodies were reacted with radiolabeled goat anti-rabbit IgG; (+): anti-AChR antibodies bound after EL exposure; (•): no EL exposure;

These results not only demonstrate that the immature thymocyte cell surface is a common target for EL and anti-AChR antibodies but also strengthen the hypothesis that the action of EL on EAMG may be due to its ability to counterbalance the deleterious effects of anti-foreign AChR antibodies on thymus function.

Two important questions remain however unanswered. The first concerns the demonstration of the role of the "AChR-like" antigen in the pathogenesis of EAMG. The second concerns the elucidation of the natural function of the thymic electrolectin in the control of the immune response. Clarification of these questions may shed light on a new thymic hormone and provide new prophylactic and therapeutic approaches to myasthenia gravis. Independently of the above issues, the fact of the presence of electrolectin in non-immune tissues and its almost exclusive expression in embryonic or tumor tissues cannot be set aside and requires some consideration. One may propose, in view of the modulatory effects produced by EL on the immune response to AChR, that electrolectins would be part of a general mechanism of immunity very much like that involving interferons. Tissue injuries, viral infections or transformation would induce the local production of electrolectins which, once released, would trigger and modulate the immune response.

Acknowledgements: This work was supported by grants from the Muscular Dystrophy Association of America and the Minerva Foundation. V.I.T. holds the Louis and Florence Katz-Cohen professorial chair in neuropharmacology.

References

1. Teichberg, V.I., I. Silman, D.D. Beitsch, and G. Resheff. 1975. Proc. Natl. Acad. Sci. 72, 1383-1387.

2. Teichberg, V.I. 1978. Methods in Enzymology Vol. L. part C (Victor Ginsburg, ed.) pp. 291-302.

3. Levi, G. and V.I. Teichberg. 1981. J. Biol. Chem. 256, 5735-5740.

4. Nowak, T.P., P.L. Haywood and S.H. Barondes. 1976. J. Biol. Chem. 252, 6026-6030.

5. Den, H. and D. Malinzak. 1977. J. Biol. Chem. 252, 5444-5448.

6. Pitts, M.J. and D.H. Yang. 1981. Biochem. J. 195, 435-439.

7. Barak-Briles, E., W. Gregory, P. Fletcher and S. Kornfeld. 1979. J. Cell Biol. 81, 528-537.

8. Beyer, E.C., S.E. Zweig, and S.H. Barondes. 1980. J. Biol. Chem. 255, 4236-4239.

9. de Waard, A., S. Hickman, and S. Kornfeld. 1976. J. Biol. Chem. 251, 7581-7587.

10. Powell, J.T. 1980. Biochem. J. 187, 123-129.

11. Childs, R.A. and T. Feizi. 1979. Biochem. J. 183, 755-758.

12. Kobiler, D., E.C. Beyer, and S.H. Barondes. 1978. Develop. Biol. 64, 265-272.

13. Beyer, E.C., K.T. Tokuyasu, and S.H. Barondes. 1979. J. Cell Biol. 82, 565-571.

14. Simpson, D.L., D.R. Thorne and H.H. Loh. 1977. Nature 266, 367-369.

15. Gremo, F., D. Kobiler, and S.H. Barondes. 1978. J. Cell Biol. 79, 491-499.

16. Eisenbarth, G.S., F.R. Ruffols, F.S. Walsh and M. Nirenberg. 1978. Biochem. Biophys. Res. Commun. 83, 1246-1252.

17. Kitamura, K. 1980. J. Embryol. exp. Morph. 59, 59-69.

18. Kitamura, K. 1981. J. Embryol. exp. Morph. 65, 41-56.

19. Powell, J.T. and P.L. Whitney. 1980. Biochem. J. 188, 1-8.

20. Raz, A. and R. Lotan. 1981. Cancer Res. 41, 3642-3647.

21. Nowak, T.P., P.L. Haywood, and S.H. Barondes. 1976. Biochem. Biophys. Res. Commun. 68, 650-657.

22. Gartner, T.K. and T.R. Podleski. 1975. Biochem. Biophys. Res. Commun. 67, 972-978.

23. MacBride, R.G. and R.J. Przybylski. 1980. J. Cell Biol. 85, 617-625.

24. Podleski, T.R. and I. Greenberg. 1980. Proc. Natl. Acad. Sci. USA 77, 1054-1058.

25. Kaufman, S.J. and M.L. Lawless. 1980. Differentiation 16, 41-48.

26. Den, H., D.A. Malinzak and A. Rosenberg. 1976. Biochem. Biophys. Res. Commun. 69, 621-627.

27. Childs, R.A. and T. Feizi. 1979. FEBS Lett. 99, 175-179.

28. Harrison, F.L. and C.J. Chesterton. 1980. Nature 286, 502-504.

29. Matsutani, E. and T. Yamagata. 1982. Develop. Biol. 92, 544-548.

30. Zalik, S.E., N. Milos, and I. Ledsham. 1982. Cell tissue Res. 225, 223-228.

31. Kobiler, D. and S.H. Barondes 1977. Devel. Biol. 60, 326-330.

32. Betz, H., J.P. Bourgeois and J.P. Changeux. 1977. FEBS Lett. 77, 219-224.

33. Levi, G., R. Tarrab-Hazdai and V.I. Teichberg. 1983. Eur. J. Immunol. 13, 500-507.

34. Patrick, J. and J. Lindstrom. 1973. Science 180, 871-874.

35. Fuchs, S. 1980. Trends Biochem. Sci. 5, 259-261.

36. Brenner, T. and O. Abramsky. 1981. Immunol. Lett. 3, 163-167.

37. Toyka, K.V., D.B. Drachman, D.E. Griffin, H. Pestonk, J.A. Winkelstein, K.H. Fishbeck and I. Kao. 1977. N. Engl. J. Med. 296, 125-130.

38. Castleman, B. and E.M. Norris. 1949. Medicine (Baltimore) 28, 27-58.

39. Fuchs, S., I. Schmidt-Hopfeld, G. Tridente and R. Tarrab-Hazdai. 1980. Nature 287, 162-164.

40. Engel, W.K., J.L. Trotter, D.E. McFarlin and C.L. McIntosh. 1977. Lancet. 1, 1310-1311.

41. Levi, G. and V.I. Teichberg. 1983. Immunol. Lett. 7, 35-39.

42. Lennon, V.A., J.M. Lindstrom and M.E. Seybold. 1976. Ann. N.Y. Acad. Sci. 274, 283-296.

43. Buckingham, J.M., F.M. Howard, P.E. Bernatz, W.S. Payne, E.G. Harrison, P.C. O'Brien and L.E. Weiland. 1976. Ann. Surg. 184, 453-457.

44. Reisner, Y., M. Linker-Israeli and N. Sharon. 1976. Cell. Immunol. 25, 129-134.

45. Madyastha, K.R. and M.A. Bucker and R.N. Taub. 1980. Transplantation 29, 252-254.

46. Levi, G. and V.I. Teichberg. 1982. FEBS Lett. 148, 145-148.

Regulation of Acetylcholine Receptors on Muscle *in vitro* by Central Nervous System Factors

T.R. Podleski, I Greenberg, D. Knaack, K. Neugebauer and M.M. Salpeter

Section of Neurobiology and Behavior, Seeley G. Mudd Hall, Cornell University, Ithaca, New York 14853 U.S.A.

Introduction

Throughout Professor Nachmansohn's scientific career he was fascinated by the organization of molecules into functional biological structures. Of course, the focus of his attention was directed towards the origin of bioelectric potentials and the organization of the relevant molecules into excitable membranes. It is an honor for us to describe briefly our work on some of the factors which regulate the distribution of acetylcholine receptors in embryonic muscle in this symposium dedicated to Professor Nachmansohn's memory.

In embryonic skeletal muscle, acetylcholine receptors (AchRs) are distributed over the surface membrane (1). *In vitro* studies have shown that during the formation of the neuromuscular synapse, AchRs move in the surface of the muscle membrane to form a localized high density cluster at the site of the synapse (2). This cluster of receptors is presumably the precursor for the organization of receptors seen in the adult neuromuscular junction where high receptor site densities are characteristically found (3). The initial development of a high density AchR cluster at regions of nerve-muscle contacts is dependent on the proximity of the nerve to the muscle, although once established such high density clusters remain in the adult even after denervation (4,5).

In addition to the clusters which form at synapses, some clusters are present in the absence of neurons (6,7,8), and

Molecular Basis of Nerve Activity

others form in response to neurons but are not necessarily located at the site of neuron-muscle contact (9,10).

Some muscle cells also show a large, 3-5 fold, increase in the average AchR site density as measured by the increase in ^{125}I α-bungarotoxin binding (IBTX) over the entire muscle under *in vitro* coculture conditions identical to those which produce an increase in receptor clusters. This increase in average receptor site density has been found for the cloned L6 cell line (9), and for chick primary muscle (10). Again, as is true for the formation of receptor clusters, the increase in average receptor site density is not limited to cells where there is contact between neurons and muscle.

These observations suggested to us that a substance, or substances, released into the medium might be responsible for both the formation of receptor clusters and the increase in average site density over entire muscle. Support for this suggestion was found when soluble components (the supernate from a 100,000xg centrifugation) prepared in physiological salt solutions from fetal rat spinal cord or brain, chick brain or neuroblastoma-glioma hybrids caused an increase in the number of receptor clusters in L6, L5, and rat or chick primary muscle cells (9,11,12), as well as an increase in total AchRs per myotube area in L6, L5 or chick muscle (9,12,13,14).

I. Separation and characterization of two factors from fetal or newborn brain

A. Factors which increase AchR number. We have found that when newborn or fetal rat brain extracts are separated on Sephadex G-200, the fractions causing the increase in AchR site density on L5 cells elute in two regions of the chromatogram (13). One, which increases the number of sites by 20-50%, is located in the void volume and has an apparent molecular weight of >140,000 daltons. The second, which increases the overall site density 2-4 fold, is found in the region of about 700 daltons. When assayed on rat primary muscle cells under identical conditions this active low molecular weight fraction increased the site density no more than 20-30% (13). Therefore, there is a marked difference in the sensitivity of the rat primary and L5 muscle systems to this factor.

Further purification of a similarly prepared low molecular weight fraction from fetal calf brain in the presence of reducing agents has led to the isolation of a component which also increases overall AchR site density about 3 fold (16). A variety of chemical tests have identified this component as ascorbic acid. Figure 1 shows that the dose-response curves of the purified component from fetal calf brain and commercial ascorbic acid are identical.

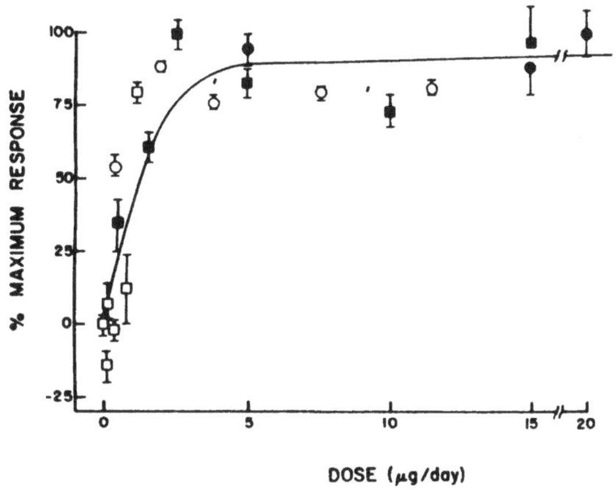

Figure 1. Dose-Response curves of IBTX binding versus ug/day of ascorbic acid or purified fraction from fetal calf brain. This figure represents the composite data from two experiments: ascorbic acid -●-, -■- , and purified fraction -○-, -□-. These data were normalized to the maximum response obtained in each experiment (16).

Additional comparisons of the biological effects of ascorbic acid and the pH 4 crude homogenate of fetal calf brain are shown in Table I (14). Both are mitogenic to myoblasts of the L5 cells, but this effect is not directly responsible for the effects on AchR number, since the mitogenic effect can be blocked by cytosine-arabinofuranoside without inhibiting the effects of either ascorbic acid or crude brain factor on AchR site density. Neither ascorbic acid nor crude brain factors have an effect on a cytoplasmic enzyme, creatinine phosphokinase (CPK), which is regulated in a similar developmental pattern as AchRs. Similarly, there is only a 20-30% increase in total protein which can be largely accounted for by the increase in the number of mononucleated cells which are generated by the mitogenic effect of ascorbic acid. We conclude that not only is ascorbic acid responsible for causing a specific increase in AchR site density, but that this effect is identical to an activity in the fetal calf brain.

Table I. pH 4.0 Fetal Calf Brain Extract and Ascorbic Acid[1]: A Comparison of Their Effects on L_5 Muscle Cells (14)

Parameter Measured	Ascorbic Acid/Control (± S.E.M.)	Crude Brain Extract/Control (± S.E.M.)
Total protein[2]	1.25±0.04 (n = 4)	1.39±0.04 (n = 5)
CPK activity	1.18±0.10 (n = 4)	1.10±0.10 (n = 5)
^{3}H Thymidine incorporation Day 6	2.66±0.45 (n = 4)	3.10±0.50 (n = 4)
^{125}I α-bungarotoxin binding	3.50±0.45[3] = 5)	4.10±0.80 (n = 5)

[1]The values given are the mean values for each specific assay after treatment with extract or ascorbic acid, divided by the value for untreated controls. The cells were dosed on days 4 and 5, and assayed on day 7 unless otherwise specified.

[2]The level of total protein in ascorbic acid treated cultures was significantly lower ($P \geq .05$) than that in extract treated cultures by "t" test.

[3]If cytosine-1-β-D-arabinofuranoside (Ara-C) was added 24 hours before ascorbic acid, the IBTX binding was no different from this value, but the ^{3}H thymidine incorporation was reduced 10 fold and was then equal in control and ascorbic acid treated cells.

B. Factors which stimulate the formation of AchR clusters. When the two fractions identified above by Sephadex G-200 chromatography were studied for their effects on the formation of AchR clusters on L5 cells and rat primary muscle, another striking difference between the two fractions was observed. An increase in the number of clusters was found with both fractions, Figure 2, but in this assay the high molecular weight fraction was more active than was the low. It caused more clusters to be formed, and the clusters formed had a higher ratio of cluster to intercluster site density than those formed in the presence of the low molecular weight factor, Figure 3. In addition, on rat primary muscle, the site densities obtained in the clusters formed in response to the crude brain extract and high molecular weight factor often exceeded 10,000 sites/um^2 (15,19).

The high molecular weight components which cause the formation of these increased AchR site densities within clusters are probably proteins, and we have partially purified and characterized several active fractions based on their properties on molecular exclusion and ion exchange chromatography. On Sephadex G200 the major activities are in fractions >200,000 and between 100,000 to 135,000 daltons. These fractions have very little effect on the overall receptor site density (unpublished).

II. Mechanism of the increase in overall AchR number

Two general classes of mechanisms can account for an increase in AchR number per myotube unit area: (1) a decrease in the rate of degradation or (2) an increase in the rate of AchR incorporation into the surface membrane. Treatment of L5 cells with crude newborn rat brain extract increases the rate of incorporation of AchRs into the muscle surface membrane 2.6 fold, and there is very little, if any effect on the rate of AchR degradation (13). Similar observations have been made with ascorbic acid (14). It is clear, therefore, that both

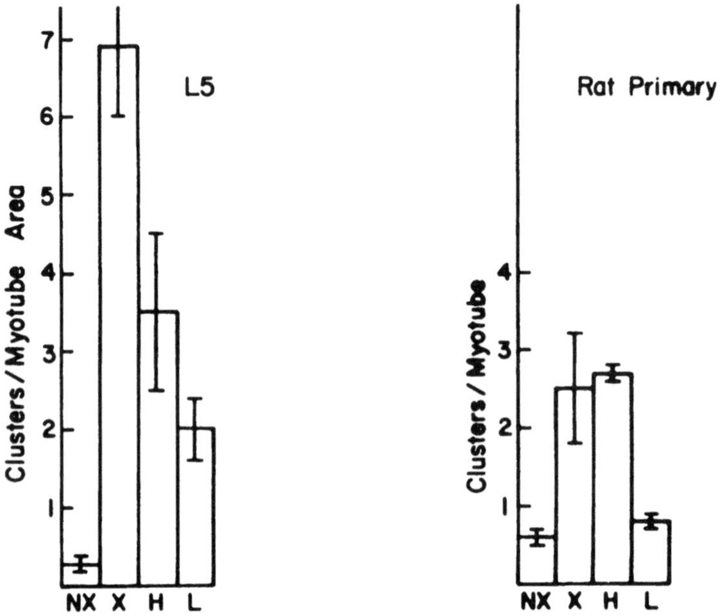

Figure 2. Number of AchR clusters obtained on L5 cells and rat primary cells after treatment with crude newborn rat brain factors (X), or high (H) or low (L) molecular weight fractions normalized to control (NX). The number of clusters was determined by overexposed light autoradiographs (11).

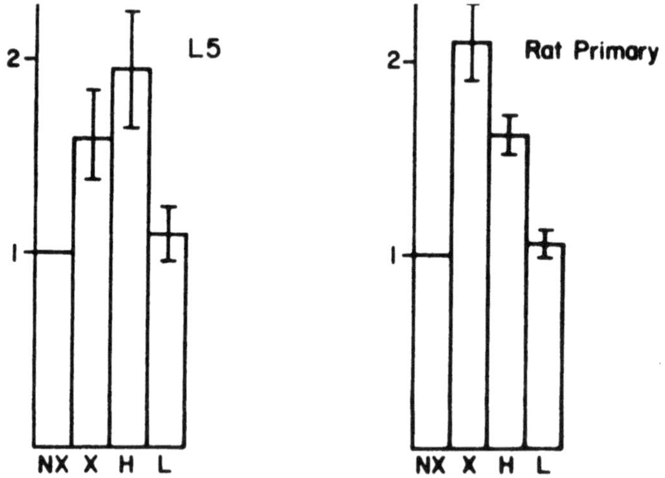

Figure 3. Peak AchR clusters in L_5 and primary cells: The cluster with the highest site density on each myotube was identified, and a ratio of cluster to intercluster site density (13) plotted relative to control (NX). Crude extract (X); high molecular weight fraction (H) and low molecular weight fraction (L).

the brain extract and ascorbic acid have a highly specific effect on the rate at which AchRs are incorporated into the surface membrane of the L5 cells. Chick brain extract also increases the average AchR number per myotube area on chick muscle without altering the rate of receptor degradation (12).

III. Mechanisms of cluster formation

Three possible mechanisms for the formation of AchR clusters are (1) selective decrease in the rate of degradation of receptors in the clusters, (2) selective insertion of new receptors into the cluster, and (3) redistribution of receptors in the surface membrane to form the cluster. We have shown by quantitative autoradiography that the receptors in clusters are degraded at the same rate as receptors not in clusters (19). No evidence has yet been presented in favor of selective insertion, but we and others have reported that prelabeled AchRs redistribute in rat primary muscle to form clusters after the addition of brain factors (19,20). We have now made similar observations on L5 cells (unpublished). The time necessary to detect the redistribution depends upon the technique used, but it is clear that it is not longer than a few hours (19), and it may be as short as 0.5 hours (20). Therefore, the clustering factors mimic the presence of the neuron in causing the receptors to redistribute in the surface membrane to form clusters.

By transmission electron microscopic autoradiography we have studied the ultrastructural features of receptor clusters formed in the presence of brain factors (15,19). These studies have not revealed an absolute correlation between an ultrastructural feature and the presence of receptor clusters. However, with or without brain factors, AchR clusters have more associated basal lamina than do non-clustered regions. Furthermore, there is, an increase in the amount of basal lamina on the top surface of myotubes in the presence of brain factors. This is due mainly to an increase in the number of clusters and an increase in the amount of basal lamina over each cluster region. Another general feature of clusters is an increase in the electron density of the plasma membrane and a cytoplasmic "fuzziness" which is not as intense as at adult synapses (15,19). The latter suggests that a cytoskeletal component may play a role in the formation or stabilization of the clusters in addition to the extracellular matrix.

IV. Stabilization of AchRs to Triton X100 extraction after treatment with brain factors

The cluster-inducing factors cause the stabilization of receptors to extraction in Triton X-100, as expected if the receptors become anchored to a cellular element such as the cytoskeleton or extracellular matrix (7,15,19,21,22,23,24,25, 26,27). Prives et al (24) found that the extraction curves of AchR in Triton X-100 were composed of two exponentials; they also provided evidence for the slowest form being AchR's from clusters. We have also found that the extraction curves on rat primary muscle cultures are composed of two exponentials: one with a half-time of about 5 min and a second of 40 min for the untreated cells, see Figure 4 (unpublished). Within 24 hours after adding the brain factors the curves were the same shape but the half-times were 7 min and 40 min. These results show that the extraction rate the rapidly extracted receptors is slowed by almost a factor of two within 24 hours after the addition of brain factors. Autoradiographic data show that AchRs in clusters are extracted more slowly than AchRs in intercluster regions, and that not all slowly extracting AchRs are initially in clusters (unpublished).

We have measured the rate at which AchRs become stabilized to Triton extraction. The AchRs were saturated with IBTX and crude brain extract was then added. At various times after labeling, the cells were extracted with Triton and the percentage of solubilized receptors was compared to the

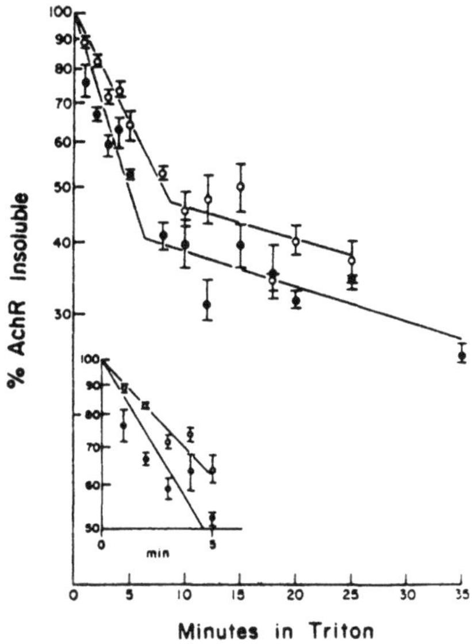

Figure 4. Extraction rates of AchRs in control (-0-) and brain factor treated (-●-) rat primary muscle *in vitro*. Nonspecific IBTX binding was determined by pretreating cells with 10^{-7}M non-iodinated BTX, washing the cells, and then adding $2x10^{-8}$M IBTX. The nonspecific binding was subtracted from the total to determine the specific binding. Insert lower left corner is with an expanded time scale. 0.25% Triton X-100 was used to extract the cells at the indicated times. The IBTX bound to cells after extraction was determined by solubilizing cells in 2N NaOH.

percentage obtained on those cells which were extracted immediately after being labeled. AchRs in both the control and brain extract treated cultures became resistant to Triton extraction with time, and the rate of stabilization is higher in brain extract treated cells than controls, see Figure 5. These results show that the stabilization of AchRs takes place after the receptors are in the surface membrane. Using this procedure we have been able to detect differences between treated and untreated cells within 3-6 hours after labeling, which is similar to measurements we made with autoradiography for the movement of AchRs into clusters. Therefore, AchRs are becoming stabilized to Triton extraction at a rate which is comparable to that found for the formation of clusters. A role for the cytoskeleton, protein synthesis and cellular metabolism in this stabilization process is implicated, since the process is inhibited by cytochalasin A or B at a low dose and for short times by cycloheximide and sodium azide (unpublished, see also ref. 25).

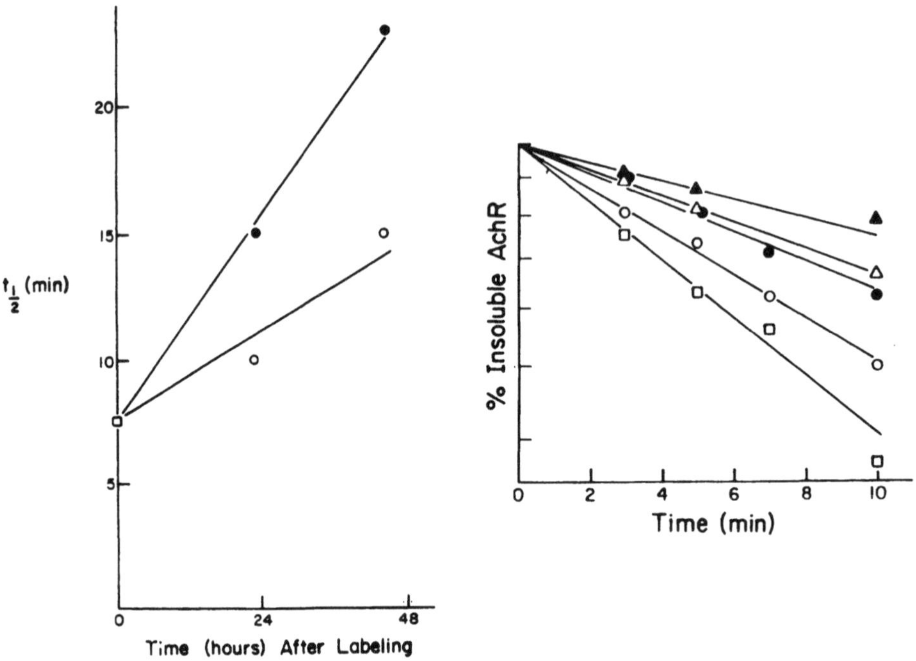

Figure 5. Newborn rat brain factors accelerate the stabilization of AchRs to Triton X-100 extraction. Extraction rates were determined by saturating the AchRs with IBTX. Specific IBTX binding was determined as described in Figure 4. Left panel: One set of cells was extracted immediately after labeling and the average half-time of the extraction during 10 minutes was determined (-□-). the other sets of cells were either treated with brain factors (-●-) or phosphate buffered saline which was the buffer used in the preparation of the factors (○). these cells were assayed 24 and 48 hours later; the extraction rates were determined precisely as before. Right panel: Extraction rates used in left panel were derived from this figure. (□) at 0 time (○,△) control at 24 and 48 hours after labeling with IBTX, (●,▲) factor treated cells 24 and 48 hours after labeling. Standard errors were 1-5% and the extraction rates were significantly different at $P \geq .01$.

Conclusions

It has been shown that AchRs redistribute in the surface membrane of skeletal muscle to form a cluster at the synapse between neurons and skeletal muscle (2). This event is considered to be a critical step in the formation of the synapse, since AchRs are highly localized at this synapse in the adult. Clusters of receptors also develop in response to soluble central nervous system components (9,11,17,18) and to purified components of basal lamina (26,27,32). These latter results suggest that the extracellular matrix plays an important role in regulating the distribution of AchRs. It should be noted that the factors from the central nervous system which cause AchR clusters to form are soluble in physiological salt solutions whereas the basal lamina components are not. This result suggests that the components isolated from the basal lamina are probably different from those isolated from brain. We have also reported an increase in basal lamina, as defined morphologically, in response to the brain factors, and it is possible that brain factors influence cluster formation by way of basal lamina components (15,19). More work, however, is necessary on this point.

It also has been shown that clusters form at contact points between inert physical structures and the muscle membrane, such as at the junction between the cell and the tissue-culture dish (7,8,23), and with polystyrene beads (29). Our work has emphasized a quantitative difference between clusters. Clusters which form in response to crude or partially purified high molecular weight components of the rat brain have a higher site density relative to that in the intercluster region than do those which form in contact with the tissue-culture dish or in response to low molecular weight factors (13,15, 19). The highest site densities within clusters, achieved with the high molecular weight factors on rat primary cells, approach those found at the adult neuromuscular junction.

A simple model has been proposed for the formation of AchR clusters in other studies (33). This model has been called the passive diffusion trap model. When applied to the present results this model suggests that components in the brain preparation induce the formation of nucleation sites for the trapping of AchRs. AchRs moving in the plane of the surface membrane randomly encounter these nucleation sites and bind to them. Once mobile AchRs are bound to the nucleation site, they are trapped and unable to move further.

Our results suggest that the stabilization of AchR to Triton extraction by brain factors requires an interaction with the cytoskeleton (see also refs. 25 and 34) as well as cellular metabolism and protein synthesis. In the framework of the passive-diffusion trap model the cytoskeleton and metabolism may be involved initially in forming and maintaining the trapping sites. Earlier studies have also implicated the cytoskeleton (22,23,24), and cellular metabolism (22,23) in maintaining spontaneously formed clusters. Since we have shown that brain factors promote and enhance stabilization of AchR against Triton extraction such a neurally controlled mechanism may be involved in stabilizing and concentrating AchR trapping sites rendering them progressively more resistant to dispersal in response to other agents as well (35).

While the potential role for clustering factors in the formation of synapses is clear, the role of a factor which stimulates the incorporation of receptors into the muscle membrane is less clear. In our work this category of components is represented by ascorbic acid (14,16). There is also evidence that ascorbic acid may play a role in regulating the amount of AchR in skeletal muscle *in situ*. Following denervation, when the synthesis of AchR is known to be increased the level of ascorbic acid increases (30, Knaack and Levitt-Gilmour, unpublished). While these results are indirect, a causal relationship *may* exist.

There may be other components that stimulate AchR insertion, e.g., those factors identified by the Fischbach group (17,18). The fact that their factors are retained on C-18 reversed-phase chromatography in the absence of ion pairing agents, suggests that these factors are not ascorbic acid (16). Whatever the physiological importance of either ascorbic acid or other stimulatory factors may be in regulating the number of AchRs, it will be interesting to determine the mechanism of their effects.

Kalcheim et al (28) have shown that ascorbic acid increases the level of collagen in primary muscle cultures. This effect is presumably related to the well-known effect of ascorbic acid on collagen synthesis in other systems. Evidence has been presented for the effect of ascorbic acid being on the transcription and stabilization of m-RNA for procollagen (31). It would be of interest to determine whether the effect on AchR synthesis is also by way of m-RNA. These results, plus our report on the specificity of the action of ascorbic acid lead us to believe that future work on the role of ascorbic acid in regulating muscle AchR is likely to be significant for our understanding of muscle development.

Acknowledgements: We wish to thank Judy Rosenthal and Cynthia Steiner for technical assistance. Supported by National Institutes of Health grant NS14679 to T.R. Podleski, and GM10422 to M.M. Salpeter.

References

1. Fambrough, D.M. 1979. Ann. Rev. Physiol. 59, 165.

2. Anderson, M.J. and Cohen, M.W. 1977. J. Physiol. (London) 268, 757.

3. Fertuck, H.C. and Salpeter, M.M. 1976. J. Cell Biol. 69, 144.

4. Porter, C.W. and Barnard, E.A. 1975. Exp. Neurol. 48, 542.

5. Loring, R.H. and Salpeter, M.M. 1980. Proc. Natl. Acad. Sci. USA 77, 2293.

6. Nirenberg, S.V., Vogel, Z., Sytkowski, A.J. and Nirenberg, M.W. 1972. Proc. Nat. Acad. Sci. USA 69, 3180.

7. Axelrod, D., Ravdin, P., Koppel, D.E., Schlessinger, J., Webb, W.W., Elson, E.L. and Podleski, T.R. 1976. Proc. Nat. Acad. Sci. USA 73, 4594.

8. Land, B.R., Podleski, T.R., Salpeter, E.E. and Salpeter, M.M. 1977. J. Physiol. (London) 269, 155.

9. Podleski, T.R., Axelrod, D., Ravdin, P., Greenberg, I., Johnson, M.M. and Salpeter, M.M. 1978. Proc. Natl. Acad. Sci. USA 75, 2035.

10. Fischbach, G.D. and Cohen, S.A. 1973. Dev. Biol. 31, 147.

11. Christian, C.N., Daniels, M.P., Sugiyama, H., Vogel, Z., Jacques, L. and Nelson, P.G. 1978. Proc. Natl. Acad. Sci. USA 75, 4011.

12. Jessel, T.M., Siegel, R.E. and Fischbach, G.D. 1979. Proc. Natl. Acad. Sci. USA 76, 5397.

13. Neugebauer, K., Salpeter, M.M. and Podleski, T.R. Brain Res. (submitted).

14. Knaack, D., Shen, I., Salpeter, M.M. and Podleski, T.R. Devel. Biol. (submitted).

15. Salpeter, M.M. and Podleski, T.R. 1982. In: Muscle Development: Molecular and Cellular Control (M.L. Pearson and H.F. Epstein, eds.). Cold Spring Harbor Laboratory.

16. Knaack, D. and Podleski, T.R. Proc. Natl. Acad. Sci. USA (in press).

17. Connolly, J.A., St. John, P.A. and Fischbach, G.D. 1982. J. Neurosci. 2, 1207.

18. Buc-Caron, M.-H., Nystrom, P. and Fischbach, G.D. 1983. Dev. Biol. 95, 378.

19. Salpeter, M.M., Spanton, S., Holley, K. and Podleski, T.R. 1982. J. Cell Biol. 93, 417.

20. Olek, A.J., Pudimat, P.A. and Daniels, M.P. 1983. Cell 34, 255.

21. Axelrod, D., Ravdin, P. and Podleski, T.R. 1978. Biochim. Biophys. Acta 511, 23.

22. Bloch, R.J. 1979. J. Cell Biol. 82, 626.

23. Bloch, R.J. and Geiger, B. 1980. Cell 21, 25.

24. Prives, J., Fulton, A.B., Penman, S., Daniels, M.P. and Christian, C.N. 1982. J. Cell Biol. 92, 231.

25. Connolly, J.A. 1984. J. Cell Biol. 99, 148.

26. Vogel, Z., Christian, C.N., Vigny, M. Bauer, H.C., Sonderegger, P. and Daniels, M.P. 1983. J. Neurosci. 3, 1058.

27. Wallace, B.G., Godfrey, E.W., Nitkin, R.M., Rubin, L.L. and McMahan, U.J. 1982. In: Muscle Development: Molecular and Cellular Control (M.L. Pearson and H.F. Epstein, eds.). Cold Spring Harbor Laboratory.

28. Kalcheim, C., Vogel, Z. and Duksin, D. 1982. Proc. Natl. Acad. Sci. USA 79, 3077.

29. Peng, H.B., Cheng, P.-C. and Luther, P.W. 1981. Nature (Lond.) 292, 831.

30. Graff, G.L.A., Hudson, A.J. and Strickland, K.P. 1965. Canadian J. Biochem. 43, 705.

31. Lyons, B.L. and Schwarz, R.I. 1984. Nuc. Acids Res. 12, 2571.

32. Daniels, M.P., Vigny, M., Sonderegger, P., Bauer, H.-C. and Vogel, Z. 1984. Int. J. Devl. Neuroscience 2, 87.

33. Poo, M. 1982. Nature 295, 332.

34. Axelrod, R., Bauer, H.C., Styra, M. and Christian, C.N. 1981. J. Cell Biol. 88, 459.

35. Block, R.J. and Steinbach, J.H. 1981. Devel. Biol. 81, 386-391.

USE AND DISUSE AND THE CONTROL OF ACETYLCHOLINESTERASE ACTIVITY IN FAST AND SLOW TWITCH MUSCLE OF RAT.

W-D. Dettbarn, D. Groswald, R.C. Gupta and K.E. Misulis

Vanderbilt University, Department of Pharmacology, and The Jerry Lewis Neuromuscular Disease Research Center, Nashville, TN 37212 USA

Although the role of acetylcholinesterase (AChE) in neuromuscular transmission is relatively well established, little is known of the mechanisms that regulate its synthesis and control its specific distribution in fast and slow muscle. Innervation plays an important role in the regulation of AChE and elimination of the influence of the nerve by surgical denervation results in a loss of AChE (1). The question of how these influences of the nerve are mediated has been a matter of continuing controversy. Muscle usage as well as other factors such as materials carried by axonal transport may participate in the regulation of this enzyme.

Nerve transection has two effects: by preventing nerve impulses from reaching the nerve terminal it produces disuse, and by interrupting axonal transport it eliminates the release of trophic materials from nerve to muscle. These two effects of denervation--disuse and loss of neurotrophic factors--are not necessarily independent.

Changes in impulse traffic along the peripheral nerve may

Molecular Basis of Nerve Activity

alter the rate, type or amount of materials that are axonally released. In addition, differences may be found in the muscles themselves which may control their characteristics by substances whose availability may depend on the level of contractile and metabolic activity. It is clear from recent reports that muscular activity, either neurally invoked (2,3,4) or generated in culture via electrical stimulation (3,4), plays a major role in the control of AChE. Studies using the interventions of limb immobilization (5,6), models of neuromuscular overload such as compensatory hypertrophy (6), and endurance training attest to the mutability of this component of neuromuscular function in the face of chronically altered neuromuscular demands. Previous reports on the effects of reduced activity by limb immobilization or tetrodotoxin (TTX) induced block of axonal conduction (7) have not effectively differentiated the response of molecular forms of AChE (8,9,10). An attempt to determine disuse on these distinct enzyme forms is essential, especially in light of evidence that a polymorphism of AChE exists which relates to function. In addition, the reports that these forms are differentially affected by nerve stump length after denervation (11,12) suggest variation among these AChE forms in their control by neuromuscular activity.

More recent observations of the effects of disuse in the regulation of the different forms of AChE do not agree with each other. When cordotomy (13) was used AChE of the non-endplate region was reduced while the enzyme activity of the endplate (16S) was unchanged. With limb immobilization (14) the opposite results were found such as a decrease in endplate AChE and increase in non-endplate AChE. These investigations differ, however, in regard to disuse technique, as well as to the muscle used.

Disuse as previously produced; by tenotomy (15), denervation

(16,17,18), spinal cord section, and/or dorsal root section involves either drastic changes in resting muscle tension, interruption of the reflex arc, transection of suprasequential inputs to the arc or manipulations that might cause damage to nerve and muscle. Results may be due, in large measure, to modification of intercellular relations which are unrelated to nerve impulses, or to cell injury of either muscle or nerve (19,20,21).

Disuse produced by injecting drugs such as TTX (7) beneath the perineurium of small nerves might mechanically damage an unspecifiable number of fibers. It seemed important, therefore, to use a new model for producing disuse, a procedure that does not involve manipulations in the vicinity of the lower motor neuron and nerve, such as the hindlimb suspension model of hypokinesia.

Although the rate of loss of (AChE) activity that follows denervation is similar in rat fast extensor digitorum longus (EDL) and slow soleus (SOL) muscles, the return of AChE activity during reinnervation showed a distinct difference between these two muscles (1). This observation raised interesting questions regarding differences in mechanisms that regulate the patterns of individual AChE forms in fast and slow mammalian skeletal muscle.

We report here on studies undertaken in our laboratory in an attempt to understand the mechanisms that regulate AChE and its molecular forms in two functionally different muscles.

First, we will describe the development of the AChE molecular forms during differentiation in fast and slow muscle. A developmental study of AChE and its molecular forms in EDL and SOL muscles may be helpful in explaining the distinct difference in activity and control of AChE in EDL and SOL and elucidate the mechanisms that control the physiological and biochemical properties of muscle. Although information is available on both morphological and biochemical changes and AChE activity during prenatal development, there have been no direct comparisons in the same series of animals between molecular forms of AChE during the postnatal development of EDL and SOL.

Second, experiments are presented which investigate the effect of denervation and reinnervation on AChE activity in SOL and EDL. This approach was chosen when previous data indicated that these muscles differed significantly in their AChE activity, their twitch speed and their histochemistry. Furthermore, recovery of AChE activity during reinnervation differed between these muscles. It was thought that the changes induced by loss and re-establishment of functional connection between nerve and muscle are a suitable model for studying the mechanisms that control AChE. A difference between these muscles in the rate of loss or recovery of this enzyme may help in the elucidation of the regulating mechanisms.

Third, we have investigated whether functional activity has a role in the control of AChE and its molecular forms. We will present experiments in which hindlimb suspension was used for reducing muscle activity. The differences found in AChE activity in SOL and EDL may have its origin in the different stimulation patterns these muscles normally receive. By removing the work load from both of these muscles the differences between them should be reduced and we can determine the regulatory effect of muscle work on AChE activity.

1. Developmental Alterations in the Molecular Forms of Acetylcholinesterase in Fast and Slow Twitch Muscle of Rat.

Termed pregnant Sprague-Dawley rats weighing 180-250 g were used for this study. Litter size was controlled at eight pups per mother. The pups were killed at 1 day, 1, 2, 3, 4, 5, 6 and 14 weeks after birth by decapitation and exsanguinated. Pups were sampled from different litters in order to minimize differences between mothers. The SOL and EDL muscles were removed, cleaned of the connective tissue, weighed and stored on ice. For very young rats (less than 2 weeks) a dissecting microscope was used during the isolation of muscles.

AChE activity as calculated per gram muscle or mg protein are shown in Fig. 1. At one day after birth, both muscles show similar AChE activity, and between the first and second postnatal week the SOL exhibited a rapid 2.5-fold increase in enzyme activity, followed by an equally rapid decline. Enzyme levels approached the lower adult values by 4 weeks. In the EDL, AChE activity also rose rapidly between the first and second postnatal week and remained at this level, however, never reaching the peak activity of the SOL. Within four weeks the mature level of enzyme activity was reached in both muscles.

Rapid changes in the pattern of AChE molecular forms were particularly apparent during the first four postnatal weeks. Representative velocity sedimentation experiments are shown for 1 day, 1 week, and 14 weeks post partum in Figs. 2 and 3. One day after birth, the intermediate (10S and 12S) and heavy (16S) molecular forms account for approximately 60% of the total enzyme activity and both muscles (Fig. 2A), exhibit similar patterns of molecular forms of AChE not unlike that of the mature SOL.

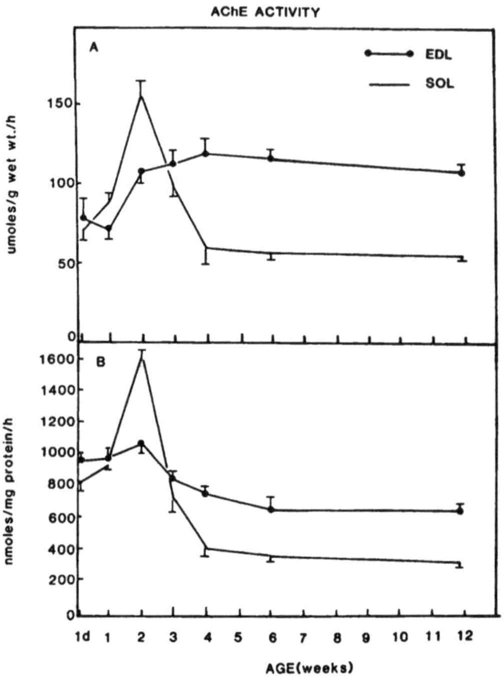

Fig. 1. Developmental changes in acetylcholinesterase(AChE) activity as a function of muscle wet weight (A) and protein (B) in extensor digitorum longus (EDL) and soleus (SOL) muscles. Values represent the mean ± SEM of 3 separate determinations for each age.

One week after birth, a relative increase of the light 4S AChE molecular form was observed in the EDL muscle (Fig. 2B). Whereas an increase of the 4S form did not occur in SOL, the relative activity of the 12S and 16S forms increased, however.

Four weeks after birth most of the postnatal alterations in AChE molecular forms in EDL and SOL muscles have occurred (not shown here) and resembled the profiles of mature muscles (Fig. 3).

As evident from Fig. 1, the most rapid alterations in total

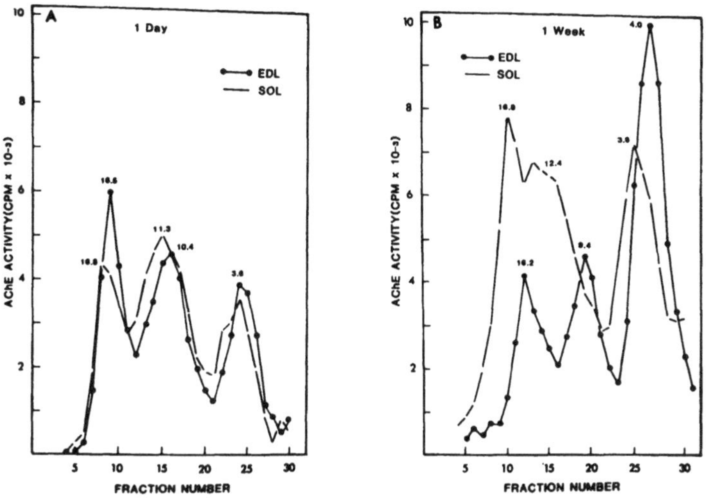

Fig 2. Developmental changes in the velocity sedimentation gradients of AChE molecular forms from: A, 1 day and B, 1 week postnatal extensor digitorum longus (EDL) and soleus (SOL) muscle. Gradients contained two hundred micrograms of a high speed supernatant from the muscles which was layered on a 5 to 20% sucrose gradient and centrifuged at 35,000 rpm for 18 hr. Estimated sedimentation constants are indicated for the separate forms. Each gradient was independently calibrated with B-galactosidase, catalase, and alkaline phosphatase. Approximately 30 fractions were collected and 0.03 ml was incubated for 4 hr in the presence of ^{3}(H)-acetylcholine and iso-OMPA (10^{-5} uM).

AChE activity occurred during the first 4 postnatal weeks. The relative percent contribution of the different AChE molecular forms, which were quantitated by comparing enzyme peak areas to the area of the total enzyme activity, are shown in Table I. One day following birth both EDL and SOL muscles exhibit similar levels of 4S, 10S and 16S AChE (27%, 43% and 30% for EDL and 22%, 54% (12S) and 24% for SOL, respectively). The marked differences of AChE molecular form patterns seen between the mature SOL and EDL muscles become

Table 1. Activities of AChE Forms During Postnatal Maturation of EDL and SOL.

Postnatal Period	EDL 4S	EDL 10S	EDL 16S	SOL 4S	SOL 10S	SOL 12S	SOL 16S
1 Day	27.0	43.0	30.0	22.0		54.0	24.0
1 Week	56.0	24.0	20.0	36.0		40.0	24.0
2 Weeks	45.0	27.0	28.0	33.0	17.0	14.0	36.0
3 Weeks	49.0	23.0	28.0	29.0		35.0	36.0
4 Weeks	51.0	33.0	16.0	21.0		62.0	17.0
14 Weeks	50.0	35.0	15.0	22.0	18.0	34.0	26.0

Activities of AChE molecular forms from EDL and SOL. AChE molecular forms were recovered from sucrose density gradients during postnatal development. Activities are expressed for each form as a percentage of the total recovered activity. Data are from three separate preparations of high speed supernatants (100,000 x g) isolated from rat muscle. Values are expressed as the mean $\pm$ SEM.

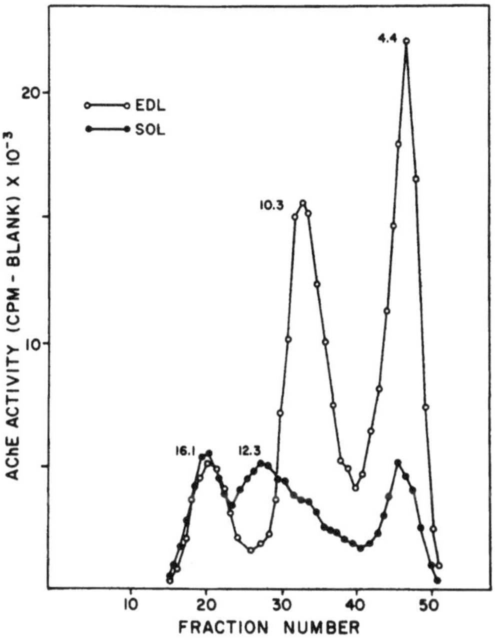

Fig. 3. Velocity sedimentation gradient separation of AChE molecular forms from EDL and SOL muscle of 14 week old rat. Exp. Neurol. 79: 519-531, 1983.

already apparent by the first postnatal week (Fig. 2B). In the EDL the 4S form accounts for approximately 50% of the total activity and remains at this level for the remaining developmental period. At no time does the EDL muscle exhibit the relative increases in the 16S AChE or decreases in 4S AChE as observed in SOL.

In the SOL muscle, an increase of the 16S molecular form is evident between 1 day and the 3rd week following birth. As the 4th postnatal week is reached the 16S has decreased from the 2 week level of activity. The decline is accompanied by an increase in the activity of the 12S molecular form of AChE.

2. Distinct Difference Between Slow and Fast Muscle in Recovery of AChE Activity After Denervation in Rats.

Within 3 days following nerve crush, AChE activity was reduced to about 40% of control in both muscles and by the end of the second week the remaining activity was 15% of control in both muscles (Fig. 4).

Following the second week, enzyme activity began to recover in both muscles; however, the rate of recovery in the SOL was much faster than in the EDL. Three weeks after crush, AChE activity in the SOL had risen to 250% of control while the activity in the EDL was only 40% of control. SOL enzyme activity returned to normal, while EDL had regained only 50% of control activity at the end of the fifth week.

The changes of total AChE activity were also studied in endplate and non endplate regions of the SOL. There was no qualitative difference in the response of AChE activity to crush and reinnervation between endplate and non-endplate regions of the SOL.

Molecular forms in the EDL, compared with their contralateral controls, exhibited marked decreases. Despite the proportional redistribution in some forms, such as in the 10S which now contributed 60% of the total activity compared with 35% in a contralateral control muscle, there remained a comparative loss in activity compared with its contralateral

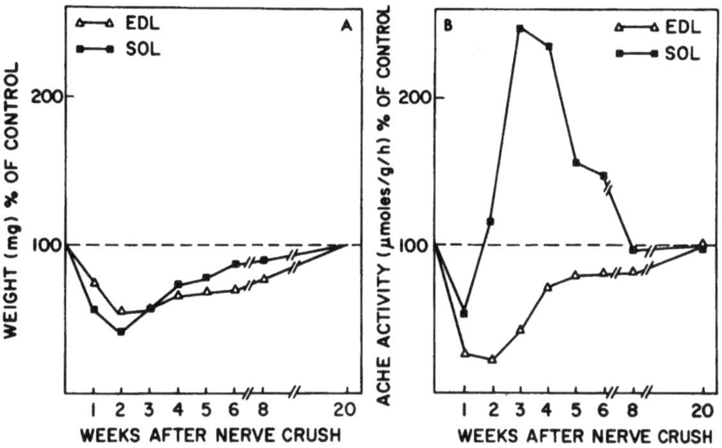

Fig. 4. Changes in muscle wet weight and AChE activity after denervation in the EDL and SOL muscles. A-EDL and SOL muscles were removed and trimmed of excess fascia before weighing. B-total AChE activity from EDL and SOL muscles after nerve crush. Twenty-microliter samples from 5% (w/v) homogenates were assayed in the presence of ^{3}H-acetylcholine for 20 min. AChE activity is expressed as the release of ^{3}H-acetate (micromoles) per gram wet weight per hour. Both wet weight and AChE activity from denervated muscles were compared with contralateral muscles and expressed as a percentage of control. Exp. Neurol. 79: 519-531, 1983.

control. Decreases in AChE molecular form activity in the SOL also were evident compared with their contralateral controls, except in the 4S form which increased above contralateral activity (Table 2).

In the SOL, 2 weeks after nerve crush all molecular forms, especially the 4S, had higher activity than those of the contralateral control muscle (Fig. 5A). Four weeks after nerve crush, as the reinnervation progressed, the molecular form pattern favored the heavier molecules in addition to the 4S form (Fig. 5B). Total AChE activity remained

TABLE 2. Activities of Molecular Forms of Acetylcholinesterase One Week After Denervation[a]

Molecular Forms	SOL Control	SOL Denervated	EDL Control	EDL Denervated
4S	25	44	50	22
10S	20	25	35	58
12S	25	17	0	0
16S	30	14	14	20

[a]Activities in AChE molecular forms 1 week after nerve crush, expressed as a percentage of the total distribution. Peaks from the distribution were weighed and divided by the sum of the peaks to give the relative amounts of AChE. Recoveries were obtained from three different preparations, each containing three muscles.

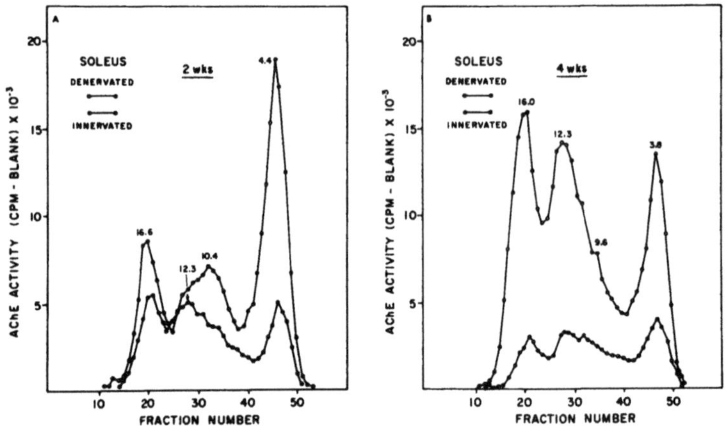

Fig. 5. Velocity sedimentation gradient separation of AChE molecular forms in the SOL muscle and contralateral control muscle after nerve crush. All gradients contained 3.5 mg protein from the SOL high-speed supernatant. A-2 weeks, B-4 weeks, after nerve crush. Exp. Neurol. 79: 519-531, 1983.

significantly higher than control. In EDL muscle, however, 2 weeks after nerve crush (Fig. 6A), the activity of individual molecular forms of AChE was reduced. In the fourth week, the 4S and 16S molecular forms had recovered (Fig. 6B), whereas the activity of the 10S AChE molecular form remained significantly below control values even at the end of the sixth week.

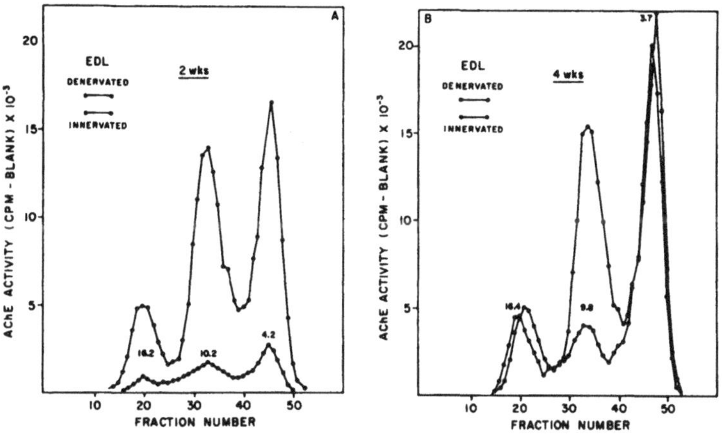

Fig. 6. Velocity sedimentation gradient separation of AChE molecular forms in the EDL muscle and contralateral control muscle after nerve crush. Gradients for the 2-week preparation contained 5 mg protein, and gradients for the 4- and 6-week preparations contained 3.5 mg protein. A-2 weeks, B-4 weeks, after nerve crush. Exp. Neurol. 79: 519-531, 1983.

As demonstrated in Table 3, two entirely different profiles for the AChE molecular forms were evident in fast and slow twitch muscle after denervation and during reinnervation. The EDL exhibited decreases in all three forms, followed by slow increases. On the other hand, the SOL muscle exhibited selective increases in the AChE activity associated with the 4S and 10S forms and decreases in the 12S and 16S forms. After reinnervation, only a gradual increase was evident in the EDL muscle. However, transient increases to several times the contralateral controls were apparent for all four

TABLE 3. Activities of the Molecular Forms of Acetylcholinesterase Expressed as a Percent of the Contralateral Controls[a]

Time after crush (weeks)	N	4S	10S	12S	16S
		Extensor Digitorum Longus			
1	3	11.0± 2.3	39.5± 9.6	0	29.6±16.3
2	5	29.9±12.2	20.8± 7.9	0	22.0±10.6
3	3	45.6± 6.7	21.6± 2.2	0	30.5± 1.3
4	5	83.1± 5.7	37.9± 9.9	0	177.8±37.3
5	3	83.8± 5.0	49.8±16.8	0	94.1±27.6
6	5	114.2± 8.1	50.1±10.8	0	96.4±10.0
		SOL			
		4S	10S	12S	16S
1	3	140± 30	85± 47	36± 10	28± 18
2	5	222± 47	157± 10	55± 17	81± 32
3	3	687±173	310± 10	248± 24	439± 39
4	5	419±101	257± 57	504±146	165±101
5	3	155± 85	262±158	219± 94	165± 23

[a]Comparisons of the AChE activity from EDL and SOL muscles after denervation for each molecular form. Peaks from the distributions between experimental muscles were compared with the peaks from their contralateral controls, to give the percentage of contralateral controls. Between three to five determinations (N) were made for each postoperative period, and there were three muscles for each group. Values are the mean ± SE. Exp. Neurol. 79: 519-531, 1984.

molecular forms of AChE in the SOL muscle. The presumed endplate form (16S) increased by more than 500% of contralateral control values, and was maximal between 3 and 4 weeks after nerve crush. The remaining AChE molecular forms exhibited a similar transient increase with respect to the contralateral control muscles between the second and fourth weeks after nerve crush.

3. Decreased Loadbearing and Its Effects on AChE and Its Molecular Forms in Slow and Fast Muscle.

Reduced loadbearing causing decreased activation of muscle units was used to examine the role of muscle activity in the regulation of AChE. This was accomplished by suspension of rat hindlimbs off the floor of animal cages for prolonged period (3 weeks). The hindlimbs so treated are not in a position to bear weight. Therefore, patterned motor unit activity producing movement of the limbs may occur, but since there is minimal load on the muscles the level of activity required to produce each movement is decreased.

There was a significant weight loss in both muscles. The weight of the EDL was reduced to 71% and that of the SOL to 40% of control weight. AChE activity in EDL showed no significant reduction when calculated as activity per g muscle or per mg protein. In SOL an increase by 420% and 537%, respectively, was seen when enzyme activity was calculated on the basis of gram per muscle or mg protein (Table 4).

These changes in enzyme activity were also reflected in the molecular forms of AChE. In SOL, hypokinesia increased

TABLE 4. Effect of Hindlimb Suspension Hypokinesia on Acetylcholinesterase Activity in Rat Extensor Digitorum Longus and Soleus Muscle.

	EDL		SOL	
	umol ACh/g/h	nmol ACh/mg protein/h	umol ACh/g/h	nmol ACh/mg protein/h
Control	109.61± 8.77 (100)	533±41 (100)	56.12± 5.91 (100)	295± 75 (100)
1 Week	93.72±11.49 (85)	470±51 (88)	71.10± 5.47 (127)	426± 69 (144)
Control	101.42± 3.89 (100)	474±18 (100)	50.54± 2.40 (100)	243± 10 (100)
2 Weeks	97.83± 3.99 (96)	440±32 (93)	102.60± 6.50[a] (206)	661± 70[a] (272)
Control	94.66± 5.97 (100)	447±35 (100)	48.28± 2.60 (100)	247± 20 (100)
3 Weeks	74.88± 5.13[a] (79)	361±20[a] (81)	126.21±10.54[a] (262)	892±100[a] (361)

Each value is the mean ± SEM of acetylcholinesterase activity, obtained from 5-10 animals, numbers in parenthesis represent percent of controls.

[a]indicate significant difference ($p < 0.05$) compared to control value at each corresponding time interval.

significantly the activity of all four major forms - 16S, 12S, 10S and 4S. In the EDL no significant change ($p < 0.05$) was observed in the 4S and 16S forms while the 10S was slightly increased.

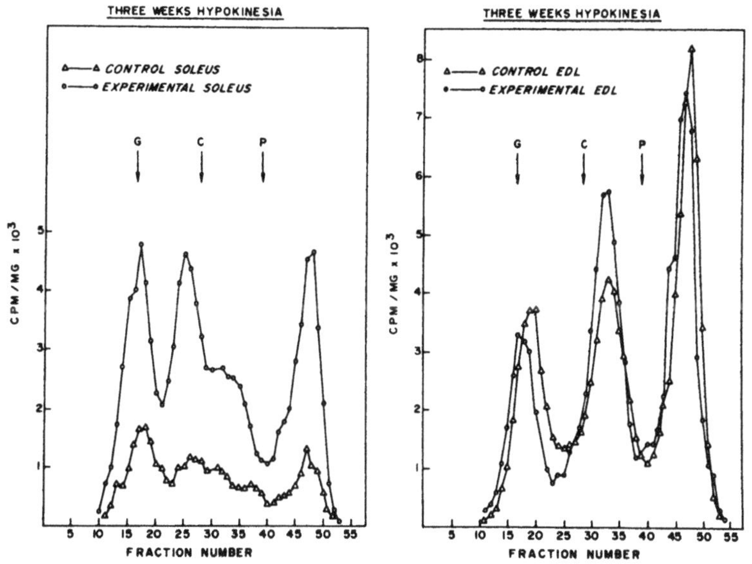

Fig. 7. Velocity sedimentation gradient separation of the acetylcholinesterase (AChE) forms from soleus (SOL) and extensor digitorum longus (EDL) muscles. A. Effects of three weeks of disuse on the SOL Δ——Δ control muscle, o——o hypokinetic muscle. B. Effects of three weeks of disuse on the EDL Δ——Δ control muscle, o——o hypokinetic muscle. Gradients contained two hundred micrograms of a high speed supernatant from the muscles. Arrows indicate position of markers, G = β-galactosidase (16.0S), C = catalase (11.1S) and P = alkaline phosphatase (6.1S).

Discussion

Adult fast EDL and slow SOL muscles have different AChE activities as well as different contributions of molecular forms to the overall AChE activities. This may be the result of innervation by distinct classes of motor neurons that transmit impulse patterns at different rates. At birth, however, AChE activity as well as the contribution of molecular forms are similar and closely approximate the characteristics of the mature SOL.

During the initial two weeks of postnatal development both muscles continue to increase AChE activity due to an endogenous program that determines synthesis of AChE. After functional innervation is fully established the fast EDL continues to synthesize AChE at the same rate, while the slow SOL synthesizes AChE at a reduced rate. During reinnervation of adult SOL, the early postnatal experience is repeated, i.e. with a reduction of functional input, SOL synthesizes AChE temporarily at a high rate, not unlike that seen during the first weeks of postnatal development (22). If the difference in AChE activity between SOL and EDL is the product of functional innervation, experiments that alter neural input should cause changes. The reduced work load placed on both muscles in the hindlimb suspension model suggests that in the absence of continuous neural stimulation and work the SOL reverts to the characteristics of the early postnatal development when innervation is established, but functional activity is not fully developed and thus synthesis of AChE is increased. AChE activity of the mature EDL muscle appears to be less affected by changes in the functional innervation (22).

Both muscles have the ability to synthesize AChE at a high rate and this may be considered a programmed event in the development of fast and slow muscle. This commitment is probably regulated by motor neuronal input through variation in impulse pattern. Thus, in SOL full functional innervation by slow motor neurons with a rate of continuous impulse frequency reduces the rate of synthesis. In case of reinnervation and hindlimb suspension the firing patterns are reduced and therefore AChE synthesis is increased. The re-establishment of muscle function appears to down regulate the intrinsic potential to synthesize AChE in slow SOL, while the EDL appears to be less dependent on functional activity.

The muscles in our hypokinesia experiments were no-loadbearing but free to contract, thus there is a need for a loadbearing function to prevent atrophy. The lesser degree of atrophy in the EDL indicates that this muscle is normally not used as an antigravity muscle during locomotion, while the SOL is normally an antigravity muscle.

These data also indicate that AChE of skeletal muscle is strongly dependent on level of loadbearing and patterns of motor unit activity. Since the control enzyme activity of the slow SOL is about half that of the fast EDL, the increase in the SOL AChE with disuse is to a level somewhat greater than the activity of the control EDL. Disuse produced little change in AChE activity in EDL.

The reasons for the SOL showing this more pronounced sensitivity to inactivity are not clear, however its greater dependency on nerve regulated influences has been observed previously. The data suggest that in the absence of normal patterns of weightbearing activity, i.e. no loadbearing and without trauma to the lower motor neuron slow SOL may revert to a faster and perhaps more immature, i.e. neonatal character. The slow SOL muscle is more dependent on activity related mechanisms than the EDL. Further studies involving axoplasmic transport, release of ACh from nerve terminals, changes in contractile characteristics and careful monitoring of electrical activity during disuse will help to more clearly elucidate the role of the nerve and nerve induced activity in the maintenance of muscle.

References

1. Dettbarn, W-D. 1981. A distinct difference between slow and fast muscle in acetylcholinesterase recovery after reinnervation in the rat. Exp. Neurol. 74, 33-50.

2. Lomo, T. 1980. What controls the development of neuro-muscular junctions. Trends Neurosci. 3, 126-129.

3. Lomo, T. and Slater, C. 1980. Control of junctional acetylcholinesterase by neural and muscular influences in the rat. J. Physiol. (London) 303, 191-202.

4. Rubin, L., Schuetze, S., Weill, C. and Fischbach, G. 1980. Regulation of acetylcholinesterase appearance at neuromuscular junctions in vitro. Nature 283, 264-267.

5. Guth, L. 1969. Effect of immobilization of sole-plate and background cholinesterase of rat skeletal muscle. Exp. Neurol. 24, 508-513.

6. Snyder, D., Rifenberick, D. and Max, S. 1973. Effects of neuromuscular activity on choline acetyltransferase and acetylcholinesterase. Exp. Neurol. 40, 36-42.

7. Butler, I., Drachman, D. and Goldberg, A. 1978. The effect of disuse on cholinergic enzymes. J. Physiol. (London) 274, 593-600.

8. Hall, Z. 1973. Multiple forms of acetylcholinesterase and their distribution in endplate and non-endplate regions of rat diaphragm muscle. J. Neurobiol. 4, 343-361.

9. Lyles, J., Silman, I. and Barnard, E. 1979. Developmental changes in levels and forms of cholinesterases in muscles of normal and dystrophic chickens. J. Neurochem. 33, 727-738.

10. Massoulie, J. 1980. The polymorphism of cholinesterases and its physiological significance. Trends Biochem. Sci. 5, 169-163.

11. Davey, B. and Younkin, S. 1978. Effect of nerve stump length on cholinesterase in denervated rat diaphragm. Exp. Neurol. 59, 168-175.

12. Fernandez, H., Duell, M. and Festoff, B. 1979. Neurotrophic control of 16S acetylcholinesterase at the vertebrate neuromuscular junction. J. Neurobiol. 10, 441-454.

13. Guth, L., Kemerer, V., Samaris, T., Warnick, J. and Albuquerque, E. 1981. The roles of disuse and loss of neurotrophic function in denervation atrophy of skeletal muscle. Exp. Neurol. 73, 20-36.

14. Gardiner, P., Lapointe, M. and Gravel, D. 1982. Exercise effects on recovery of muscle acetylcholinesterase from reduced neuromuscular activity. Muscle & Nerve 5, 363-368.

15. Vrbova, G. 1963. Effect of motoneurone activity on the speed of contraction of striated muscle. J. Physiol. 169, 513-536.

16. Eccles, J., Eccles, R and Kozac, V. 1962. Further investigations on the influence of motoneurones on the speed of muscle contraction. J. Physiol. 163, 324-340.

17. Buller, A., Eccles, J. and Eccles, R. 1960. Differentiation of fast and slow muscles in the cat hind limb. J. Physiol. 150, 399-416.

18. Buller, A., Eccles, J. and Eccles, R. 1960. Interaction between motoneurones and muscles in respect of their characteristic speeds of their responses. J. Physiol. 150, 417-439.

19. Engel, W., Brooke, M. and Nelson, P.G. 1966. Histochemical studies of denervated or tenotomized cat muscle - illustrating difficulties in relating experimental animal conditions to human neuromuscular diseases. Ann. N. Y. Acad. Sci. 138, 160-185.

20. McMinn, R. and Vrbova, G. 1964. The effect of tenotomy on the structure of fast and slow muscle in the rabbit. Q. J. exp. Physiol. 49, 424-428.

21. Tower, S.S. 1937. Trophic control of non-nervous tissues by the nervous system. A study of muscle and bone innervation from an isolated quiescent region of spinal cord. J. Comp. Neurol. 67, 241-267.

22. Groswald, D.E. and Dettbarn, W-D. 1983. Nerve crush induced changes in molecular forms of acetylcholinesterase in soleus and extensor digitorum longus muscles. Exp. Neurol. 79, 519-531.

This work was supported in part by NASA Grant No. NAGW-469. The authors wish to thank Mrs. Barbara Page for the typing of the manuscript.

THE EXPRESSION OF MICROTUBULE PROTEINS DURING THE DEVELOPMENT OF THE NERVOUS SYSTEM

IRITH GINZBURG AND URIEL Z. LITTAUER
Department of Neurobiology, The Weizmann Institute of Science, Rehovot, Israel

Microtubules are subcellular structures found in all eukaryotic cells. They participate in formation of a large variety of structures such as cytoplasmic networks, mitotic spindles, and highly ordered organelles such as centriols and flagella axoneme. Functionally microtubules are involved in a number of dynamic processes (1,2), such as chromosomal movement, cell migration, intracellular transport, formation of cell shape, and secretion. In the nervous system microtubules are present in high concentration and play a unique role in the nervous system. The neuronal cells have an asymmetric shape that governs their connectivity as well as their function in the neuronal network. Synapse formation in the developing brain is an integral part of neuronal differentiation. As such it depends not only on the establishment of axon terminals and synaptic junctions but also on the extent and degree of branching of the axons and dendrites that provide the framework on which synapse are made. All these elements are formed by a process in which morphologically and functionally distinct domains are progressively established within the neuroblast. The basic neuronal microdifferentiation, which distinguishes between axons and dendrites, depends on redistribution of microtubules. This stage occurs when the nerve cells, in developing brain, stop dividing and

Molecular Basis of Nerve Activity

enter the phase of terminal differentiation. Later in development, axon outgrowth and axoplasmic transport depend on microtubule integrity.

The dynamic nature of microtubules and the broad spectrum of cellular processes in which they are involved suggests that their assembly and growth is controlled not only temporally but also spatially in all cells. The detailed mechanisms by which microtubules regulate this process are not yet well understood. The great variety of microtubule functions can only be explained by the great number of interactions between tubulin and other proteins such as microtubule associated proteins (MAPs) and is possibly governed by the microheterogeneity of the microtubule proteins. Microtubule heterogeneity can arise from its specific assembly from its various components, i.e., the diverse tubulin isoforms and the family of MAPs. These include high molecular components MAP_1 and MAP_2 and a number of closely related proteins of 56-62K designated as TAU factors (3,4). The availability and the change in composition of these components may determine their interaction. Thus it was established that during rat brain development there is an increase in number as well as a change in the distribuiton of the various tubulin isotypes and TAU forms.

It was shown that the major differences in isotubulin and TAU forms that arise during brain development are controlled at the mRNA level (3,5). Thus, the genes coding for microtubule proteins exhibit brain-specific expression which is characterized by a wider variety of isoforms than in other tissues, as well as by a specific time of activation during brain differentiation.

Moreover, enhanced synthesis of beta-2 isotubulin species and two TAU forms was observed in the goldfish retina during optic nerve regeneration, suggesting that these isoforms are involved in the formation of the new axonal matrix (6,7). It is therefore suggested that the various isotubulin and MAP forms are not randomly distributed within a given microtubular system, rather their organization is directly coupled to their function.

In order to study the involvement of the different isotubulins and MAPs at the molecular level, specific probes are a prerequisite.

For this purpose we have constructed cDNA libraries derived from newborn and adult rat brain as well as from various neuroblastoma and glioma cells.

The specific cDNA clones together with specific antibodies will be used for immunocytochemistry and in situ hybridization studies which will determine the expression of specific microtubule components within individual cells as well as in the various regions of the central nervous system.

We have developed a specific binding assay for the interaction of [^{125}I]MAPs with tubulin or its cleavage peptides. To identify the tubulin binding domains to MAPs we have examined, in collaboration with Dr. H. Ponstingl (German Cancer Research Center, Heidelberg), the binding of rat brain [^{125}I]MAP_2 or [^{125}I]TAU factors to 46 cleavage peptides derived from pig alpha- and beta-tubulin. Our results show that MAP_2 will specifically interact with only two peptides located at the C-terminus of beta-tubulin, between positions 407-415 and 416-445. Strong binding sites for TAU factors were also located at the C-terminus of beta-tubulin between positions 416-445. In addition, TAU factors, but not MAP_2, interacted with two peptides derived from alpha-tubulin. The first peptide is located near the N-terminus of alpha-tubulin between positions 37-78 and a second binding site located between positions 314-327 (7).

Comparison of the amino acid sequences of human beta-tubulin isotypes to that of pig, chicken and yeast reveals a variable domain within the 15 carboxy-terminal amino acids. This same regions was found by our studies to include, at least in part, the binding site to MAP_2 and TAU factors. It is tempting to speculate that the variability in amino acid sequence in this region will determine the specificity of interaction of beta-tubulin isotypes with the various MAPs, thus generating functionally different microtubules.

Acknowledgements: Supported in part by grants from the United States-Israel Binational Science Foundation, from the Forchheimer Center for Molecular Genetics, from the National Council for Research and Development, Israel and from the Deutsches Krebsforschungzentrum, Heidelberg, Germany.

References

1. Kreis, T.E., T. Scherson, U.Z. Littauer, J. Schlessigner and B. Geiger 1982. J. Cell Biol. 95, 348a.

2. Scherson, T., T.E. Kreis, J. Schlessinger, U.Z. Littauer, G.G. Borisy and B. Geiger 1984. J. Cell Biol. 99, 425-434.

3. Littauer, U.Z. and I. Ginzburg 1984. In: Gene Expression in Brain (C. Zomzely-Neurath and W.A. Walker, eds.), J. Wiley and Sons, N.Y., in press.

4. Scherson, T., B. Geiger, Z. Eshhar and U.Z. Littauer 1982. Eur. J. Biochem. 29, 295-302.

5. Ginzburg, I., T. Scherson, D. Giveon, L. Behar and U.Z. Littauer 1982. Proc. Natl. Acad. Sci. USA 79, 4892-4896.

6. Neuman, D., T. Scherson, I. Ginzburg, U.Z. Littauer and M. Schwartz 1983. FEBS Lett. 162, 270-275.

7. Ginzburg, I., T. Scherson, S. Rybak, Y. Kimhi, D. Neuman, M. Schwartz and U.Z. Littauer 1984. Cold Spring Harbor Symposium on Quantitative Biology, Vol. XLVIII: Molecular Neurobiology, 783-790.

8. Ginzburg, I. and U.Z. Littauer 1984. In: Molecular Biology of the Cytoskeleton. Cold Spring Harbor Laboratory, Cold Spring Harbor, New York, in press.

V. Acetylcholinesterase and Choline Acetyltransferase

ACETYLCHOLINESTERASE OF MUSCLE AND NERVE

New molecular tools applied to an enzyme study pioneered by David Nachmansohn

W.R. Randall, Josephine Lai, E.A. Barnard
Department of Biochemistry, Imperial College of Science and Technology, London SW7 2AZ, England

Introduction

The modern study of acetylcholinesterase, it can reasonably be argued, arose in considerable measure from the observations and ideas of David Nachmansohn. He and his co-workers coined this name for the enzyme [1], discovered its high concentration in electric organs of Electrophorus [2] and of Torpedo [3,4], and at the endplates of amphibian [5] and mammalian [6] skeletal muscles, and showed that this declined on interruption of the nerve supply [6]. In his laboratory the first major purifications of acetylcholinesterase (AChE) were achieved [7,8]. It was there, also, that I.B. Wilson obtained the first evidence for the acyl-enzyme intermediate as the essential feature of the action of this and other serine esterases, and as early as 1950 he and Nachmansohn [9] depicted this catalytic mechanism of AChE in a form still accepted as correct. Likewise the principal features of the binding site and its relationship to the catalytic site were well comprehended in that early work.

The study of the relationship of AChE to muscle development

was pioneered by the measurements by Nachmansohn in 1938 [5] which showed that AChE increases greatly in chick muscle in the developing egg at the time that the embryo movements commence [10,11]. The value deduced as early as 1939 of 1.6 x 10^9 molecules of ACh split per millisecond at each motor endplate of frog muscle is of the same order as in modern determinations, and concentrated attention upon the great efficacy of AChE in cholinergic synaptic transmission, as Nachmansohn himself expounded [11,12]. Subsequently, as is well known, his interpretation of the roles of AChE and ACh departed from what is now generally accepted. However, this should not detract from our appreciation of the characteristic resourcefulness and energy which led him from the outset to the choice of appropriate zoological sources and to make many of the fundamental observations upon which later studies elsewhere were built. In truth one can say that his work focussed attention upon AChE as a fruitful topic of research, and stimulated the whole field of activity upon it which has since developed.

Results

New tools for the investigation of AChE in the biology of nerve and muscle

(i) In recent years the analysis of multiple molecular forms of AChE has become of major importance in studies of its biological role. We shall describe here, along with such analysis, the use of micro-dissection, e.g. of endplates, in locating different forms within the various muscle fibre types.

(ii) Studies have become feasible of the regulation of AChE in muscle cells by the nerve, both by trophic factors and, as exemplified below, by motor nerve activity.

(iii) The new tools of molecular genetics are in the course of application to AChE. Studies at the level of its mRNAs will be described here.

(iv) The value of monoclonal antibodies directed against AChE, both as probes and as vehicles of enzyme purification, will be considered.

This list does not exhaust the catalogue of newer tools which have been applied to AChE, with different approaches illustrated in some of the other chapters of this symposium volume.

Multiple molecular forms of AChE in muscle fibres

Avian muscles. Multiple forms of AChE were originally recognised by their different electrophoretic mobilities, but that method has become outdated: the heavy forms do not penetrate the usual gels, and the charge differences between the proteins are not predictable or of intrinsic significance in this series. The method universally used now for this purpose is sedimentation velocity on sucrose density gradients, a method introduced for this purpose by J. Massoulié and F. Rieger [13]. This separates a series of forms; in decreasing size order these are the asymmetric collagen-tailed (as denoted by subscript c) or "heavy" (H) forms, H_{2c} and H_{1c}, and the globular forms M, L_2 and L_1. An alternative nomenclature (A_{12}, A_8 and G_4, G_2, G_1, respectively) based on the apparent structures of these forms in electric organ, has been introduced by J. Massoulié and colleagues (see the article by J. Massoulié in this Volume). The H_{2c} form shows a value for $s_{20,w}$ (in 1% Triton X-100 medium) of 16 S in the mammals but 20 S in the birds. This form requires a combination of detergent and high salt concentration for its extraction and analysis; it contains a collagenous "tail" polypeptide [16]. The globular L_2 form (~5.5 S in mammals), and perhaps others, can exist in a hydrophobic version as well (as discussed by T.L. Rosenberry and by I. Silman in this Volume), so that one may need additional designations, eg. L_2 and L_{2h}, for such cases.

In those muscles of birds which are essentially pure in fast-twitch fibres (type II), typically 95% of the AChE is found, upon such extraction and analysis, to be in the H_{2c} collagen-tailed 20 S form (Fig. 1). This is true both in the mature chicken, where the pectoral muscle has 99% type IIB (glycolytic) fibres [15] and in the mature pigeon, where >90% of its fibres are also fast-twitch but of the oxidative sub-type, IIA.

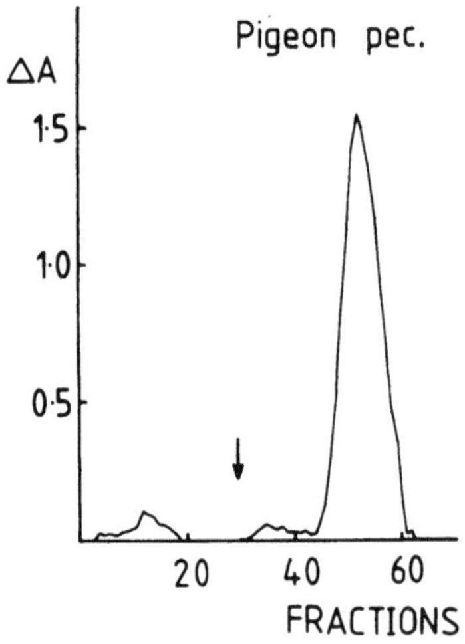

Figure 1. AChE of the pigeon pectoral muscle.
Pigeon muscle is extracted in 1% Triton X-100/1 M NaCl/10 mM EGTA (pH 7) plus a set of protease inhibitors [24]. The AChE forms present are separated on a 5-20% sucrose density gradient in a similar medium. In this and all subsequent gradient profiles, the procedure is the same, the heaviest species is on the right and the arrow denotes the reference position of catalase (11.4 S).

It is interesting to note that in developing avian muscle the lighter forms appear first, and the asymmetric heavy forms become detectable at the stage where innervation becomes significant. This is illustrated in Fig. 2 and

Table 2 for the developing chick pectoral muscle, where innervation occurs at 6-7 days *in ovo*. As has been deduced also for innervating myotubes in nerve/muscle co-cultures [16] and likewise for mixed muscles in the embryonic chick leg [17], the production of the asymmetric forms is triggered by the formation of synapses. The results shown in Fig. 2 and Table 1 and the other results just noted represent a modern confirmation of the early findings of Nachmansohn [10-12] that AChE appears in embryonic muscles when they commence movement.

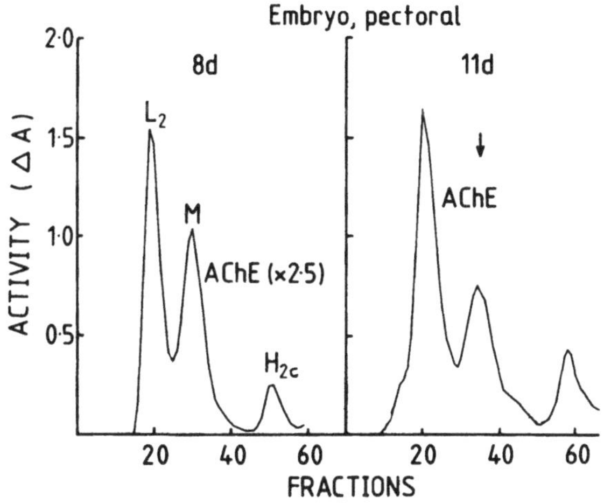

Figure 2. AChE forms of neural muscle from 8-day and 11-day chick embryos. The amount of muscle employed is 2.5 times greater in the 8-day sample due to its lower content.

Table 1. Development of AChE and Its Major Collagen-tailed Form in the Pectoral Muscle of the Chick

	Days	% H_{2c} (% of total AChE)	Total AChE activity (U/g wet weight)
In ovo	7	3	
	8	8	2.2
	11	11	
	14	19	5.5
	16	23	3.7
	19	55	7.0
Adult		90	0.06

From Barnard *et al* [18] and unpublished data.

The collagen-tailed form of AChE (16-20 S) in vertebrate muscles has for some time been regarded as the AChE type associated with the endplate. The principal evidence for this has been the selective occurence in neural, and not aneural, segments of certain muscles, and the loss of H_{2c} therefrom upon denervation and its reappearance upon reinnervation (reviewed in ref. 14). However, exceptions to those observations have been reported, as in the H_{2c} AChE seen in aneural regions of human muscle [19] and the disappearance of the H_{2c} form from chicken tonic fibre endplates [20, 21]. Moreover, the sedimentation profiles of AChE forms shown by various authors for different mammalian muscles vary considerably, and as Massoulie and Bon [14] note in their review, the asymmetric forms "generally constitute only a small fraction of the enzyme". Hence, the questions as to whether H_{2c} is the sole form of AChE at the motor endplate, and whether any is maintained outside the endplate, have not received a conclusive answer. The most direct evidence on these questions is obtainable now using the approach of micro-dissection of individual endplates, removed from the rest of the fibre (Fig. 3). These are extracted and the forms of AChE actually in the endplate are determined by sedimentation analysis using [^{3}H]ACh assay [22, 23]. In the fast-twitch fibres of chicken muscles this method has demonstrated that only the H_{2c} form of AChE is present at the endplate, and that none of this form exists outside it (Fig. 3A). This is true even in younger birds, where appreciable L_2 AChE is present extra-synaptically [22]. In the slow-tonic chicken muscles, in contrast, the AChE at the endplate is largely in the L_2 form, with only a small proportion of H forms (Fig. 3B). In these muscles the same proportions of AChE forms exist outside and inside the endplates, although the concentration of the extra-synaptic AChE is, of course, far lower than in the endplates [23].

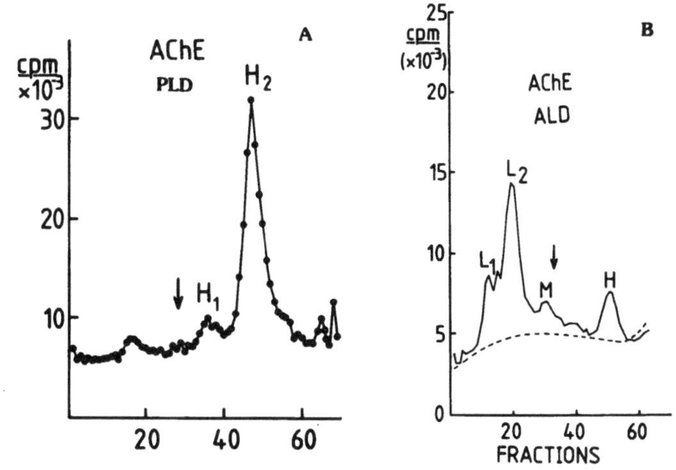

Figure 3. Forms of AChE separated from extracts of micro-dissected endplates. Endplates are free of almost all underlying muscle fibre. The broken line in B represents the background hydrolysis of the substrate as measured in a control gradient; the lowest activity level in B also represents that background level. Activity is measured in cpm of [^{3}H]ACh hydrolysed in a standard assay. For conditions, see [22]. A, endplates from PLD muscle of a 4-month chicken. (No activity was detectable in non-endplate segments from this muscle). B, endplates from the anterior latissimus dorsi (ALD) muscle (pure slow-toxic from a 4-month chicken.

<u>AChE in mammalian muscles</u>. The above-mentioned variability in the proportions of the AChE forms in reports in mammalian muscles, and the rather low contribution of the H forms there, are removed when stringent precautions are taken against proteolysis during AChE extraction. With a sufficient set of inhibitors of the whole spectrum of muscle proteases and rapid processing to the stage of gradient separation [24], profiles of the types shown in Fig. 3 are obtained with mammalian muscles, always with a majority of AChE in the 16 S (H_{2c}) form. In the fast-twitch (type II) fibres two types of behaviour are found, varying with the muscle: in some (e.g. hamster anterior tibialis) the enzyme is very largely in the H_{2c} form, whereas in others (e.g. guinea pig extensor digitorum longus, EDL), there is a considerable minority, which can approach one-half of the total, in the L forms (Fig. 3). As in the birds, there is no

essential difference between type IIA and type IIB mammalian fibres in their AChE forms.

In the slow-twitch fibres, e.g. in the guinea pig soleus muscle, the H forms greatly predominate, with the H_{1c} form (13 S) also contributing significantly in these cases (Fig. 4). When micro-dissection analysis is applied, it is found that an essential difference exists between the slow-twitch and fast-twitch fibres.

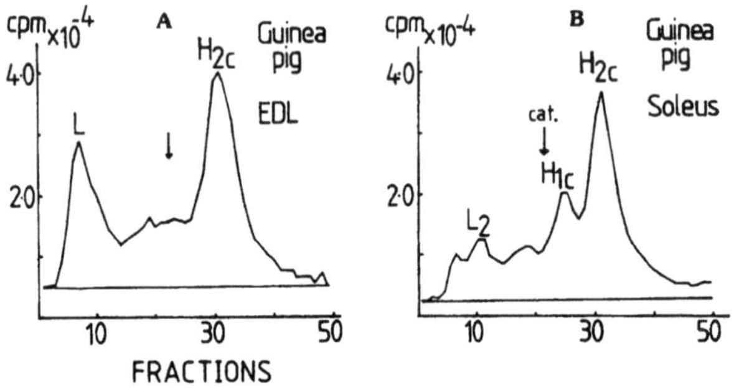

Figure 4. AChE forms from slow-twitch (A) and fast-twitch (B) guinea-pig muscles.

In the slow-twitch muscle the collagen-tailed forms are present throughout the fibre, whereas they are virtually absent extra-synaptically in the fast-twitch muscle (Fig. 5). The L and M AChE content of the muscle in all cases is due to extra-synaptic AChE.

Upon chronic denervation, in all muscle fibre types the content of H form declines dramatically [24]. This is true even in the exceptional cases (birds, rabbit, guinea pig) where the total AChE content of the muscle <u>increases</u> after denervation. However, in the slow-twitch mammalian muscles the H forms do not disappear, but reach a constant low level. Micro-dissection analysis reveals that this is due to the extra-synaptic H AChE, noted above to be present in slow-twitch fibres. In these fibres, clearly, collagen-tailed AChE is independent of neural influence where this form is extra-synaptic but is dependent upon it when synaptic.

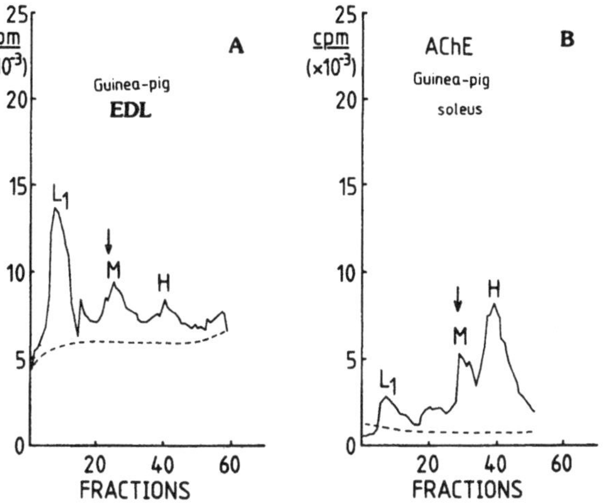

Fig. 5 AChE forms from micro-dissected Regions of Muscle Fibres from outside the Endplates.
A, guinea pig EDL muscle (fast-twitch fibres). B, guinea pig soleus muscle (slow-twitch fibres). (In both muscles the corresponding endplate specimens show only the H forms).

In the fast-twitch muscles the extra-synaptic AChE, present in L and M forms, is either constant or increases (depending on the species) upon denervation.

In conclusion, we can say that the distribution and maintenance of the forms of AChE, when analysed under fully protective conditions, is found to be determined in part by the neural contact and in part by the muscle fibre type. At the fast-twitch fibre endplates, where rapid destruction of ACh is required during motor nerve firing, the form present is the collagen-tailed H_{2c} molecule, but this is not needed at the endplates of slow tonic fibres. Maintenance of the H_{2c} AChE at the endplate is dependent upon the synaptic activity, but this is not so for the various forms present away from the endplates. The significance of the large amounts of L_2 AChE occurring extra-synaptically in certain fast-twitch fibres is unknown; whether this is a transient or stored precursor of the endplate AChE, or whether it has some role in distant regions of the muscle fibre, remains to be determined.

Regulation of muscle AChE by nerve activity

Neural control of muscle is divisible into the direct effect of neurally-secreted factors and the indirect effect via the nerve-evoked muscle action potentials [25]. We have tested for an effect of the latter component in the regulation of AChE forms, by employing prolonged repetitive stimulation of a muscle through its nerve, in situ in the living chicken. Silver electrodes were implanted in contact with the posterior latissimus dorsi (PLD) nerve, and impulses at 10 Hz were delivered to them from a totally-implanted, encapsulated printed-circuit stimulator. This neural stimulation was continuous for up to 4 weeks. The animals moved and fed freely. The stimulation produced a characteristic increase in the total AChE of the muscle (about 5-fold at the maximum) and a marked increase in the content of L_2 AChE (Fig. 7). The absolute amount of H_{2c} was also augmented.

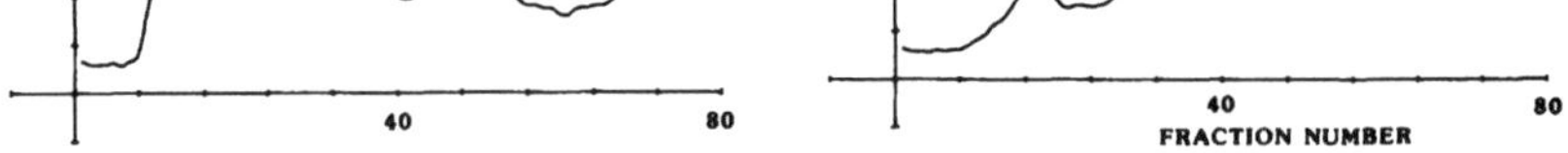

Figures 6 and 7. AChE forms present in the stimulated PLD muscle. Analysis was made of AChE in the PLD muscle of an 8-week chicken (Fig. 6) after chronic stimulation in vivo at 10 Hz for 4 weeks; or (Fig 7) in the contralateral, unstimulated PLD muscle of the same bird (using 1.7 times the amount of muscle used above). (Background not subtracted). In these muscles the total AChE activity was increased 3.2-fold by the stimulation. From [24].

Table 2. Effects of Constant Stimulation at 10 Hz on AChE in Chicken Muscle

Total AChE	Forms at endplate		Forms outside endplate		Fibre types	
Stim./Control	Stim.	Contr.	Stim.	Contr.	Stim.	Contr.
2.8	90% H_{2c}	90% H_{2c}	80% $(H_{2c}+H_{1c})$	N.D.	80% IIA	10% IIA
			20% (L_2+M)		20% IIB	90% IIB

Chickens (9-week) were stimulated through the PLD nerve for 3 weeks. Stim. = stimulated PLD muscle; Control = contralateral non-stimulated PLD muscle. The forms of AChE were separated from total AChE extracts of micro-dissected endplates, or of fibre regions far from the endplates.

The muscle changed from glycolytic to oxidative in its fibre types, and the extra-synaptic regions of the fibres accummulated AChE, primarily the collagen-tailed forms (Table 2).

These results extend to the in vivo level some parallel observations previously reported on cells in culture, e.g. that synaptic blockers such as d-tubocurarine retard the synthesis of muscle cell AChE and especially of H_{2c} [26] and that tetrodotoxin blockade of the action potentials in cultured myotubes acts in a similar direction on the H_{2c} AChE [27]. In stimulated mammalian muscle the synthesis of some enzymes is selectively increased [28], including the slow-myosin isoforms [29]. The results illustrated here suggest that electrical activity of the muscle fibre and the resultant ionic changes are a major influence on the production of the catalytic subunits of muscle AChE. The regulation of the forms of this enzyme in vivo appears to depend both upon electrical activity and factors derived from the nerve.

Note the characteristic change of the muscle form strongly glycolytic to strongly oxidative in type due to the stimulation. From data of J. Jarvis, J. Lai and P.J. Barnard.

Studies of AChE-specific mRNA activity.

Although, as discussed previously, indirect evidence suggests that the synthesis of asymmetric forms occur through assembly of lower molecular weight precursor forms [14, 16, 30], there is no direct evidence demonstrating the post-translational events in assembly and little is known about the regulation of the specific mRNAs.

We have currently been interested in developing methods to measure the levels of translatable mRNA coding for AChE in developing skeletal muscle by:

(1) monitoring mRNAs responsible for the appearance of AChE forms by their translation in microinjected Xenopus oocytes.
(2) preparing monoclonal antibodies (mAb) against components of AChE for measuring intermediates in the assembly of the asymmetric forms.

The accumulation of AChE activity was measured in Xenopus oocytes injected with poly (A^+) RNA (mRNA) which had been isolated from developing chick muscle (Fig. 8).

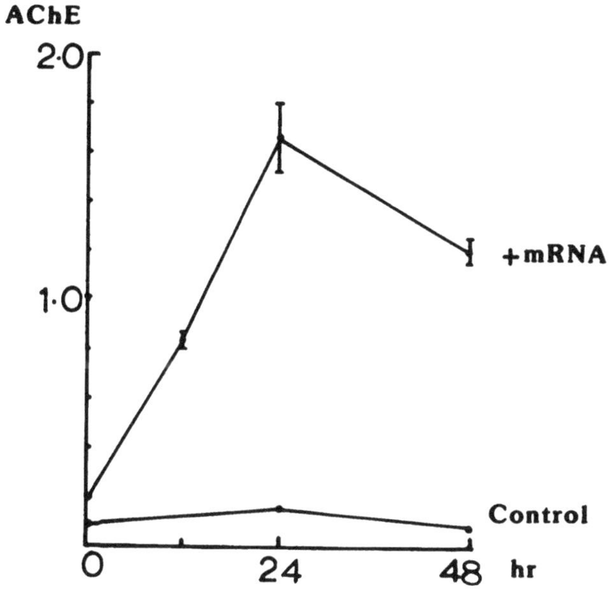

Figure 8. AChE synthesis directed by muscle mRNA: length of oocyte incubation. Poly A^+ RNA (mRNA) from 19-day embryonic pectoral muscles was prepared as described [31, 32] and microinjected (50 ng per oocyte) into stage 6 Xenopus oocytes [33]. Oocytes were incubated (10 per group) in Barth's medium (0.1 ml) + antiproteases: leupeptin (30 g/ml), aprotinin (10^{-2} μg/ml) bacitracin (1 mg/ml) and lima bean trypsin inhibitor (100 μg /ml). Oocytes were homogenized and extracted in 1M NaCl, 1% Triton-X-100, 5 mM EGTA, 10 mM HEPES, pH 7.5. The centrifuged supernatant (30,000g; 30 min) was assayed for AChE radiometrically [34] in the presence of 1 mM p-chloromercuri- phenylsulphonate and 10^{-4} M tetraisopropylpyrophosphorimide. Values are mean ± SE; N = 4.

AChE reached its highest activity in the mRNA-injected oocytes at about 24 hr after injection and then began to decrease by 48 hr. The extent of the decrease varied with the condition of the oocytes and was, in separate experiments, as high as 95% or as low as 60% of the 24-hr value. Because of seasonal differences in the oocytes [35], levels of translated message were assessed within each experimental set of oocytes. AChE activity varied as a function of the amount of mRNA injected: it reached its maximum activity at 40-50 ng of mRNA injected per oocyte.

We next examined the level of translatable AChE from muscle mRNA during the later stages of embryonic development (Table 3). The level of intracellular AChE of the injected oocyte is similar for all ages tested. Little enzyme activity is detected in the incubation medium (Table 3), suggesting that muscle mRNA directs a form of AChE that is not secreted by the oocyte.

Table 3. Effect of mRNA from Embryos at Different Ages.

Age of Embryo (days)	AChE Intracellular	AChE Secreted
13	1.55 ± 0.18	
15	1.62 ± 0.16	
17	1.71 ± 0.18	0.10 ± 0.02
19	1.50 ± 0.14	0.08 ± 0.03

Oocytes were injected, incubated, homogenized and extracted as in Figure 8. The intracellular extract and incubation medium were assayed for AChE radiometrically [34]. Values are means ± SE; N=4.

The low level of secreted AChE that was observed probably represents non-specific substrate hydrolysis, since it is nearly identical to values obtained from control oocytes injected with mRNA from non-AChE containing cultured fibroblasts (Table 4).

Table 4. Tissue Specificity of mRNA-directed AChE Synthesis

mRNA Source	AChE Intracellular	AChE Secreted
Chick muscle	3.48	0.44
Chick fibroblasts	0.40	0.30

mRNA preparation, injection and AChE assay were as in Figure 8. Values are means of two groups of ten oocytes.

Xenopus oocytes which have been injected with muscle mR translate and assemble AChE into multiple molecular forms (Fig. 9). Although the forms exhibit sedimentation coefficients similar to those seen in embryonic muscle from which the mRNA was derived, their proportions are very different. Although the collagen-tailed asymmetric (20 S) form is reproducibly detected at low levels, the monomeric (L_1, ~5 S) form appears to be the most abundant. This contrasts with the profiles seen from 19-day embryonic pectoral muscle used as source, where the majority of the enzyme is in the 20 S form (Table 1). One explanation for the apparent low level is that the oocyte may be deficient in cofactors necessary for collagen synthesis. However, no detectable increase in the 20 S form was observed after either co-injection of ascorbate (10^{-5}M), α-ketoglutarate ($5x10^{-5}$M) and iron-bound conalbumin ($5x10^{-5}$M) with the mRNA or incubation of injected oocytes in Barth's medium containing those cofactors.

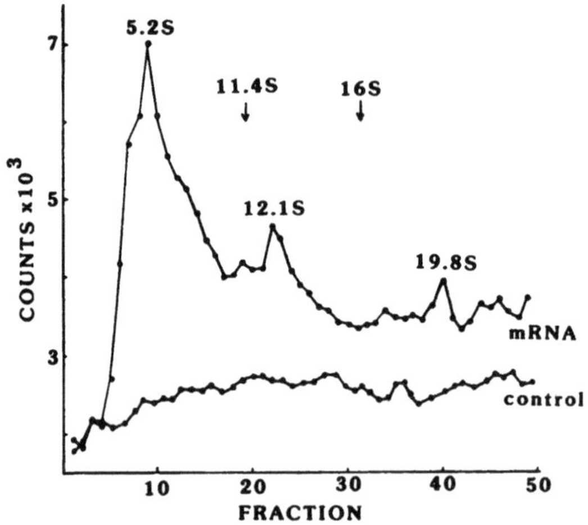

Figure 9. Molecular forms of AchE translated in injected Xenopus oocytes. Injected oocytes were homogenized and extracted as in Figure 1. Extracted proteins were centrifuged in 5%-20% sucrose gradients for 16 hours at 39,000 rpm in a Beckmann SW 40 rotor. Collected fractions were assayed for AChE radiometrically. Catalase (11.45) and -galactosidase (16 S) were used as markers.

Detection of the low levels of oligomeric forms (especially the higher molecular weight forms) has been restricted because of large dilution factors inherent in the use of sucrose gradients. One way to circumvent this problem is to measure directly the newly synthesized forms using mAbs that are specific for one or a subclass of the enzyme forms.

Monoclonal Antibodies as a tool for the Study of Chick AChE

In the first instance, mAbs have been prepared using chick brain AChE as the antigen: this contains the hydrophobic dimeric and (primarily) tetrameric forms. AChE from 1-day chick brain was purified using three consecutive chromatographic steps on; immobilized Concanavalin A (Con A), DEAE ion-exchange resin and N-methylacridinium affinity column (Fig. 10a). The product showed a major single polypeptide (M_r 110,000) on sodium dodecyl sulphate/polyacrylamide gel electrophoresis (SDS-PAGE) when stained by Coomassie blue (Fig. 10b, track B). On pretreatment of the native product with 10^{-4} M [^{3}H] diisopropylfluorophosphate (DFP), the M_r 110,000 band was specifically labelled; this reaction was blocked when 10^{-4} M eserine was present with the [^{3}H]DFP. Hence, chick brain AChE has a single catalytic subunit of M_r 110,000.

(a)

Chick brain (1 day old) homogenate
↓
Wash in 140 mM NaCl
↓
Extract in 0.5 M NaCl/0.5% Triton X-100
↓
Concanavalin A-Agarose column
↓
a-methyl mannoside elution, dialysis
↓
DEAE-Sephacel column (0.5% Triton, 10 mM phosphate)
↓
0-250 mM NaCl gradient elution
↓
peak (75 mM +) dialysed
↓
N-methyl-acridinium (MAC)-Sepharose* column 0.05% Triton, 10 mM phosphate)
↓
10 mM Decamethonium elution
↓
Dialysis and concentration

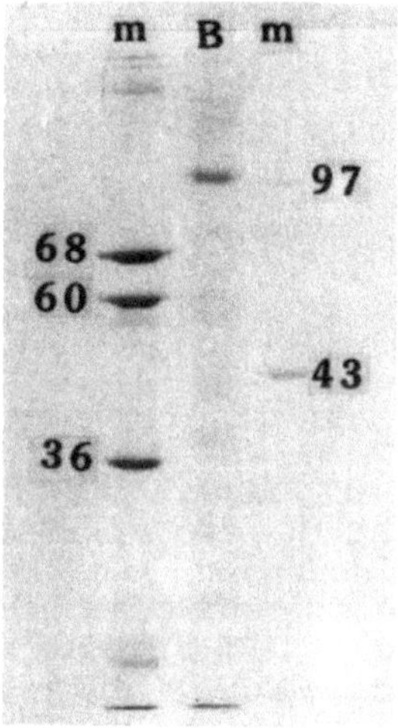

(b)

Figure 10. Purification of AChE from 1-day chick brain. Markers (in K_d) are shown in m tracks; purified AChE in B.

This product was employed to immunize Balb/C mice for the production of mAbs.

Three mAbs which bind brain AChE have been isolated, and characterised on nitrocellulose blots (Fig. 11). One mAb (ACB-1) specifically recognizes SDS-denatured AChE from both chick brain and muscle extracts; this mAb does not show any cross-reactivity with extracted oocyte proteins. A minor second band which stains in the brain extract represents a proteolytic fragment of AChE, since it becomes greatly enhanced when antiproteases are omitted from the extract preparation. ACB-1 does not recognize chicken serum pseudocholinesterase (ψChE), Brain AChE showed no decrease in catalytic activity when bound to ACB-1 at saturating concentrations, suggesting that the mAb does not alter the accessibility or structure of the catalytic site.

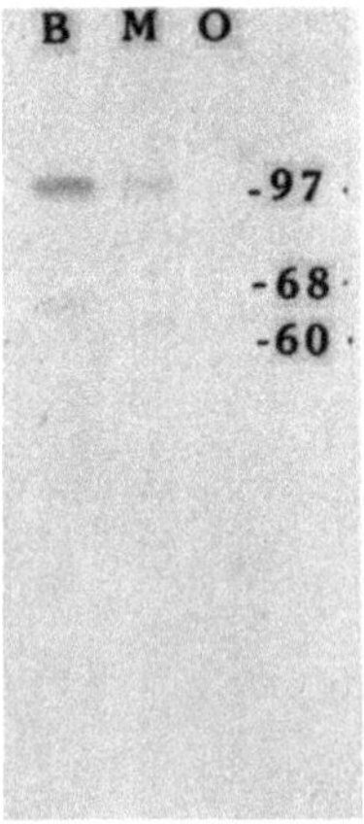

Figure 11. Specificity of monoclonal antibody (ACB-1) to chick AChE. Nitrocellulose-blotted proteins from SDS-PAGE were incubated with anti-AChE ACB-1 and stained after binding to peroxidase-linked anti-AChE anti-mouse Ig second antibody. ACB-1 binds to brain (B) and muscle (M) proteins at approximately 110 Kd from crude extracts but does not bind to extracted oocyte proteins (o).

Table 5. Immunoprecipitation of AChE from <u>Xenopus</u> Oocytes Injected with Chick Muscle mRNA.

mRNA	Immunoprecipitate CPM	Exp./Control
Muscle	3024	2.02
Fibroblast	1683	1.22
Control	1500	-

^{35}S-methionine (2 μCi/oocyte) was co-injected with the mRNA, followed by incubation in Barth's medium and antiproteases for 24 hours. Oocytes (10/group) were homogenized, extracted as described [36] and incubated with ACB-1 (approximately 20 μg/group) for 24 hours at 4°C. Antibody-antigen was insolubilized using protein-A-Sepharose 4B (20 μl/group), filtered through a glass wool plug, washed and solubilized with SDS and β-mercaptoethanol. 10 μl aliquots were blotted onto glass fibre filters and counted. Muscle = 1.9 day, embryonic pectoral; fibroblast = cultured chick fibroblasts; Control = HEPES buffer (10mM pH 7.5); Exp./control = experimental/control.

ACB-1 was used to immunoprecipitate AChE extracted from oocytes which had been co-injected with ^{35}S-methionine and the mRNA (Table 5). The immunoprecipitate was significantly labelled above that from controls, and contained approximately 0.9% of the total ^{35}S-methionine incorporated into the injected oocytes.

In summary, when oocytes execute the mRNA directed synthesis of AChE using chick muscle as the source, the time-course is similar to that found when Torpedo mRNA was used [37]). A notable characteristic of the muscle-mRNA-dependent AChE accumulation is that there is very little enzyme secreted into the medium. This is in agreement with a report showing similar localization in oocytes injected with ascidian muscle total RNA [38] and contrasts with the results found for AChE synthesis directed by rat brain mRNA or Torpedo electric organ mRNA, where the majority of the enzyme was secreted into the medium [37]. Proteolytic degradation cannot account for this difference since when an antiprotease mixture, was used, also, by others with Torpedo mRNA [39], the same result was obtained. Experiments are in progress to determine if chick brain mRNA-directed AChE activity is primarily secreted.

Oocytes will assemble, therefore, globular [40] and asymmetric oligomeric forms of AChE, albeit at low efficiency for the latter. One possibility for the observed low levels of the latter is that the mRNA for the "collagen-like" peptide is not translated to the same extent as the mRNA for the catalytic subunit of AChE. It is clear that injected foreign mRNA must compete with endogenous mRNA for polysomes [41, 42] but it is not known what factors influence the translatability of each specific mRNA. Another possibility is that post-translational modification is necessary prior to assembly and is not efficiently performed in the oocyte. However, supplementing the oocyte with cofactors necessary for procollagen synthesis does not enhance 20 S AChE appearance, suggesting that these are not limiting. The synthesis of procollagen has been reported after injection of mouse 3T6 mRNA into oocytes, indicating that foreign collagen proteins can be made in the oocyte [43]. However it was unclear from that study whether conversion from procollagen to collagen occurs correctly. Although glycosylation of foreign protein occurs in oocytes [44, 45] it is not always correct [46] or complete [47] and incomplete processing may affect the transport and metabolism of the collagen-tailed forms made in the oocyte. Monoclonal antibodies to AChE were developed, in part, to use for immunoprecipitation of chick-mRNA-induced AChE. We have demonstrated that ACB-1 does not cross-react with oocyte proteins. We have obtained several such anti-AChE mAbs, which recognise all of the forms of chick muscle AChE.

The isolation of mAbs which recognize molecular forms of AChE can now be combined with the immunoprecipitation method to extend the analysis of the components of the chicken muscle enzyme synthesized in the injected oocyte. A second application for such mAbs is immobilisation on a support and immuno-affinity purification of the AchE forms. This can lead to the isolation in pure form of all of the AChE forms discussed above.

Acknowledgements

W.R.R. holds a Postdoctoral Research Fellowship of the Muscular Dystrophy Association. J.L. holds a Croucher Foundation Studentship. This work was supported in part by the Muscular Dystrophy Group of Great Britain. We thank Prof. J. Massoulie (Paris) for providing the MAC-Sepharose gel.

References

1. Augustinsson, K.B. D. Nachmansohn. 1949. Science 110, 98-99.

2. Nachmansohn, D. C.W. Coates, R.T. Cox. 1941. J. Gen. Physiol. 25, 75-88.

3. Nachmansohn, D. E. Lederer. 1939. Bull. Soc. Chim. Biol. (Paris) 21, 797-808.

4. Marnay, A. 1937. C.R. Soc. Biol. (Paris) 126, 573-574.

5. Marnay, A., D. Nachmansohn. 1938. J. Physiol. 92, 37-47.

6. Couteaux, R. D. Nachmansohn. 1940. Proc. Soc. Exp. Biol. (N.Y.) 43, 177-181.

7. Rothenberg, M.A., D. Nachmansohn. 1947. J. Biol. Chem. 168, 223-231.

8. Lawler, H.C. 1959. J. Biol. Chem. 234, 799-801.

9. Wilson, I.B., D. Nachmansohn. 1950. J. Biol. Chem. 186, 781-790.

10. Nachmansohn, D. 1938. C.R. Soc. Biol. (Paris) 128, 599-673.

11. Nachmansohn, D. 1939. J. Physiol. 95, 29-35.

12. Nachmansohn, D. 1959. Chemical and Molecular Basis of Nerve Activity. Academic Press, New York.

13. Massoulié, J. F. Rieger. 1969. Eur. J. Biochem. 11, 441-455.

14. Massoulié, J., S. Bon. 1982. Ann. Rev. Neurosci. 5, 57-106.

15. Barnard, E.A., J.M. Lyles, J.A. Pizzey. 1982. J. Physiol. 331, 333-354.

16. Koenig, J., M. Vigny. 1978. Nature 271, 75-77.

17. Kato, A.C., A. Vrachliotis, B. Fulpius, Y. Durant. 1980. Dev. Biol. 76, 222-228.

18. Barnard, E.A., J.M. Lyles, I. Silman, J. Jedrzejczyk, P.J. Barnard. 1982. Reprod. Nutr. Develop. 22, 261-273.

19. Carson, S., M. Bon, M. Vigny, J. Massoulie, M. Fardeau. 1979. FEBS Lett. 97, 348-352.

20. Lyles, J.M., E.A. Barnard. 1980. FEBS Lett. 109, 9-12.

21. Lyles, J.M., I. Silman, L. Di Giamberardino, J.Y. Couraud, E.A. Barnard. 1982. J. Neurochem. 38, 1007-1021.

22. Jedrzejczyk, J., Silman, I., Lyles, J.M., Barnard, E.A. 1981. Biosci. Rep. 1, 45-51.

23. Jedrzejczyk, J., Silman, I., Lai, J., Barnard, E.A. 1984. Neuroscience Letters, 46, 283-289.

24. Barnard, E.A., Barnard, P.J., Jarvis, J., Jedrzejczyk, J., Lai, J., Pizzey, J.A., Randall, W.R. 1984. In: Cholinesterases (M. Brzin, E.A. Barnard, and D. Sket eds.) Walter de Gruyter, Berlin, pp. 49-71.

25. Vrbova, G., Gordon, T., Jones, R. 1979. Nerve-Muscle Interactions. Chapman and Hall, London.

26. Rubin, L.L., Schuetze, S.M., Weill, C.L., Fischbach, G.D. 1980. Nature 283, 264-267.

27. Rieger, F., Koenig, J., Vigny, M. 1980. Develop. Biol. 76, 358-365.

28. Pette, D., Smith, M.E., Staudte, H.W., Vrbova, G. 1973. Pflugers Arch. 338, 257-272.

29. Heilig, A., Pette, D. 1983. FEBS Lett. 151, 211-214.

30. Cisson, C.M., C.H. McQuarrie, J. Sketelj, M.G. McNamee, B.W. Wilson. 1981. Develop. Neurosci. 4, 157-164.

31. Chirgwin, J.M., A.E. Przybyla, R.J. MacDonald, W.J. Rutter. 1979. Biochemistry 18, 5294-5299.

32. Aviv, H., P. Leder. 1972. Proc. Nat. Acad. Sci. USA 69, 1408-1412.

33. Dumont, J.N. 1972. J. Morphol. 136, 153-179.

34. Johnson, C.D., R.L. Russell. 1975. Analyt. Biochem. 64, 229-238.

35. Soreq., H. 1984. Crit. Rev. Biochem. (in press).

36. Opdenakker, G., H. Weening, P. Collen, A. Billian, P. Se Somer. 1982. Eur. J. Biochem. 121, 269-274.

37. Soreq., H., R. Pavoni, I. Silman. 1982. Proc. Nat. Acad. Sci. USA 79, 830-834.

38. Meedel, T.H., J.R. Whittaker. 1983. Proc. Nat. Acad. Sci. USA 80, 4761-4765.

39. Soreq., H., R. Miskin. 1981. FEBS Lett. 128, 305-310.

40. Soreq., H., D. Zevin-Sonkin, N. Razon. 1984. EMBO. J. 3, 1371-1375.

41. Laskey, R.A., A.D. Mills, J.B. Gurdon, G.A. Partington. 1977. Cell 11, 345-352.

42. Richter, J.D., D.C. Evers, L.D. Smith. 1983. J. Biol. Chem. 258, 2614-2620.

43. Lane, C.D. and J.S. Knowland. 1975. In: The Biochemistry of Animal Development. (R.A. Weber, ed). Vol.3. Academic Press, New York. pp. 145-181.

44. Coleman, A., J. Morser, C.D. Lane, J. Besley, C. Wylie, G. Valle. 1981. J. Cell Biol. 91, 770-780.

45. Sumikawa, K., M. Houghton, J.S. Emtage, B.M. Richards, E.A. Barnard. 1981, Nature 292, 862-864.

46. Mous, J.M., B.L. Peeters, W.J. Heyns, W.S. Rombouts. 1982. J. Biol. Chem. 257, 11822-11828.

47. Kruppa, J., D. Darmes, H. Kaltaff, D. Richter. 1983. Eur. J. Biochem. 129, 537-542.

EXPRESSION OF ACETYLCHOLINESTERASE IN MURINE NEURAL CELLS IN VIVO AND IN CULTURE

Jean Massoulié+, Marc Vigny+ and Monique Lazar++

+ Laboratoire de Neurobiologie, Ecole Normale Supérieure, 46 rue d'Ulm, 75005 Paris, France.
++ Laboratoire de Biochimie Cellulaire, Collège de France, Place Marcelin Berthelot, 75005 Paris, France.

The polymorphism of acetylcholinesterase (E.C. 3.1.1.7) (for reviews, see Massoulié and Bon, 1982, Brimijoin, 1983, Massoulié et al., 1984) raises a number of intriguing questions regarding for example the genetic and biosynthetic relationships between the different globular and asymmetric molecular forms, and their subcellular localization.

We have examined the expression of acetylcholinesterase molecular forms in cultures of a murine neural cell line, T28, a neuroblastoma X sympathetic ganglion cell hybrid (Greene and Rein, 1977). These cells produce only globular forms of acetylcholinesterase, mostly the monomeric (G1) and tetrameric (G4) forms, with a minor proportion of dimers (G2) (Lazar and Vigny, 1980). This is similar to the situation observed in the central nervous system of mammals, where G1 and G4 are the major molecular forms, and asymmetric, collagen-tailed forms exist only in trace amounts (Grassi et al. , 1982). Moreover, the evolution of acetylcholinesterase during in vitro differenciation of T28 cultures is remarkably parallel to that observed in vivo during the maturation of the central nervous system in the rat (Rieger and Vigny, 1976, Muller et al. , in press). When the cells become confluent or when they are deprived of serum in the culture medium, they extend neurites, and their acetylcholinesterase content increases, as previously observed in the case of neuroblastoma cells (Vimard et al. , 1976; Kimhi et al. , 1976). This increase in acetylcholinesterase specific activity is accompagnied by a striking change in the proportion of the G1 and G4

Molecular Basis of Nerve Activity

forms. In dividing cells, in the logarithmic phase, the G1 form is largely predominant. Morphologically differentiated cells, in the stationary phase, possess approximately equal amounts of G1 and G4 forms. T28 cell cultures therefore appeared to represent an excellent experimental model for a study of the expression of acetylcholinesterase in mammalian nerve cells.

Endocellular and ectocellular compartments

In a first step, we examined the distribution of acetylcholinesterase molecular forms in intracellular and extracellular compartments. These compartments were defined experimentally by the sensitivity of the enzyme to non-permeant compounds. Phospholine, a phosphorylating inhibitor possessing a quaternary ammonium group, was found to react equally well with the G1 and G4 forms of acetylcholinesterase in a detergent solubilized cellular extract. When whole, living cells are incubated with this compound, the inhibition reaction proceeds in a more complex manner : a fraction of the enzyme is inactivated as rapidly as in solution, and another fraction is either totally resistant, or inactivated at a slower rate, depending on the concentration of inhibitor and on the incubation temperature. The first phase corresponds to the phosphorylation of the extracellularly accessible enzyme, and the second phase corresponds to the penetration of the inhibitor through the plasma membrane, and the inactivation of intracellular enzyme. After treatment with phospholine in carefully controlled conditions, it is thus possible to analyze the molecular forms of the non-inhibited intracellular enzyme by sedimentation in a sucrose gradient. We found that in the stationary phase the intracellular compartment contained most of the G1 form (72 to 80%) and only a small fraction of G4 (6 to 21%) (Lazar and Vigny, 1980). We obtained similar results in the case of dividing cells (unpublished results).

In T28 cells, therefore, G1 is essentially intracellular, while G4 represents the bulk of the enzyme that is associated with the plasma membrane, with an externally accessible catalytic site. This observation shows without ambiguity that the different forms correspond to really distinct molecular entities in the cells.

Solubilization and cellular localization

When stationary phase T28 cells are extracted in a detergent-free medium, about 1/4 of the acetylcholinesterase activity may be solubilized. This extract contains very little G4 (5% of total G4) but a large fraction of G1 (37% of total G1). A quantitative solubilization of the remaining particulate enzyme may be achieved in the presence of detergent, e.g. 1% Triton X100. This is very similar to the results obtained with the adult rat brain (Rieger and Vigny, 1976; Muller *et al.* , in press) or with the bovine caudate nucleus (Grassi *et al.* , 1982) : in both cases the detergent-free extract, which contains about 10% of the total activity, is considerably enriched in the monomeric G1 form. The solubilization characteristics of acetylcholinesterase are thus similar in T28 cells and in adult brain.

We obtained rather different results, in the case of embryonic rat brain (15 day-old embryos): a larger fraction of the enzyme (about 60%) may be solubilized in the absence of detergent, and the soluble extract contains the same proportion of G1 and G4 molecular forms as a total detergent extract. This may be related to recent findings, obtained in collaboration with Y. Berwald-Netter, on the endo and ectocellular components of acetylcholinesterase forms in primary cultures of neurons from embryonic rat brain (14 day old embryos): we found that the G1 and G4 forms were equally distributed between the two compartments, in contrast with the results obtained with T28 cells (unpublished results).

These results suggest that the enzyme that may be solubilized in the absence of detergent probably corresponds to a fraction of cytoplasmic acetylcholinesterase: in embryonic neurons, in culture or *in vivo* G1 and G4 appear to be equally distributed intra and extracellularly, and equally well solubilized; in the T28 cell cultures, on the contrary, the proportion of G1 is higher in the intracellular compartment, and the detergent-free extract is enriched in this form. In this respect, T28 cells resemble adult rather than embryonic brain tissue, even when they are in a non-differentiated state. It is therefore not possible to establish a valid parallelism between *in vitro* cellular differentiation of T28 cells and brain maturation.

Solubility, amphipathic character and molecular structure

Rosenberry and his collaborators have recently showed, in the case of acetylcholinesterase from human erythrocyte membranes, that the enzyme possesses a small hydrophobic peptide (about 2000 daltons) which may anchor the enzyme at the external surface of the plasma membrane, or bind a detergent micelle (e.g. Triton X100). This peptide may be labelled by a lipid-soluble reagent, and may be cleaved by proteolytic digestion (Dutta-Choudhurry and Rosenberry, 1984; Rosenberry and Scoggin, 1984). Brodbeck and cowokers have reached a similar conclusion in the case of *Torpedo* membrane-bound dimers (Stieger *et al.* , 1984). The digestion of membrane vesicles with proteases such as papain or pronase solubilizes an active acetylcholinesterase lytic form, which no longer interacts with non-denaturing detergents (Li and Bon, 1983 ; Dutta-Choudhury and Rosenberry, 1984).

It is important to note that detergent-interacting, amphipathic molecules which possess a hydrophobic peptide and may be converted by proteolytic treatment into lytic, detergent-insensitive molecules, may be solubilized in the absence of detergent. It is possible that such molecules exist in loose association with intracellular membranes, e.g. in the endoplasmic reticulum. Amphipathic acetylcholinesterase molecules also occur in soluble form in extracellular media (see next section).

Are amphipathic and detergent-insensitive acetylcholinesterase molecules entirely different proteins, as suggested by several authors, or do they differ only by the presence of a small hydrophobic peptide? In order to partially answer this question, we have studied their immunological reactions with antisera raised against pure detergent-soluble G4 acetylcholinesterase from rat brain (Marsh *et al.*, 1984). We found no difference between the buffer-soluble and detergent-soluble fractions. These populations of enzyme molecules may therefore derive from the same gene (or genes), by differential splicing of a pre-messenger RNA, or by post-transcriptional maturation processes.

Secretion

Acetylcholinesterase is normally secreted by nerve and muscle cells. It is for example released from cerebral structures, particularly from dopaminergic nigral cells, in an activity-regulated manner (Greenfield et al., 1980, 1983). It is also released from nerve terminals at neuromuscular junctions, upon stimulation of motoneurons (Skau and Brimijoin, 1978). In this case, the secreted enzyme is mostly G4, with a small proportion of G1.

The T28 cells actively secrete the three globular forms of acetylcholinesterase. We found that in the stationary phase, the accumulation of enzyme in the culture medium was linear for at least two days, with a rate of secretion of about 1.4 % of the cellular activity released per hour.

We studied the effect of Triton X100 on the secreted forms, and found that some of the secreted enzyme shows a clear shift in its sedimentation : for example the secreted G1 form sedimented at 4.6 S (± 0.2 S) in the absence of detergent and 3.8 S (± 0.2 S) in the presence of Triton X100. The G2 form was also detergent-sensitive, and the G4 form was composed of detergent sensitive and detergent-insensitive components, in variable proportions (Lazar et al., 1984). It is interesting however that the secreted enzyme appeared to differ slightly from the soluble, detergent-interacting cellular enzyme, which sedimented at 3.4 S (+ 0.2 S) in the presence of Triton X100. In any case, this observation shows very clearly that amphipathic molecules may occur in truly soluble form. Naturally soluble amphipathic acetylcholinesterase has also been characterized in vivo, in Torpedo plasma (Toutant et al., in preparation).

Metabolic pools

T28 cells offer an interesting and favourable material for

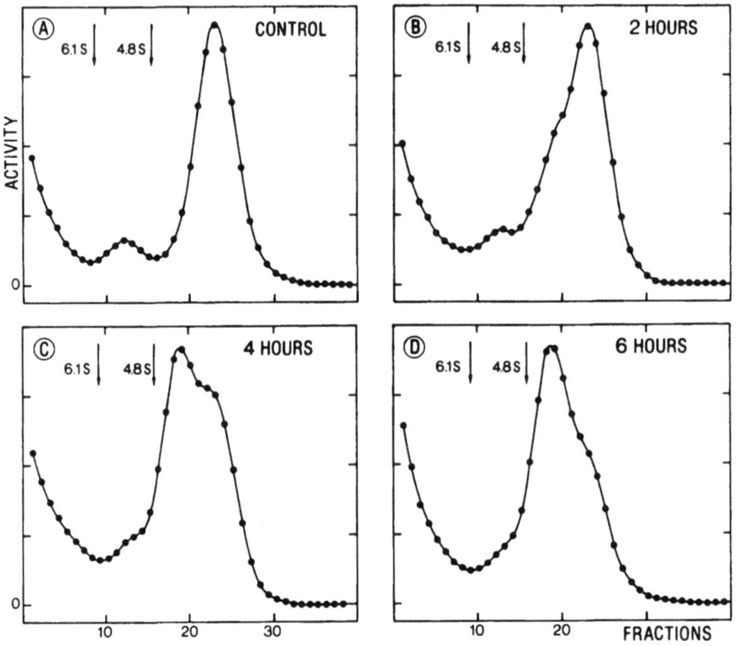

Figure 1 - Evolution of the sedimentation pattern of the cellular G1 acetylcholinesterase forms after transfer into the heavy medium

Cellular extracts were obtained after incubation of the cells for the indicated periods of time in the heavy medium and were centrifuged in detergent-saline sucrose gradients, at 40,000 rpm for 48 hours. The arrows indicate the positions of the internal sedimentation standards, E. coli alkaline phosphatase (6.1 S) and horse liver alcohol dehydrogenase (4.8 S). A, control; B, 2 hours; C, 4 hours; D, 6 hours. (Reproduced from Lazar et al., 1984).

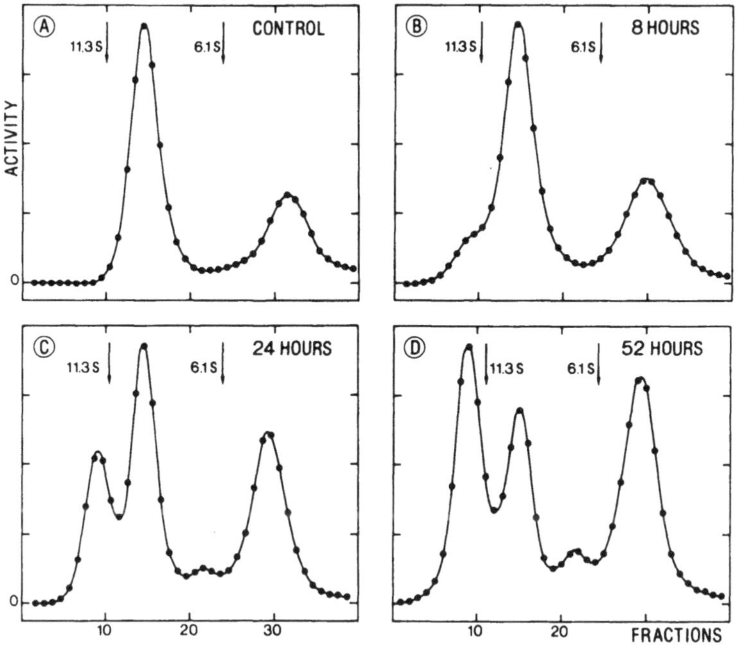

Figure 2 - Evolution of the sedimentation pattern of the cellular G4 acetylcholinesterase form after transfer into the heavy medium

The experiment was performed as indicated in figure 1 except that the gradients were centrifuged for 24 hours. The internal sedimentation standards were catalase (11.3 S) and *E.coli* alkaline phosphatase (6.1 S). *A,* control; *B,* 8 hours; *C,* 24 hours; *D,* 52 hours. (Reproduced from Lazar *et al.* , 1984).

investigating the metabolic relationships between the various acetylcholinesterase molecular forms, for several reasons : they constitute a homogenous cellular population, producing and secreting mostly G1 and G4 forms of the enzyme, and segregating these forms in a simple manner, with predominantly intracellular G1 and predominantly extracellular, membrane-bound G4. These cells exhibit neuronal-like morphological and biochemical differentiation in vitro .

We analyzed the renewal of acetylcholinesterase in these cells by the method of heavy isotope labelling (Lazar et al. , 1984). This method, introduced by Fambrough and his colleagues to determine the rate of metabolic turnover of the acetylcholine receptor molecule (Devreotes et al. , 1977; Gardner and Fambrough, 1979) appeared particularly well suited to analyze the renewal rates of the different forms of acetylcholinesterase. It makes it possible to separate physically light and heavy molecules by differential sedimentation or by equilibration in a density gradient and to determine their distribution by a specific assay (ligand-binding or enzymatic activity). In the case of acetylcholinesterase we may thus take advantage of the high sensitivity of detection of its catalytic activity in total cellular extracts or in culture media. We wished to analyze simultaneously the different molecular forms of the enzyme and we found that this was possible by sucrose gradient centrifugation. We obtained a satisfactory resolution of the light and heavy components of the G1 and G4 forms, by choosing correct sedimentation conditions.

We studied in this manner the replacement of pre-existing light enzyme molecules by newly synthesized heavy molecules, after transferring the cells from normal to heavy medium. We verified that the cells maintained constant levels of cellular enzyme forms and a constant rate of secretion, so that it was possible to attempt a steady state analysis of metabolic pools.

The most obvious observation is that the G1 form is renewed much more rapidly than the G4 form, as shown in figures 1 and 2. A more detailed analysis of the data was made possible by estimating precisely the relative proportions of the light and heavy components of the G1 and G4 forms (we did not attempt to analyze the minor G2

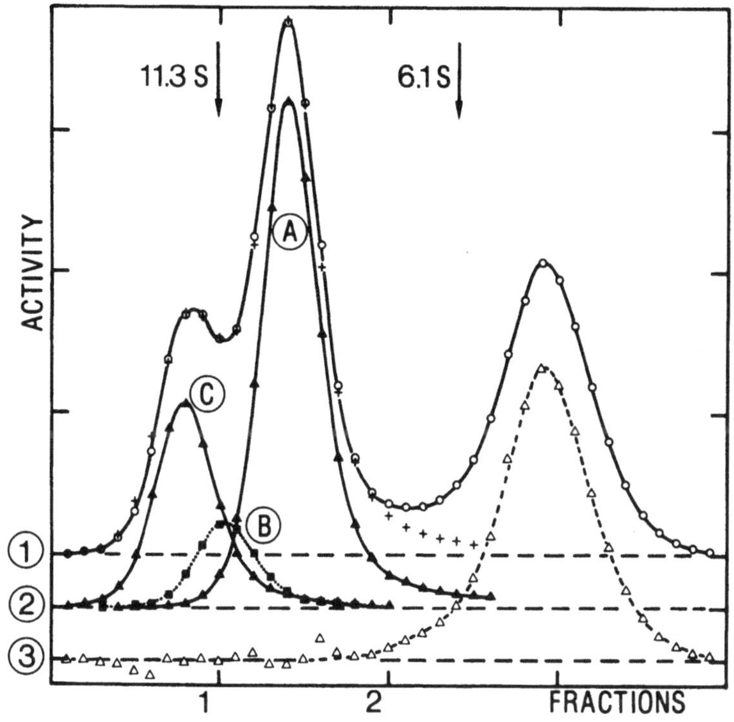

Figure 3 - Example of analysis of a G4 sedimentation pattern

Sedimentation profile of the G4 form corresponding to a cellular extract obtained after 18 hours in the heavy medium.

○——○ experimental curve

+- - -+ reconstructed curve (sum of elementary distributions)

△——△ difference between experimental and reconstructed curves

The elementary distributions correspond respectively to:

A (▲——▲), light G4 component

B (■——■) intermediate (hybrid) G4 component

C (▲——▲), heavy G4 component

(Reproduced from Lazar et al., 1984).

form), with the help of a computer reconstitution procedure. In this method, we assumed that the distribution functions of the light and heavy components of a given form in a gradient were identical, except for a difference in their sedimentation rates, i.e. in their positions along the tube. We observed however that the distributions of G1 and G2 forms were different in their half-widths, so that in our attempts to reconstitute the multi-component sedimentation profiles, we used for elementary distributions the experimental curves obtained for purely light enzyme forms in control gradients. In this manner, we were able to obtain a satisfactory fit between the experimental and the reconstituted curves.

In the case of G1, the sedimentation profiles could all be fitted with two components, representing purely heavy and purely light molecules. This showed that the replacement of the amino acid pool could be considered as instantaneous, in comparison with the rate of renewal of acetylcholinesterase molecules.

In the case of the G4 form, we had to include in some cases an intermediate component of intermediate sedimentation coefficient (figure 3). We verified the existence of such hybrid heavy-light polymers, by reanalyzing the corresponding fractions in sucrose gradients. In agreement with numerous observations showing that the catalytic subunits are not in a dissociation-reassociation equilibrium, this observation shows that the more complex forms are produced by specific biosynthetic processes. We found that hybrids were not formed when a mixture of purely heavy and light G4 forms was analyzed in the same conditions.

The data obtained by analyzing experimental sedimentation profiles such as those shown in figures 1 and 2 may be summarized by a series of curves representing the evolution of the various light and heavy molecules in cellular extracts and in the culture medium, after transferring the cells from a light to a heavy medium (figure 4). These curves could be fitted with theoretical models which showed a number of interesting common features :

1) Acetylcholinesterase appears to be first synthesized as an inactive precursor, and rapidly converted into active enzyme (half-life of about 30 minutes).

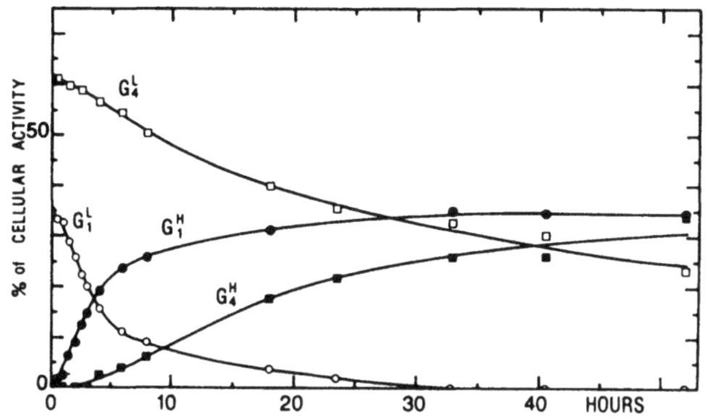

Figure 4 - Evolution of light and heavy molecular forms in cellular extracts after transfer into the heavy medium

The activities of the light G1 form (GL1,○), heavy G1 form (GH1, ●), light G4 form (GL4, □) and heavy G4 form (GH4, ■) were calculated from the overall proportions of G1 and G4 and from the proportions, within each form, of its heavy and light components. They are expressed as fractions of the total cellular activity.

(Reproduced from Lazar et al., 1984).

2) The monomeric form G1, which is rapidly renewed, consists in fact of several metabolic pools. Our models included three pools with half-lives on the order of 1, 5 and 10 hours (fast, medium and slow : G1f, G1m, G1s).

3) Most of the synthesized G1 form is degraded (80%), only a minor fraction being converted into G4 and exposed at the outer cellular surface, or secreted.

4) The G4 form derives from a rapidly renewed precursor, such as G1s (or possibly a distinct inactive precursor).

5) The G4 form consists of at least two components with half-lives of about 10 and 40 hours.

6) Heavy molecules appear in the culture medium approximately 3 hours after the change of medium.

7) The time course of the continued secretion of light molecules indicates that they derive from at least two cellular pools.

a) G4 derives partially from a slowly renewed G4 pool.

b) G1 and part of G4 appear to derive from the medium G1 pool (G1m), which is exhausted within a few hours, in the heavy medium.

8) The fact that secreted G4 may derive from a cellular G1 pool suggests that a polymerization step is coupled or closely linked with secretion.

These various aspects of acetylcholinesterase metabolism indicate that complex cellular processes are involved in the synthesis, transport, localization and secretion of the enzyme. The main features present a remarkable convergence with the results obtained by Rotundo (1984), who recently studied the metabolism of acetylcholinesterase in primary cultures of chick muscle cells, by radio-labelling and immunoprecipitating the newly formed enzyme. In this case also the enzyme was first produced as an inactive precursor, and again, 90% of the synthesized protein was rapidly degraded. (In the chick cells, degradation occurred at the level of inactive precursors; the method used in our study of T28 cells allowed us to observe only active enzyme: we noted a rapid degradation of G1, but we have no information about the inactive precursors).

Rotundo (1984) also studied the binding of lectins (concanavalin A, wheat germ agglutinin and ricin) as an indication of the sequential acquisition of N-acetyl-galactosamine and terminal galactose by a fraction of the enzyme. This showed that the formation of globular forms occurred before the enzyme reached the Golgi and acquired complex sugars, whereas asymmetric forms were assembled in the Golgi. As observed previously (Rotundo and Fambrough, 1980) the enzyme reached the cell's surface about 3 hours after its synthesis, i.e. approximately the period observed for the appearance of secreted heavy molecules in the culture medium of T28 cells.

One of the most puzzling aspect of these findings is the wastefulness of the biosynthetic process, in which only a small fraction of the enzyme reaches the cellular surface where it is expected to exert its physiological role. This possibly allows the cells to rapidly respond, if submitted to a physiological demand, by providing the adequate molecular forms of acetylcholinesterase, and direct them to the relevant subcellular sites.

References

-Brimijoin, S. (1983) Molecular forms of acetylcholinesterase in brain, nerve and muscle : nature, localization and dynamics. Progress in Neurobiology, 21 , 291-322.

-Devreotes, P.N., Gardner, J.M. and Fambrough, D.M. (1977) Kinetics of biosynthesis of acetylcholine receptor and subsequent incorporation into plasma membrane of cultured chick skeletal muscle. Cell, 10, 365-373.

-Dutta-Choudhury, T.A. and Rosenberry, T.L. (1984) Human erythrocyte acetylcholinesterase is an amphipathic protein whose short membrane-binding domain is removed by papain digestion. J. Biol. Chem., 259 , 5653-5660.

-Gardner, J.M. and Fambrough, D.M. (1979) Acetylcholine receptor degradation measured by density labeling. Effects of cholinergic ligands and evidence against recycling. Cell, 16 , 661-674.

-Grassi, J., Vigny, M. and Massoulié, J. (1982) Molecular forms of acetylcholinesterase in bovine caudate nucleus and superior cervical ganglion: solubility properties and hydrophobic character. J. Neurochem., 38 , 457-469.

-Greene, L.A. and Rein, G. (1977) Dopaminergic properties of a somatic cell hybrid line of mouse neuroblastoma X sympathetic ganglion cells. J. Neurochem., 29 , 141-150.

-Greenfield, S., Chéramy, A., Leviel, V. and Glowinski, J. (1980) In vivo release of acetylcholinesterase in the cat substantiae nigrae and caudate nuclei. Nature, 284 , 355-357.

-Greenfield, S.A., Chéramy, A. and Glowinski, J. (1983) Evoked release of proteins from central neurons in vivo. J. Neurochem., 40, 1048-1057.

-Kimhi, Y., Palfrey, C., Spector, I., Barak, Y. and Littauer, U.Z. (1976) Maturation of neuroblastoma cells in the presence of dimethylsulfoxide. Proc. Natl. Acad. Sci. U.S.A., 73 , 462-466.

-Lazar, M. and Vigny, M. (1980) Modulation of the distribution of acetylcholinesterase molecular forms in a murine neuroblastoma X sympathetic ganglion cell hybrid cell line. J. Neurochem. 35, 1067-1079.

-Lazar, M., Salmeron, E., Vigny, M. and Massoulié, J. (1984) Heavy isotope-labeling study of the metabolism of monomeric and tetrameric acetylcholinesterase forms in the murine neuronal-like T28 hybrid cell line. J. Biol. Chem., 259 , 3703-3713.

-Li, Z.I. and Bon, C. (1983) Presence of a membrane-bound acetylcholinesterase form in a preparation of nerve endings from Torpedo marmorata electric organ. J. Neurochem., 40 , 338-349.

-Marsh, D.J., Grassi, J., Vigny, M. and Massoulié, J. (1984) An immunological study of rat acetylcholinesterase: comparison with acetylcholinesterase from other vertebrates. J. Neurochem. 43 , 204-213.

-Massoulié, J. and Bon, S. (1982) The molecular forms of cholinesterase and acetylcholinesterase in vertebrates. Ann. Rev. Neurosci., 5 , 57-106.

-Massoulié, J., Bon, S., Lazar, M., Grassi, J., Marsh, M., Méflah, K., Toutant, J.P., Vallette, F. and Vigny, M. (1984) The polymorphism of cholinesterases: classification of molecular forms; interactions and solubilization characteristics; metabolic relationship and regulations, in "Cholinesterases : Fundamental and Applied Aspects" (Brzin, M., Kiauta, T. and Barnard, E.A., eds) Walter de Gruyter, Berlin, 93-97.

-Muller, F., Dumez, Y. and Massoulié, J. Molecular forms and solubility of acetylcholinesterase during the embryonic development of rat and human brain. Brain Res., in press.

-Rieger, F. and Vigny, M. (1976) Solubilization and physico-chemical characterization of rat brain acetylcholinesterase: development and maturation of its molecular forms. J. Neurochem., 27 , 121-129.

-Rosenberry, T.L. and Scoggin, D.M. (1984) Structure of human erythrocyte acetylcholinesterase. Characterization of inter-subunit disulfide bonding and detergent interaction. J. Biol. Chem., 259, 5643-5652.

-Rotundo, R.L. and Fambrough D.M. (1980) Secretion of acetylcholinesterase: relation to acetylcholine receptor metabolism. Cell, 22 , 595-602.

-Rotundo, R.L. (1984) Asymmetric acetylcholinesterase is assembled in the Golgi apparatus. Proc. Natl. Acad. Sci. USA, 81 , 479-483.

-Stieger, S., Brodbeck, U., Reber, B. and Brunner, J. (1984) Hydrophobic labeling of the membrane binding domain of acetylcholinesterase from Torpedo marmorata. FEBS Lett., 168 , 231-234.

-Skau, K.A. and Brimijoin, S. (1978) Release of acetylcholinesterase from rat hemidiaphragm preparations stimulated through the phrenic nerve. Nature, 275 , 224-226.

-Vimard, C., Jeantet, C., Netter, and Gros, F. (1976) Changes in the sedimentation properties of acetylcholinesterase during neuroblastoma differentiation. Biochimie, 58 , 473-478.

THE INVOLVEMENT OF PHOSPHATIDYLINOSITOL IN THE ANCHORING OF HYDROPHOBIC FORMS OF ACETYLCHOLINESTERASE TO THE PLASMA MEMBRANE

A.H.Futerman[1], P.L.Barton[1*], R.M.Fiorini[1**], M.G.Low[2], W.R.Sherman[3] and I.Silman[1]

[1]Dept. of Neurobiology, Weizmann Institute of Science, Rehovot 76100 (Israel), [2]Cardiovascular Research Program, Oklahoma Medical Research Foundation, Oklahoma City, OK 73104 (USA), and [3]Dept. of Psychiatry, Washington University School of Medicine, St.Louis, MO 63110 (USA).

Introduction

Acetylcholinesterase (Acetylcholine hydrolase; EC 3.1.1.7., AChE) exists in multiple molecular forms which can be characterized by their sedimentation coefficients, reflecting differences in quaternary structure, and by their solubility characteristics, reflecting differences in their mode and degree of attachment to cellular structures (1). In *Torpedo* electric organ, a substantial part of the AChE occurs as a hydrophobic, catalytic subunit dimer (2-4) which displays the solubility characteristics of an integral membrane protein inasmuch as it is only soluble in the presence of detergent (DSAChE). Much of the DSAChE in *Torpedo* is associated, presumably through hydrophobic interactions, with the external surface of the presynaptic plasma membrane, where it is the only molecular form of AChE detected (5-7).

We have recently shown that this hydrophobic form of AChE is selectively solubilized, in the absence of detergent, both from the particulate fraction and from a purified synaptosomal preparation of *Torpedo* electric organ by a phosphatidylinositol-specific phospholipase C (monophosphatidylinositol phosphodiesterase, EC

Permanent addresses:
* Dept. of Biochemistry, University of Bath, England, (U.K.);
** Institute of Biochemistry, Faculty of Medicine, University of Ancona, (Italy).

Molecular Basis of Nerve Activity

3.1.4.10., PIPLC) from Staphylococcus aureus (8-10). This raises the possibility that the catalytic subunit of AChE is anchored to the membrane via a specific interaction with the head-group(s) of one or more PI molecules. In this report, we present evidence to support this idea, and suggest that PI is indeed tightly attached to DSAChE and may be responsible for anchoring the enzyme to the plasma membrane.

Solubilization of AChE by PIPLC

On addition of PIPLC to presynaptic plasma membranes from Torpedo ocellata electric organ, AChE is solubilized in a concentration-dependent manner (Fig. 1), with almost quantitative solubilization being achieved at the highest PIPLC concentrations.

In order to confirm that the highly purified PIPLC preparation which we used in this study is devoid of proteolytic activity (11), we examined its ability to degrade ^{125}I-casein. PIPLC shows no detectable proteolytic activity towards ^{125}I-casein, even at levels far above those at which it solubilizes membrane-bound AChE, whereas various proteases tested extensively degrade the ^{125}I-casein under similar conditions(10). Thus, the release of AChE by PIPLC cannot be due to a proteolytic contaminant and must presumably be ascribed to its phospholipase activity.

Detergent-soluble forms of AChE exist in a number of tissues. However, although PIPLC is able to quantitatively solubilize Torpedo DSAChE (Fig. 1) as well as AChE from intact bovine and rat erythrocytes (Fig. 2), without any concomitant haemolysis, it solubilizes AChE only partially from mouse and human erythrocytes (Fig. 2), and not at all from various brain regions of four different mammalian species(10).

In Torpedo electric organ, PI accounts for ca. 2% of the total phospholipids (13) and it can be readily calculated from the amount of AChE in this tissue that PI is in huge excess on a molar basis with respect to the enzyme. In human erythrocytes, which are a much less rich source of AChE, PI also accounts for ca. 2% of the total PI (14);

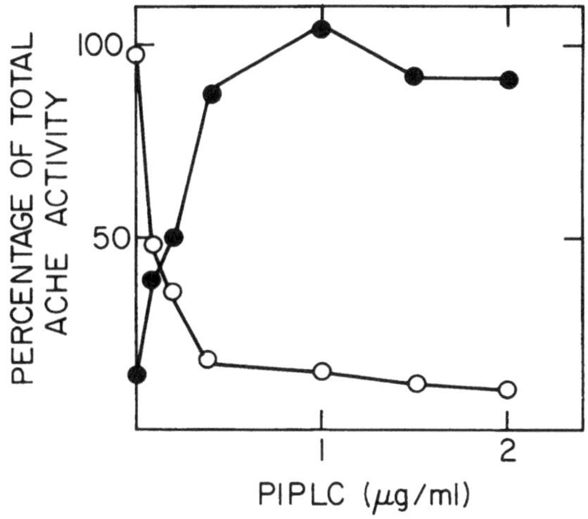

Fig. 1. Solubilization of AChE from presynaptic plasma membranes of Torpedo ocellata electric organ by PIPLC.

Synaptic membranes (ca. 1.2mg protein/ml) from *Torpedo ocellata* electric organ, suspended in 10mM Tris, pH 8.0, were incubated with increasing concentrations of PIPLC at 37°C. After 30min, samples were centrifuged at 100,000g for 10min in a Beckman airfuge and AChE activity was measured by a radiometric assay (12) in the pellet (○) and the supernatant (●).

it is, therefore, unlikely that the partial solubilization of AChE from human erythrocytes is due to a limiting amount of PI. Three possibilities may be considered for the differential susceptibility of AChE to PIPLC: a) The susceptible bond in the PI molecule may not be accessible to PIPLC due to steric reasons deriving from the molecular topography of the membrane; b) Part of the PI may be chemically modified (e.g. by phosphorylation) in a fashion which renders it resistant to PIPLC; c) Part of the AChE may be anchored by a totally different mechanism. With respect to this last possibility it is worth noting that whereas *Torpedo* and erythrocyte DSAChE correspond to 'globular' G_2 dimers, the bulk of the particulate AChE in mammalian brain corresponds to a G_4 tetramer (15-17). Although we have not made a detailed study of the mode of action of PIPLC on skeletal muscle AChE, a small but significant solubilization of particulate AChE was

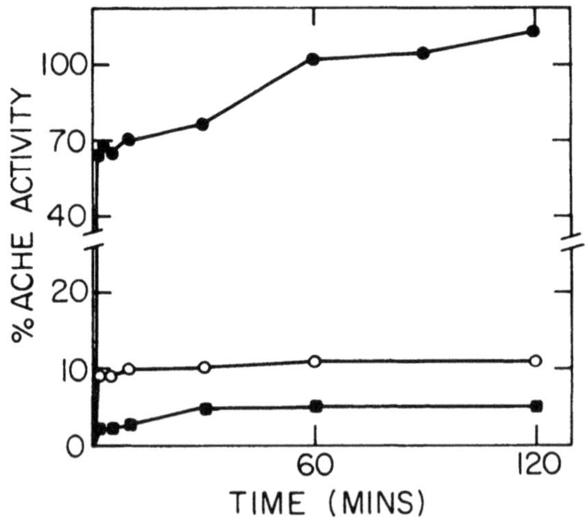

Fig. 2. Solubilization of AChE from erythrocytes by PIPLC. Washed erythrocytes (100µl) were incubated with PIPLC (8µg/ml) at 37°C. At appropriate times, aliquots were removed and centrifuged as described for Fig. 1. AChE activity was measured in the supernatant and taken as the percentage activity of that solubilized in Triton X-100 (100%). This DSAChE corresponds to greater than 95% of the total AChE activity in all erythrocytes examined. ●——● rat erythrocytes, ○——○ human erythrocytes, ■——■ mouse erythrocytes. All three curves present data from which background release in the absence of PIPLC (less than 1% of total activity) has been subtracted.

found in non-end plate regions of rat diaphragm (Futerman, Low and Silman, unpublished results), where lighter globular forms of AChE predominate (18).

Earlier literature showed that only a limited number of membrane-bound enzymes are solubilized by bacterial PIPLC (reviewed in ref. 19), and our present data confirm the selectivity of its action, since in *Torpedo* synaptic membranes the polypeptide corresponding to AChE is the only one visualized which is detectably solubilized(10). Furthermore, no other phospholipases tested, including a non-specific phospholipase C from *Staphylococcus aureus*, were able to solubilize DSAChE from *Torpedo* homogenates or synaptic membranes, mammalian brain or erythrocytes, pointing to a special mode of anchoring of AChE to the membrane, involving PI (or a PI derivative)(10). Additionally, it has been shown (20), that in rat hepatocytes the greater part of the alkaline phosphatase on the surface membrane could be solubilized by PIPLC with hydrolysis of as little as 1% of the total PI, suggesting that release of certain ectoenzymes by PIPLC is not due to a non-specific degradation of membrane PI.

Similarly to its action in releasing AChE quantitatively from the external surface of intact erythrocytes, PIPLC could almost completely solubilize the G_2 form of AChE from intact Torpedo synaptosomes (10). This is in agreement with the demonstration of Morel and Dreyfus (6) that the enzyme is on the outer surface of the synaptic membrane, and the observation of Li and Bon (7) that it is possible to remove up to 98% of the AChE activity from intact synaptosomes with pronase, without disrupting the integrity of the synaptosomes. Our observations suggest that PIPLC releases AChE from the outer surface of the synaptosome by attack on PI located on the same surface, just as in the case of the intact erythrocyte (10,21,22). Although the presence of PI in synaptosomal membranes of Torpedo electric organ has not been directly established, it is present in significant amounts in the whole electric organ as well as in synaptic vesicles (13,23). Interestingly, much of it is present in these latter organelles in the outer leaflet of the vesicle bilayer (23).

Physicochemical behaviour and structural characteristics of PIPLC-solubilized AChE

Torpedo DSAChE has an apparent sedimentation coefficient which depends strongly on the detergent employed, ranging from ca. 5.0S in Brij 96 up to ca. 7.3S in cholate. The apparent sedimentation coefficient (s) of DSAChE solubilized by PIPLC in the absence of detergent is 7.2S (8). Addition of various detergents to the PIPLC-solubilized AChE has no influence on its sedimentation coefficient. For instance, AChE solubilized with Brij 96 or Triton X-100 has sedimentation coefficients of 5.0S and 6.3S respectively, in agreement with earlier reports (2,3), whereas the sedimentation coefficient of

PIPLC-solubilized AChE remains at ca. 7.2S even in the presence of either of these detergents. The effect of PIPLC thus appears similar, in physicochemical terms, to that of Proteinase K, which was reported earlier (2) to render DSAChE insensitive to detergent by this criterion, a finding which we have confirmed.

DSAChE purified by affinity chromatography after cholate solubilization has a sedimentation coefficient of 7.3S (Fig. 3A); if this enzyme is diluted 100-fold into Brij 96, the sedimentation coefficient decreases to 5.4S (Fig.3B), i.e. to a value similar to that obtained by direct solubilization with Brij 96. However, if incubation of the cholate-purified DSAChE with either PIPLC (Fig. 3C) or Proteinase K (Fig. 3D) precedes dilution of the enzyme into Brij 96, no such decrease is observed.

These results suggest that one or more molecules of PI must be cosolubilized with DSAChE in the presence of cholate and remain tightly attached to the enzyme throughout purification, since we have clearly demonstrated modification by PIPLC of the physicochemical behaviour of the purified enzyme. Thus, after digestion of cholate-purified DSAChE with PIPLC, the sedimentation coefficient of DSAChE is no longer sensitive to detergent (Fig. 3).

DSAChE, purified after cholate solubilization, binds almost completely to liposomes composed of a PC/PI mixture (Fig. 4B), whereas it fails to interact after prior treatment with PIPLC (Fig. 4D). AChE purified after Proteinase K solubilization also fails to interact with such liposomes (Fig. 4C). DSAChE, purified after cholate solubilization, binds similarly to liposomes composed of PS and PC, but AChE directly solubilized with PIPLC is unable to bind to PS and PC liposomes (24).

It is of particular interest that DSAChE treated with PIPLC does not bind even to liposomes containing PI. The effect of PIPLC suggests that it either eliminates a site on the enzyme responsible for binding of DSAChE to the plasma membrane or cleaves a phosphate-glycerol bond in such a manner as to leave a fragment of PI (or a derivative thereof) adhering to the enzyme, which blocks reconstitution. Similarly, alkaline phosphatase solubilized from pig kidney microsomes by PIPLC is also unable to bind to phospholipid vesicles, even if they contain PI, whereas alkaline phosphatase

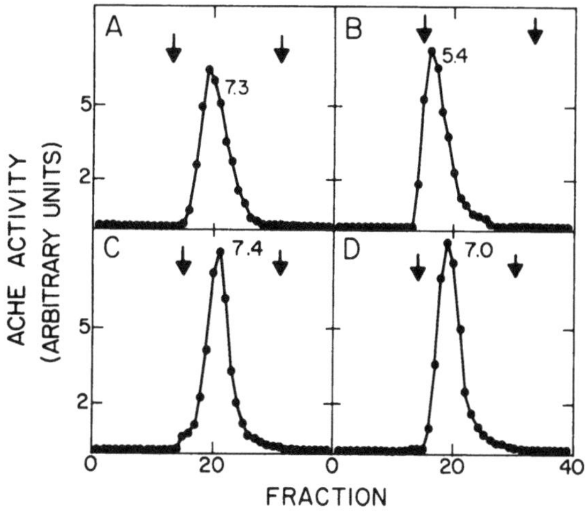

<u>Fig. 3. Sedimentation profiles of cholate-purified DSAChE before and after treatment with PIPLC or Proteinase K, and subsequent dilution into Brij 96.</u>

Cholate-purified DSAChE (see 24), (2µl, ca. 380µg/ml) was diluted into 200µl of 40mM $MgCl_2$-0.1M NaCl-10mM Tris, pH 8.0, containing either 1% cholate or 1% Brij 96; after 1h at room temperature, the whole sample was layered on top of a sucrose gradient containing the same detergent used for dilution. Centrifugation was for 16h at 39,000rpm in a Beckman L8-70 ultracentrifuge using an SW40Ti rotor. A) Cholate-purified AChE diluted into 1% cholate. B) Cholate-purified AChE diluted into 1% Brij 96. C) Cholate-purified AChE after incubation with PIPLC (25µg/ml, 37°C, 30min) diluted into 1% Brij 96. D) Cholate-purified AChE after incubation with Proteinase K (50µg/ml, 25°C, 30min) diluted into 1% Brij 96. The arrows denote the position of the marker proteins, alcohol dehydrogenase (5.0S) and catalase (11.4S), with the top of the gradients at the left. The sedimentation coefficient of AChE is given in each panel.

solubilized by butanol does bind to liposomes (25). The failure of PIPLC-treated AChE and PIPLC-treated alkaline phosphatase to bind to liposomes thus favours the latter possibility that a fragment of PI remains tightly attached to the enzyme after PIPLC treatment. The similar loss of detergent-sensitivity and ability of AChE to bind to liposomes after treatment with Proteinase K might then be ascribed to proteolytic detachment of a peptide to which the PI residue(s) would be strongly attached.

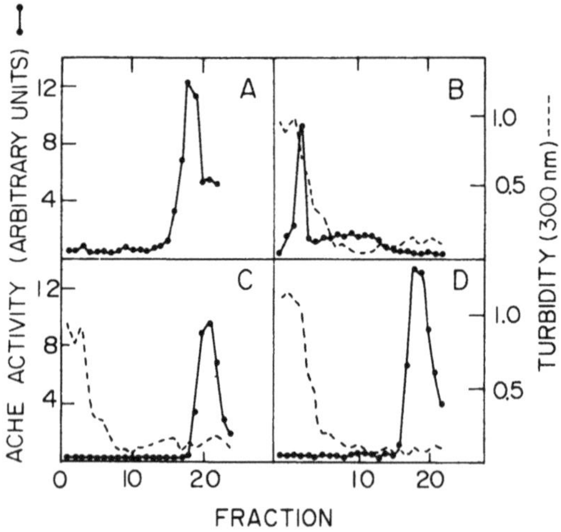

Fig. 4. Interaction of DSAChE with liposomes of defined composition
Flotation gradients (see 24) were prepared as follows: 3µl of cholate-purified AChE was mixed with 280µl of sonicated liposomes. Step gradients consisted, from the bottom up, of 0.5ml of 2M sucrose; 0.7ml of the AChE-liposome mixture in 1.2M sucrose, 3.1ml of 1M sucrose, 0.5ml buffer. The buffer, in which sucrose solutions were also made up, was 0.1M NaCl-0.01M phosphate, pH 7.0. A) Cholate-purified DSAChE in the absence of liposomes. B) Cholate-purified DSAChE with PC/PI liposomes. C) AChE, purifed after solubilization with Proteinase K, with PC/PI liposomes. D) Cholate-purified DSAChE after PIPLC treatment (25µg/ml, 37°C, 30min) with PC/PI liposomes.

In order to determine the size of the Proteinase K susceptible peptide, AChE samples labelled with ^{3}H-diisopropyl phosphorofluoridate (^{3}H-DFP) were coelectrophoresed with a mixture of ^{14}C-methylated proteins. However, SDS-polyacrylamide gel electrophoresis analysis revealed no significant difference in apparent molecular weight (M_r) between intact DSAChE and the enzyme modified with either PIPLC or Proteinase K (Table 1). The catalytic polypeptide derived from high-salt-soluble AChE migrated almost together with the bovine serum albumin marker, and was clearly ca. 2Kda heavier than the catalytic polypeptide of DSAChE. This observation confirms that made by Witzemann and Boustead (26) who similarly compared the corresponding forms of AChE from <u>Torpedo marmorata</u>.

Table 1: M_r of AChE after solubilization by various methods.
The M_r of ^{3}H-DFP-AChE was determined by comparison with the electrophoretic mobility of the ^{14}C-methylated protein markers. The results are shown as the means ± S.E.M..

	Average M_r
High-salt-soluble AChE	65.6±0.4(3)
Low-salt-soluble AChE	63.8±0.9(6)
Detergent-soluble AChE	63.2±0.25(5)
PIPLC-solubilized AChE	63.8±0.3(5)
Proteinase K purified AChE	63.2±0.6(3)

Furthermore, it shows that any similar change in M_r induced by Proteinase K would have been readily detectable, and sets an upper limit of ca. 1.5Kda (i.e. 10-15 amino acids) for the putative peptide removed by its action. Moreover, of 6 homogeneous proteases tested only one, namely Proteinase K, modifies the detergent-sensitivity of DSAChE, also suggesting that a relatively small polypeptide sequence is accessible to proteolytic attack.

In studies of membrane-bound enzymes, Arrhenius plots of the temperature-dependence of enzymic activity have often been used to correlate the activity of the enzyme with the phase state of the membrane lipids. We adopted this approach to determine the effect of membrane lipids on the activity of DSAChE in Torpedo and erythrocyte membranes. Figure 5A shows that membrane-bound AChE in Torpedo gave a non-linear Arrhenius plot, with a transition temperature (T_c) at 16°C and activation energies (E_A) of 20kJ/mol and 15kJ/mol below and above the break point, respectively. Linear Arrhenius plots were found for enzyme solubilized with either PIPLC (Fig. 5B) or proteinase K (Fig. 5C). Both of the solubilized enzymes had an E_A of ca. 18kJ/mol, a value intermediate between those obtained for the membrane-bound enzyme. Similarly, a linear Arrhenius plot was obtained for AChE purified after proteinase K solubilization (27). A much sharper discontinuity in the Arrhenius plot was found for rat erythrocyte ghost AChE (Fig. 6A), with a T_c at 14°C and E_A of 25kJ/mol below the break point and 15kJ/mol above the break point. Upon solubilization with either PIPLC (Fig. 6B) or papain (Fig. 6C), the enzyme still gave a sharp break point with similar transition temperatures and activation energies to that of the membrane-bound enzyme. These results strongly suggest that the kinetic properties of DSAChE are not significantly altered when it is released from the membrane, as none of the three *Torpedo* forms studied show appreciable changes in E_A and nor do the membrane-bound and soluble forms in rat erythrocyte ghosts.

The break points in the Arrhenius plots of the PIPLC-solubilized and the papain-solubilized enzymes from rat erythrocytes were surprising since non linear Arrhenius plots are usually expected for membrane-bound enzymes and are commonly attributed to phase separation phenomena in membrane lipids which have been shown to occur near the transition temperature of the enzyme (i.e. see refs. 28 and 29). However, this interpretation is almost certainly too simplistic, since many cases can be found in the literature of non linear Arrhenius plots for soluble enzymes (i.e. 30,31). Indeed, a recent report (22) shows that bovine erythrocyte AChE, whether solubilized with Triton X-100 or PIPLC from *Bacillus thuringiensis*, shows break points in its Arrhenius plots, with E_A values nearly identical to those obtained in this study. Thus, even though great care must be taken in interpreting non-linear Arrhenius plots as being due solely to the influence of membrane lipids, our results suggest that the temperature-dependence of the catalytic activity of DSAChE in *Torpedo* and rat erythrocytes is not significantly influenced by its membrane environment. These data support our model (see Fig. 7), in which DSAChE is bound to the plasma membrane via discrete and specific interactions with one or more PI molecules, and not via relatively non-specific interactions with membrane lipids as postulated in the fluid-mosaic model (32).

Discussion

It is currently believed that integral membrane proteins are usually positioned in the plasma membrane by one or more transmembrane hydrophobic polypeptide sequences (33). In many cases these proteins are amphipathic molecules and proteolytic digestion permits separation of the functional part of the protein, as a water-soluble entity, from the hydrophobic anchor. This anchor can be derived from COOH-

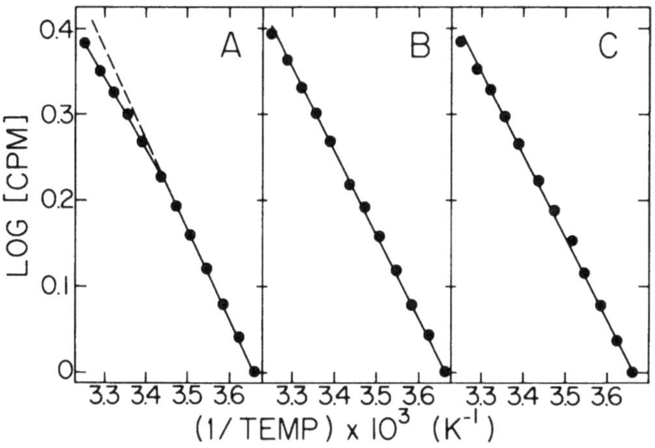

Fig. 5. Arrhenius plots of Torpedo AChE activity.
AChE activity was assayed in triplicate by the radiometric assay (12), over the temperature range 0°C to 35°C using an accurately controlled thermostated water bath. Each panel shows results from five experiments, normalized to 0°C. One-line and two-line fits were made to each data set by minimizing χ^2; the two-line fit was chosen for the cases in which χ^2 was substantially better than for the one-line fit. A) Membrane-bound AChE. B) PIPLC-solubilized AChE. C) Proteinase K-solubilized AChE.

terminal, NH_2-terminal or intermediate sequences of the polypeptide chain (see, for example, 34-35, and references therein). Such sequences must contain at least 20-25 amino acids in order to span the bilayer, and thus should have a minimal molecular weight of ca. 2500 ; indeed typical reports in the literature describe hydrophobic fragments ranging in size from 3Kda upwards (for literature see 34,35).

In recent years, however, considerable evidence has accumulated for another mode of attachment of integral membrane proteins, which involves attachment via covalently bound fatty acids (see, for example, 36-39). A particularly well studied protein in this respect is the amphiphilic membrane-bound penicillinase from *Bacillus licheniformis* (40). This enzyme is anchored to the plasma membrane via

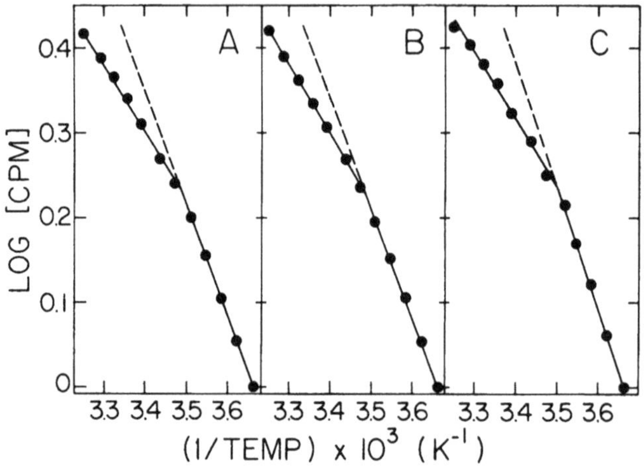

Fig. 6. Arrhenius plots of rat erythrocyte ghost AChE activity.
Experiments were carried out as described in the legend to Fig. 5. A) Membrane-bound AChE. B) PIPLC-solubilized AChE. C) Papain-solubilized AChE.

its NH_2-terminal region (41,42) and a water-soluble exo-form can be produced by cleavage of the NH_2-terminal octapeptide (43). Its amino-acid composition shows this short peptide to be hydrophilic, and it cannot, therefore, be responsible for the amphipathic character of the membrane-bound form or for its membrane anchorage. Evidence has been produced (44,45) that this is achieved by covalent modification of both the sulphydryl and amino groups of the NH_2-terminal cysteine by a diglyceride and a fatty acid respectively. The evidence we have presented above suggests that tightly bound PI (or a derivative thereof) may similarly suffice to anchor AChE to the plasma membrane in Torpedo electric organ, since PIPLC is able to remove the enzyme swiftly and quantitatively, and the studies with proteases suggest that only a small polypeptide is involved. It is worth noting that both for intestinal alkaline phosphatase (46) and human erythrocyte AChE (47) relatively small anchoring peptides, 3Kda and 2Kda respectively, have been reported; furthermore, both alkaline phosphatase (11) and erythrocyte AChE (10,21) are solubilized by PIPLC, although only partially in the case of human erythrocyte AChE.

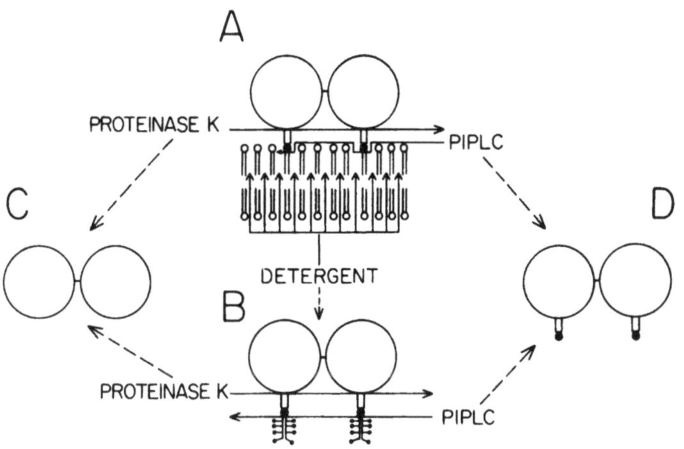

Fig. 7. Schematic representation of modes of solubilization and enzymic transformation of the hydrophobic dimer of AChE.

The model assumes that both subunits of the AChE dimer are associated with one or more molecules of PI (or a PI derivative) which serve to anchor the enzyme in the membrane (A) and are responsible for its interaction with detergent upon detergent solubilization (B). Proteinase K can directly solubilize membrane-bound AChE (A) by digestion of a small peptide to produce a form which is unable to interact with detergents (C). PIPLC can similarly solubilize membrane-bound AChE directly by digestion of one or more PI molecules to yield a form that is also unable to interact with detergents (D). Both Proteinase K and PIPLC are able to act on detergent-soluble AChE (B) to yield the same detergent-insensitive forms (C and D) as produced by direct solubilization with these enzymes.

If the analogy with the bacterial penicillinase holds fully, a stoichiometric amount of PI might be covalently attached to the catalytic subunit of AChE as shown in the model in Fig. 7. We are currently investigating this possibility by direct chemical analysis, using a GC/MS technique (48).

The results presented so far strongly support the involvement of tightly bound phosphoinositides in the anchoring of DSAChE to the synaptic membrane of *Torpedo* electric organ and to the plasma membrane of mammalian erythrocytes. Preliminary data have revealed the presence of significant amounts of *myo*-inositol in affinity-purified *Torpedo californica* DSAChE. This *myo*-inositol is probably covalently attached to the enzyme molecule since it is detected only after acid hydrolysis of enzyme samples. *Torpedo* AChE samples purified after Proteinase K solubilization contained less *myo*-inositol than the DSAChE samples purified after cholate solubilization, as predicted from our model (Fig. 7). Interpretation of the data is, however, complicated by low recovery in both samples and suitable controls. Furthermore, substantial amounts of *chiro*-inositol, an inositol isomer usually found in seagrass, were detected both in DSAChE affinity-purified after cholate-solubilization and after Proteinase K-solubilization, as well as in *Torpedo* homogenates.

Acknowledgments

A.H.F. was supported by the Science and Engineering Research Council of Great Britain and R.M.F. by the Italian C.N.R. A grant to I.S. from the Muscular Dystrophy Association of America is gratefully acknowledged as is the support of the Washington University Mass Spectrometry Resourse, NIH grant RR-00954. We thank Dafna Raviv, Esther Roth and Karen Eckerman for skillful technical assistance.

References

1. Massoulie, J. and Bon, S. (1982) Ann. Rev. Neurosci. 5, 57-106.
2. Bon, S. and Massoulie, J. (1980) Proc. Natl. Acad. Sci. U.S.A. 77, 4464-4468.
3. Viratelle, O.M. and Bernhard, S.A. (1980) Biochemistry 19, 4999-5007.
4. Witzemann, V. (1980) Neurosci. Lett. 20, 277-282.
5. Witzemann, V. and Boustead, C. (1981) Neurosci. Lett. 26, 313-318.
6. Morel, N. and Dreyfus, P. (1982) Neurochem. Int. 4, 283-288.
7. Li, Z.Y. and Bon, C. (1983) J. Neurochem. 40, 338-349.
8. Futerman, A.H., Low, M.G. and Silman, I. (1983) Neurosci. Lett. 40, 85-89.
9. Futerman, A.H., Fiorini, R.M., Roth, E., Michaelson, D.M., Low, M.G. and Silman, I. (1984) in Cholinesterases, Fundamental and Applied Aspects. Walter de Gruyter, Berlin, 99-113.
10. Futerman, A.H., Michaelson, D.M., Low, M.G. and Silman, I. (1985b),submitted.
11. Low, M.G. and Finean, J.B. (1977b) Biochem. J. 167, 281-284.
12. Johnson, C.D. and Russell, R.L. (1975) Analyt. Biochem. 64, 229-238.
13. Deutsch, J.W. and Kelly, R.B. (1981) Biochemistry 20, 378-385.
14. Ways, P. and Hanahan, D.J. (1964) J. Lipid. Res. 5, 318-328.
15. Grassi, J., Vigny, M. and Massoulie, J. (1982) J. Neurochem. 38, 457-469.
16. Sorensen, K., Gentinetta, R. and Brodbeck, U. (1982) J. Neurochem. 39,1050-1060.
17. Razon, N., Soreq, H., Roth, E., Bartal, A. and Silman, I. (1984) Exptl. Neurol., 84, 681-695.
18. Hall, Z.W. (1973) J. Neurobiol. 4, 343-361.
19. Shukla, S.D. (1982) Life. Sci. 30, 1323-1335.
20. Shukla, S.D., Coleman, R., Finean, J.B. and Michell, R.H. (1980) Biochem. J. 187, 277-280.
21. Low, M.G. and Finean, J.B. (1977a) FEBS Lett. 82, 143-146.
22. Taguchi, R., Suzuki, K., Nakabayashi, T. and Ikezawa, H. (1984) J. Biochem. (Tokyo) 96, 437-446.
23. Michaelson, D.M., Barkai, G. and Barenholz, Y. (1983) Biochem. J. 211, 155-162.
24. Futerman, A.H., Fiorini, R.M., Roth, E., Low, M.G. and Silman, I. (1985a) Biochem. J., in press.

25. Low, M.G. and Zilversmit, D.B. (1980) Biochemistry 19, 3913-3918.

26. Witzemann, V. and Boustead, C. (1983) EMBO J. 2, 873-878.

27. Barton, P.L., Futerman, A.H. and Silman, I. (1985), in preparation.

28. Lee,. A.G., Birdsall, N.J., Metcalfe, J.C., Toon, P.A. and Warren, G.B. (1974) Biochem. 13, 3699-3705.

29. Thilo, L., Trauble, H. and Overath, P. (1977) Biochem. 16, 1283-1290.

30. Massey, V., Curti, B. and Ganther, H. (1966) J. Biol. Chem. 241, 2347-2357.

31. Dean, W.L. and Tanford, C. (1978) Biochem. 17, 1683-1690.

32. Singer, S.J. and Nicolson, G.L. (1972) Science 175, 720-731.

33. Capaldi, R.A. (1982) Trends in Biochem. Sci. 7, 292-295.

34. Macnair, R.D.C. and Kenny, A.J. (1979) Biochem. J., 179, 379-395.

35. Kensil, C.R., Hediger, M.A., Ozols, J. and Strittmatter, P. (1983) J. Biol. Chem. 258,14656-14663.

36. Schlesinger, M.J. (1981) Ann. Rev. Biochem. 50, 193-206.

37. Sefton, B.M., Trowbridge, I.S., Cooper, J.A. and Scolnick, E.M. (1982) Cell 31, 465-474.

38. Henning, R. and Lange-Mutschler, J. (1983) Nature 305, 736-738.

39. Ferguson, M.A.J. and Cross, G.A.M. (1984) J. Biol. Chem. 259, 3011-3015.

40. Lampen, J.O. (1978) Symp. Soc. Gen. Microbiol. 28, 231-247.

41. Simons, K., Sarvas, M., Garoff, H. and Helenius, A. (1978) J. Mol. Biol. 126, 673-690.

42. Izui, K., Nielsen, J.B.K., Caulfield, M.P. and Lampen, J.O. (1980) Biochemistry 19, 1882-1886.

43. Chang, C.N., Nielsen, J.B.K., Izui K., Blobel G. and Lampen J.O. (1982) J. Biol. Chem. 257, 4340-4344.

44. Nielsen, J.B.K., Caulfield, M.P. and Lampen, J.O. (1981) Proc. Natl. Acad. Sci. U.S.A. 78, 3511-3515.

45. Lai, J.S., Sarvas, M., Brammer, W.J., Neugebauer, K. and Wu, H.C. (1981) Proc. Natl. Acad. Sci. U.S.A. 78, 3506-3510.

46. Colbeau, A. and Maroux, S. (1978) Biochim. Biophys. Acta 511, 39-51.

47. Dutta-Choudhury, T.A. and Rosenberry, T.L. (1984) J. Biol. Chem. 259, 5653-5660.

48. Sherman, W.R., Packman, P.M., Laird, M.H. and Boshans, R.L. (1977) Analyt. Biochem. 78, 119-131.

THE HYDROPHOBIC DOMAIN OF HUMAN ERYTHROCYTE ACETYLCHOLINESTERASE CONTAINS NONAMINO ACID COMPONENTS

Terrone L. Rosenberry, Robert Haas, William L. Roberts
and Benedict H. Kim

Department of Pharmacology, Case Western Reserve University
Cleveland, Ohio 44106

Introduction

Acetylcholinesterases (AChE[1]; EC 3.1.1.7) in vertebrate tissues are classified either as asymmetric if they include a collagen-like tail structure or as globular if they are devoid of such a structure (see 1, 2). In addition to soluble and secreted forms, the globular AChEs include a class of integral membrane AChE forms with a hydrophobic domain that interacts with cell plasma membranes. These globular membrane-bound AChEs are the predominant AChEs in mammalian brain and in muscle cells outside the neuromuscular junctions. A dimeric (G_2) form in this class is the only AChE present on human erythrocytes, and we are studying this enzyme (RBC AChE) as a general model for integral membrane AChEs. These forms require nonionic detergents for extraction and continue to interact with detergents following solubilization (1). Highly purified RBC AChE associates with detergent in micelles (3, 4) but forms soluble aggregates on removal of detergent (5). RBC AChE is an amphipathic protein because its small hydrophobic domain may be cleaved from the remaining hydrophilic enzyme by papain digestion of either enzyme-detergent micelles or soluble enzyme aggregates (6). The hydrophilic enzyme fragment has virtually the same polypeptide size as the intact amphipathic enzyme and retains full enzyme activity, but it no longer associates with detergents or forms aggregates in the absence of detergent.

Molecular Basis of Nerve Activity

In this report we focus on the hydrophobic domain fragment cleaved from RBC AChE by papain digestion. We show that this hydrophobic domain is essential for the binding of RBC AChE to the phospholipid membranes of liposomes (7), and we demonstrate the radiolabeling of this domain both by a hydrophobic photoactivated reagent (8) and by reductive methylation (9,10). Reductive methylation reveals that the domain contains nonamino acid components and that it is covalently linked to the C-terminus of the subunit polypeptides (10).

Results and Discussion

Release of RBC AChE from liposomes by papain digestion. The reconstitution of RBC AChE into liposomes in this study was based on the cholate dialysis procedure of Kagawa and Racker (11). Inclusion of the competitive inhibitor edrophonium chloride in the initial reconstitution mixture was necessary to protect the enzyme against inactivation (7). The size of the reconstituted RBC AChE liposomes, determined by gel exclusion chromatography on Sepharose CL-4B, was controlled by the conditions under which the stock phospholipid was sonicated prior to reconstitution. In initial experiments, large liposomes were desired so that successful reconstitution of RBC AChE could be monitored by a shift in enzyme activity to the large liposome fraction that was eluted in the column void volume. Large liposomes were obtained by treating the stock phospholipid only by bath sonication and by carrying the highly turbid reconstituted liposome mixture through a freeze-thaw cycle. Intact RBC AChE could be reconstituted into these liposomes but the papain-disaggregated RBC AChE could not (7).

More vigorous probe-type sonication of the stock phospholipid prior to reconstitution resulted in small liposomes which were eluted from Sepharose CL-4B at a partially included volume. Fig. 1A shows a preparation that contained a mixture of large and small RBC AChE liposomes, but preparations that were almost exclusively small liposomes were frequently obtained. The relative elution volume of the

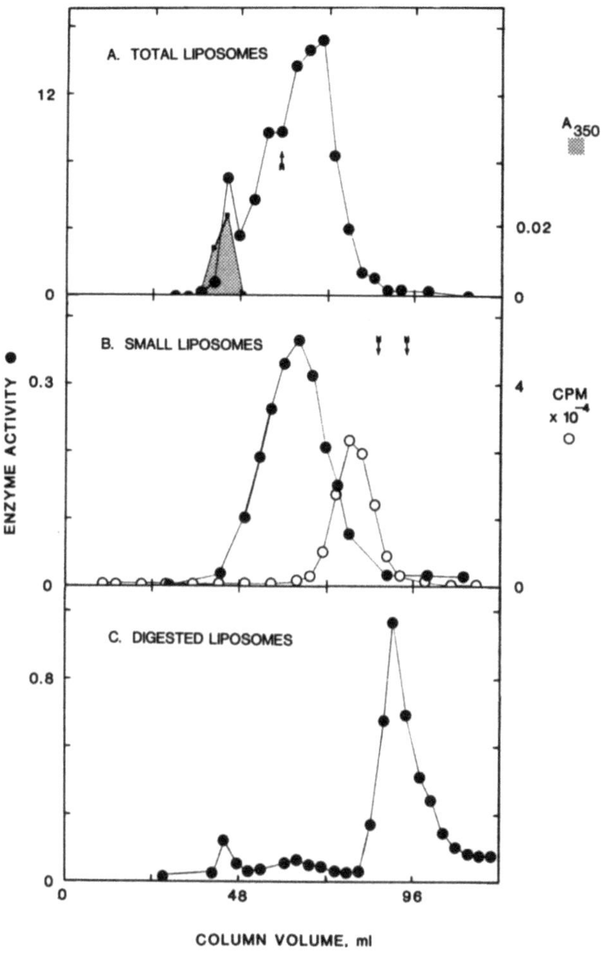

Fig. 1. Release of RBC AChE from small liposomes by papain. Samples were analyzed by chromatography on Sepharose CL-4B equilibrated with 20 mM Tris-Cl (pH 7.4). The reconstitution mixture (1.0 ml) included 200 units of RBC AChE, 2 mg of sonicated phospholipid (Sigma Type IX-E), 1 mM edrophonium chloride, and 1.4% sodium deoxycholate in buffer and was dialyzed at 4^0 C. Enzyme activity (●) is in units. The arrows in panel B indicate the elution positions of β-galactosidase and catalase. A. Elution profile of the entire liposome reconstitution mixture. Turbidity ($A_{350\ nm}$) is shaded. B. Rechromatography of small liposomes. A 1.0 ml aliquot of the fraction indicated by the arrow in panel A was mixed with 30 μl of [^{3}H]DFP-labeled control RBC AChE. (O) ^{3}H-cpm. C. A 2.0 ml aliquot of the fraction indicated by the arrow in panel A (5 units) was mixed with activated (6) soluble papain (0.14 mg), incubated for 1 hr at 25° C, and chromatographed. For further details, see (7).

small RBC AChE liposomes varied slightly from preparation to preparation but was always less than that of the smaller soluble AChE aggregates. This point was demonstrated in Fig. 1B, where a selected fraction of small RBC AChE liposomes was rechromatographed with aggregated [^{3}H]DFP-labeled RBC AChE. The elution volume of the RBC AChE liposomes was unchanged and clearly distinct from the elution volume of the labeled RBC ACHE, which corresponded precisely to the 10 nm Stokes radius previously observed for unlabeled RBC AChE aggregates (4). Fig. 1B also demonstrated that aggregated RBC AChE did not interact with preformed liposomes.

The hypothesis that the hydrophobic domain removed from purified RBC AChE aggregates by papain digestion is responsible for the interaction of RBC AChE with phospholipid membranes was supported by the effects of papain on RBC AChE liposomes. Small RBC AChE liposomes were incubated with papain in the presence of 0.1 mM edrophonium chloride to prevent papain inactivation of the RBC AChE (6), and the digest was chromatographed on Sepharose CL-4B (Fig. 1C). More than 80% of the recovered RBC AChE activity had been released from the liposomes and was eluted precisely at the position of the papain-disaggregated RBC AChE obtained from digests of RBC AChE aggregates (6, 7). Thus the site of papain cleavage that releases the hydrophobic domain from the active hydrophilic domain of RBC AChE is accessible to papain in these small RBC AChE liposomes. The importance of size fractionating the RBC AChE liposomes to obtain this result became apparent when a similar papain digestion was conducted on large RBC AChE liposomes like those in the turbid void volume fractions in Fig. 1A (7). Less than 15% of the enzyme activity was released from these large liposomes, perhaps in part because large liposomes tend to be multilamellar and could exclude soluble papain from access to the RBC AChE.

To investigate the site of papain cleavage that releases enzyme activity from RBC AChE liposomes, we characterized intact and papain-released enzyme by SDS-PAGE[1]. Residual soluble papain rapidly degrades RBC AChE samples exposed to SDS, but this degradation can be avoided if digestion is conducted with papain resin (6). RBC AChE liposomes were purified by affinity chromatography on acridinium resin (7) and digested with papain resin. Released and liposomal enzyme in the digest supernatant were recovered by further affinity chromatography and separated by gel exclusion chromatography on Sepharose CL-4B (7). SDS-PAGE revealed that the intact subunit mass of about 75 kDa was decreased by about 2 kDa following release of the RBC AChE from the liposomes by papain (7). This apparent decrease in subunit mass is identical to the one observed when papain cleaved purified RBC AChE aggregates to disaggregated RBC AChE (6).

Radiolabeling the hydrophobic domain with a photoactivated reagent. A useful probe for radiolabeling the hydrophobic domain of RBC AChE is [^{125}I]TID.[1] This reagent partitions selectively into the hydrophobic phase of membranes, liposomes or protein-detergent complexes and, on photoactivation, reacts covalently with molecules in the hydrophobic phase (12). RBC AChE could be labeled with [^{125}I]TID either in liposomes, in Triton X-100 micelles, or simply in soluble RBC AChE aggregates; the aggregates showed the highest incorporation of radiolabel (8). The labeled enzyme was repurified from the reaction mixture by affinity chromatography and incorporated into liposomes or Triton X-100 micelles. The ^{125}I label was shown to be localized exclusively in the hydrophobic domain by Sepharose CL-4B chromatography of a partial papain digestion of small [^{125}I]TID-labeled RBC AChE liposomes (Fig. 2A). About one-half of the enzyme activity was released by the digestion as soluble G_2 RBC AChE, but none of the ^{125}I label remained associated with the released enzyme. The cleaved ^{125}I label instead remained in the small liposome pool together with residual intact ^{125}I-labeled RBC AChE. The released hydrophobic domain fragment could be separated from residual intact enzyme by rechromatography of the small liposome pool on Sephadex LH-60 in a formic acid-ethanol solvent. Three radiolabeled peaks were observed on the LH-60 column in Fig. 2B. The earliest peak in Fig. 2B at about 20 ml (the column void volume) corresponded to intact control RBC AChE with or without ^{125}I label. Recovery of the intact enzyme tended to be low, probably because of its poor solu-

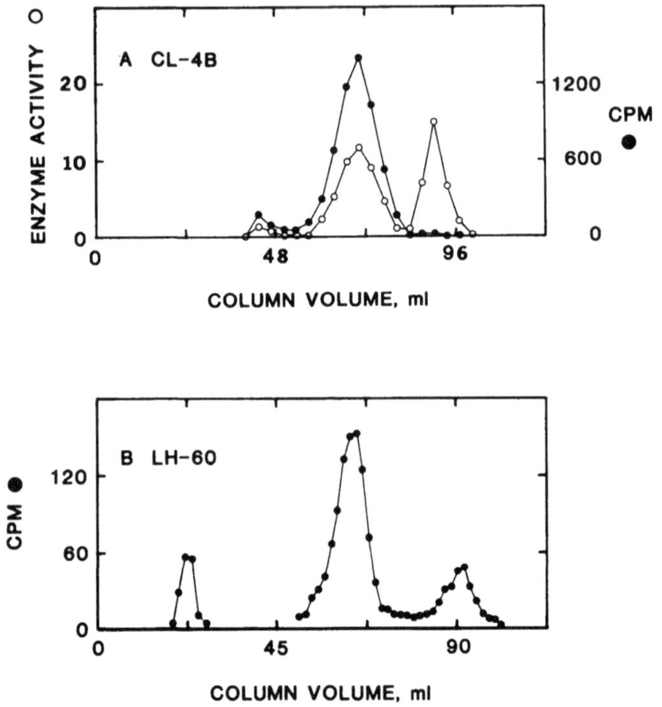

Fig. 2. Isolation of the [^{125}I]TID-labeled hydrophobic domain generated from RBC AChE liposomes by papain digestion. Purified detergent-free RBC AChE (4800 units in 0.5 ml 0.2 M sodium phosphate, pH 7) was mixed with [^{125}I]TID (50 µCi) in a sealed cuvette and irradiated at 350 nm in a Farrand MK1 Spectrofluorometer (150 W Xenon lamp) for 15 min at ambient temperature. The mixture was dialyzed, purified by affinity chromatography on acridinium resin (4) to a level of 2400 ^{125}I cpm/unit (7.6% radiolabel incorporation), and dialyzed again. Radiolabeled RBC AChE (45 units) was mixed with unlabeled RBC AChE (2000 units) and phospholipid (15 mg/ml) and reconstituted and fractionated into small liposomes as in Fig. 1A. A. Selected fractions containing small RBC AChE liposomes (125 units, 10,000 ^{125}I cpm, 2.6 ml) were digested with papain (0.14 mg) for 24 hr at 4° C and chromatographed as in Fig. 1C. Enzyme activity (O) in units and ^{125}I cpm (●) are indicated. B. The small liposome fraction with the peak ^{125}I radioactivity in panel A was dried, resuspended in formic acid (0.2 ml), and applied to a Sephadex LH-60 column equilibrated with 25% formic acid, 75% ethanol. The recovery of ^{125}I cpm in panel B was 70% of the onput, with the loss selectively involving intact ^{125}I-labeled RBC AChE in the peak eluting at 20 ml. For further details, see (8).

bility in the formic acid-ethanol solvent. The major hydrophobic domain fragment was eluted from the LH-60 column at about 60 ml, while an additional and usually quite minor peak appeared near the approximate solvent volume of about 90 ml.

Papain digestion of RBC AChE in either aggregates, liposomes, or Triton X-100 micelles demonstrated the exclusive localization of the [^{125}I]TID-label in the released hydrophobic domain (8). Furthermore, the LH-60 column elution volume of the labeled major hydrophobic domain fragment released by all digestion procedures was highly reproducible. Thus we conclude that a single hydrophobic domain in RBC AChE is responsible for the formation of enzyme-detergent micelles, aggregation in the absence of detergents, and incorporation into liposomes.

Amino acid composition of the isolated hydrophobic domain fragment. Attempts to further purify the major ^{125}I-labeled hydrophobic domain fragment by reverse-phase high pressure liquid chromatography under conventional peptide fractionation procedures were unsuccessful because no radioactivity could be recovered from C-8 or C-18 columns. Therefore, we examined the amino acid composition of the major peak from the LH-60 column. Amino acid compositions of the hydrophobic domain fragment generated by papain digestion of RBC AChE liposomes as in Fig. 2 included unacceptable levels of background amino acids apparently derived from contaminants in the stock phospholipid used to prepare the liposomes. Background amino acid levels were much lower when the hydrophobic domain fragment was prepared by the following four-step procedure (8): 1) papain resin digestion of RBC AChE in Triton X-100 micelles; 2) affinity chromatography of the digest on acridinium resin to separate nonretained hydrophobic domain fragment in Triton X-100 micelles from retained intact RBC AChE and hydrophilic G_2 RBC AChE generated by the papain cleavage; 3) gel exclusion chromatography of the nonretained affinity column pool on Sepharose CL-6B in 0.1% Triton X-100 to fractionate hydrophobic domain fragment in Triton X-100 micelles from small peptide fragments produced by minor nonselective papain degradation of a small portion of the RBC AChE; and 4) gel exclusion chromatography of the hydrophobic domain fragment on Sephadex LH-60 as in Fig. 2B. Calibration of the LH-60 column with peptide standards indicated that the 60-ml peak corresponded to a 3000 dalton peptide (8), and thus some 25-30 moles of presumably hydrophobic amino acids would be expected per mole of hydrophobic domain fragment. Molar quantities of hydrophobic domain fragment were estimated from the radiolabel content either of ^{125}I from the $[^{125}I]$TID reaction described above or of ^{14}C from the reductive methylation described below. In contrast to the expectation, however, only two moles of amino acids per mole of hydrophobic domain fragment were found: one mole of histidine and one mole of glycine (8). No other amino acid exceeded 0.15 moles per mole of fragment. This observation was our first experimental suggestion that the hydrophobic domain was composed largely of nonamino acid components.

Radiolabeling of RBC AChE by reductive methylation. Reductive methylation is a convenient and useful way to label amine groups in proteins. The reaction leaves RBC AChE substantially unaltered by several criteria, including enzymatic activity, gel exclusion chromatography and SDS-PAGE (9). Mono- and dimethylated amines produced by reductive methylation are stable to acid hydrolysis and may be identified by comparison to methylated standards with slight extensions of conventional amino acid analysis as outlined under Table I. Although the presence of two methylated derivatives per amine resulted in greater complexity in the amino acid analyzer elution patterns, it provided a more definite identification of the unknown amine.

The use of reductive methylation to determine lysine residues and N-terminal amino acids in RBC AChE is summarized in Table I. The major N^{α}-methylated amino acid from hydrolyzates of RBC AChE which had been radiomethylated at 0.2 mM HCHO coincided with dimethyl-Glu, and the assignment of Glx as an N-terminus was confirmed by the presence of a minor peak comigrating with monomethyl-Glu. Corresponding Glu peaks also coincided when a hydrolyzate of ^{3}H-methylated

Table I

Methylation Stoichiometry for Reductive Methylation

of RBC AChE at Two Concentrations of [^{14}C]HCHO.

Formaldehyde concentration mM	Residues methylated per RBC AChE subunit[a]		
	N^{α}-methyl		N^{ϵ}-methyl
	Glu	Arg	Lys
0.2	0.56	0.36	0.96
10.0	0.66	0.34	7.40

[a]Reductive methylation was carried out by incubating RBC AChE (0.1-2 mg; 0.1-1.0 mg/ml) with 0.2-10 mM [^{14}C]HCHO (Amersham, 14 mCi/mmol, or ICN, 40 mCi/mmol) and 50 mM $NaCNBH_3^-$ (Aldrich, recrystallized) in 20-200 mM sodium phosphate (pH 7) for 15 minutes at 37° C. The labeled enzyme was dialyzed against 1-20 mM sodium phosphate for 24-72 hr at 4° C to remove unincorporated radioactivity and, in some cases, repurified by affinity chromatography on acridinium resin. Methylated RBC AChE was hydrolyzed in 6 N HCl at 115° C for 16 hr.

Mono- and dimethylated amino acids were quantitated by scintillation counting 0.4 ml fractions of effluent from a Beckman 119CL Amino Acid Analyzer. Methylated amino acids were resolved on a four-buffer elution schedule and identified by comparison to the elution times of 16 hydrolysis-stable methylated standard amino acids. The methyl group specific activity (S. A.) was calculated from the analyzer elution profile of hydrolyzed methylated Glu standards.

The pmols of residues methylated (N) was calculated from the formula N = [cpm"/2 + cpm']/(S. A.), where cpm" and cpm' refer to the radioactivity of the dimethylated and monomethylated forms of a given amino acid. The pmols of RBC AChE subunit was calculated from the protein determined by amino acid analysis assuming a subunit equivalent weight of 70,000 g protein. Data presented are averages of 2-3 determinations of AChE labeled at 0.2 mM HCHO and 6 determinations at 10 mM HCHO with coefficients of variation of the mean less than 6% of the mean. For further details, see (9).

RBC AChE was mixed with ^{14}C-methylated Glu in a single run. The N-terminal Glx as well as a minor N-terminal species Arg were essentially fully methylated by 0.2 mM HCHO, whereas methylation of the less reactive Lys residues increased nearly 8-fold at 10 mM HCHO. This pattern is consistent with previous observations that α-amino groups in amino acids are more reactive than ϵ-amino groups and thus become fully dimethylated at lower formaldehyde concentrations (13).

Additional radioactive peaks corresponding to nonamino acid components in RBC AChE were also observed and are considered in greater detail below. Recovery of radioactivity following hydrolysis averaged 97% of the radiolabel incorporated into native RBC AChE, and 87% of the radiolabel applied to the amino acid analyzer was recovered in identified peaks. Thus any significant label incorporation into hydrolysis unstable amino acids can be ruled out. The identification of Glu as the predominant N-terminal amino acid was established by manual Edman determination (9). Six manual Edman cycles were conducted with intact RBC AChE and gave the sequence Glu-Gly-Ala-Glu-Asp-Ala... Small amounts of Arg also were detected in the first Edman cycle, but a second minor sequence was not detected.

The presence of N^{α}-methylated Arg in Table I is puzzling. The mono- and dimethylated species were found consistently at 45-65% of the level of methylated Glu, a ratio not changed either in methylated RBC AChE repurified by affinity chromatography, in RBC AChE methylated in the presence of 1% SDS, or in nonreduced or disulfide-reduced subunit bands of methylated RBC AChE that were cut from SDS-PAGE gels. More extensive sequence analyses of RBC AChE are required to establish whether the methylation stoichiometries in Table I imply that dimeric RBC AChE has N-terminal homogeneous subunits (1.0 residues of Glu per subunit, 0.0 residues Arg) or heterogeneous subunits (0.5 residues each of Glu and Arg). Only one previous report of possible AChE catalytic subunit heterogeneity has appeared (14), and that involved the unusually large subunits of chicken AChE. The deviation of the methylation stoichiometries for Glu and Arg in Table I from either of these alternatives could arise from an error of 10-15% in the subunit molecular weight due to uncertainty in the carbohydrate content or from incomplete methylation caused by protein aggregation or folding, competing side reactions, or endogenous N-terminal blockade (9).

Localization of the hydrophobic domain to the C-terminus of RBC AChE. A considerable advantage of radiomethylation is that it permits labeling of polypeptide N-termini for protein cleavage studies. Since papain cleavage of the hydrophobic domain fragment caused only a slight decrease in the residual subunit mass of RBC AChE, we had concluded that the hydrophobic domain must occur at the N- or C-terminus of the subunit polypeptide (6). We reported above that the hydrophobic domain fragment contained only one residue of His and one residue of Gly, neither of which was identified as an N-terminal amino acid by reductive methylation in Table I. To confirm this indication that the hydrophobic domain is not located at the N-terminus, we digested radiomethylated RBC AChE with papain. Samples of RBC AChE were reductively methylated with 10 mM [^{14}C]HCHO as in Table I or iodinated with [^{125}I]TID as in Fig. 2, reconstituted into liposomes, purified on acridinium affinity resin, and mixed with papain. The digestion mixture was applied to acridinium resin, and the hydrophobic domain fragment in nonretained liposomes was isolated on Sepharose CL-4B and Sephadex LH-60 as in Fig. 2. RBC AChE retained on the acridinium resin and eluted with decamethonium bromide was fractionated on Sepharose CL-4B to obtain the papain-released hydrophilic enzyme fragment as in Fig. 2A. Radiomethylated amines in intact RBC AChE and in the hydrophilic and hydrophobic domain fragments were resolved on the amino acid analyzer in Fig. 3. Labeled peaks were identified by comparison to reductively methylated standards as in Table I. The predominant species in intact RBC AChE (Fig. 3A) included N^{ϵ},N^{ϵ}-dimethyl-Lys (K"), dimethyl-Glu (E"), and dimethyl-Arg (R") as indicated in Table I. These labeled amino acids were quantitatively retained in the hydrophilic enzyme fragment following papain digestion (Fig. 3B) but were totally absent from the isolated hydrophobic domain fragment (Fig. 3C). Since digestion of radiomethylated RBC AChE released the hydrophobic

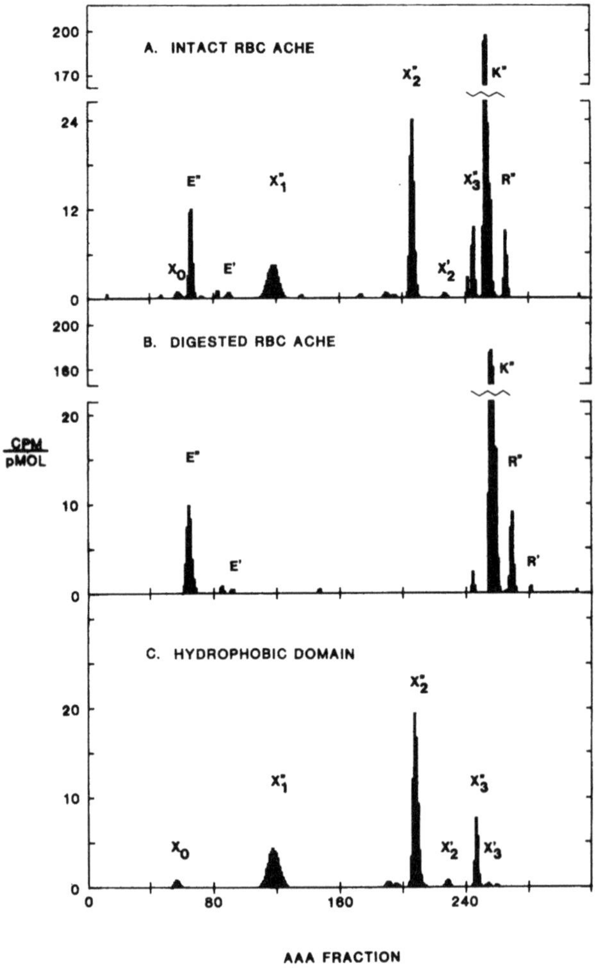

Fig. 3. Profile of methylated amine components of intact RBC AChE and of its hydrophilic and hydrophobic domain fragments generated by papain. Radiolabeled RBC AChE (12,800 units, 36 x 10^6 ^{14}C cpm, 14,000 ^{125}I cpm) and phospholipid (50 mg/ml) were reconstituted into liposomes as in Fig. 1, purified by affinity chromatography, and digested with papain (2.7 mg in 25 ml) for 2 hr at 25° C as in Fig. 1. Samples corresponding to control intact ^{14}C-labeled RBC AChE (panel A), G_2 RBC AChE released by papain and isolated by affinity and Sepharose CL-4B chromatography (panel B), and the major hydrophobic domain fragment obtained following Sephadex LH-60 chromatography (panel C) were hydrolyzed. Radioactive peaks from the amino acid analyzer (AAA) were identified and normalized to subunit pmols as in Table I. E designates Glu; K, Lys; R, Arg; and X, nonamino acids identified in Fig. 4. Primes and double primes designate mono- and dimethylated derivatives, respectively. For details, see (10),

domain but did not remove labeled N-terminal Glu or Arg from the hydrophilic enzyme fragment, we conclude that the hydrophobic domain must be at the subunit C-terminus. Reductive methylation of the isolated hydrophobic domain fragment produced N^{α}-methylated His and suggested that papain cleaved a C-terminal dipeptide attached to the hydrophobic domain with the sequence His-Gly (10).

Identification of nonamino acid components in the hydrophobic domain. Although radiomethylation of intact RBC AChE and isolation of the hydrophobic domain fragment in Fig. 3C revealed no methylated amino acid components, several peaks labeled X in Fig. 3 were apparent only in the intact enzyme and the hydrophobic domain fragment. Possible identities for these methylated components were suggested in a report by Campbell et al. (15) on the hydrophobic membrane-binding structure linked to the C-terminus of the Thy-1 glycoprotein of rat brain. Amino acid analysis of the C-terminal tryptic peptide of Thy-1 gave ninhydrin staining peaks that indicated glucosamine and galactosamine and suggested ethanolamine. We prepared ^{3}H-methylated standards of these amines, hydrolyzed mixtures of each of the standards with ^{14}C-methylated RBC AChE hydrophobic domain fragment, and compared the amino acid analyzer elution profiles in Fig. 4. Mono- and dimethylated amine derivatives were distinguished by comparing samples of a standard methylated to different extents; methylation with a lower HCHO concentration increased the monomethyl peak relative to the dimethyl. The profile in Fig. 4A indicates that the X_2" and X_2' peaks from the hydrophobic domain fragment corresponded to di- and monomethylated ethanolamine. The X_1" component coincided with the rather broad peak of dimethyl glucosamine in Fig. 4B. Although very little X_1' is apparent in the hydrophobic domain fragment profile in Fig. 4B, samples with higher levels of X_1' eluted at the position of monomethyl glucosamine in Fig. 4B. This correspondence clearly distinguished the X_1 peaks from the somewhat later mono- and dimethyl galactosamine peaks in Fig. 4C. The X_3 peaks arose as acid degradation products and coeluted with identical degradation products generated from both methylated glucosamine and methylated galactosamine. The X_3 peaks were greatly

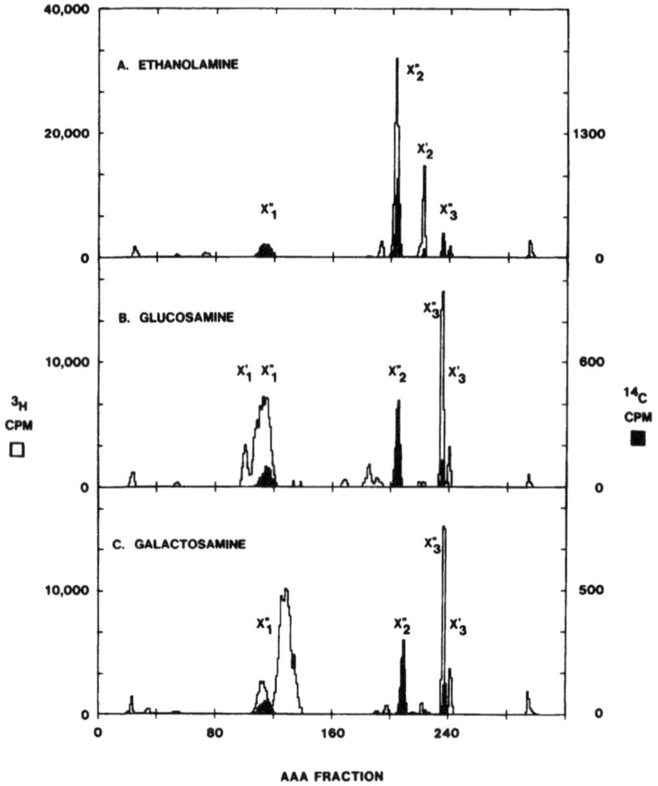

Fig. 4. Identification of nonamino acid components of the RBC AChE hydrophobic domain by reductive methylation and amino acid analysis. Ethanolamine, glucosamine, and galactosamine were reductively methylated with 1 mM [^{3}H]HCHO as in Table I, and excess unreacted HCHO was removed either by drying under vacuum or by quenching with a known amino acid. Aliquots of each ^{3}H-labeled standard were mixed with samples of ^{14}C-labeled hydrophobic domain fragment that was produced by papain digestion of radiolabeled RBC AChE in Triton X-100 micelles and isolated by the four-step procedure outlined in the text. The mixed samples were dried, hydrolyzed and analyzed by amino acid analysis as in Fig. 3. The profiles of ^{3}H (open histograms) and ^{14}C (shaded histograms) radioactivity determined by dual channel scintillation counting are shown for each of the three standards. ^{3}H-Labeled peaks derived from quenching amino acids have been omitted for clarity. For further details, see (10).

reduced in samples hydrolyzed for shorter times or at lower concentrations of HCl, and these peaks were completely absent in nonhydrolyzed methylated standards of glucosamine and galactosamine. Nevertheless, the correspondence of the X_3 peaks with the degradation products of the methylated hexosamine standards further confirms the identification of glucosamine in the hydrophobic domain fragment. Quantitative analysis as in Table I indicated that ethanolamine and glucosamine were present in identical amounts of about 1.2 moles per mole of hydrophobic domain fragment (10).

The hydrophobic domain of RBC AChE thus contains both ethanolamine and glucosamine with unblocked primary amino groups. These components are incorporated through covalent linkages into an unusual, possibly glycolipid, structure at the C-terminus of RBC AChE that currently is being explored in our laboratory. The fact that a similar structure appears linked to the C-terminus of the Thy-1 glycoprotein suggests the important new concept that a nonamino acid hydrophobic domain can be linked to an otherwise soluble protein in a post-translational modification designed to produce a membrane-bound protein.

[1]Abbreviations: AChE, acetylcholinesterase; RBC AChE, human erythrocyte acetylcholinesterase; SDS, sodium dodecylsulfate; PAGE, polyacrylamide gel electrophoresis; DFP, diisopropylphosphoryl fluoride; $[^{125}I]$TID, 3-(trifluoromethyl)-3-(*m*-$[^{125}I]$iodophenyl)diazirine.

We thank Michael Hodges, Patricia Brandt and Jonathon Knight for their technical assistance. This investigation was supported by Grants NS-16577 and NS-07118 from the National Institutes of Health and by grants from the Muscular Dystrophy Association U.S.A.

References

1. Massoulié, J. and S. Bon. 1982. Ann. Rev. Neurosci. 5, 57-106.

2. Rosenberry, T.L. 1984. In: The Enzymes of Biological Membranes, Volume IV, Second edition, Martonosi, A. (editor), Plenum Press, New York, in press.

3. Wiedmer, T., C. Di Francesco and U. Brodbeck. 1979. Eur. J. Biochem. 102, 59-64.

4. Rosenberry, T.L. and D.M. Scoggin. 1984. J. Biol. Chem. 259, 5643-5652.

5. Ott, P. and U. Brodbeck. 1978. Eur. J. Biochem. 88, 119-125.

6. Dutta-Choudhury, T.A. and T.L. Rosenberry. 1984. J. Biol. Chem. 259, 5653-5660.

7. Kim, B.H. and T.L. Rosenberry. 1985. Biochemistry, submitted.

8. Roberts, W.L. and T.L. Rosenberry. 1985. J. Biol. Chem., submitted.

9. Haas, R. and T.L. Rosenberry. 1985. Analyt. Biochem., submitted.

10. Haas, R., P.A. Brandt, J. Knight and T.L. Rosenberry. 1985: to be published.

11. Kagawa, Y. and E. Racker. 1971. J. Biol. Chem. 246, 5477-5487.

12. Brunner, J. and G. Semenza. 1981. Biochemistry 20, 7174-7182.

13. Sherman, G., T.L. Rosenberry and H. Sternlicht. 1983. J. Biol. Chem. 258, 2148-2156.

14. Rotundo, R.L. 1984. J. Biol. Chem. 259, 13186-13194.

15. Campbell, D.G., J. Gagnon, K.B.M. Reid and A.F. Williams. 1981. Biochem. J. 195, 15-30.

CHOLINESTERASE: TWO SURPRISING INHIBITORS

I.B. Wilson

Department of Chemistry, University of Colorado,
Boulder, CO 80309

I am pleased to be represented in this volume honoring the memory of David Nachmansohn. I spent early years with David in his laboratory at Columbia University and even after I had attained an independent position at Columbia and rose through the professional ranks, we worked together and were friends.

David always emphasized the broad view of science but still he was very curious and had a keen interest in solving problems of a narrower nature, if they were unusual. I think he would have enjoyed this paper.

Introduction

The inhibition of acetylcholinesterase and butyrylcholinesterase by arsenite and by fluoride has been known for many years (1-4). Since these two substances inhibit other enzymes (5), why is it surprising that they inhibit cholinesterase? The remarkable fact is that cholinesterase does not have the structural features that enable some enzymes to bind arsenite and other enzymes to bind fluoride.

Arsenite is a potent inhibitor of enzymes that contain lipoic acid (5,6). Arsenite (arsenious acid, $As(OH)_3$, at pH < 9.2) forms stable cyclic dithioesters with 1,2 and 1,3 dithiols, such as lipoic acid (5,6). It is a potent poison because it inhibits the pyruvic dehydrogenase complex of brain. The inhibition is reversible and activity can be restored by

Molecular Basis of Nerve Activity

dithiols, notably BAL (6), which is an antidote for arsenical poisons.

Enzymes that are specifically inhibited by fluoride contain metal ions, especially iron, or require metal ions for activity as, for example, enolase which needs magnesium. Enzymes that contain iron porphorin compounds such as cytochrome oxidase are often inhibited by fluoride because fluoride is bound as a coordination complex.

Cholinesterase contains no lipoic acid nor even free sulfhydryl groups (1,7,8). It contains no metal ion and needs no metal ion for activity (9). Enzymes that are inhibited by arsenite or fluoride are not inhibited by both. It is suprising that cholinesterase is inhibited by arsenite and by fluoride, isn't it?

Arsenite Inhibition

When I visited Israel Silman in 1975, we discussed arsenite inhibition of acetylcholinesterase and came up with a theory. The theory proved to be wrong, but in our study we did delineate the kinetic aspects of arsenite inhibition (10). We found that arsenite inhibits acetylcholinesterase (from David's favorite animal - the electric eel) in a second order reaction with a rate constant of $140M^{-1}min^{-1}$ and an equilibrium dissociation constant of $1.4 \times 10^{-5}M$. To give some feeling for the pace of the inhibition process, we note that the half time for inhibition with $10^{-3}M$ arsenite is five minutes. The potency of inhibition is about the same as for lipoic acid containing enzymes.

Because the dissociation of the arsenite-enzyme with concomitant restoration of enzyme activity is very slow, the half time is minutes, we have called arsenite a quasi-irreversible inhibitor. Although there are quantitative differences, the inhibition of butyrylcholinesterase (human serum) is essentially the same (11). Enzyme activity can be restored rapidly by oxidizing agents such as iodine or permangenate which oxidize bound and free arsenite to arsenate.

Another interesting observation concerning arsenite inhibition is that some substances such as 2-PAM, and Tris very greatly increase the rate of inhibition and the rate of dissociation of arsenite-enzyme without much change in the equilibrium constant. The rates can be increased by as much as 200 fold (10). The fundamental question is "what are the binding interactions between arsenite and the enzyme?" There is some help from the chemical literature. Arsenite reacts weakly and spontaneously with diols to form diesters (12,13). The equilibrium constants for the hydrolysis of the diesters (dissociation of the complex) are relatively large, 500 times larger than the arsenite-enzyme. One of the most stable diesters, is formed with o-catechol as a 1:1 complex. Thus, given the limited possibilities for binding presented by the amino acid side chains, it seemed reasonable to investigate whether arsenite binds to cholinesterase by forming a diester with two tyrosine residues, fortuitously so oriented that a very stable diester can form. Since phenols can be nitrated by mild nitrating agents such as tetranitromethane but their esters cannot (14,15), it seemed that a comparative study of the nitration of acetylcholinesterase and the arsenite enzyme might show that two tyrosines are protected in the arsenite enzyme. Fuchs et al (16) had shown that acetylcholinesterase is inhibited by TNM as far as hydrolysis of acetylthiocholine is concerned, but the hydrolysis of the very poor substrate indophenyllacetate was not greatly affected. They had also found that many of the tyrosines were nitrated and that 2-PAM greatly accelerates the inhibition of the enzyme.

With this background information, we undertook a careful quantitative measurement of the number of tyrosine residues (17). We found that 12 of the 21 tyrosines per monomer were nitrated by TNM in the uninhibited enzyme but only 10 in the arsenite enzyme. In the presence of 2-PAM, nitration was much more rapid but the results were the same (Figure 1).

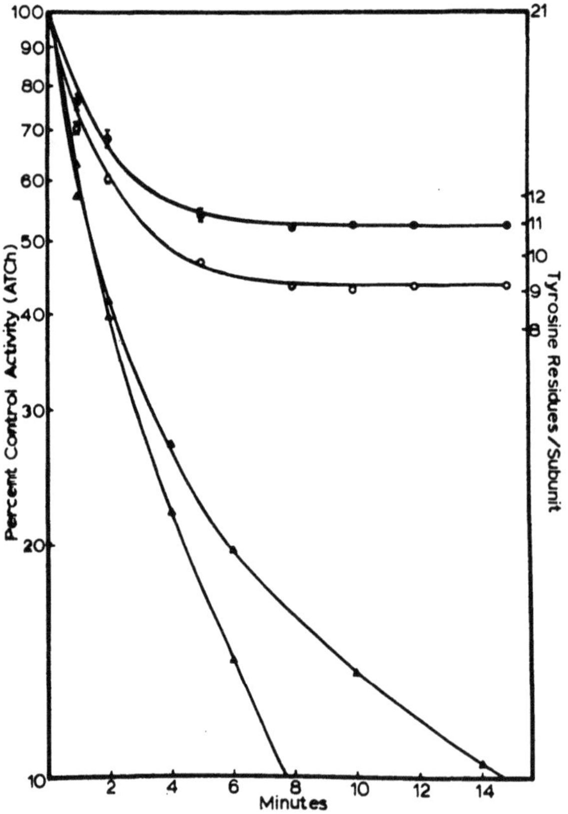

Figure 1 - The loss of enzyme activity and tyrosine residues during nitration with tetranitro methane. Tyrosine residues O without arsenite; ● with arsenite. Enzyme activity △ without arsenite; ▲ with arsenite. 0.1mM 2PAM was present.

It is thus apparent that the molecular basis for the binding of arsenite is the formation of a diester with two properly oriented tyrosine residues. Why is the formation of the diester so favorable? It may be that the tyrosines are so disposed that their geometry is just perfect for the formation of the diester.

There is another possibility. The formation of an arsenite diester with diols is measured by the acid production that occurs because the diester is more acidic than arsenious acid. The ionization of the diester helps to pull the reaction. We can imagine that in the enzyme there might be various stabilizing interactions. For example, the ionized diester might be stabilized and further bound by a neighboring positive charge of the enzyme

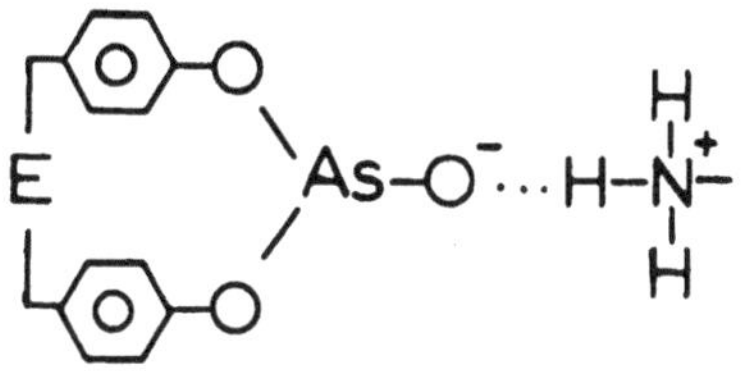

It appears that the two tyrosines play a special role in maintaining the three-dimensional structure of cholinesterase because they have persisted through the independent evolution of acetylcholinesterase and butyrylcholinesterase from a common ancestral gene.

Fluoride Inhibition

Although fluoride inhibition of cholinesterase has been known for a long time, its kinetic effects have not been clearly delineated. Some studies have indicated that fluoride combines with the enzyme, enzyme substrate complex and acylenzyme and inhibits acylation and deacylation. But there is contrary evidence from studies with the carbamylenzyme that suggests that fluoride does not affect deacylation (18). Methylcarbamyl choline (and other compounds) reacts slowly with acetylcholinesterase to produce a carbamylenzyme derivative analogous to the acetylenzyme (19). But the carbamylenzyme hydrolyzes slowly, in minutes, and therefore can be studied as a separate process. Fluoride has no effect on the rate of hydrolysis of the carbamylenzyme which indicates that either fluoride does not bind to the carbamylenzyme or it does so with no effect. Thus, it would appear that fluoride should have no effect on deacetylation of the acetylenzyme intermediate.

Fluoride has the peculiar effect of enhancing or inducing marked substrate inhibition even with substrates like phenylacetate that do not show substrate inhibition in its absence. This phenomenon has not yet been explained but one may speculate that fluoride and substrate bind in a highly cooperative manner to the acetylenzyme so that in the presence of fluoride substrate is much better bound to the acetylenzyme and the bound substrate prevents the hydrolysis of the acetylenzyme. This possibility can be tested with the carbamylenzyme derivative.

In the course of some experiments involving the rapid labeling of acetylcholinesterase, it appeared that fluoride might not react rapidly with the enzyme. We therefore decided to investigate the rate of reaction of fluoride with the enzyme on a milliscecond time scale using a stopped-flow apparatus. We used phenylacetate as a chromogenic substrate and the hydrolysis of this substrate was monitored by an increase in absorbance at 270mm (20). When fluoride was added to the substrate syringe in an amount that would cause marked inhibition of the hydrolysis of phenylacetate, the slow and progressive development of inhibition could be seen during the first 300 milli-seconds, which confirmed that the reaction was slow. The changing rate of increase in absorbance in the presence of fluoride relative to the control without fluoride was converted into a plot of inhibited enzyme, i.e. enzyme with bound fluoride as a function of time (Figure 2).

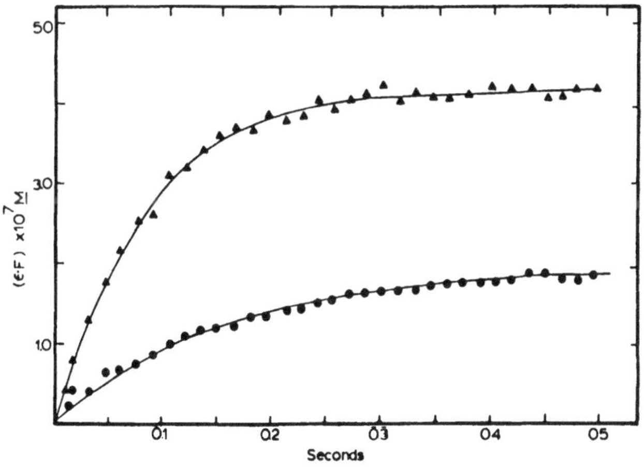

Figure 2 - The formation of fluoride inhibited enzyme, $\mathcal{E}\cdot F$, as a function of time. Upper curve: 2mMF^-, 1mM phenylacetate, and 5.1×10^{-4}mN enzyme. Lower curve: 1mMF^-, 1mM phenylacetate, and 3.2×10^{-4}mN enzyme. The medium was 500mM NaCl, 10mM phosphate pH 7.0, 23.5°C.

Our results for the reaction

$$\mathcal{E} + F^- \underset{k_{-1}}{\overset{k_1}{\rightleftharpoons}} \mathcal{E}\cdot F$$

are given in Table 1.

Table 1

PA	F^-	$k_1 \times 10^{-3}$	k_{-1}	$K_I \times 10^4$
mM	mM	$M^{-1}sec^{-1}$	sec^{-1}	M
1.0	2.0	4.9	2.5	5.2
1.0	2.0	4.0	2.2	5.5
1.0	2.0	4.9	2.4	4.9
0.5	1.0	5.8	3.9	6.6
0.2	0.5	8.7	6.3	7.3

You will notice that the constants depend on the concentration of phenyl acetate, PA(K_M = 1mM). This is because there are actually three different forms of the enzyme that may react with fluoride; free enzyme, enzyme-substrate complex and acetylenzyme. The ratio of the last two is independent of substrate concentration. The distribution of these enzyme forms depends upon the concentration of substrate. It is evident that fluoride reacts decidedly faster with the free enzyme than with the other forms. It is also evident that free enzyme with bound fluoride dissociates more rapidly than the other forms of enzyme with bound fluoride. The dissociation constant also varies with substrate concentration in a manner that suggests there is some cooperativity in the binding of fluoride and substrate. Independent measurements of the binding of fluoride in the absence of substrate give a value of 9 x 10^{-4}M for the dissociation constant.

At first glance it appears that fluoride is much stronger bound by buryrylcholinesterase than by acetylcholinesterase because when butyrylcholinesterase is assayed with its commonly used substrate, benzoylcholine, much lower concentrations of fluoride suffice to inhibit the enzyme. Actually, it turns out that fluoride is bound with about equal strength by both enzymes. The apparent difference arises from a very high degree of cooperativity in the binding of fluoride and substrates with butyrylcholinesterase (11). The cooperativity is 30 fold with benzoylcholine, 21 fold with butyrylthiocholine, 8 fold with propionylthiocholine and only 1.5 fold with acetylthiocholine. This remarkable potential for cooperativity is manifest with many ligands. It can be measured by the affect of the ligand with and without fluoride, on the equilibrium binding of arsenite. These experiments are easy to do because arsenite is a quasi irreversible inhibitor of the enzyme and enzyme activity is not restored when the arsenite-enzyme is extensively diluted for assay. With this technique we find the dissociation constant for binding of a ligand to free enzyme and to arsenite enzyme. It turns out that fluoride binds only to the free enzyme but other ligands bind to both. We found that two characteristic inhibitors of butyrylcholinesterase, dibucaine and procainamide bind very cooperatively with fluoride 13 fold and 17 fold. It turns out that the cooperativity in the binding of ligands and fluoride as with substrates is determined not so much by the substituted amine functions as by the acyl group.

There is an interesting genetic variant of human serum butyrylcholinesterase called atypical butyrylcholinesterase that does not readily hydrolyze succinyl dicholine (21). The atypical enzyme is also less readily inhibited by fluoride (22). The fluoride number is the percent inhibition by 5×10^{-5}M fluoride as measured with 5×10^{-5}M benzoylcholine. Its value is $\sim$ 60% for the usual enzyme and $\sim$15% for the atypical enzyme. The simplest explanation for this observation is that fluoride is bound weaker by the atypical enzyme. However, this is not the case; we have measured fluoride binding using the arsenite-enzyme technique and find that both the usual and atypical enzymes bind fluoride equally.

Then it must be that the atypical enzyme manifests less cooperativity in the binding of fluoride and other ligands. Again that is not the case. The explanation is probably related to the fact that the K_M is poorer (higher) for the atypical enzyme.

We have seen that fluoride reacts slowly with acetylcholinesterase in a second order reaction with a rate constant on the order of $10^4 M^{-1} sec^-$. This rate is some 10^4 times slower than for the substrate. It is surprising that such a simple symmetric ion should react so slowly. The reason may be that fluoride does not find a preexisting binding site but must create one. The structural nature of the fluoride binding site is not known but given the nature of a protein, it seems reasonable to expect that hydrogen bonds and ionic interactions are involved. This would be similar to the interactions in NH_4F crystals. Further, it seems reasonable to suppose that the hydrogen bond donors of the enzyme are already hydrogen bonded. Then as a speculation we might consider types of reactions such as:

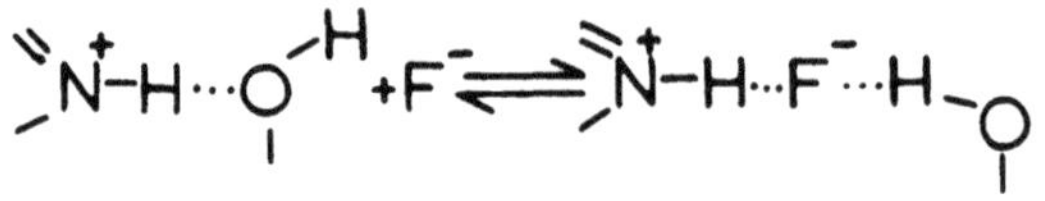

or even a rate controling proton transfer from a relatively strong base to a weak base (23):

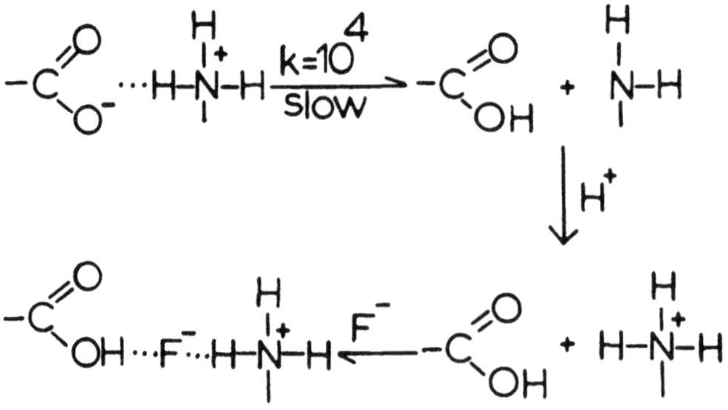

References

1. Mounter, L.A. and V.P. Whittaker. 1953. Biochem. J. 53, 167-173.
2. Thompson, R.H.S. 1947. J. Physiol. 105, 370-381.
3. Wilbronn, E. 1965. Acta Chem. Scand. 19, 1333.
4. Krupka, R.M. 1966. Mol. Pharmacol. 2, 558-569.
5. Boyer, P.D., H. Lardy and K. Myrback, eds. 1963. The Enzyme. 2nd ed., Academic Press, New York.
6. Peters, R.A. 1949. Symp. Soc. Exp. Biol. 3, 36-59.
7. Rosenberry, T.L. 1975. Advan. Enzymol. 43, 103-218.
8. Karlin, A. 1967. Biochim. Biophys. Acta 139, 358-362.
9. Wilson, I.B. and E. Cabib. 1954. J. Amer. Chem. Soc. 76, 5154.
10. Wilson, I.B. and I. Silman. 1977. Biochemistry 16, 2701-2708.
11. Page, J.D., I.B. Wilson and I. Silman. 1984. (Submitted for publication).
12. Roy, G.L., A.L. Laferriere and J.D. Edwards. 1957. J. Inorg. Nucl. Chem. 4, 106-114.
13. Antikainen, P.J. and E. Huttunen. 1973. Suomen Kemist, B. 46, 184-190.
14. Bruice, T.C., M.J. Gregory and S.L. Walters. 1968. J. Amer. Chem. Soc. 90:6, 1612.
15. Riordan, J.F., M. Sokolovsky and B.L. Vallee. 1967. Biochemistry 6, 3609-3617.
16. Fuchs, S., D. Guarari and I. Silman. 1974. Arch. Biochem. Biophys. 165, 90-97.
17. Page, J.D. and I.B. Wilson. 1984. J. Biol. Chem. (In press).
18. Greenspan, C.M. and I.B. Wilson. 1970. Mol. Pharmacol. 6, 266-272.
19. Wilson, I.B., Harrison, M.A. and S. Ginsburg. 1961. J. Biol. Chem. 236, 1498-1500.

20. Froede, H.C. and I.B. Wilson. 1984. (Submitted).

21. Kalow, W., K. Genest and N. Staron. 1956. Can. J. Biochem. Physiol. 34, 637.

22. Brown, S.S., W. Kalow, W. Piley, M. Whittaker, and C.L. Woronick. 1981. Adv. Clin. Chem. 22:1.

23. Eigen, M. in Nobel Symposium 5. Interscience (1967).

CORRELATION OF ASSUMED FUNCTIONS OF ACETYLCHOLINESTERASE WITH ITS CELLULAR AND SUBCELLULAR DISTRIBUTION

M. Brzin

Institute of Pathophysiology, School of Medicine, E. Kardelj University, 61 105 Ljubljana, Yugoslavia

Introduction

The localization of acetylcholinesterase (AChE) activity at the postsynaptic membrane in the vicinity of the cholinreceptor was postulated at the very beginning as one of the basic requirements of the neurohumoral theory of cholinergic transmission (1). No other but synaptic function was provided for AChE by the theory, therefore, the presence of the enzyme outside of synapses in excitable tissues was considered irrelevant. Furthermore, new biochemical, histochemical and electrone microscopic-cytochemical evidence which was since then accumulating steadily over the decades and which revealed a wide spectrum of diverse structures, cells and subcellular locations bearing AChE activity (2,3,4), it was given selective acceptance and the rest was largerly neglected. It is likely that such atmosphere in the field has been the main reason for the Nachmansohn's membrane theory of nerve function (5,6), although it was the first attempt to incorporate biochemical and molecular correlates into the mechanisms of nerve activity, was often widely opposed or ignored. The latter, in na extended form, may still represent the most likely basis on which the large body of experimental evidence could be accomodated within an integrated future view of nerve functions.

With time, more detailed information about the localization of AChE in various structures became available, contributing to the conclusion that nothing like a general cholinergic synapse exists either functionally (7, 8) or in regard to AChE activity and localization. Thus, the mostly explored cholinergic synapse, muscle endplate seem to represent exceptional

Molecular Basis of Nerve Activity

case of the cholinergic mechanisms characterized by an extremely high concentration of AChE activity in the junctional space between presynaptic and postsynaptic membranes (11, 12).

Parallel to the discoveries of profound functional differences between cholinergic structures it became evident that the transmitters can functione also as modulators (13,14). In the peripheral and central nervous structures either a synaptic or a non-synaptic release of transmitters can act on a variety of receptor-bearing sites, modulating its own or other transmitter release from nerve endings or affecting directly the responses of neurons (14,15,16). Large amounts of transmitters, including ACh, can be released without an electrical impulse and such silent release excedes considerably the spontaneous quantal release of the transmitter (17,18,19). A transmitter can also have several different actions on the same or on different postsynaptic cell regarding the duration of electrical responses (20,21,22), or triggering metabolic changes in the cell (23), which may or may not show in measurable electrical responses of postsynaptic membrane (24). Thus, the action of transmitters may not always be linked to electrical signals; moreover, functionally involved neurones exist without ever propagating an electrical impulse (25).

The accumulated evidence shows that the field of neurotransmission is much more complex than originally assumed by the now outmoded concept of simple chemical synaptic transmission (14). A new and general functional scheme, unifying into a logical framework the different transmitters and a variety of their actions and interactions, inclusive non-impulsive communications among neurons, has yet to emerge, hopefully in the near future.

Different functional requirements imposed on the cholinergic system should also be reflected on its components, and it seems justified to assume that AChE activity, distribution and ultrastructural localization would be designed to match specifically to the requirements of each of different functions. Therefore new information about subcellular localization of AChE is particularly important as a complement to the future interdisciplinary research of cholinergic functions in CNS. To achieve this goal much higher resolution, accuracy and precision will be essential Resolution now offered by current cytochemical procedures is not sufficient.

<u>Present state of procedures for ultrastructural localization of AChE.</u> The methods based on thiocholine esters as substrates for localization of sites of cholinesterase activity (26,27) have contributed considerably to our present knowledge about the distribution of cholinergic elements in peripheral excitable tissues (28,29,30) and in CNS, where in addition, the cholinergic pathways have been revealed and elaborated on the light microscopic level (2,3,31,32,33).

The drive toward further increase of spatial resolution of AChE localization by using electrone microscope was natural and various existing histochemical approaches were adopted for ultrastrucutral localization of the enzyme (3,4,34,35,36). The new information gained by this upgraded approach was worth the effort particularly in regard of intracellular distribution of the enzyme but some disagreement appeared in observations of different authors about the details of ultrastructural localization, most notably related to the presence of AChE activity in junctional clefts of ganglionic or CNS synapses (3,4,9). The problems stemm in the inherent inadeguacy of methods to meet the requirements imposed by the complexity of conditions essential for a reliable localization at ultrastructural dimensions (4). As pointed by Koelle there is no gain to be expected by increasing just the resolution of observation without reducing the diffusion of the reaction product and other artefacts as well (37).

Considering the high substrate specifity and the frequency of use, as judged by the past and currently published observations, only the procedures based on thiocholine esters will be critically evaluated here. The Copper-thiocholine histochemical reaction of Koelle and Friedenwald (26) followed by the sulphide conversion step is the basis for the majority of cytochemical methods (3,34,35). At present several important points can already be listed explaining why this group of methods, which has contributed large body of new information on the light microscopic level, can not yield more satisfactory results in regard to details of ultrastructural localization of AChE:

1. The chemistry of reaction: In 1970 (38,39) it was demonstrated that the precipitated reaction product is not the cupric thiocholine sulphate, as originally assumed, but cuprous copper-thiocholine-iodide complex and

that iodide is absolutely essential for the precipitate formation. This finding was the potential basis for the rational modifications of the procedures. However, the methods in use were not adopted accordingly and the essentiality of iodide, which must be present in sufficient concentration in the medium is not taken into consideration, creating experimental conditions for a delayed capturing reaction and diffusion artefacts.

2.The conversion of the primary reaction product. The procedures consist of two steps: first is the incubation resulting in precipitation of the primary product presumably close to the sites bearing AChE activity, and the second is the conversion of already formed reaction product into CuS by $(NH_4)S$ or Na_2S. Considering the dimensions of structures under examination and the resolution desired and obtainable in an electrone microscope, such conversion of primary precipitate is hardly acceptable for two main reasons: a)during conversion the primary precipitate is first dissolved and then precipitated as CuS, but not essentialy on the same location as originally (40,41); b)due to the difference in densities of the two precipitates (2.2 and 5.5 respectively) (42), the specific volume of CuS is only about 40 % of the specific volume of the primary precipitate, meaning, that even if there is no translocation by previous diffusion, the secondary precipitate could not be located on exactly the same sites as the original primary reaction product.

3.The length of incubation: All Cu-Thiocholine procedures modified for ultrastructural localization of AChE use extremely long incubation times which are often longer than one hour for nervous tissue and 10 minutes or more for muscle endplates. Such long incubation is not only unnecessary but harmful and can not be reconciled with ultrastructural distinction of actual locations of AChE activity. From the quantitative data about AChE activity in the sample extrapolated considering the decrease of activity in histochemical experimental conditions (fixation, low pH, low temperature, Copper ion concentration, diffusion distance for substrates) and the available space on ultrastructure, it could be calculated that the incubation time used, is long enough for an enormous axcess of the reaction product to be formed (4,43). The spaces adjacent to the sites of enzyme activity are occupied first so that the reaction product, which is being steadily produced by the enzyme, diffuses and precipitates on remote locations, usually forming large crystals and aggregates (43, this paper). The prescribed (34, 35) prolonged rinsing of samples before and after conversion and particularly the prolonged conversion step in sulphide solutions, which often contain palysulphides, eliminates largerly the excess of the precipitate, whereas the residual reaction product remains attached in intercellular clefts and intracellular membranes (43). It is possible that the sites of enzyme activity, where the first reaction product was initially formed, remain labeled even after the long postincubational treatments, but it is even more probable that a good part of excessive precipitate on structures without AChE activity also remains, giving a false localization. This point has been experimentally demonstrated on caudate nucleus of rat, by first observing the samples directly after incubation and then after the generally recomended prolonged rinsing-conversion-rinsing treatment (43).

The relation between the primary and the secondary reaction product in muscle motor endplates, where the additional complication is created by an extremely high focal activity of AChE, is presented in this paper.

Materials and Methods

The rat diaphragm muscle was first shortly (20 min) fixed in situ (4% paraformaldehyde in Ringer-phosphate buffer, pH 7.2 then the rib case with the diaphragma still attached was directed and additionally fixed for 4 to 8 hours at 2-4°C. A ribbon of muscle, about 3 mm wide and containing most of endplates, was excised along the branches of the phrenic nerve and sliced parallel to the muscle fibres to pieces 100-200 um thick. The samples were rinsed in Ringer for two hours and processed for visualization of AChE activity (44) first by preincubation in the basic medium without substrate for 20 min and then by incubation in the same medium sontaining 3 mM AThCh Cl for 30-120 sec. Preincubation and incubation were performed at 2-4°C.

The composition of the basic medium: glycine 20 mM, $CuSO_4$ 10 mM, and acetate buffer 10 mM, pH 5. To supply the capturing anions, necessary for reaction product precipitation, one of the following salts was included: KJ, KCN or KCNS, at a final concentration of 5 mM. This basic medium, with AThCh-Cl added, serves for the subsequent incubation of samples. In the incubation solution containing AThCh-Cl, but without a capturing anion (either J^- or CN or CNS) no end product is precipitated in the presence of AChE. Each of those anions forms with cuprous ion and thioline its specific reaction product which can be readiliy recognized in electrone microscope by the characteristic appearance (44).

After incubation for desired time (incubation times given in legends to the figures), the samples are rinsed in cold destiled water for 5 min and postfixed in 1% $KMnO_4$ solution for 1 hour (2-4°C). Some samples incubated in the presence of J^- were additionally treated with Na_2S (1%, 15 min + 5 min rinsing in Ringer's). After dehydration and embedding in Epon, the samples are thin-selectioned and photographed in a Jeol 100 B electrone microscope.

The postfixation in $KMnO_4$ is essential for this one-step procedure (36) because it stabilizes all three reaction products at the sites where they were formed. This is in opposit with postfixation with OsO_4 which dissolves the primary reaction product (36) or after solubilization forms a secondary product which precipitates randomly in the intracellular spaces (43). The solubilization of the primary reaction product by osmic acid, which is a standard fixative for electrone microscopy, may be the main reason that the conversion into copper sulphide, used in light microscopic histochemistry was maintained by most researchers working on ultrastructural localization of AChE activity.

Results

One-step Cu-thiocholine procedure with iodide as the capturing ion (present as the moiety of the commercial substrate) gave satisfactory results on tissues with moderate or low AChE activity (35,45,46), but failed completely when applied to muscle motor endplates which have an extremely high focal AChE activity. In all experimental conditions tested, even at short incubation times (30 sec) the reaction product in the endplate was precipitated in large elongated crystals, individually scatered or in clusters, suggesting crystallic growth unrelated to the sites of origin of end-product molecules (Fig. 1 and 3).

Such characteristics of the end-product exclude the use of iodide as the capturing anion for high resolution localization of AChE in neuromuscular junctions not only when one-step procedure is employed, but also in two--steps procedures where the primary reaction product is subsequently converted to CuS. The fact remains that secondary precipitate is formed by solubilization of the primary reaction product, therefore the localization of the former can not be more precise than the localization of the latter. Due to partial solubilization, redistribution and a lower specific volume of CuS mentioned in the introduction, the localization of smaller microcrystalline particles as the endplate looks satisfactory (Fig. 2 and 4). But only from an aesthetic point of view, because the localization of the secondary product is an artefact caused by an alteration and translocation of the primary product caused by the conversion. In addition, the conversion causes diffusion of the second reaction product as shown on Fig. 2 and 4 by small particles scatered in the nerve ending and muscle which were absent before conversion (Figs. 1 and 3).

As shown on Figs. 5-9 a considerable progress in regard to resolution was achieved by eliminating J^- from the medium and substituting it by CNS^- or CN^- as capturing anions. One-step Cu-thiocholine procedure in the presence of CNS^- yields a primary precipitate which is exclusively distributed in synaptic clefts and in nerve ending-Schwann cell interspace (Fig. 8). At a higher magnification the impression is that the activity of AChE is more densely deposited close to synaptic membranes, notably presynaptic, and that the synaptic cleft substance (basement membrane) exhibits less AChE activity (Fig.9). However, in the presence of CN^- (Figs.5 and 6), which gives smaller reaction product the membrane affiliation of particles is much less pronounced, suggesting, that basal lamina in the synaptic space is the main structure bearing AChE activity. Such localization is particularly evident in some samples incubated for a short time (Fig.5 and 6, 30 sec), where in junctional regions, remote from the surface, the reaction seems somewhat delayed due to a longer diffusion path for the substrate. At longer incubation times (90 sec) the synaptic cleft is uniformly filled with the product and even in such conditions, no diffusion artefacts can be observed (Fig. 7).

Discussion

Recently, one-step procedure with CNS^- as capturing anion was applied for localization of AChE in developmental studies of muscle and sympathetic ganglia (4,59,60). The results show a clear improvement in relyability and resolution when compared with the two-steps methods, where

Fig. 1 Motor endplate of adult rat diaphragm. Capturing anion J^-, incubation time 30 sec. One-step procedure; conversion to sulphide omitted. Large elongated crystals of primary reaction product in the synaptic cleft (arrows) and Schwann cell-nerve ending interspace (crossed arrows). NE: nerve ending; M: muscle fiber. The resolution of localization is extremely poor but no diffusion of the primary product outside clefts can be observed. Bar: 0.5 um.

Fig. 2 Motor endplate of adult rat diaphragm. Capturing anion J^-, incubation time: 30 sec. Two-steps procedure; primary reaction product converted into CuS to demonstrate transformation, translocation and altered localization of the product. The secondary precipitate is formed by dispersion of the primary reaction product (see Fig. 1) and reprecipitated in small particles in the synaptic clefts (arrows). NE: nerve ending. M: muscle fiber. The conversion caused diffusion of secondary product as indicated by scatered small particles in nerve ending and muscle fiber. Bar: 0.5 um.

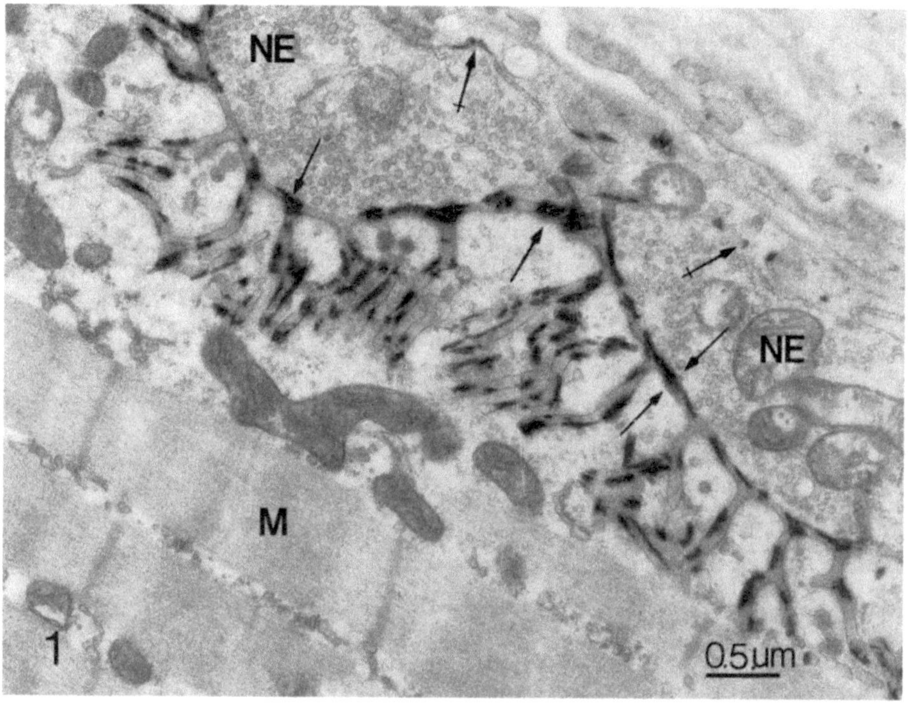
NE
NE
M
1
0.5µm

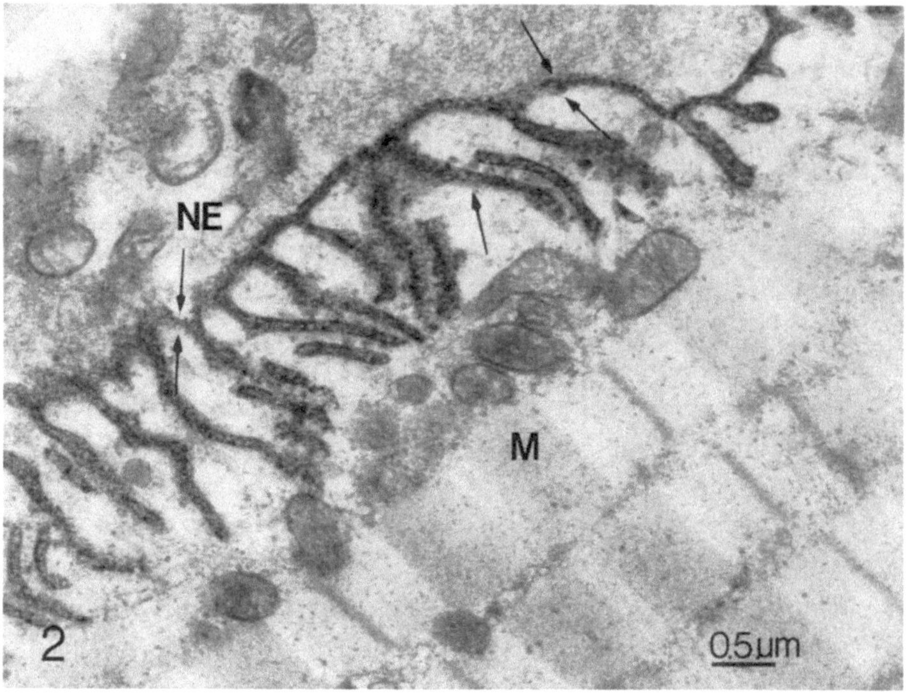
NE
M
2
0.5µm

the translocation of the reaction product during conversion step makes the distinction of ultrastrucutral sites of AChE activity impossible. The precision and relyability of ultrastructural localization of AChE is becoming important because the results are increasingly used to complement the interpretation of quantitative data in functional terms. Considering the new and broader concepts of cholinergic functions, which are emerging the accurate infcrmations about ultrastructural localization of AChE activity could be of great value for the studies of the cholinergic system which serves different specific functions in CNS and in other tissue. It seems justified to expect that a distinct spatial organization, distribution and density of AChE, if revealed cytochemi cally, may indicate domains where either synaptic transmission or various modulatory function predominate, characterized by their different requirements for the rate of ACh elimination and for mutual biochemical separation by AChE. With a precise cytochemical method, which gives a high level of resolution it might be possible to reach this goal.

Fig. 3 The same as in Fig. 1, but the incubation time was 90 sec. Larger quantity of the primary product has accumulated due to prolonged incubation. The resolution is even worse than in Fig. 1, but still no diffusion artefacts can be seen. Bar: 0.5 um.

Fig. 4 The same as Fig. 2, but the incubation was prolonged to 90 sec, resulting a larger quantity of deposited primary product (see Fig. 1). After conversion, the primary product is again dispersed and the secondary product in synaptic cleft is reprecipitated and evenly distributed in small particles. The impression is that the localizatoon is satisfactory, but in fact it is an artefact (compare Fig. 3 - localization of the primary product). In addition, diffusion of the secondary product in nerve ending and muscle cell is clearly visible. Bar: 0.5 um.

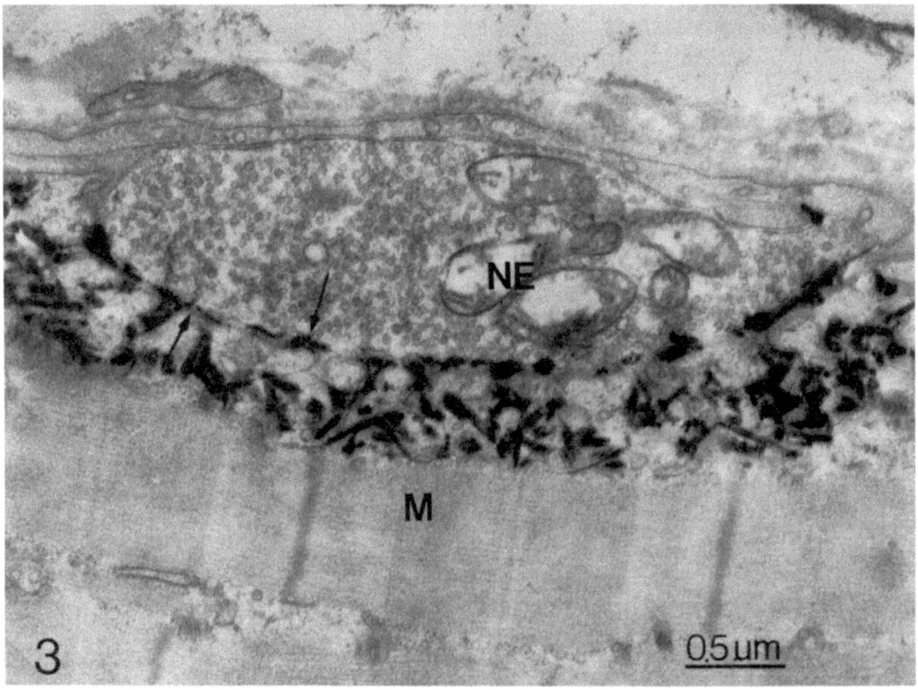
NE
M
3
0.5 μm

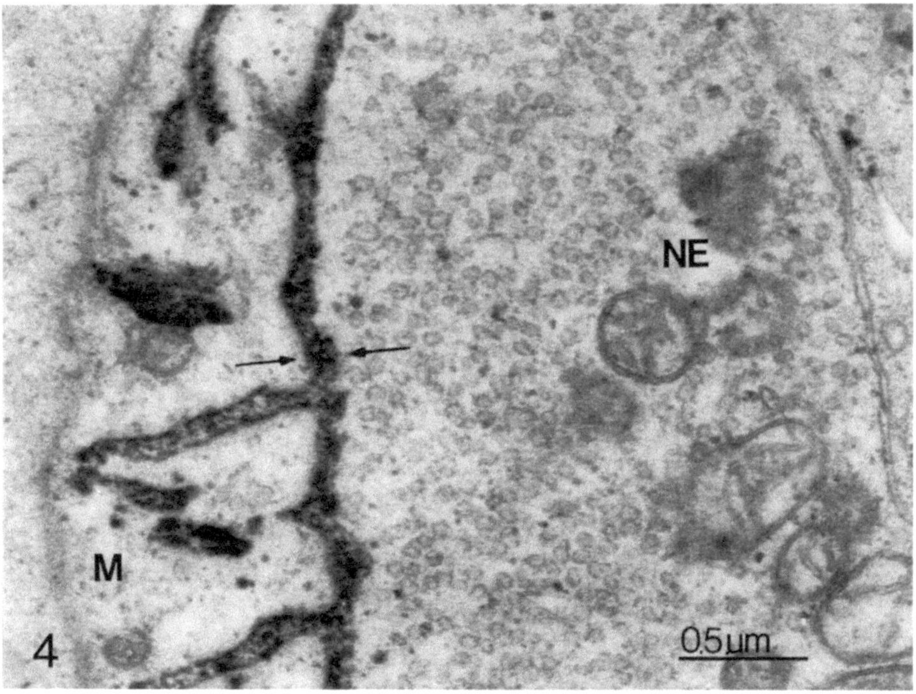
NE
M
4
0.5 μm

Furthermore, the precise data on ultrastructural localization should also help in resolving the recent puzzle of functional significance of AChE release or secretion by neurons and muscle (47,48,49,50). An interesting example is the AChE containing neurons (52) in substantia nigra where the enzyme and dopamine are released upon stimulation and purified AChE applied in vivo has a long lasting effect on the discharge of neurons, behavior of animals and release of dopamine (53,54,55,56). Another area for application of cytochemistry is indicated by the suggestion that the released AChE in CNS may act as a peptidase implying the involvment of the enzyme in regulation of brain peptides (57,58).

Fig. 5 Motor endplate of adult rat diaphragm. Capturing anion CN^-, incubation time: 30 sec. One-step procedure without conversion. Particles of the primary reaction product distributed in the basal lamina of synaptic clefts (arrows). Synaptic space closer to the fiber surface exhibits a larger density of particles, due to a shorter diffusion path for substrate. The resolution is high and the localization within the cleft can be distinguished. NE: nerve ending; MN: muscle nucleus. No diffusion artefacts are present. Bar: 0.5 um.

Fig. 6 The same as Fig. 5, at larger magnification. Bar: 0.5 um.

Fig. 7 The same as in Fig. 5, but the incubation time was prolonged to 90 sec. The primary reaction product formed in the presence of CN^- is evenly distributed in synaptic clefts (arrows) and Schwann cell-nerve ending interspace (crossed arrows). The appearance of the precipitate is amorphous with scatered nuclei of small particles. In spite of longer incubation time, no diffusion outside clefts is visible. Bar: 0.5 um.

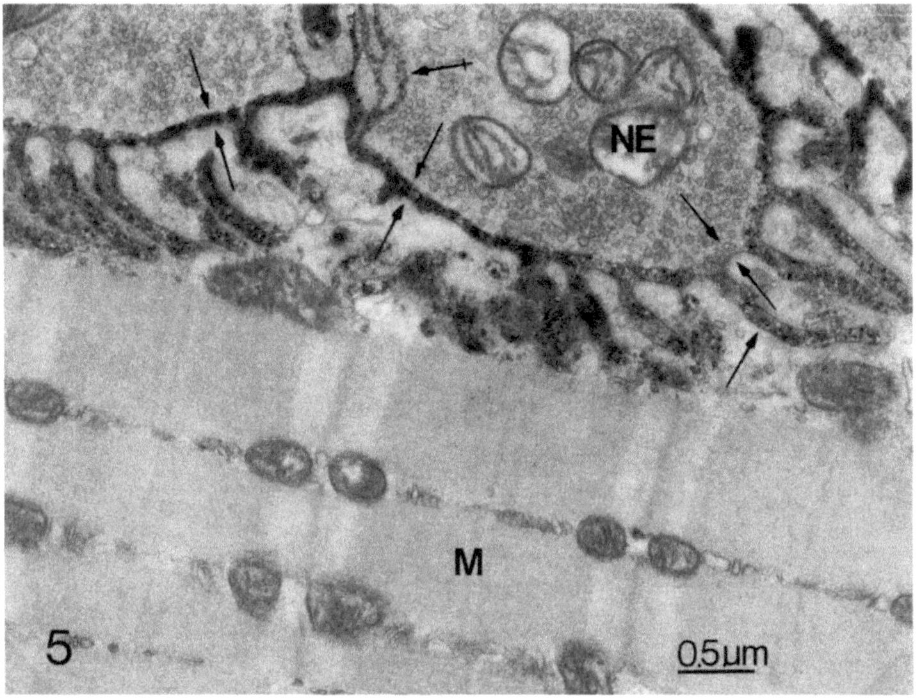
NE
M
5
0.5 µm

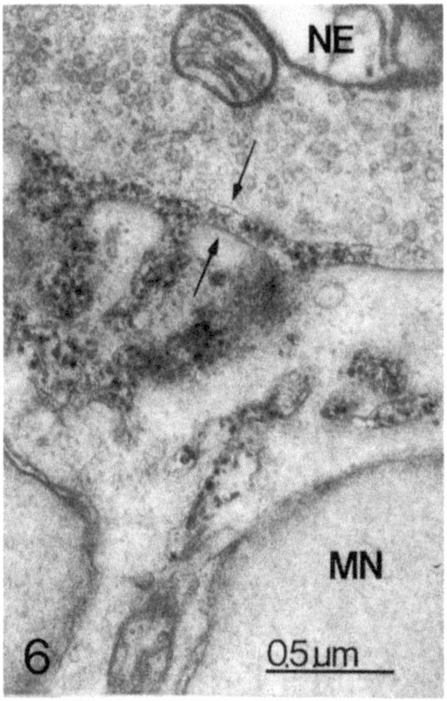
NE
MN
6
0.5 µm

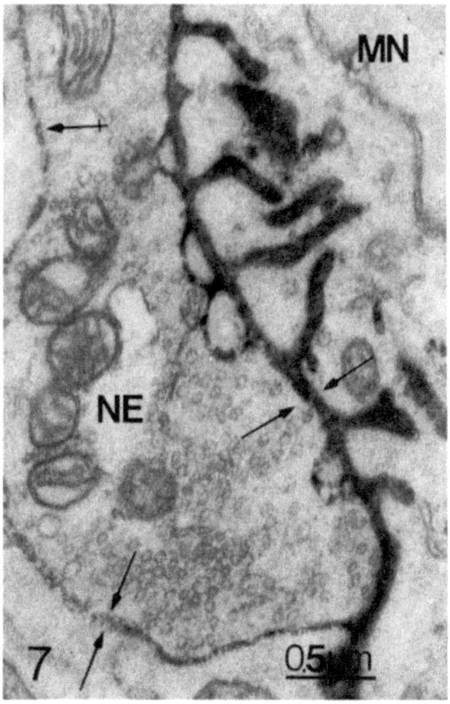
MN
NE
7
0.5 µm

The intention of this presentation was to draw the attention of neurochemists and physiologists to the inherent difficulties, unsolved problems but also for the not properly exploited potentialities of the complex field of electrone microscopic cytochemistry of AChE. The main errors and crutial points in contemporary procedures pointed out and the directions for the further improvement of one-step approach was indicated. And if cytochemical observations are going to be accepted in the domain of neurochemistry on equal terms as quantitative assays, then the neurochemists should take a more critical and more active standpoint to the ways and means how these results are obtained. The path toward a better method should follow the strict criteria for precission and accuracy as valid for quantitative procedures.

Acknowledgements

This work was supported by a grant of Slovenian Research Community. The assistance of Mrs. Simona Pucihar and Mrs. Duša Čuček with cytochemistry and electrone microscopy, and of Mrs. Belinda Lampret with the preparation of the manuscript is greatfully acknowledged.

Fig. 8 Motor endplate in diaphragm of three weeks old rat. Capturing anion CNS^-, incubation time: 90 sec. One-step procedure without conversion. Low magnification overview of the whole endplate. Particles of primary reaction product localized in synaptic clefts (arrows), Schwann cell-nerve endings interspace and between Schwann cell processes (crossed arrows). During the period of muscle growth in young rats, AChE activity in Schwann cells can still be visualized (double arrow). NE: nerve endings; M: muscle cell; MN: muscle nucleus; SN: Schwann cell nucleus. Bar: 1.0 um.

Fig. 9 The same as in Fig. 7, but a larger magnification. Primary reaction product in synaptic cleft preferentially localized close to synaptic membranes, notably presynaptic (compare with Fig. 6). Bar: 0.5 um.

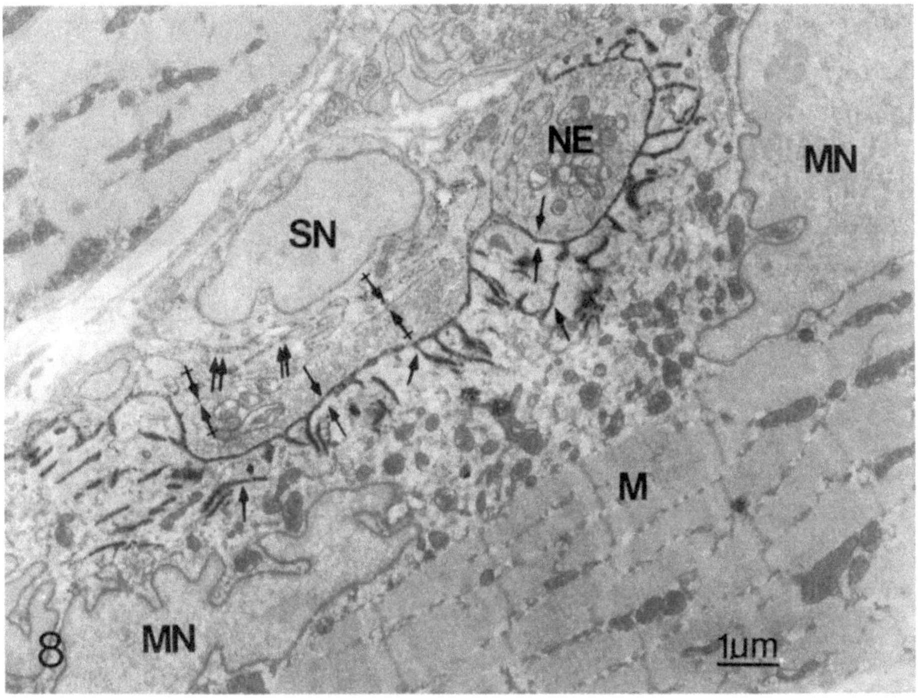
NE
MN
SN
M
MN
8
1µm

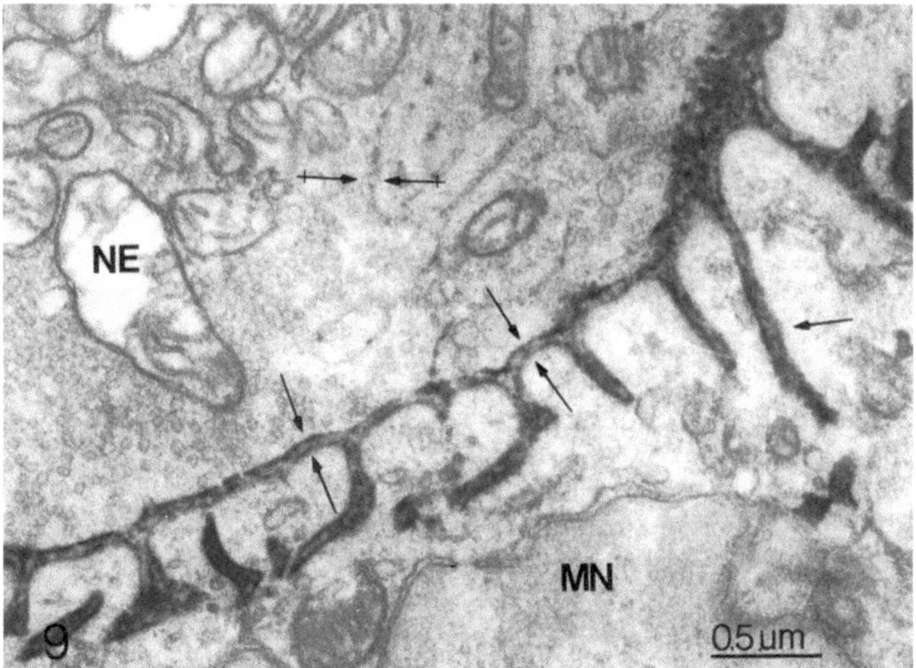
NE
MN
9
0.5µm

References

I.Katz, B.. 1966. Nerve, Muscle and Synapse. McGraw Hill, New York.

2.Koelle, G.B.. 1963. In: Cholinesterases and Anticholinesterase agents (G.B. Koelle, ed.). Springer Verlag, Berlin. pp. 187-298.

3.Silver, A.. 1974. The Biology of Cholinesterases. North Holland, Amsterdam.

4.Brzin, M., J. Sketelj and B. Klinar. 1983. In: Handbook of Neurochemistry. Vol. 4. (A. Lajtha, ed). Plenum Press, New York - London. pp.251-292.

5.Nachmansohn, D.. 1959. Chemical and Molecular Basis of Nerve Activity. Academic Press, New York - London.

6.Nachmansohn, D. and Newman, E.. 1975. Chemical and Molecular Basis of Nerve Activity. Academic Press, New York - London.

7.Krnjević, K.. 1969. Federetion. Proc. 28, 113-120.

8.Krnjević K., R. Pumain and L. Renard. 1971. J. Physiol. 215, 247-268.

9.Koelle, G.B., R. Davis, E.G. Smyrl and A.V. Fine. 1974. J. Histochem. 22, 252-259.

10.Klinar, B. and M. Brzin. 1979. Experientia, 35, 754-756.

11.Brzin, M. and J. Zajiček. 1958. Nature, 181, 626 only.

12.Barnard, E.A.. 1974. In: The Peripheral Nervous System. (J.I. Hubbard, ed.). Plenum Press, New York, pp. 201-224.

13.Florey, E.. 1960. Fed.Proc. 26, 1164-1178.

14.Florey, E.. 1984. Neurochem. Res. 9, 413-427.

5.Weiner, N.. 1979. Fed. Proc. 36, 2193-2202.

16.Vizi, E.S. and J. Knoll. 1971. J. Pharm. Pharmacol. 23, 918-925.

17.Katz, B. and R. Miledi. 1977. Proc. R. Soc. London, B 196, 59-72.

18.Miledi, R., P.C. Molenaar, R.L. Polak, J.W.M. Tas and T. Van der Laakes. 1982. Proc. R. Soc. London, B, 214, 153-168.

19.Katz, B. and R. Miledi. 1981. Proc. R. Soc. LOndon, B, 212, 131-137.

20.Hartzebl, H.C., S.W. Kuffler, R. Stickgold and D. Yoshikami. 1977. J. Physiol. 271, 817-846.

21.Purves, R.D.. 1974. Brit. J. Pharmacol. 52, 77-86.

22.Weight, F.F., Schulman, J.A., P.A. Smith and N.A. Busis. 1979. Long Lasting synaptic Potentials and Modulation of Synaptic Transmission. Fed. Proc. 38, 2084-2094.

23.Greengard, P.. 1976. Nature, 291, 101-108.

24.Hill-Smith I. and R.D. Purves. 1978. J. Physiol. 279, 31-54.

25.Roberts, A. and B.M.H. Bush. 1981. Neurones Without Impulses. Cambridge University Press, London. 290.

26.Koelle, G.B. and J.S. Friedenwald. 1949. Proc. Soc.exp.Biol. 70,617-622.

27.Karnovsky, M.J. and L. Roots. 1964. J. Histochem. Cytochem. 12, 219-221.

28.Couteaux, R. and J. Taxi. 1952. Arch.Anat.Microscop.Morphol.Exp. 41, 352-392.

29.Couteaux, R. 1955. Intern.Rev. Cytol. 4, 335-375.

30.Eränkö, O. 1967. Ann.Rev. Pharmacol. 7, 203-222.

31.Shute, C.C.D. and P.R. Lewis. 1963. Nature. 199, 1160-1163.

32.Shute, C.C.D. and P.R. Lewis. 1965. Nature. 205, 242-246.

33.Shute, C.C.D. and P.R. Lewis. 1967. Brain. 90, 497-520.

34.Lewis, P.R. and C.C.D. Shute. 1969. J. Microsc.Sci. 89, 181-193.

35.Lewis, P.R. and Shute, C.C.D. 1966. J. Cell Sci. 1, 381-390.

36.Brzin, M., V.M. Tennyson and P. Duffy. 1966. J. Cell Biol. 31, 215-242.

37.Koelle, G.B. 1970. In: Drugs and cholinergic Mechanisms in the CNS (E. Heilbronn and A. WInter, eds.). Research Institute of National Defense, Stockholm, pp. 431-439.

38.Souchay, P. and S. Tsuji. 1970. Ann. Histochim. 15, 263-271.

39.Tsuji, S. 1974. Histochemistry. 42, 99-110.

40.Zajiček, J., B. Sylven and N. Datta. 1954. J. Histochem. Cytochem. 2, 115-121.

41.Holmsted, B. 1957. Acta Physiol. Scand. 40, 331-337.

42.Grubić, Z. and M. Brzin. 1978. Histochemistry. 56, 213-220.

43.Klinar, B. and M. Brzin. 1977. Histochemistry. 50, 313-318.

44.Brzin, M. and S. Pucihar. 1976. Histochemistry. 48, 283-292.

45.Tennyson, V.M. and M. Brzin. 1970. J. Cell Biol. 46, 64-80.

46.Tennyson, V.M., M. Brzin and P. Slotwiner. 1971. J. Cell Biol. 51, 703-721.

47.Wilson, S.H., B.K. Schrier, J.L. Farber, E.J. Rosenberg, E.J. Thompson, A.J. Blume and M. Nirenberg. 1972. J. Biol. Chem. 247, 3159-3169.

48.Vogel, Z., A.J. Sytkowsky and M. Nirenberg. 1972. Proc. Natl. Acad. Sci. USA, 69, 3180-3184.

49.Meunier, J.C., R. Sealock, R. Olsen and J.-P. Changeaux. 1974. Eur. J. Biochem. 45, 371-394.

50.Prives, J., I. Silman and A. A sterdam. 1976. Cell. 7, 543-550.

51.Gisiger, V. and M. Vigny. 1977. Febs Letters. 84, 253-256.

52.Butcher, L.L. and K. Talbot. 1977. In: Cholinergic-monominergic interactions in the brain.(Butcher, L.L., ed.). Academic Press, London. pp. 25-95.

53.Greenfield, S.A. and S.G. Shaw. 1982. Neuroscience. 7, 2883-2893.

54.Greenfield, S.A., J.F. Stein, J.F. Hodgson and I.W. Chubb. 1981. Neuroscience, 6, 2287-2295.

55.Greenfield, S.A., I.W. Chubb, R.A. Grünwald, Z. Henderson, J. May, S. Portnoy, J. Weston, and M.C. Wright. 1984. Exp.Brain Res. 54, 513-520.

56.Greenfield, S.A. 1984. In: Cholinesterases, Fundamental and Applied Aspects (Brzin M., E. Barnard and D. Sket, eds.). W. de Gruyter, Berlin, New York, pp. 289-303.

57.Chubb, I.W., A.J. Hodgson and G.H. White. 1980. Neuroscience 5, 2065-2072.

58.Chubb, I.W. 1984. In: Cholinesterases, Fundamental and Applied Aspects (Brzin, M. E. Barnard and D. Sket, eds.). W. de Gruyter, Berlin, New York, pp. 345-459.

59.Brzin, M., J. Sketelj, V.M. Tennyson, T. Kiauta, M. Budininkas--Schoenebeck. 1981. Muscle&Nerve 4, 505-513.

60.Klinar, B. and M. Brzin. 1979. Experientia 35, 754-756.

THE MECHANISM OF ACTION OF CHOLINE ACETYLTRANSFERASE

Henry G. Mautner

Department of Biochemistry and Pharmacology. Tufts University School of Medicine, Boston, Massachusetts, 02111, USA.

Introduction

Choline acetyltransferase (ChA) was discovered by Nachmansohn and Machado more than forty years ago (1). Nachmansohn's group also showed that a dialyzable cofactor was required for cell-free ATP-dependent acetylation to take place (2). This cofactor was later named CoA (3) and found to be of ubiquitous importance in acyl group transfer reactions (4).

Nachmansohn, in addition to postulating the central role of acetylcholine (Ach) in modulating membrane cation permeability through attachment to a protein receptor (5), recognized that the Ach receptor and the enzymes responsible for the hydrolysis and synthesis of Ach had to function in a coordinated cyclic process, in which ChA, the enzyme catalyzing the synthesis of Ach, provided the essential link between the metabolism of the cell and the operations of the Ach cycle.

In recent years, the great usefulness of antibodies to ChA in mapping cholinergic neurons (6) as well as the possible link of Alzheimer's disease and ChA (7) has added further interest to studies of this enzyme. It is, therefore, surprising that so many questions regarding the structure of ChA, the nature and topography of its binding sites and the regulation of its action remain unanswered.

The Ordering of the Choline Acetyltransferase Reaction

Ach synthesis: ChOH + CoASAc $\rightleftharpoons$ ChoAc + CoASH

involves two substrates (a thiolester and an alcohol) and two products (an ester and a thiol). Since Ach is a "high energy" compound with a $G^{0'}$ for hydrolysis only slightly lower than that of AcCoA, it is not surprising that, under physiological conditions, the acetylation of choline by acetyl CoA shows an equilibrium constant of only 12 (8) making this a readily reversible reaction. It has been postulated that

Molecular Basis of Nerve Activity

the fact that Ach is a "high energy" ester can be related to the gauche conformation of this compound with the coulombic interaction between the trimethylammonium group and the acyloxy oxygen of Ach activating the carbonyl carbon to nucleophilic attack (9, 10).

Studies of the ordering of the ChA reaction suggested a Theorell-Chance mechanism in which the steady-state concentration of the ternary complex is very low (11, 12): More recently, Hersh and his coworkers (13) proposed a random Theorell-Chance mechanism in which acetyl CoA is generally the leading substrate to be attached to ChA and CoA is generally the last product to depart:

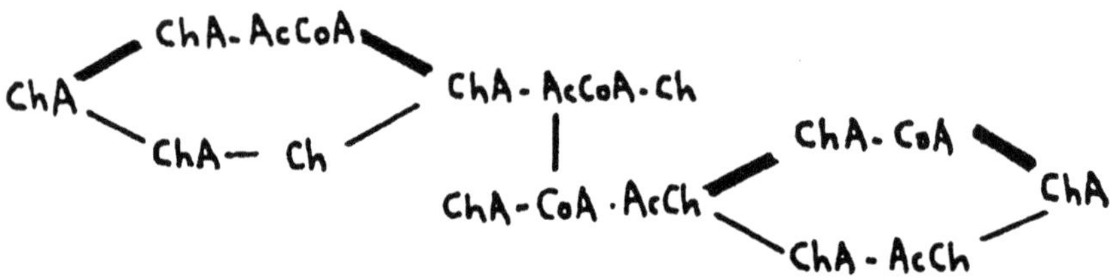

Kinetic Mechanism of ChA Reaction as redrawn from (13) Predominant pathway is indicated by heavy line; concentration of ternary complex is assumed to be low.

ChA catalysis is complicated by salt effects. Addition of salt alters the binding constant of the substrates and reactants and increases the maximum velocity of the reaction (14, 15). It could be shown that, at low salt concentration, the rate-limiting step is CoA release while, at higher salt concentration, the interconversion of the ternary complex becomes rate-limiting. Considerable controversy surrounded the ion specificity of these salt effects. Rossier (16) made the suggestion that, in a presynaptic enzyme complex in which ChA is membrane-bound, enzyme activation occurs through a depolarization-induced increase in chloride concentration. However, the ion specificity reported for "fresh and unpurified enzyme" (17) could not be confirmed using highly purified ChA (18). However, the possibility that ChA may become membrane-bound under physiological as well as under experimental conditions (19, 20) cannot be excluded. It is at least possible that ion-specific activation of membrane-bound ChA might occur.

The Binding Sites of Choline Acetyltransferase

The Anionic Binding Site

Nachmansohn and his coworkers (21) proposed the presence of an anionic binding site in ChA to which the cationic portion of ChA would be attached. The substrate specificity of this site is relatively low. Thus, the maximal Ch -catalyzed acetylation rate for the dimethylsulfonium analog of choline is as high as that for choline (22). That the alignment of the cationic portion of the substrate on the anionic binding site is relatively unimportant is also shown by the relatively small differences in the acetylation rates of the enantiomers of α-methylcholine.

$(CH_3)_3\overset{+}{N}\overset{\alpha}{C}H_2\overset{\beta}{C}H_2OH$	$(CH_3)_2\overset{+}{S}CH_2CH_2OH$	$(CH_3)_3\overset{+}{N}CH_2CH_2CH_2OH$
CHOLINE	DIMETHYLSULFONIUM ANALOG OF CHOLINE	HOMOCHOLINE

The alignment of the hydroxy group of choline and of the thiolester group of acetyl CoA is, however, very important. This is shown by the very high stereospecificity of the acetylation of choline substituted on the β-carbon, the carbon adjacent to the hydroxy group (23), and by the lack of ChA catalyzed acetylation of homocholine (24) in which three rather than two carbons separate the hydroxy and trimethylammonium groups.

The Coenzyme A Binding Site

CoA and acetyl CoA are bound much more tightly to ChA than either choline or Ach. Studies of the specificity of Co -ChA interactions show that dethiation of CoA has little effect on cofactor binding. The similarity of the binding of CoA and $1,N^6$-etheno CoA shows that the 6-amino group on the purine ring affects ChA binding relatively little. On the other hand, the 3'-phosphate on the ribose ring is most important for tight binding; 3'-dephospho CoA is attached only relatively loosely to ChA (25, 26, 27).

It is interesting to note that a series of aromatic dyes, such as Congo Red or Cibacron Blue F3GA, with little obvious structural resemblance to CoA, are excellent competitive antagonists of this cofactor both in ChA (28) and in carnitine acetyltransferase.

Thiol Groups

The inhibition of ChA by thiol reagents was noted long ago (29, 30). We investigated the essentiality of the thiol groups of ChA by means of a series of alkyl alkanethiolsulfonates, a group of reagents capable of forming disulfides with biopolymer thiol groups (31). We found methyl methanethiolsulfonate to be an effective inhibitor of ChA with inhibition showing a lag period. This was due to the inhibitor reacting not with an essential thiol of ChA but rather with the thiol group of the reaction product CoA. This was confirmed by the synthesis of a series of alkyl disulfides of CoA. These were potent, competitive inhibitors of ChA showing no lag period (32, 33). This work was later extended by Hersh (34) who concluded that ChA contains two classes of reactive thiol groups with neither located in the active sites of the enzyme.

Histidine Residues

The ChA catalyzed acetylation of choline and aminocholine shows a considerable D_2O isotope effect, greatest at the pH (pD), near neutrality, at which enzyme activity is maximal (24). This led to the postulate that the imidazole of a histidine residue was involved in rate-limiting proton transfer. In many hydrolytic enzymes,, imidazole through such general-base interaction with the hydroxy group of a serine residue, enhances the nucleophilicity of the alcohol moiety (35). What is unusual in the case of ChA is that the imidazole seems to interact, not with the hydroxy group of a serine, but rather with the hydroxy group of the substrate choline attached to its binding site (24). That imidazole is indeed essential to the functions of ChA could be confirmed by the observation that rose bengal inactivates ChA in the presence, but not in the absence, of light (24). Rose bengal is a relatively specific reagent for the inactivation of histidines, presumably through the light catalyzed generation of singlet oxygen (36). The esssentiality of histidine to the functions of ChA was confirmed by Roskoski (37) and Malthe-Sørenssen (38), who used ethoxyformic anhydride as histidine-directed reagent.

Arginine Residues

We found the arginine-specific reagents 2,3-butanedione, camphorquinone-10-sulfonic acid and phenylglyoxal to be irreversible inhibitors of ChA (39). Protection against inactivation by these reagents was provided by CoA, however, NAD^+, NADH, $NADP^+$, NADPH and choline provided no protection even at high concentration. 3'-Dephospho CoA, lacking the phosphate essential for tight binding to ChA, provided less protection against arginine reagents than CoA. These results suggest the interaction of the guanidinium group of an active-site arginine with the 3'-phosphate of the cofactor.

Partial protection against ChA inactivation by arginine-specific reagents was provided by salt. Presumably, the 3'-phosphate of CoA and salt anions compete for attachment to the same arginine. Sodium chloride, bromide and iodide as well as cesium chloride wre equiactive in protecting ChA against arginine inactivation. This agrees with the lack of ion specificity seen in studies of the effect of salt on the binding of substrates and products of the ChA reaction (18). Our data also agree with the observation that the binding of CoA to ChA is much more sensitive to salt concentration than the binding of 3'-dephospho CoA (13).

Hydrophobic Binding

The first suggestion that ChA contains lipophilic binding sites was provided by the ability of this enzyme to be retained tightly by alkyl- and aryl-substituted chromatographic columns (26, 27, 40,). Lenghtening of the substituent chains leads to tighter binding. Not surprisingly, hydrophobic columns can provide a useful step during the purification of ChA.

Elution patterns of ChA from columns carrying aromatic dyes, found to be competitive antagonists of CoA binding (28), as well as the elution of ChA from CoA columns, suggest the presence of a lipophilic component in the binding of CoA to ChA. However, lipophilic interactions outside the CoA binding site also seem likely. Elucidation of this problem may throw light on the question whether ChA, generally thought of as a cytoplasmic enzyme, may interact with membranes under physiological conditions. As a rule, highly lipophilic enzymes tend to be membrane-associated. In contrast to the other components of the Ach Cycle - acetylcholinesterase and the AchR - ChA is not a glycoprotein. Therefore, if ChA is membrane-associated at any time, this association is likely to involve intra-cytoplasmic membranes.

A Model for the Substrate and Cofactor Interactions of Choline Acetyltransferase

Since so far no one has obtained enough ChA to make a serious attempt to crystallize the enzyme, x-ray diffraction studies are likely to be several years away. While the procedures for the purification of ChA, from various sources, have been greatly improved during the last decade (41) the pure enzyme is still available only in miniscule amounts and the application of physicochemical methods to studies of the topography of ChA has not been possible. The information about the binding sites of ChA, derived from studies already discussed can be summarized as follows:

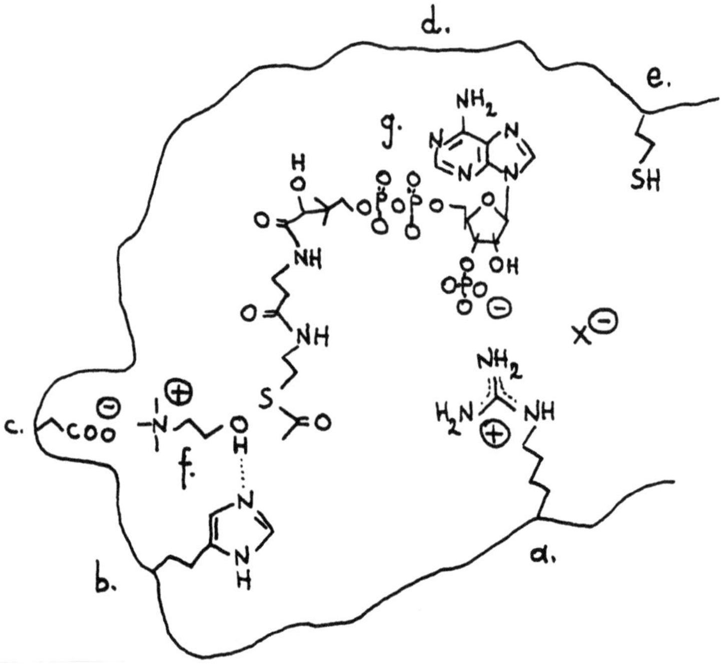

PROPOSED BINDING SITE FOR CHOLINE ACETYLTRANSFERASE

a. Active-site arginine
b. Active-site histidine
c. Anionic binding site
d. Hydrophobic binding site
e. Cysteine
f. Choline
g. Acetyl CoA
X Salt anion

Kinetic studies show acetyl CoA to be generally, but not obligatorily, the leading substrate in the ChA catalyzed synthesis of Ach (11, 12, 13). Similarly, CoA is the leading substrate in the reverse reaction. It seems likely that attachment of the leading substrate brings about a conformational change in ChA affecting binding of the second substrate. The availability of aromatic, fluorescent dyes, competitively attached to the CoA binding sites of ChA (28), should permit direct studies of such conformational alterations.

The Control of Acetylcholine Synthesis

A sodium dependent high-affinity system for choline transport has been reported by several laboratories (42, 43). It was also suggested that the high-affinity choline transport system is linked functionally and even structurally to ChA. As pointed out in a recent thoughtful review by Tuček (44), the evidence for specific coupling of choline transport and Ach synthesis is not yet convincing nor is it clear whether choline transport or Ach synthesis is the rate-limiting step in the synthesis of Ach. Tuček has also reviewed the complicated question of the origins of the acetyl groups used for Ach synthesis (44). Acetyl groups used for the synthesis of Ach seem to be transferred from mitochondria to the cytosol in the form of citrate and in the form of acetylcarnitine.

In view of the deficiency of ChA activity in Alzheimer's dementia, several attempts have been made to increase Ach synthesis by feeding massive amounts of choline. It should be noted, however, that the binding of choline and Ach is looser by almost two orders of magnitude than the binding of acetyl CoA or CoA. It is very unlikely, therefore, that either choline or Ach levels control Ach synthesis. It seems much more likely that acetyl CoA or CoA have a rate-limiting effect on Ach synthesis. In view of the aforementioned ability of salt anions to compete with the 3'-phosphate of CoA for attachment to an active-site arginine, the role of depolarization-linked changes in salt concentration affecting Ach synthesis cannot be excluded.

The Roles of the Cholinergic System

More than thirty years have passed since David Nachmansohn, in a brilliant extrapolation from the data then available, proposed the Ach Cycle and the role of Ach in regulating the cation permeability of electrically excitable membranes. During the following decades a great deal of work was centered on studies of the presence or absence of components of the Ach Cycle in various neuronal preparaions with the assumption that their presence or absence would prove or disprove a role for Ach in the conduction of the nerve impulse.

It is clear, however, that since ChA and acetylcholinesterase are found in a wide variety of non-neuronal cells, their presence cannot be used as proof of Ach being involved in neuronal function. ChA is found, in varying amounts, in many electrically excitable tissues. However, it is also found in placentae (a very rich source), in nettles, a wide variety of other plants and in unicellular organisms (45, 46, 47). Even though some of the older literature bears reinvestigation because of possible confusion between ChA and carnitine acetyltransferase and the ability of the latter enzyme to acetylate choline (48), the presence of ChA in non-neuronal cells is well established. It has been reported that light regulates Ach levels in plants (46); trophic effects have been ascribed to Ach (45) and a possible link of Ach to chemotaxis has been proposed (26).

We assume that Ach plays an essential role in intercellular communication. However, it seems likely that, as with other extracellular chemical messengers (49), intercellular functions of Ach were, early in evolution, preceded by intracellular functions, possibly in regulating membrane potential. David Nachmansohn proposed that Ach can exert intracellular effects. It seems likely that here, as in so many other things, he was right.

Bibliography

1. D. Nachmansohn and A.L. Machado (1943). The Formation of Acetycholine, a New Enzyme, "Choline Acetylase". J. Neurophysiol., 6, 397-403.

2. D. Nachmansohn, H.M. John, H. Waelsch (1943). Effect of Glutamic Acid on the Formation of Acetylcholine. J. Biol. Chem., 150, 485-486.

3. F. Lipmann (1945) Acetylation of Sulfanilamide by Liver Homogenates and Extracts. J. Biol. Chem., 160, 173-190.

4. L. Jaenicke and F. Lynen (1960) Coenzyme A. In: The Enzymes, 3B, 2. Ed. Ed.: P.D. Boyer, H. Lardy, K. Myrback; Acad. Press, New York, pp. 1-103.

5. D. Nachmansohn (1954) Metabolism and Function of the Nerve Cell, Harvey Lectures, 1953-1954, pp. 57-99.

6. B.H. Wainer et al., (1984) Cholinergic Systems in Mammalian Brain Identified with Antibodies Against Choline Acetyltransferase. Neurochem. Internat., 6, 163-182.

7. For a review, see, J.T. Coyle, D.L. Price, M.R. De Long (1983) Alzheimer's Disease: A Disorder of Cortical Cholinergic Innervation. Science, 219, 1184-1190.

8. J.R. Pieklik and R.W. Guynn (1975) Equilibrium Constants of the Reactions of Choline Acetyltransferase, Carnitine Acetyltransferase and Acetylcholinesterase. J. Biol. Chem., 250, 4445-4450.

9. H.G. Mautner, D.D. Dexter, B.W. Low (1972) Conformation Requirements of Acetylcholine Analogues and Local Anaesthetics, 238, 87-88.

10. H.G. Mautner (1980) Acetylcholine and Choline Acetyltransferase. Neurochem. Internat., 2, 123-134.

11. D. Morris, A. Maneckjee, C. Hebb (1971) The Kinetic Properties of Human Placental Choline Acetyltransferase. Biochem. J., 125, 857-863.

12. B. V. Rama-Sastry and G.I. Henderson (1972) Kinetic Mechanisms of Human Placental Choline Acetyltransferase. Biochem. Pharmacol., 21, 787-902.

13. L.B. Hersh (1982) Kinetic Studies of the Choline Acetyltransferase Reaction Using Isotope Exchange at Equilibrium. J. Biol. Chem., 257, 12820-12825.

14. L.B. Hersh and M. Peet (1978) Effect of Salts on the Physical and Kinetic Properties of Human Placental Choline Acetyltransferase. J. Neurochem., 30, 1087-1093.

15. L.B. Hersh, L.A. Barker, B. Rush (1978) Effect of Sodium Chloride on Changing the Rate-Limiting Step in the Human Placental Choline Acetyltransferase Reaction. J. Biol. Chem., 253, 4966-4970.

16. J. Rossier (1977) Choline Acetyltransferase, Review with Special Reference to Its Cellular and Subcellular Localization. Int. Rev. Neurobiol., 20, 283-337.

17. J. Rossier, Y. Spantidakis, P. Benda (1977) The Effect of Cl^- on Choline Acetyl-transferase Kinetic Parameters and a Proposed Role for Cl^- in the Regulation of Acetylcholine Synthesis. J. Neurochem., 29, 1007-1012.

18. L.B. Hersh (1980) Studies of the Kinetic Mechanism and Salt Activation of Bovine Brain Choline Acetyltransferase. J. Neurochem., 34, 1077-1081.

19. F. Fonnum (1968) Choline Acetyltransferase Binding to and Release from Membranes. Biochem. J., 109, 389-398.

20. C.G. Benishin and P.T. Carroll (1983) Multiple Forms of Choline-O-Acetyltransferase in Mouse and Rat Brain: Solubilization and Characterization. J. Neurochem., 41, 1030-1039.

21. R. Berman, I.B. Wilson, D. Nachmansohn (1953) Choline Acetylase Specificity in Relation to Biological Function. Biochim. Biophys. Acta, 12, 315-324.

22. L. Frankenberg et al. (1973) Biochemical and Pharmacological Studies on the Sulfonium Analogues of Choline and Acetylcholine. Eur. J. Pharmacol., 23, 37-46.

23. B.A. Hemsworth and J.C. Smith (1970) Enzymic Acetylation of the Steroisomers of α- and β-Methyl Choline. Biochem. Pharmacol., 19, 2925-2927.

24. S.F. Currier and H.G. Mautner (1974) On the Mechanism of Action of Choline Acetyltransferase. Proc. Nat. Acad. Sci. SA, 71, 3355-335.

25. S.F. Currier and H.G. Mautner (1977) Interaction of Analogues of Coenzyme Acetyltransferase. Biochem. J., 321-326.

26. H.G. Mautner (1977) Choline Acetyltransferase. CRC Crit. Revs. Biochem., 4, 341-370.

27. J. Rossier (1977) Acetyl Coenzyme A and Coenzyme A Analogues. Their Effect on Rat Brain Choline Acetyltransferase. Biochem. J., 321-326.

28. H.G. Mautner, et al., (1981) Interaction of Aromatic Dyes with the Coenzyme A Binding Site of Choline Acetyltransferase. J. Med. Chem., 24, 1534-1537.

29. R. Berman-Reisberg (1957) Sulfhydryl Groups of Choline Acetylase. Biochim. Biophys. Acta, 14, 442-443.

30. B. Mannervik and B. Sörbo (1970) Inhibition of Choline Acetyltransferase from Bovine Caudate Nucleus by Sulfhydryl Reagents - Reactivation of the Inhibited Enzyme. Biochem. Pharmacol., 19, 2509-2516.

31. D.J. Smith, E.T. Maggio, G.L. Kenyon (1975) Simple Alkane Thiol Groups for Temporary Blocking of Sulfhydryl Groups of Enzymes. Biochem., 14, 755-771.

32. S.F. Currier and H.G. Mautner (1976) Evidence for a Thiol Reagent Inhibiting Choline Acetyltransferase by Reacting With the Thiol Group of Coenzyme A Forming a Potent Inhibitor. Biochem. Biophys. Res. Commun., 69, 431-436.

33. S.F. Currier and H.G. Mautner (1977) The Interaction of Analogs of Coenzyme A with Choline Acetyltransferase. Biochem., 16 1944-1948.

34. L.B. Hersh, R.V. Nair, D.J. Smith (1979) The Reaction of Choline Acetyltransferase with Sulfhydryl Reagents. Methoxycarbonyl Coenzyme A Disulfide as an Active-Site Directed Reagent. J. Biol. Chem., 254, 11988-11992.

35. For a review see, D.M. Blow (1976) Structure and Mechanism of Chymotrypsin. Accts. Chem. Res., 9, 145-152.

36. A.P. Schaap et al. (1975) Polymer-Based Sensitizers for Photooxidations. J. Amer. Chem. Soc., 97, 3741-3745.

37. R. Roskoski (1974) Choline Acetyltransferase and Acetylcholinesterase: Evidence for Essential Histidine Residues. Biochem., 13, 5141-5144.

38. D. Malthe-Sørensen (1976) Choline Acetyltransferase - Evidence for Acetyl Transfer by a Histidine Residue. J. Neurochem., 27, 873-881.

39. H.G. Mautner, A.A. Pakula, R.E. Merrill (1981) Evidence for the Presence of an Arginine Residue in the Coenzyme A Binding Site of Choline Acetyltransferase. Proc. Nat. Acad. Sci. USA, 78, 7749-7752.

40. H.G. Mautner and S.F. Currier (1976) Studies on the Mechanism of Action of Choline Acetyltransferase. In: Cholinergic Mechanisms and Psychopharmacology Ed.: D.J. Jenden, Plenum Press, New York, pp. 197-206.

41. For a recent review see, P.L. McGeer, E.G. McGeer, J.H. Peng (1984) Choline Acetyltransferase: Purification and Immunohistochemical Localization. Life Sci., 34, 2319-2338.

42. H.J. Yamamura and S.H. Snyder (1972) Choline: High Affinity Uptake by Rat Brain Synaptosomes. Science, 178, 626-628.

43. M.J. Kuhar and L.C. Murrin (1978) Sodium-Dependent High-Affinity Choline Uptake. J. Neurochem., 30, 15-21.

44. S. Tuček (1984) Problems in the Organization and Control of Acetylcholine Synthesis in Brain Neurons. Progr. Biophys. Mol. Biol., 44, 1-46.

45. For a review, see B.V. Rama-Sastry and C. Sadavongvivad (1978) Cholinergic Systems in Non-Nervous Tissue. Pharmacol. Revs., 30, 65-132.

46. G. Hartman and H. Kilbinger (1974) Gas-Liquid Chromatographic Determination of Light-Dependent Acetylcholine Concentrations in Moss Callus. Biochem. J., 137, 249-252.

47. B.N. Smallman and A. Maneckjee (1981) The Synthesis of Acetylcholine by Plants. Biochem. J., 194, 361-364.

48. D.R. Goodman, F.K. Adatsi, R.D. Harbison (1984) Evidence for the Extreme Overestimation of Choline Acetyltransferase in Human Sperm, Seminal Plasma and Rat Heart: A Case of Mistaking Carnitine Acetyltransferase for Choline Acetyltransferase. Chem. Biol. Interacts., 49, 39-53.

49. J. Roth et al. (1982) The Evolutionary Origins of Hormones, Neurotransmitter and Other Extracellular Chemical Messengers. Implications for Mammalian Biology. New England. J. Med., 306, 523-527.

ACETYLCHOLINESTERASE IN REGENERATING SKELETAL MUSCLES

J. Sketelj, A. Blinc, M. Brzin

Institute of Pathophysiology, School of Medicine, University E. Kardelj, 61 105 Ljubljana, Yugoslavia

Introduction

Acetylcholinesterase (AChE) is an important functional protein of skeletal muscles. The polymorphism of its molecular forms probably reflects processes of intracellular synthesis, transport and attachement of AChE to functional sites (1,2). The asymmetric molecular forms of AChE in muscles are particularly interesting since their activity is strongly influenced by motor nerves and they practically disappear after denervation (3,4). However, it seems that different mechanisms control the asymmetric AChE forms in mature or immature muscles, at least as far as their localization (5) or their rate of synthesis and degradation is concenrned. Namely, immature muscle cells in culture do accumulate AChE forms even in the absence of nerves (6,7) but it is difficult to discard the possibility that this is due to some nonphysiologic conditions in tissue culture. Another approach to this problem was sought and, for this reason, investigations of AChE in muscles during regeneration <u>in vivo</u> were undertaken. Muscle regeneration mimics to a certain extent the normal ontogenetic development of muscles from mcnonuclear precursor cells to muscle fibres (8). Regenerating muscle is more easily amenable to experimentation than the embryonic muscles and, in comparison with cultured muscle cells, the regenerating muscle cells grow in a much more physiologic environment. Autotransplantation of skeletal muscles was performed in rats, causing ischaemic degeneration which is followed by consecutive regeneration. This is currently the best experimental model of muscle regeneration (8).

Molecular Basis of Nerve Activity

Materials and methods

Male rats, weighing 160 - 180 g were used in experiments.

Muscle regeneration. Leg muscles were permanently denervated by high axotomy of the ischiadic nerve. A combined ischaemic and toxic degeneration of extensor digitorum longus (EDL) muscles was produced as described in (9,10). In addition, the isolated muscles were incubated in 0.2 mM phospholine for 15 min at room temperature before autotransplantation. Rats were sacrificed 8-10 days, 14-15 days or 21-22 days after surgery.

In experiments designed for histochemical analysis of AChE, soleus muscle was excised from one leg, incubated in 10 µM soman (pinacolylmethylphosphonofluoridate) for 20 min at room temperature before incubation in bupivacaine (Marcaine) and sutured in place of the EDL muscle in the other leg. Soman was chosen because of its fast ageing reaction which makes its complex with AChE practically impossible to reactivate (11).

Velocity sedimentation. Isolated muscle regenerates were immediately homogenized in ice cold medium (1 mg/10 µl) containing 1 M LiCl, 50 mM Tris-HCl, pH 7.3, 1 % Triton X-100 and a set of antiproteases (10 mM EGTA, 2 mM benzamidine, 5 mM ethylmaleimide, 1 mg/ml bacitracine). Further procedure was essentially the same as described earlier (12).

Histology and histochemistry. Muscle regenerates were isolated and fixed overnight at $4^{o}C$ in a buffered 4 % formaldehyde solution and standard hematoxyline-eosine stained histological preparation were prepared. For histochemical demonstration of AChE, serial cryotome sections were made throughout the length of muscle regenerates and incubated for 30 min at room temperature in a histochemical medium according to (13). Normal soleus muscles were used as controls.

Results

A fairly ordered muscle structure is found in 8-11 days old muscle consisting mostly of myotubes. In 14-15 days old regenerates muscle fibres predominate, showing typical cross-striations and polygonal shape (Fig. 1).

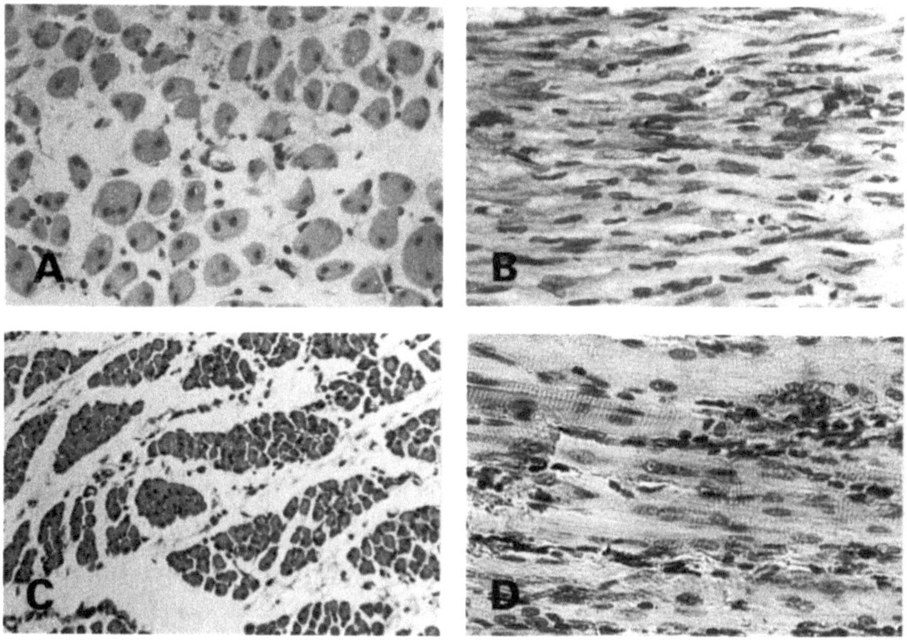

Figure 1.Histology of regenerating EDL muscle. A, B: 11 days old regenerate; C, D: 14 days old regenerate. Hematoxyline-eosine.

Four main molecular forms of AChE (4 S, 6 S, 10 S and 16 S) are present in EDL muscle of adult rats from which muscle regenerates were prepared (Figure 2). The 4 S form predominates. Activity of all AChE forms in this muscle is much reduced after denervation and the 16 S form cannot be reliably identified anymore.

The patterns of AChE molecular forms in noninnervated regenerates of EDL muscles are shown in Figure 3. The 16 S molecular form of AChE is clearly present in these regenerates. Its relative activity is somewhat higher in 8-11 days old regenerates (11 $\pm$ 2 % of total activity, n=9) than in more mature ones (4$\pm$ 2 %, n=5, and 5$\pm$ 2 %, n=6, in 14-15 days old and 21-22 days old regenerates, respectively). The 4 S molecular form of AChE predominated in 8-11 days old regenerates but its activity decreased and that of the 10 S form much increased in more mature two or three weeks old muscle

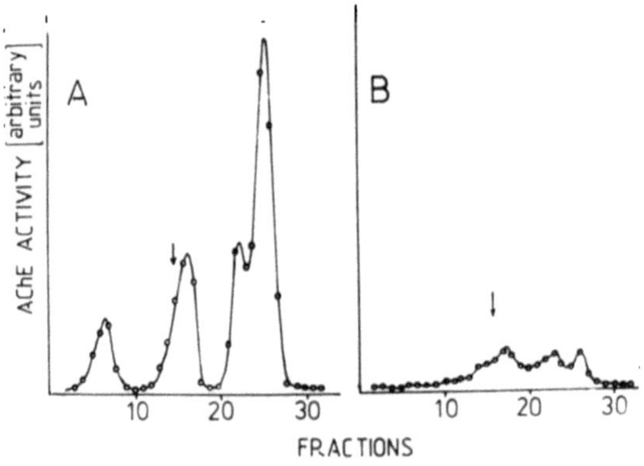

Figure 2.Velocit) EDL muscle of
adult ra denervation (B).
The top e. Arrows show
the pos s.

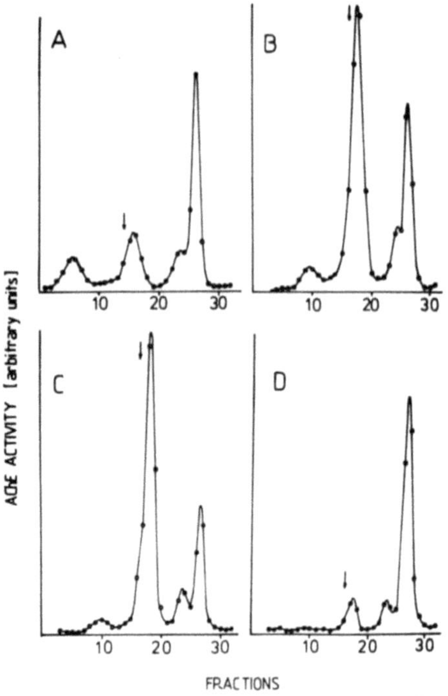

Figure 3.Velocity sedimentation profiles of AChE in 8 days (A), 15 days (B), and 22 days (C; D: unusual pattern) old regenerating muscles. The top of the gradients is on the right hand side. Arrows show the position of catalase (11.4 S) in the gradients.

regenerates. Essentially the same patterns of AChE molecular forms were found in regenerates prepared from muscles denervated 14 days before surgery. An unusual pattern of AChE molecular forms characterized by a great preponderance of the 4 S form over heavier forms (Figure 3) was found in 3 out of 22 investigated regenerates.

Preliminary histochemical analysis of AChE localization showed focal accumulations of AChE activity on the sarcolemma of some muscle fibres in 15 days old regenerates prepared with special precaution to eliminate the preexistent AChE in basal lamina in transplanted muscles (Figure 4). In control muscles, treated by an organophosphate inhibitor in the same way as the transplanted ones, less than 1 % of AChE activity was left and AChE in motor endplates cannot be demonstrated histochemically even after twice longer incubation time than that used with the regenerates. AChE accumulations in regenerates of soleus muscle were present only in the middle third of the regenerate, thus mimicking the distribution of the motor endplates in normal soleus muscle (Figure 5).

Discussion

Old basal lamina is not destroyed during the process of muscle degeneration and some preexistent AChE in the basal lamina of motor endplates is usually preserved in muscle regenerates (14,15). In order to obviate this problem we treated the isolated muscles by a potent irreversible organophosphate inhibitor of AChE (16) before transplanting them back. Essentially all AChE analysed in our muscle regenerates was therefore synthesized by the noninnervated regenerating muscle cells. In comparison with standard procedure (9,17) this additional step produced no major change in forthcoming muscle regeneration: myotubes were present in 8-11 days old regenerates and matured into muscle fibers by the day 14 after surgery.

There is no doubt that regenerating myotubes developing in the absence of motor nerves are capable to synthesize and accumulate the asymmetric 16 S molecular form of AChE (A_{12}). This has already been demonstrated in muscle cell culture (6,7) and some earlier negative results have been explained by the influence of some substances in culture medium (18,19). Our results

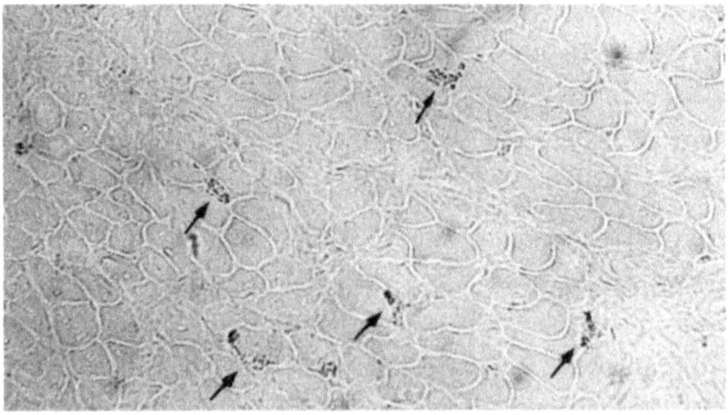

Figure 4. Focal accumulations of AChE activity in 15 days old regenerating muscle fibers. Cryotome section, histochemical method by Koelle and Friedenwald (13).

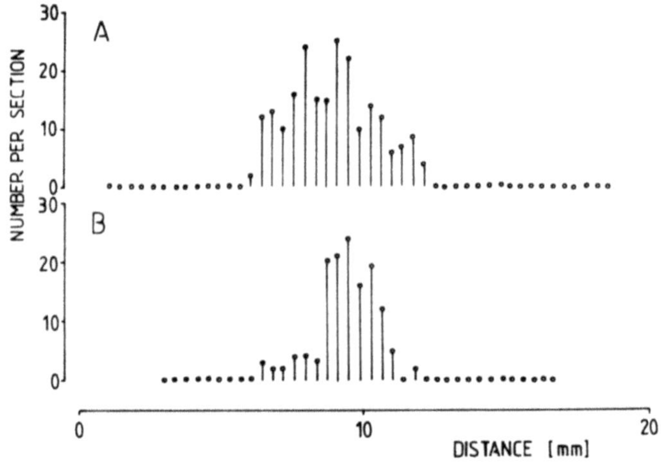

Figure 5. Distribution of motor endplates in normal soleus muscle (A) and of AChE accumulations in 15 days old muscle regenerate prepared from tne soleus muscle (B).

strongly support the hypothesis that the ability to synthesize the asymmetric AChE forms represents a basic feature of immature muscle cells (myotubes) which is not directly dependent upon innervation and is not due to some unidentified factors in tissue culture media. However, it seems that motor nerves indirectly influence the expression of this property of myotubes. Destruction of the spinal chord in chicken embryo greatly diminished although did not prevent the accumulation of A_{12} AChE in muscles (19). Muscle cells in cultures prepared from embryonic muscles which have not yet been contacted by the motor nerve do not contain the asymmetric AChE forms (7,19).

Several kinds of myoblasts appear during early stages of myogenesis: 'early myoblasts' are later superseded by 'late myoblasts' (19,20). Only myotubes produced by the fusion of the late myoblasts are able to synthesize the asymmetric AChE forms while those originating from early myoblasts are not (19). The presence of motor nerve in developing muscle is essential for proper development and final predominance of a certain group of late myoblasts from which further development of muscles takes place (21).

What kind of cells do the regenerating myotubes originate from? It is generally accepted that the 'satellite cells' represent the stem cells for most mononuclear precursor cells from which the regenerating myotubes are formed (8). The ability of regenerating rat myotubes to synthesize the asymmetric AChE forms suggests that most of the satellite cells in rat EDL muscle originate from late myoblasts. This seems to be in accordance with recent investigations of myogenesis in chicken muscles (20). These cells can still express this information in spite of long dormant period before and many divisions during the regeneration. Continuous presence of motor nerves is not necessary for this ability to appear since muscle regenerates prepared from previously denervated muscles also produced the 16 S AChE form.

The regulation of the asymmetric molecular forms in muscles seems to change irreversibly by maturation. These forms are present throughout the length of immature muscle cells (5,22), and are not directly dependent upon innervation while in mature muscle fibres they are highly concentrated only in the motor endplate region and critically dependent upon innervation (3).

In parallel with the appearance of cross-striations in regenerating muscle fibres a significant change in the pattern of AChE molecular forms occurs: the activity of the 10 S form (G_4) increases and that of the 4 S form (G_1)

decreases. This pattern was further accentuated in three weeks old regenerates. This has not been described to occur in cultured rat muscle cells. The pattern of AChE molecular forms in cultured myotubes was very similar to that found in 8-11 days old regenerates (6,7). It seems as if muscle cells in culture cannot reach this stage of maturation or they start peeling off the culture dish and degenerate when they do. The pattern of AChE molecular forms in more mature muscle regenerates is similar to that in muscles denervated during early postnatal period while they are still immature (23). AChE in mature rat muscles, however, behaves quite differently since there is no particular absolute or relative increase of 10 S AChE after their denervation.

In a few muscle regenerates the 4 S form by far exceeded the activity of heavier forms. Biochemical maturation might have been impeded by some unfavourable local conditions in these regenerates or, in these animals, the satellite cells may have originated from early myoblasts. Myotubes, formed from such myoblasts in cell culture namely produce mainly the 4 S AChE form (19).

Preliminary results of hisoochemical analysis showed that spots of relatively concentrated AChE activity are present in the middle third of 15 days old soleus muscle regenerates. This AChE must have been synthesized de novo by regenerating muscle cells since practically all preexistent AChE, including that in basal lamina of motor endplates, had been irreversibly inhibited. Similar distribution of AChE spots in regenerates and of former motor endplates suggests that basal lamina of old motor endplates may have induced the accumulation of AChE in regenerating muscle fibres. This process may run in parallel with the accumulation of the cholinergic receptor molecules which are also concentrated in places of old motor endplates (24,25). However, more detailed ultrastructural analysis of AChE localization in muscle regenerates is necessary in order to support or refute this hypothesis.

Summary

Muscle regeneration has been used as a model to study the regulation of AChE during muscle maturation. Irreversible inhibition of preexistent AChE which made possible to analyse only the enzyme synthesized by regenerating muscle cells did not influence the course of muscle regeneration. Noninnervated regenerating myotubes are able to synthesize and accumulate the asymmetric AChE forms. This indicates that the satellite cells from which the muscle regeneration starts originate from the late myoblasts. The satellite cells do not require continuous presence of motor nerves for this property to be expressed. In noninnervated regenerating myotubes the 4 S AChE form shows the highest activity but later, in parallel with the appearance of cross-striations in muscle fibres, the 10 S AChE form becomes predominant which has not been observed in muscle cell cultures or in adult denervated muscles. Focal accumulations of AChE appear in regenerating muscle fibres in the region of the former motor endplates. It is tentatively suggested that components of basal lamina of old motor endplates may induce these accumulations of AChE, however, further research is necessary to clarify this point.

Acknowledgment

This research was supported by a grant from the Research Community of Slovenia. Technical assistance of Mrs.Nina Zupančič is gratefully acknowledged. Standard histological preparations were made at the Dpt. of Pathology, School of Medicine, Ljubljana.

References

1. Masoulié, J., S. Bon. 1982. Ann.Rev.Neurosci. 5, 57.

2. Brzin, M., J. Sketelj, B. Klinar. 1983. In: Handbook of Neurochemistry, Vol. 4 (A.Lajtha, ed.). Plenum Press, New York, p. 251.

3. Hall, Z.W. 1973. J.Neurobiol. 4, 343.

4. Sketelj, J., M.G. McNamee, B.W. Wilson. 1978. Exp.Neurol. 60, 624.

5. Sketelj, J., M. Brzin. 1980. Neurochem.Res. 5, 653.

6. Sugiyama, K. 1977. FEBS Letters 84, 257.

7. Koenig, J., M. Vigny. 1978. Nature 71, 75.

8. Carlson,B.M. 1978. Physiol.Bohemosl. 27, 387.

9. Carlson,B.M. 1976. Exp.Neurol. 52, 421.

10. Bader, D. 1980. Dev.Biol. 77, 315.

11. Fleisher, J.H., L.W. Harris. 1965. Biochem.Pharmac. 14, 641.

12. Klinar, B., J. Sketelj, D. Sket, M. Brzin. 1983. J.Neurosci.Res. 9, 437.

13. Koelle, G.B., J.S. Friedenwald, 1949. Proc.Soc.Exptl.Biol.Med. 70,617.

14. Gruber, H., G. Freilinger, J. Holle, R. Mayr. 1974. Experientia 30, 1191.

15. Carlson, B.M., K.R. Wagner, S.R. Max. 1979. Muscle&Nerve 2, 304.

16. Welsch, F., W.-D. Dettbarn. 1972. Biochem.Pharmacol. 21, 1o39.

17. Carlson, B.M., E. Gutmann. 1976. Exp.Neurol. 53, 82.

18. Bulger, J.E., W.R. Randall, P.S. Nieberg, G.T. Patterson, M.G. McNamee, B.W. Wilson. 1982. Dev.Neurosci 5, 474.

19. Toutant, J.P., M. Toutant, M.Y. Fiszman, J. Massoulie. 1983. Neurochem. Internat. 5, 751.

20. Rutz, R., S. Hauschka, 1982. Dev.Biol. 91, 103.

21. Bonner, P.H., T.R. Adams. 1982. Dev.Biol. 90, 175.

22. Koenig, J., F. Rieger. 1981. Dev. Neurosci. 4, 249.

23. Gautron, J., F. Rieger, B. Blondet, M. Pincon-Raymond. 1983. Biol.Cell. 49, 55.

24. Burden, S.J., P.B. Sargent, U.J. McMahan. 1979. J. Cell Biol. 82, 412.

25. Bader, D. 1981. J. Cell Biol. 88, 338.

Distribution and regulation of acetylcholinesterase forms in chick myotubes : comparison with rat myotubes.

M. VIGNY and F. VALLETTE

Laboratoire de Neurobiologie
Ecole Normale Superieure
46, rue d'Ulm
75005 Paris, France.

Abbreviations:
- AChE: Acetylcholinesterase
- HS: Horse serum
- FCS: Foetal calf serum
- TTX: Tetrodotoxin

Introduction

Acetylcholinesterase (AChE) is a polymorphic enzyme. In vertebrate tissues it presents two classes of molecular forms : the globular forms (G), corresponding to the monomer G1, dimer G2, and tetramer G4, and the asymmetric forms (A), corresponding to one, two or three tetramers linked to a collagen-like tail (A4, A8, A12) (for a review, see Massoulié and Bon, 1982).

The distribution of the collagen-tailed and globular forms along muscle fibers varies according to the animal species and the physiological state. In normal adult rat muscle the A12 form, which represents the major collagen-tailed form, is localized almost exclusively in the endplate region whereas in rat embryo this form is distributed over the entire length of the muscle fiber. After denervation it almost completely disappears from rat and chicken muscles. After ectopic reinnervation of a denervated rat muscle it reappears both at the ectopic site and at the old end-plate. During

Molecular Basis of Nerve Activity

embryogenesis its appearance coincides with the establishment of neuromuscular contacts (see references in Massoulié and Bon, 1982).

In culture, muscle cells obtained from the hind limb of 18 day-old rat embryos synthesize the A12 form in absence of nerve cells. With chick muscle cultures rather contradictory results have been obtained. On one hand, it was reported that chick embryonic muscle cells seeded without nerve cells do not synthesize any asymmetric form (Rotundo and Fambrough, 1979, Rubin et al., 1980). Chick muscle cells were thus considered to require the presence of neural cells to accumulate the A12 form. On the other hand, Kato et al., (1981) and Toutant et al., (1983) observed a synthesis of A12 form in aneural cultures. Toutant et al. (1983) showed that the A12 form is expressed when the culture medium is supplemented with foetal calf serum (FCS) but not with horse serum (HS). They found that a component which is present in horse serum but not in foetal calf serum binds the globular and asymmetric forms of chicken AChE, forming aggregates of variable sizes.

Rat myotubes exhibit action potentials and contract spontaneously (Koenig and Vigny, 1978). In the presence of tetrodotoxin (TTX), which blocks the sodium channels responsible for the propagation of the action potentials, the A12 form almost completely disappears (Rieger et al.,1980, Brockman et al.,1984). This effect is reversible. Veratridine, which maintains the sodium channels in the open state, provokes a partial depolarization leading to the blockade of the action potentials and the paralysis of the myotubes. In rat myotubes, veratridine induces a dramatic increase in total AChE activity and in the relative proportion of the A12 form. By itself, a depolarization obtained by a change of the concentration of potassium ions in the culture medium, does not produce the same effect (De la Porte et al., in press). Therefore the effect of veratridine cannot be explained simply by a decrease of the cellular resting potential. It probably implies a modification of the permeability of the cell membrane to calcium ions, although this is not fully understood (De la Porte et al., in press).

In similar embryonic rat myotubes, Brockman et al. (1982)

analyzed the cellular localization of the different molecular forms. Their data indicate that the external enzyme represents about 30% of the total cellular activity and that intracellular and external AChE have different compositions. All forms however are distributed in the two compartments. Particularly the A12 form was 40% extracellular and 60% intracellular. In a recent work, we essentially confirmed these data (Vallette et al. , in preparation).

In this paper we summarize the data concerning the distribution and regulation of acetylcholinesterase in chick myotubes. We compare these results with those previously reported in the case of the rat myotubes.

Results and discussion

1) Growth and morphological differentiation of chick muscle cells in culture : influence of the culture medium .

We analyzed the growth and differentiation of chick muscle cells in three culture media. In one case the medium was supplemented with 10% horse serum (HS) (HS medium), in the second it was supplemented with 5% foetal calf serum (FCS medium) and in the third case no serum was added and the medium was supplemented with a mixture of completely defined compounds, according to the formula proposed by Bottenstein and Sato (1979) (defined medium).

Whatever the medium used, the mononucleated myoblasts attached to the dish and fused two days later. The morphological appearance of the myotubes was very similar if not identical in media supplemented with horse and with foetal calf serum. In contrast, the myotubes formed in the defined medium looked thinner and longer. The absence of one or several serum factors is probably responsible for this morphological difference.

We also attempted to culture rat muscle cells in the completely defined medium. In this case, the myoblasts did not attach. Addition of purified horse fibronectin did not improve the attachment. However, after the fusion has occurred in a serum supplemented medium, a switch to the defined medium does not kill the cells nor change the subsequent differentiation of the rat myotubes. Therefore, the attachment and fusion of myoblasts require serum factors but serum is not necessary for the survival of mature rat myotubes.

We noticed another major difference between the two types of myotubes. Mature chick myotubes in the three media exhibited very few if any contractions whereas in the same conditions rat myotubes contracted vigourously.

2) AChE content and molecular forms .

Whatever the medium used, the AChE activity reached similar maximum values. However the time-course of evolution was different. In FCS cultures the maximal value was attained as soon as the third day, but not before the fifth or the sixth day in HS or defined medium cultures. We noticed no obvious correlation between the morphological appearance and the AChE content of the myotubes. For example, as described above the aspect of the cells in HS and FCS cultures was roughly similar, whereas the time-course of the evolution of the AChE activity was quite distinct in these two conditions.

As previously described by Toutant *et al.* (1983) the sedimentation pattern of AChE of chick myotubes cultured in the presence of horse serum reflects an interaction of the enzyme with a component present in this serum. The G2 form is shifted from 7 S to 9 S, the G4 originally at 11 S is converted into a broad peak at 15-17 S and the A12 form (20 S) is no longer visible and gives aggregates of about 30 S. We investigated the nature of the HS factor(s) responsible for these effects. We purified the IgG fraction from the serum by DEAE chromatography. This fraction produced the same shift of S values as the complete serum, indicating that the observed interaction was actually due to an IgG population present in the serum. In contrast, the horse serum does not modify the sedimentation profiles of AChE solubilized from rat, guinea pig or quail tissues. Horse serum therefore appears to contain specific immunoglobins against chicken AChE.

In FCS cultures, we never detected any fast sedimenting polydisperse aggregates, and the chick myotubes express all molecular forms, including the A12 form (Fig.1). The relative proportion of this form increased as a function of the age of the cultures to reach about 10% after the seventh day of culture. In defined medium cultures, chick myotubes also synthesize the A12 form. It is noteworthy that the proportion and therefore the total content of the A12 form was much higher in this case than in FCS cultures (Fig.1). Addition of FCS to the defined medium led to a decrease of the relative proportion of A12. Therefore the presence of FCS had an inhibitory effect on the expression of the A forms. The mechanism(s) and the factor(s) responsible for this effect remain unknown.

3) Cellular localization of the molecular forms of acetylcholinesterase .

The relative amounts of AChE in the intracellular and external compartments of chick myotubes were determined in FCS and defined medium cultures. We used two different methods. In the first method we used echothiophate, a cationic phosphorylating agent to inactivate external AChE selectively (Lazar and Vigny, 1980; Brockman et al. , 1982). In the second method we used a mixture of DFP, which is a lipid soluble irreversible inhibitor, and a non-permeant reversible inhibitor such as BW284C51. This reversible inhibitor protects from inactivation by DFP those enzyme molecules to which it has access, i.e. the external compartment (Inestrosa et al. , 1982). The two methods gave similar results. In young cultures the internal compartment was predominant, whereas the reverse was true in older cultures. We also analyzed the endo or ectocellular distribution of the different molecular forms by sucrose gradient sedimentation. Except in the case of the G1 form which is always completely intracellular, the other forms are divided among the two compartments. In young cultures the G2 form was preferentially intracellular whereas A12 and G4 were almost equally intra and extracellular. In old cultures the G2 form became more externalized and the G4 and A12 forms were predominantly found in the external compartment. Thus, the subcellular localization of the different forms depends upon the age of the culture. This result may explain some of the discrepancies found in the litterature. For example, Inestrosa et al. (1982)

reported that in the C2 cell line, the A12 form exist only in the ectocellular compartment, whereas Brockman *et al.* (1982) showed that 60% of this form is endocellular in primary rat muscle cells. In addition to the difference in the nature of the cells used by these authors it is possible that the age, the degree of differentiation and/or differences in culture media may be partly responsible for the difference of localization reported.

4) Effect of tetrodotoxin and veratridine on AChE in muscle cell cultures

In the C2 cell line, TTX treatment led to a 50% reduction of all AChE forms, (Inestrosa *et al.*, 1983) and in rat myotubes this agent induced the almost complete disappearance of the A forms (Rieger *et al.*, 1980).

We investigated the effects of tetrodotoxin and veratridine on AChE in cultures of chick myotubes. Veratridine was used at a concentration of 20 µM and tetrodotoxin at a concentration of 1 µM. The drugs were added at day 5 to defined medium cultures and allowed to act for two days. After exposure to tetrodotoxin (TTX), the AChE activity and the distribution of the molecular forms were not modified (Fig.2). Thus chick myotubes do not respond to TTX treatment in the manner described for mammalian myotubes.

When chick myotubes were treated with veratridine, the total AChE activity was not significantly changed but the relative proportion of the different molecular forms was modified. The percentage of the A12 form was increased by a factor of 2 or 3 (Fig. 2). It is noteworthy that the morphological aspect of the myotubes was similar in control and in veratridine-treated cultures. The effect of veratridine was completely abolished in the presence of tetrodotoxin. This probably results from a blockade of the sodium channels by TTX, even in the presence of veratridine.

Veratridine acts as a depolarizing agent by maintaining the sodium channels in the open state. The presence of 20 mM potassium ions in the culture medium lowered the resting potential of the myotubes to a value similar to that obtained in the presence of 20 µM veratridine. In the case of chick myotubes, depolarization induced by a high potassium external concentration produced the same effect as veratridine. Therefore the effect of veratridine can be explained simply by a decrease of the resting potential.

In rat myotubes (see introduction) veratridine induces a dramatic increase in total AChE activity and in the relative proportion of the A12 form. Depolarization by itself does not produce the same effect. Once again, chick and rat myotubes respond differently to the veratridine treatment and to the depolarization induced by a high potassium concentration in the external medium.

Thus one must be aware that the regulation of the biosynthesis of AChE molecular forms differs between different species.

References

-Bottenstein, J.E., Sato, G.H. (1979). Growth of a rat neuroblastoma cell line in serum-free supplemented medium. Proc. Natl. Acad. Sci. USA, 11 , 514-517.

-Brockman, S.K., Przybylski, R.F., Younkin, S.G. (1982). Cellular localization of the molecular forms of acetylcholinesterase in cultures of embryonic rat myotubes. J. Neurosci., 2 , 1775-1785.

-Brockman, S.K., Younkin, L.H. and Younkin, S.G. (1984). The effect of spontaneous electromechanical activity on the metabolism of acetylcholinesterase in cultured embryonic rat myotubes. J.Neurosci., 4, 131-140.

-De la Porte, S., Vigny, M., Massoulié, J., Koenig, J. (1984). Action of veratridine on acetylcholinesterase in cultures of rat muscle cells. Dev. Biol., in press.

-Inestrosa, N.C., Silberstein, L., Hall, Z.W., (1982). Association of the synaptic form of acetylcholinesterase with extracellular matrix in cultured mouse muscle cells. Cell 29 , 71-79.

-Inestrosa, N.C., Miller, J.B., Silberstein, L., Ziskind-Conhaim, L., Hall, Z.W. (1983). Development and regulation of 16S acetylcholinesterase and acetylcholine receptors in a mouse muscle cell line. Exp. Cell Res., 147 , 393-406.

-Koenig, J., Vigny, M. (1978). Neural induction of the 16 S acetylcholinesterase in muscle cell cultures. Nature, 271 , 75-77.

-Lazar, M., Vigny, M. (1980). Modulation of the distribution of acetylcholinesterase molecular forms in a murine neuroblastoma X sympathetic ganglion cell hybrid cell line. J. Neurochem., 35 , 1067-1079.

-Massoulié, J., Bon, S. (1982). The molecular forms of cholinesterase and acetylcholinesterase in vertebrates. Ann. Rev. Neurosci., 5 , 57-106.

-Rieger, F., Koenig, J., Vigny, M. (1980). Spontaneous contractile activity and the presence of the 16 S form of acetylcholinesterase in rat muscle cells in culture. Reversible suppressive action of tetrodotoxin. Dev. Biol., 76 , 358-365.

-Rotundo, R.L., Fambrough, D.M. (1979). Molecular forms of chicken embryo acetylcholinesterase in vitro and in vivo , isolation and characterization. J. Biol. Chem., 254 , 4790-4799.

-Rubin, L.L., Schuetze, S.M., Weill, C.L. Fischbach, G.D. (1980). Regulation of acetylcholinesterase appearance at neuromuscular junction in vitro . Nature, 283 , 264-267.

-Toutant, J.P., Toutant, M., Fiszman, M.Y., Massoulié, J. (1983). Expression of the A12 form of acetylcholinesterase by developing avian leg muscle cells in vivo and during differentiation in primary cell cultures. Neurochem. Int., 5 , 751-762.

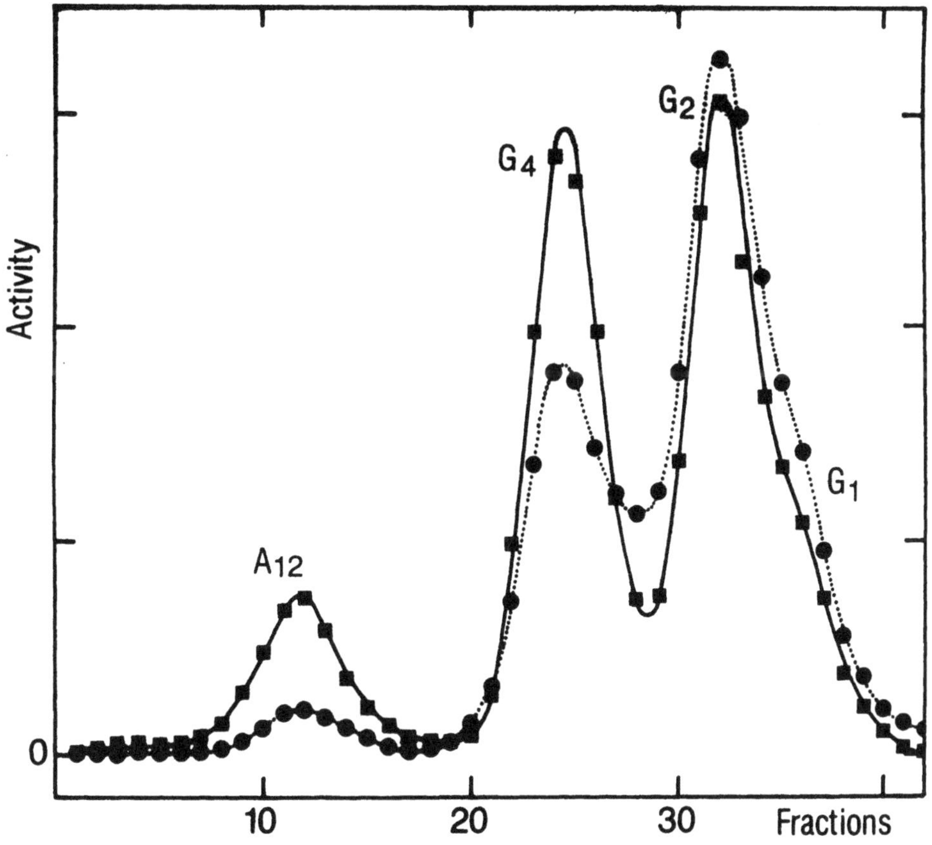

Fig. 1 : Sedimentation analysis of acetylcholinesterase of 6 day-old chick myotubes.

●- -● defined medium cultures

■——■ Foetal calf serum cultures.

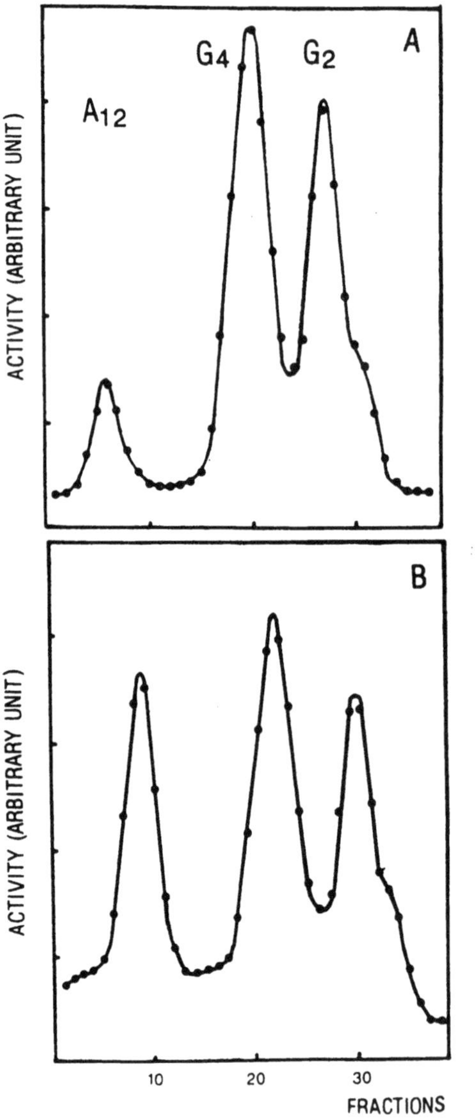

Fig 2 : 7 day-old chick myotubes cultivated in the defined medium.

A : Control myotubes or myotubes treated for 2 days in presence of 1 µM Tetrodotoxin.

B : Myotubes in the presence of 20 µM veratridine or 20 mM K^+ in the medium for 2 days.

ACETYLCHOLINESTERASE IN MAMMALIAN SKELETAL MUSCLE AND SYMPATHETIC GANGLION CELLS. EXTRA AND INTRACELLULAR HYDROPHILIC AND HYDROPHOBIC ASYMMETRIC FORMS.

P. Dreyfus, M. Verdière, D. Goudou, L. Garcia and F. Rieger

Groupe de Biologie et Pathologie Neuromusculaires, INSERM U 153 17, rue du Fer-à-Moulin 75005 Paris, France

Introduction

Mammalian skeletal muscle and superior cervical ganglion neuronal cells contain multiple molecular forms of acetylcholinesterase (acetylcholine acetylhydrolase, AChE, EC 3-1-1-7). A detailed analysis of the sedimentation profiles of mouse muscle AChE by gaussian decomposition into individual sedimentation peaks has revealed up to 8 distinct molecular forms (1). In rat superior cervical ganglia (SCG), at least 5 different molecular forms of AChE have been recognized in vivo as well as in vitro, in different relative proportions depending on the developmental stages reached by the neuronal cells (2). The main molecular forms in mouse muscle, and also in rat SCG neurons, are 4 S, 6.5 S, 10 S, 12 S and 16 S. It is now well established that AChE forms differ by their quaternary structure. There are globular or asymmetric forms, first studied after extraction from fish electric organs (3, 4, 5, 6). In mouse muscle, the collagenase-sensitive 16 S and 12 S AChE belong to the asymmetric type (7, 8, 1). However, a detailed biochemical analysis is still lacking to propose a precise structural model including all identified AChE forms in mammalian tissues (1). In particular, it has recently been demonstrated that additional non-catalytic proteic components may contribute to the AChE molecule (9) and hydrophilic and hydrophobic

Molecular Basis of Nerve Activity

variants exist not only for globular forms (10, 11, 12) but also for asymmetric forms as well (8, 13). In view of such a diversity of AChE the question of the physiological significance of AChE is now widely open. We have recently begun a study of the subcellular localization of the various molecular forms of AChE and their variants. In this paper, we describe the results obtained studying AChE localization with two main strategies : the use of cell permeant and impermeant reversible or irreversible inhibitors (14) and the use of subcellular fractionation (8).

Material and Methods

1) Permeant and impermeant inhibitors of AChE.

- Phospholine or ecothiopate iodide (Promedica, France) is a charged, cell impermeant, irreversible inhibitor of the ectoenzyme.
- O-ethyl-S_2-diisopropylaminoethylmethylphosphorothiolate (MPT ; a gift from Dr F. Leterrier, Hopital Percy, Clamart) is cell permeant and irreversible. Used alone, it inhibits both the endo- and the ectoenzyme.
- 1,5-bis(4-allyldimethylammoniumphenyl)-pentane-3-one dibromide (BW 284 C 51 ; Burroughs Wellcome) is cell impermeant and reversible. It inhibits the ectoenzyme, and the inhibition is reversed by simple washing. It protects the ectoenzyme against irreversible inhibition by the MPT compound.

2) Subcellular fractions from mouse muscle and isolated neuromuscular junctions.

The subcellular fractionation of mouse muscle was performed as described in (8) using 3 week old mouse intercostal muscles. A fraction enriched in isolated neuromuscular junctions (C_1) and a fraction containing microsomal membranes (A), enriched in sarcolemmal membranes and transverse tubules (L. Garcia, J.M. Laval, P. Dreyfus and F. Rieger in preparation) were studied. All incubations were performed in the presence of Krebs medium (8). Phospholine was used at a concentration of 10^{-5} M at 20°C and for 5, 15 and 30 min. For the protection experiments, the muscles were preincubated in the presence of 10^{-4} M BW for 15 min and incubated with both 10^{-4} M BW and 1.5 10^{-7} M MPT during 15 min. After a first wash in the Krebs medium plus BW during 5 min, 3 other washes in the Krebs medium followed and subsequent AChE analysis was performed without any delay as described in (8).

3) Superior cervical ganglion in vivo study.

We used near term (embryonic day 21 ; E21) rat embryo superior cervical ganglia (SCG). Incubations with inhibitors were performed in a culture medium composed of basal medium (Eagle MEM and 199, 3 : 1; GIBCO) with 200 mM glutamine. Phospholine at a concentration of 10^{-5} M was applied for 3 min. After 3 washes in the basal medium, the ganglia were further analyzed for AChE (15) and proteins (16). In the protection experiments, 10^{-4} M BW was applied for 15 min, followed by 10^{-4} M BW and 1.5 10^{-7} M MPT for 15 min and 3 washes followed, the first one in the presence of the BW compound. Subsequent AChE analysis was performed as described in (2).

4) Hydrophilic and hydrophobic properties of AChE.

We used sequential extractions of AChE from the various tissues or tissue fractions studied in the presence of detergent (hydrophobic pool) or in the presence of high salt (hydrophilic pool) as reported in (8). Detergent was 1 % Triton X-100 ; High salt was 1 M NaCl.

A) Extra and intracellular AChE of mouse intercostal muscle in isolated neuromuscular junctions and other cellular membranes

Muscles still attached to the costal grid were exposed either directly to 10^{-5} M phospholine (Echothiopate) or to BW followed by BW + MPT and cell fractionation performed as reported in the Methods Section. Fig 1 shows typical AChE sedimentation profiles obtained from control, phospholine and BW + MPT treated muscle fractions. In the fraction C_1, which is enriched in isolated neuromuscular junctions (see 8), most of AChE activity is protected by the BW compound (fig 1a). The phospholine treatment considerably decreases the AChE activity recovered in the C_1 fraction and affects all AChE peaks. In the fraction A, all AChE peaks are significantly decreased by 10-20 % after the BW + MPT

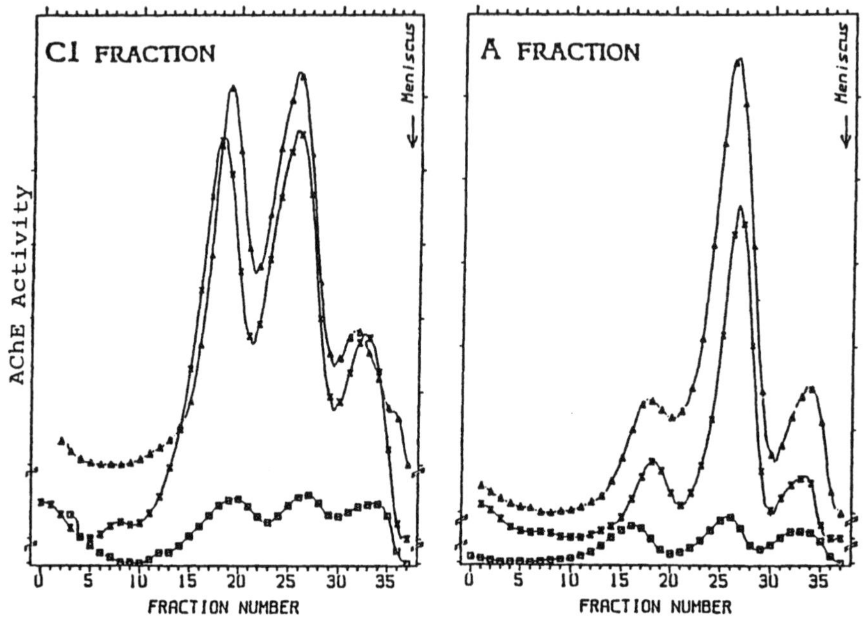

Fig 1 Ecto-and Endoforms of AChE in Subcellular Fractions of Mouse Muscle. Control muscle (Δ) ; after treatment with phospholine (□) and with BW + MPT (⧗).

treatment (fig 1b) and are much more reduced after phospholine. In particular, "10 S" AChE is the peak the most affected by the phospholine treatment in both C_1 and A fractions. Thus, "10 S" AChE is mostly external. In the C_1 fraction, "16 S" AChE is more affected by the phospholine action (11 % of residual activity) than in the A fraction (39 % of residual activity). These results indicate that "16 S" AChE is mainly extracellular in the isolated neuromuscular junctions as well as "10 S" and "4 S" AChE. In the A fraction, "16 S" AChE has an important intracellular component, in contrast to "10 S" and "4 S" which are mainly extracellular.

We know from previous work (8. 13) that most if not all AChE forms present in mouse muscle possess hydrophobic and hydrophilic variants characterized by their components extracted respectively in the presence or in the absence of detergent. We sequentially extracted the fractions C_1 and A from control and phospholine treated muscles, using detergent in a first step followed by high salt in a second step. AChE extracted by detergent or AChE extracted by high salt are both decreased after 5 and 15 min of phospholine in a similar way than total (detergent + high salt) AChE in both the C_1 and the A fractions (data not shown). This result suggested that there is no privileged extra- or intracellular location of (total) hydrophilic or hydrophobic AChE. We then wondered whether any of the hydrophobic or hydrophilic variants of the main AChE peaks "4 S", "10 S" or "16 S" could present a privileged inside-outside cellular localization. Fig 2 shows that all AChE peaks detected in detergent or

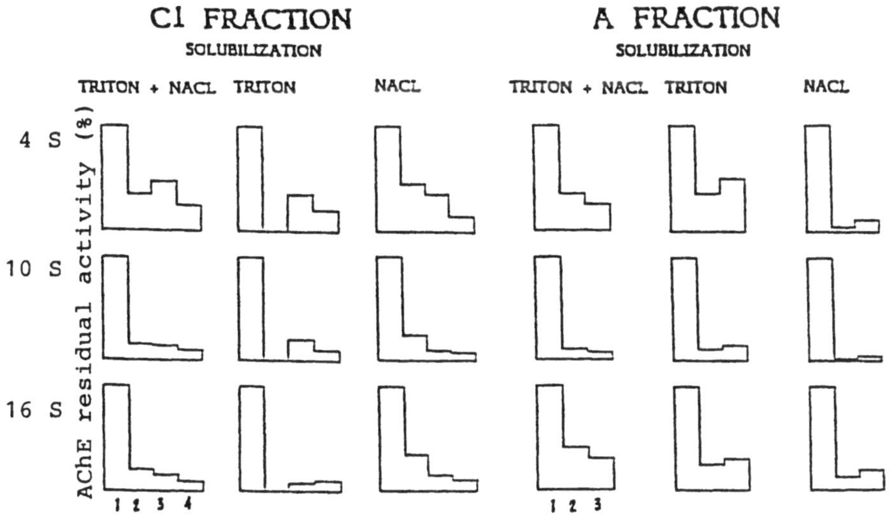

Fig 2 Sequential Extractions of AChE "Forms" from Control and Phospholine-treated Mouse Muscle. Phospholine incubation time : 0 (1), 5 (2), 15 (3) and 60 (4) min.

high salt extracts in both the C_1 and the A fractions were similarly decreased after the phospholine treatment. The only exception is "4 S" AChE in the C_1 fraction : hydrophilic "4 S" is more affected by the phospholine treatment than hydrophobic "4 S" and is thus more external than hydrophobic "4 S".

B) Extra- and intracellular AChE in superior cervical ganglion SCG neurons in vivo at birth

We applied the same phospholine or BW + MPT treatments than above to near term (E21) rat SCG kept as described in Methods. AChE was extracted by homogenization in the presence of complete (detergent + high salt) medium. Table 1 gives the total residual AChE activities per ganglion in control, phospholine or BW + MPT homogenates and demonstrates similar intracellular and extracellular AChE pools. Fig 3

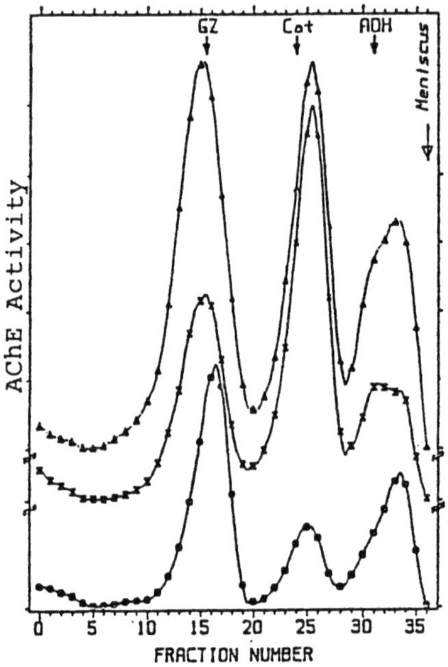

Table 1. Intracellular and Extracellular AChE in SCG Neurons in Vivo.

	Total activity per ganglion	% control values
Control	4.8 ± 0.5	100
Phospholine (inside AChE)	2.8 ± 0.4	61.3 ± 6.4
BW + MPT (outside AChE)	2.2 ± 0.2	53.7 ± 4.1

Fig 3 Ecto-and Endoforms of AChE in Rat E21 SCG. Control (▲) ; after treatment with BW + MPT (x), with phospholine (□).

shows the sedimentation patterns of AChE from control, phospholine and BW + MPT treated tissues. The phospholine treatment results in an important decrease of "10 S" and only a small decrease in "4 S" and "16 S" AChE, suggesting a privileged internal location of these two forms. The BW + MPT treatment mainly confirms that "10 S" AChE is predominantly extracellular in SCG tissues. Table 2 gives the proportions

Table 2. Intracellular and Extracellular Molecular Forms of AChE in SCG Neurons in Vivo. Results are expressed as percentages of control value.

		"4 S"	"10 S"	"16 S"
PHOSPHOLINE (inside AChE)	n=4	70.3 ± 3.5	35.5 ± 3.5	76.1 ± 12.7
BW + MPT (outside AChE)	n=4	43.0 ± 5.4	74.3 ± 4.7	42.7 ± 4.2

of residual activities in SCG tissues in vivo after phospholine and BW + MPT for each main AChE peak. We found dissimilar extra- and intracellular pools of "4 S", "10 S" and "16 S" AChE. "10 S" AChE is about 65 % extracellular and both "4 S" and "16 S" AChE are about 75 % intracellular.

We also determined the hydrophobic and hydrophilic contributions to the extra- and intracellular total AChE using the sequential detergent followed by high salt extractions. Table 3 shows the percentages of

Table 3. Hydrophobic and Hydrophilic AChE in Control, Phospholine and BW + MPT Treated SCG from Near-Term E21 Rat Embryos.

Extraction Medium		Total Activity OD/hr/ganglion	% total AChE *
Detergent Medium	CONTROL	4.94±0.38	93.2±1.15
	PHOSPHOLINE	2.03±0.37	91.2±1.15
	BW+MPT	1.82±0.06	89.2±1.4
High salt Medium	CONTROL	0.35±0.04	6.7±1.2
	PHOSPHOLINE	0.20±0.05	8.6±0.9
	BW+MPT	0.22±0.03	10.8±1.4

* "Total AChE" is the sum of AChE activities extracted by detergent and by high salt in each experiment.

detergent (hydrophobic) or high salt (hydrophilic) extracted AChE in control, phospholine and BW + MPT treated ganglia. We find roughly the same proportions of hydrophilic or hydrophobic AChE in the phospholine or BW + MPT extracts than in the control extracts. Thus, intra- and extracellular AChE are composed of hydrophobic and hydrophilic variants in a similar 90 versus 10 % proportion. AChE in these experiments was further analyzed by sucrose gradient centrifugation and hydrophobic and hydrophilic "4 S", "10 S" and "16 S" AChE determined in each control, phospholine and BW + MPT experiments. All AChE peaks present roughly the same hydrophobic-hydrophilic ratio in both their intra- and extracellular locations (Table 4 ; "4 S" AChE percentages were very similar to "10 S" percentages).

Table 4. Hydrophobic and Hydrophilic Molecular Forms of AChE in Control, Phospholine and BW + MPT Treated SCG from Near-Term E21 Rat Embryos.

Extraction Medium		"10 S" activity	"10 S" % total* "10 S" AChE	"16 S" activity	"16 S" % total* "16 S" AChE
Detergent Medium	Control	1.93±0.19	95.60±0.50	0.82±0.04	88.30±5.80
	Phospholine	0.38±0.08	88.31±1.40	0.33±0.03	82.38±2.65
	BW+MPT	0.90±0.06	92.89±1.03	0.20±0.003	71.90±2.24
High salt	Control	0.09±0.003	3.70±0.90	0.17±0.04	18.10±3.80
	Phospholine	0.05±0.01	11.69±1.40	0.07±0.02	17.62±2.65
	BW+MPT	0.07±0.01	7.12±1.03	0.08±0.01	28.10±2.24

* Total "10 S" or "16 S" AChE is the sum of AChE activities extracted by detergent and by high salt in each experiment.

Discussion

Our work deals with the cellular distribution of AChE and its main sedimentation peaks "4 S", "10 S" and "16 S" in extra- and intracellular compartments, in both muscle and neuronal cells. Our results bring a few important points concerning of the cellular biology of AChE : 1) There are important intracellular pools of AChE 2) "10 S" AChE is mainly extracellular in both mouse muscle and rat sympathetic neuronal cells 3) Asymmetric 16 S AChE is not only external but also internal to the muscle fiber or the SCG neuron and 4) Hydrophobic and hydrophilic variants of all forms ("4 S", "10 S" and "16 S") are present both inside and outside the cells.

The analysis of AChE and its molecular forms in the present contribution was performed by reference to 1) the definition of an internal and an external cell compartment through the use of cell permeant and impermeant AChE reversible or irreversible inhibitors (14) and 2) the characterization of socalled "hydrophobic" and "hydrophilic" variants of AChE (8, 13) which suggest the possible occurence of various AChE membrane associations in situ and thus differential localizations. The limitations of our methodological approach are the following 1) No one of the irreversible inhibitor (phospholine) or the reversible inhibitor (BW 284 C 51) can be considered as perfectly non permeant entities. Some unknown quantities of phospholine or BW will cross the cell membrane and respectively inactivate or protect intracellular AChE in an unknown proportion 2) Hydrophobic and hydrophilic variants of AChE forms have been operationally defined, and correspond to the AChE pools extracted from tissues by homogenization in the presence of, respectively, anionic detergent and high salt (8). There is no known correlation of hydrophobic/hydrophilic ratios with particular cell membranes.

In mouse intercostal muscle, the study of the subcellular fractions C_1 and A shows that a major part of particulate AChE is external to the cell, although "16 S" AChE in fraction A has an important intracellular component (40 %). This latter result suggests the presence of "16 S" AChE associated to sarcoplasmic reticulum membranes in fraction A and offers the possibility of isolating and studying internal membranes rich in "16 S" AChE. In the superior cervical ganglion in vivo we found similar amounts of AChE inside and outside the cell. However, "16 S" AChE and also "4 S" AChE are essentially located inside the neuronal cells, in contrast to "10 S" AChE which is mainly (60-70 %) located outside the cell. Thus, in neuronal cells, "10 S" AChE is prevalent outside the cell, and may have a physiological role in cholinergic as well as non-cholinergic mechanisms (17). It is more difficult to imagine a physiological role for the internal "16 S" AChE pool. The presence of such a high proportion of the cell "16 S" AChE in an internal localization suggests that all AChE forms are synthesized and preassembled intracellularly (8, 18, 19).

In mouse intercostal muscle we found that the ratios of hydrophilic versus hydrophobic AChE are similar outside and inside the cell. Thus, there is no preferential localization of hydrophilic AChE outside the muscle cell, in association to the basal lamina. In the near-term rat sympathetic ganglia in vivo, all AChE forms possess hydrophobic and hydrophilic variants both inside and outside the cell, in a similar 90/10 proportion.

In conclusion, the techniques of cell fractionation, of AChE protection or irrevesible inhibition and differential extraction allow to gather detailed data on AChE localization. It will be of great value in the future to directly determined the localization of the different forms of AChE and their variants at the light of our results by entirely different methodologies, such as specific antibodies against each type of forms and immunocytochemistry.

Acknowledgements

This work was partially supported by a grant from DRET (82-83.29).

References

1 - Dreyfus, P., A. Friboulet, H.L. Tran, F. Rieger. 1984. Biol. Cell. 51, 35.

2 - Verdière, M., M. Derer, F. Rieger. 1982. Dev. Biol. 89, 509.

3 - Rieger, F., S. Bon, J. Massoulié, J. Cartaud. 1973. Eur. J. Biochem. 31, 539.

4 - Dudaï, Y., M. Herzberg, I. Silman. 1973. Proc. Natl. Acad. Sci. U.S.A. 70, 2473.

5 - Cartaud, J., F. Rieger, S. Bon, J. Massoulié. 1975. Brain Res. 88, 127.

6 - Massoulié, J., S. Bon. 1982. Ann. Rev. Neurosci. 5, 57.

7 - Rieger, F., M. Ruberg, M.L. Shelanski. 1979. Brain Res. 170, 568.

8 - Dreyfus, P., F. Rieger, M. Pinçon-Raymond. 1983. Proc. Natl. Acad. Sci. U.S.A. 80, 6698.

9 - Lee, S.L., P. Taylor. 1982. J. Biol. Chem. 257, 12292.

10- Bon, S., J. Massoulié. 1976. Febs Letters 67, 99.

11- Viratelle, D.M., S.A. Bernhard. 1980. Biochemistry 19, 4999.

12- Tran, H.L., F. Rieger. 1984. Eur. J. Biochem. (submitted).

13- Nicolet, M., F. Rieger, M. Pinçon-Raymond. 1984. J. Cell. Biol. (submitted).

14- McIssac, R.J., G.B. Koelle. 1959. J. Pharmac. Exp. Ther. 126, 9.

15- Ellman, G.L., K.D. Courtney, V. Andres, R.M. Featherstone. 1961. Biochem. Pharmacol. 7, 89.

16- Lowry, O.H., N.J. Rosebrough, A.L. Farr, R.J. Randall. 1951. J. Biol. Chem. 193, 265.

17- Chubb, I.W., A.J. Hodgson. 1983. 2nd Int. Meet on Cholinesterases. Bled, Yugoslavia (Sept. 17th to 21st), 9.

18- Rotundo, R.L. 1984. Proc. Natl. Acad. Sci. U.S.A. 81, 479.

19- Fernandez, H.L., J. Stiles. 1984. Neurosc. Lett. 44, 187.

INHIBITION AND PHOTOAFFINITY LABELING OF ACETYLCHOLINESTERASE BY PHENCYCLIDINE AND TRIPHENYLMETHYLPHOSPHONIUM

Jan Verdenhalven, Ferdinand Hucho
Institut für Biochemie, Freie Universität Berlin, Thielallee 63, 1000 Berlin 33

Summary:

Acetylcholinesterase was extracted by high salt treatment or proteolytic procedures and isolated by affinitychromatography from the electric organ of *Torpedo marmorata*. Both phencyclidine (PCP) and triphenylmethylphosphonium ($TPMP^+$) could be shown to be potent inhibitors of the enzyme. The mechanism of inhibition of the two substances was investigated by enzymekinetics and photoaffinity labeling of the purified enzyme. PCP shows a pure competitive inhibition in the Lineweaver-Burk-plot with acetylthiocholine as substrate. This was confirmed by photoaffinity labeling of the enzyme with ^{3}H-PCP, which was specific for the active site and could be blocked by acylation of the active center. In contrast $TPMP^+$ shows a mixed competitive/uncompetitice type of inhibition. There are two principal possibilities to explain these data; either the inhibitor binds to different sites on the esterase molecule or it does not differentiate between the free and the acylated active center. The second possibility was supported by the specific photoreaction of ^{3}H-$TPMP^+$ with the native and the acylated enzyme.

Introduction:

The acetylcholinesterase is a typical serine hydrolase. The catalytic center is termed the esteratic site since a serine-

Molecular Basis of Nerve Activity

OH is acylated and deacylated during the catalytic cycle. An anionic binding site for the positively charged choline moiety of the acetylcholine molecule is located in the vicinity of the serine residue. Both sites together constitute the active center of the enzyme. Moreover there is evidence for an additional peripheral anionic site, that is only seen at very low salt concentrations and therefore seems to be without any significance in vivo (1).

To investigate the catalytic mechanism of the acetylcholine-esterase a number of cationic inhibitors have been used. They are mostly aromatic ammonium compounds positively charged at physiological pH-values. Here we present two novel esterase inhibitors, which seem to be useful as tools for the elucidation of the active site of the enzyme. Both were recently successfully used for the characterization of the noncompetitive binding site of the acetylcholine receptor (2,3) and are commercially available as tritium labeled compounds.

Phencyclidine (PCP) is positively charged at pH 7.4 showing pronounced psychotropic effects probably mediated by a high affinity binding site in the central nervous system (4,5). But some side effects may be due to interactions with different neurotransmitter receptors (6,7).Though the effects on the acetylcholine receptor are well investigated the interaction of PCP with the acetylcholinesterase has been neglected so far (3,5).

The lipophilic cation triphenylmethylphosphonium ($TPMP^+$) is widely used as a sensor of membrane potentials. In addition it is a noncompetitive blocker of the acetylcholine receptor (3). This and the fact that $TPMP^+$ is a congener of the esterase inhibitor triphenylmethylammonium suggested that $TPMP^+$ might be an esterase inhibitor itself.

Materials and Methods:

Preparation of an acridinium column: 9-Aminopropylamino acridine was prepared from 9-Cl-acridine and 1,2 Diaminopropane according to (8). The oily product was recristallised twice from benzene, m.p. 133° C. It yields two spots on thin layerchromatography which are both ninhydrine-positive. This shows that in fact a mixture of the two possible isomers of aminopropylacridines were obtained. The product was coupled to activated CH-Sepharose (Pharmacia) according to the producer instructions at a concentration of 2,2 mg/ml of aminopropylaminoacridine. The resin was washed with 1 liter of 1 M Tris/HCl pH 9 per ml of gel. The extent of substitution was calculated from the absorption of aminopropylacridine in a washbuffer using an extinction coefficient e_{412} = 8ooo (cm^2/Mol). Usually an incorporation of o.2 µMol acridine residues/ml gel was obtained.

Purification of the acetylcholinesterase: The asymmetric species of acetylcholinesterase were extracted from Torpedo marmorata electric tissue as described (9) with the only modification that 1 M NaCl was used instead of 2 M MgCl. Trypsinolytic release of the enzyme was performed as described. The crude solubilised esterase preparations from 1oo g of electric tissue were shaken for two hours with 24 ml of acridinium sepharose in either 1 M NaCl, 1o mM Tris/HCl, o.1 mM EDTA, pH 7.4 (salt extracted enzyme) or 1o mM Tris/HCl, pH 7.4 (lytic enzyme). The resin was washed with 1oo ml of buffer and poured in a column (1,6 cm diameter). The esterase was eluted at 5 ml/h with 1o mM decamethonium in the appropriate buffer. The fractions were tested for esterase activity and the enzyme-containing fractions were pooled and chromatographed on Sephadex G 25 to remove the decamethonium.

Esterase activity was determined according to Ellman (1o) and the enzyme activity is ecpressed in Ellman-Units (11). One Unit corresponds to 1,o3 µMol of acetylcholine hydrolysed per min and ml(11). SDS-polyacrylamide gel electrophoresis was performed according to Laemmli (12) with a 3% acrylamide upper and a 1o% acrylamide lower gel. Photoaffinity labeling was carried our in analogy to (13) with 1o min irridation.

Cl-acridine was obtained from Aldrich or prepared from anthrone (Fluka) according to (14). 1,2-diaminopropane was from Fluka, $TPMP^+$ bromide from Sigma, ^3H-$TPMP^+$ iodide, ^3H-PCP and ^3H-DPF were from NEN. Trypsin (Merck) was dissolved in 1o mM Tris/HCl pH 7.4. Torpedo marmorata was obtained from Prof.Whittaker, Göttingen. All other reagents were of the highest purity commercially available.

Results:

Purification of different species of acetylcholinesterase: The various species of acetylcholinesterase can be purified in a

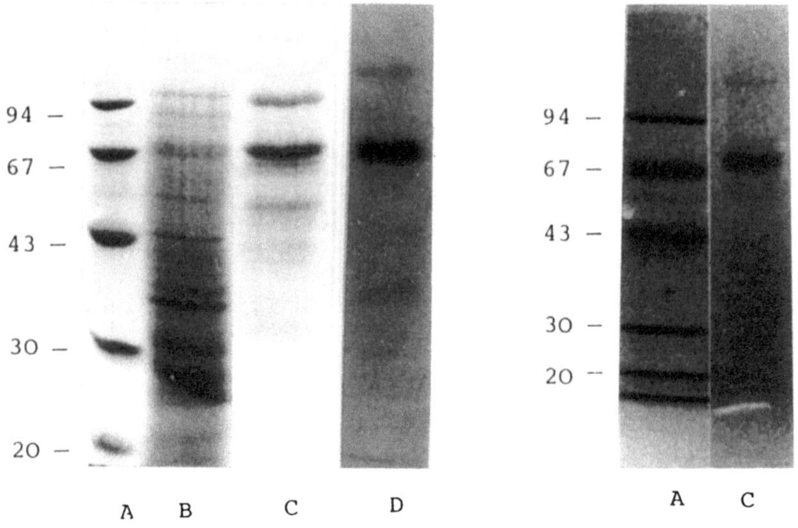

Fig.1: SDS-polyacrylamide gel electrophoresis of purified acetylcholinesterase species; left: 17+13 S esterase, right: lytic (12 S) esterase, A: Molecular weight standards (Mr x 10^{-3}), B: Crude high-salt extract from electric tissue, C: Purified acetylcholinesterase, D: ^3H-diisopropylfluorphosphate labeled 17 + 13 S acetylcholinesterase; A-C: Coomassie-stained gels, D: Fluorography.

single step to homogeneity by means of the acridinium column (Fig.1). From 1oo g electric tissue about 3.5 mg asymmetric and 2.8 mg lytic (12 S) acetylcholinesterase were obtained.

Quaternary structure: Asymmetric acetylcholinesterase (AChase) is composed of three polypeptide chains of Mr 1oo ooo, 72 ooo, 55 ooo, in a molar ratio of about 1:3:1 as determined by densitometry of the coomassie-stained SDS polyacrylamide gel. Only the Mr 72 ooo polypeptide chain appears to contain an active site since this is the only polypeptide chain which could be labeled with ^{3}H-diisopropylfluorophosphate to a significant extent (Fig.1, track D). By sucrose density centrifugation two components can be separated one of about 17 S and one of 13 S, at a ratio of 3:1 (Fig.2 A). Lytic AChase can be shown by SDS-polyacrylamide gel electrophoresis to contain only the Mr 72 ooo polypeptide chain. After partial reduction this form sediments on the sucrose density gradient slightly slower (12 S).

Fig.2,A: purified 17+13 S esterase

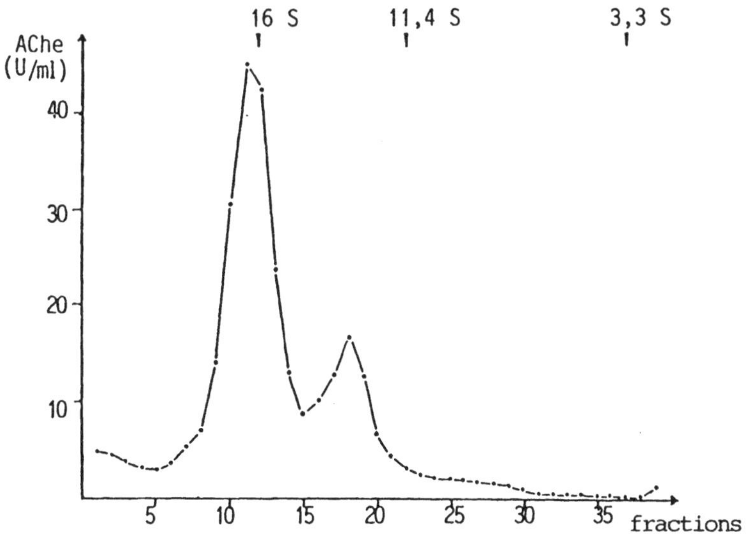

B)purified lytic (12 S) esterase after reduction and alkylation

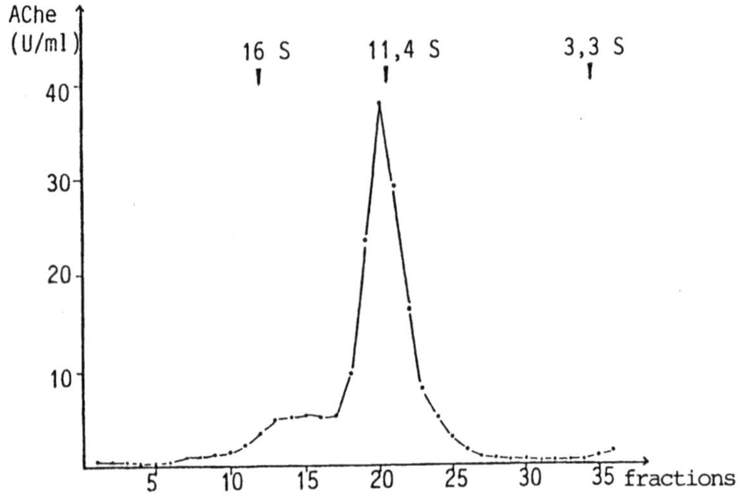

Fig.2: Sedimentation analysis of purified esterase species. A sample of the purified enzyme was sedimented in a 5-2o% sucrose gradient containing 1.o M NaCl, 1o mM Tris/HCl pH 7.4,using a SW-41 Ti rotor at 4o ooo rpm for 15 h, at 4° C. ß-Galactosidase (16 S), beef liver catalase (11.4 S) and bovine carbonic anhydrase B (3.3 S) were used to estimated sedimentation coefficients.

Inhibition by triphenylmethylphosphonium ($TPMP^+$) and phencyclidine: 17 S, 13 S, and 12 S acetylcholinesterase are similarly inhibited by $TPMP^+$ and PCP (Fig.3). The apparent inhibitor

Inhibitor	1 M NaCl	1o mM Tris/HCl	Type of inhibitor
$TPMP^+$	15	o.5	mixed competitive/ uncompetitive
PCP	4o	12	competitive

Table 1: Inhibitor constants and type of inhibition for $TPMP^+$ and PCP (K_i-values in μM).

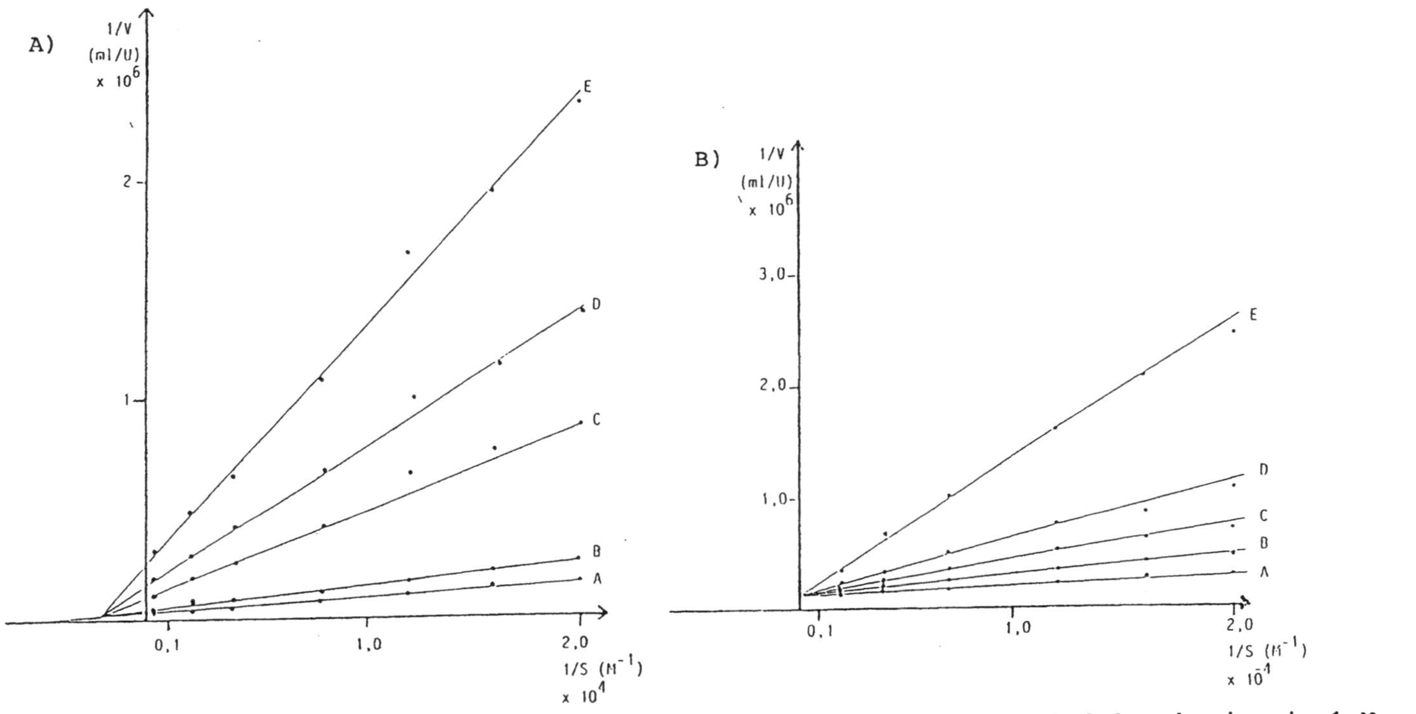

Fig.3,A):Inhibition of 17+13 S acetylcholinesterase by triphenylmethylphosphonium in 1 M NaCl;A: No TPMP$^+$;B:5o μM TPMP$^+$;C:125 μM TPMP$^+$;D:25o μM TPMP$^+$;E:5oo μM TPMP$^+$.B):Inhibition of 17+13 S acetylcholinesterase by phencyclidine in 1 M NaCl;A: No Phencyclidine; B: 5o μM Phencyclidine; C: 125 μM Phencyclidine; D: 25o μM Phencyclidine; E: 5oo μM Phencyclidine.

constants K_i are strongly dependent on the ionic strength (Table 1).

Photoaffinity labeling of 12 S AChase by ^3H-$TPMP^+$. Irradiation of the AChase/^3H-$TPMP^+$ complex with UV light causes incorporation of radioactivity into the enzyme (Fig.4). Up to o.5 %

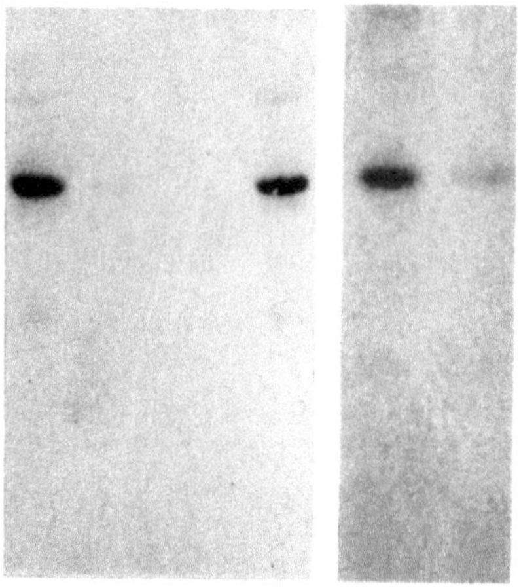

Fig.4: Photoaffinity labeling of purified 12 S acetylcholinesterase with ^3H-triphenylmethylphosphonium; A: 1 uM ^3H-$TPMP^+$ B: 1 uM ^3H-$TPMP^+$ + 1oo uM decamethonium, C: 1 uM ^3H-$TPMP^+$ + 1oo uM propidium, D: 1 uM ^3H-$TPMP^+$ + 1oo uM diisopropyl fluorophosphate, E: 1 uM ^3H-$TPMP^+$ + 1oo uM edrophonium.

of the AChase molecules were labeled under our experimental conditions. The reversible AChase inhibitors decamethonium, propidium and edrophonium prevent photolabeling completely (Fig.4,track B,C,E). But the inhibitor of the esteratic subsite diisopropylfluorophosphate causes only partial inhibition of photolabeling (Fig.4, track D).

Similarly, irradiation of an AChase/^3H-PCP cpmplex results in covalent radioactive labeling of the Mr 72ooo polypeptide chain of the enzyme (Fig.5). Interestingly even more radioactivity is incorporated into a polypeptide chain of an apparent

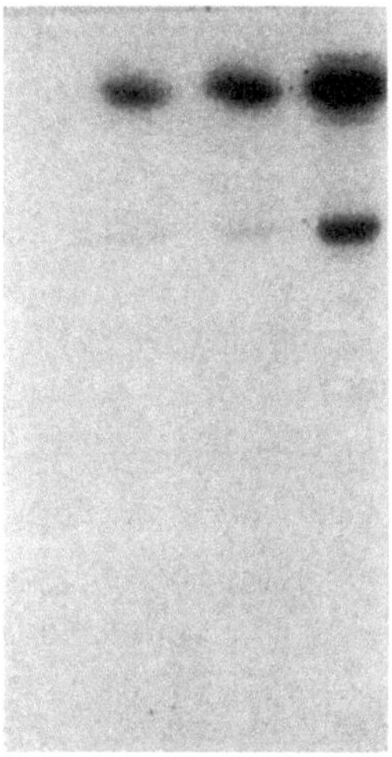

Fig.5: Photoaffinity labeling of purified 12 S acetylcholinesterase with ^{3}H-phencyclidine; A: 1 µM ^{3}H-PCP + 1oo µM diisopropylfluorophosphate, B: 1 µM ^{3}H-PCP + 1oo µM propidium, C: 1 µM ^{3}H-PCP + 1oo µM decamethonium, D: 1 µM ^{3}H-PCP

Mr 22o ooo. Edrophonium, propidium and decamethonium decrease photolabeling by ^{3}H-PCP significantly, DFP inhibits labeling of both bands completely.

Discussion:

Asymmetric AChase purified from *Torpedo marmorata* electric tissue by affinity chromatography consists of three types of polypeptide chains. In addition to the known Mr 72 ooo (15) and Mr 55 ooo (16) polypeptide chains a further polypeptide of Mr 1oo ooo is present in our preprarations. A similar component was described recently in AChase from *Torpedo californica* (9) as an extremely protease sensitive protein which could be found only if isolation was performed rapidly and in the presence of Trasylol and EDTA. It contains probably no catalytically active center because it is neither labeled by ^{3}H-DPF nor by our photolables ^{3}H-TPMP$^+$ and ^{3}H-PCP. Furthermore the lytic (12 S) AChase does not contain the Mr 1oo ooo polypeptide chain; its catalytic activity is identical to that of the asymmetric form.

The kinetics of $TPMP^+$ and PCP inhibition is the same for the 17 S, 13 S and 12 S forms, respectively. But PCP is a purely competitive inhibitor of all three AChase species while $TPMP^+$ is not. PCP appears to bind to the free active site, and it covers both the anionic and the esteratic subsite of the AChase active center. Blocking of the latter with DFP prevents photolabeling of the AChase with PCP probably because it inhibits reversible binding of the photolabel.

A problem is posed by the observation that both edrophonium, a specific ligand of the active site (17), and propidium, a ligand supposed to be specific for the peripheral site of the AChase (18), prevent photolabeling of AChase with ^{3}H-PCP. This seems to contradict the purely competitive nature of PCP inhibition. One explanation for this could be that propidium at the concentrations applied in the photolabeling experiments (10^{-4}M) binds to a significant extent to the active site as well.

$TPMP^+$ is a mixed competitive/uncompetitive inhibitor as evidenced by the Lineweaver-Burk-plots (Fig.3). This type of inhibition is indicative of inhibitor binding to the acylated enzyme intermediate (thereby slowing down the deacylating step) (19) in addition to the free enzyme. Similar inhibition kinetics is also expected if $TPMP^+$ binds to both the active and the peripheral binding site of the enzyme. On the basis of the present data we cannot discriminate between these alternatives. For this direct binding studies would be required.

Acknowledgement: This work was supported by the Deutsche Forschungsgemeinschaft, the Land Berlin (FGS Biomembranen) and the Fonds der Chemischen Industrie.

References:

1. Rosenberry, T.L.; Adv.Enzymol.43, 1o3 218 (1975)

2. Eldefrawi,M.E., Eldefrawi,A.T., Aronstam,R.S., Maleque, M.A., Warnick,J.E., Albuquerque,E.X.; Proc.Natl.Acad. Sci.USA 77, 7458-7462 (198o).

3. Lauffer,L., Hucho.F., Proc.Natl.Acad.Sci.USA 79, 24o6-24o9 (1982).

4. Vincent,J.P., Kortalovski,B., Geneste,P., Kameka,J.M., Lazdunski,M.; Proc.Natl.Acad.Sci.USA 76, 4678-4682 (1979)

5. Altura,B.T., Quirion,R., Port,L.B., Altura,B.M.; Proc. Natl.Acad.Sci.USA 8o, 865-869 (1983)

6. Albuquerque, E.X., Tsai,M.G., Aronstam,R.S., Witkop,B., Eldefrawi,A.T., Eldefrawi, M.E.; Proc.Natl.Acad.Sci.USA 77, 1224-1228 (198o)

7. Domino,E.F., in: Phencyclidine (PCP) Abuse; Petersen,E.F. & Stillmaryk, eds., 18-34 (1978)

8. Dudai,Y., Silman,I.; Methods Enzymol. 73, 571-591 (1974)

9. Lee,S.L., Heinemann,S., Taylor,P.; J.Biol.Chem.257, 12283-12291 (1982).

1o. Ellman,G.L., Courtney,K.P., Andres,V., Featherstone; Biochem.Pharmacol.7, 88-95 (1961).

11. Massoulié,J., Bon,S.;Eur.J.Biochem.68, 531-539 (1976).

12. Laemmli,U.K.; Nature 15, 68o-685 (197o).

13. Muhn,P., Hucho,F.; Biochemistry 22, 421-425 (1983)

14. Albert,A., Ritchie,B.; Org.Synthesis 22, 5-8 (1948).

15. Rieger,F., Bon,S., Massoulié,J., Cartaud,J., Picard,B., Benda,P.; Eur.J.Biochem.68, 513-521 (1967).

16. Anglister,L., Silman,I.; J.Mol.Biol. 125, 293-311 (1978)

17. Wilson,J.B., Quan,C.; Arch.Biochem.Biophys.73, 131-143 (1958).

18. Taylor,P., Lwebuga-Mukasa,J., Lappi,S., Rademacher,J.; Mol.Pharmacol.1o, 7o3-7o8 (1974).

19. Krupka,R.M.; Biochemistry 3, 1749-1755 (1964).

A COMPARATIVE STUDY OF THE AGEING OF DFP-INHIBITED SERINE HYDROLASES BY MEANS OF ^{31}P-NMR AND MASS SPECTROMETRY

Alfons C.M. van der Drift
Medical Biological Laboratory TNO, P.O. Box 45, 2280 AA Rijswijk, The Netherlands

Introduction

Therapeutic treatment of organophosphate intoxication may be impeded by a spontaneous and gradual conversion of the phosphylated acetylcholinesterase (AChE; EC 3.1.1.7) into a non-reactivatable form (1-3). This in vivo process, called ageing, involves the release of an oxygen-bound alkyl group from the inhibiting organophosphorus moiety according to first-order kinetics (4-7). The rate of ageing depends on the structure of both enzyme and inhibitor (2,4,7,8), on temperature (2,8-11), pH (5,6,8-12) and ionic strength (10,11,13); besides, it may be influenced by non-covalent binding of particular ligands (14-16). The molecular events underlying the dealkylation and leading to the non-reactivatability of phosphylated enzyme are partly understood.

Ageing may be envisaged to occur via C-O or P-O scission (17) by a nucleophilic substitution reaction in which either a unimolecular (S_N1) or a bimolecular (S_N2) character may dominate (18,19). For electric eel AChE inhibited with pinacolyl methylphosphonofluoridate (soman), ageing has been demonstrated to result almost exclusively from C-O scission by an S_N1 type reaction (5). However, it is not known to what extent this ageing mechanism will be operative after inhibition with other organophosphorus compounds such as diisopropyl phosphorofluoridate (DFP).

The native structure of the protein plays an important part in ageing but its exact involvement is unclear. For instance, it is still unsettled whether the protonated group involved in ageing of phosphylated AChE is an imidazolium group or a carboxylic acid (10,11,20). It has been suggested that elements of the protein structure responsible for the

Molecular Basis of Nerve Activity

hydrolase activity also play an essential role in ageing (4). If so, ageing might be a common phenomenon among phosphylated serine hydrolases. Indeed, spontaneous dealkylation has also been observed in phosphylated butyrylcholinesterase (EC 3.1.1.8) (4,10,11) and phosphorylated α-chymotrypsin (4), but not for soman-inhibited α-chymotrypsin (5). Since only the three-dimensional structure of the latter is well known, it cannot be decided to what extent ageing is dependent on common structural features of the active sites of these enzymes. Further, it is still uncertain whether the refractoriness of aged phosphylated enzymes to reactivation results entirely from the intrinsic stability of the remaining phosphorus moiety or is partly due to a conformational change occurring in the enzyme structure upon ageing (21,22).

The present report concerns a comparative study of the ageing of three DFP-inhibited serine hydrolases, which are unrelated from an evolutionary point of view - each is a representative of a different superfamily - but whose active centres show a close structural similarity. It shows that structurally well-characterized enzymes, which allow the application of spectroscopic techniques to which AChE is hardly accessible, viz. ^{31}P-NMR and mass spectrometry, may serve as suitable model systems to resolve some of the above-mentioned questions. A more detailed account of the present work has been given elsewhere (23).

Outline of the experimental approach

α-Chymotrypsin from bovine pancreas (Chymo; EC 3.4.21.1), subtilisin A (Sub; EC 3.4.21.14) and atropinesterase from Pseudomonas putida PMBL-1 (AtrE) were investigated. Chymo and Sub are proteases with well-known three-dimensional structures; AtrE is a true esterase, like AChE, and has an active centre that contains the same structural elements as that of the proteases, viz. an apolar substrate binding pocket, a so-called charge-relay system and an oxy-anion hole. The side chain of the serine residue in the active centre of these enzymes was phosphorylated with a diisopropyl phosphoryl (DP) group by incubation with DFP. The DP-enzymes were studied in buffered aqueous solution under various conditions with

^{31}P-NMR spectrometry before and after being subjected to ageing at 36 °C. Buffers used were 0.1 M Tris-HCl (pH $\geq$ 7.6) or 0.1 M citric acid-sodium citrate (pH < 7.6). Besides, some model compounds of low molecular weight bearing a DP or a monoisopropyl phosphoryl (MP) group were studied under identical conditions and in various solvents for the purpose of identification and interpretation of the NMR data. Mass spectrometry was used to analyse products resulting from ageing in the presence of $H_2^{18}O$.

Results

^{31}P-NMR spectra of the DP-enzymes showed a single line, the location of which as characterized by the chemical shift δ differed with the enzyme (Fig. 1). In all three cases lowering of the pH from 9 to 5 resulted in

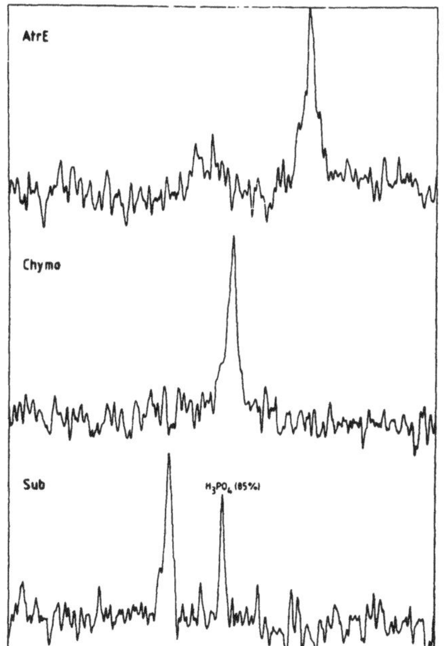

FIGURE 1: ^{31}P-NMR spectra of DP-enzymes in buffered aqueous solution at 4°C. AtrE = atropinesterase (pH = 7.76), Chymo = α-chymotrypsin (pH = 7.60), Sub = subtilisin A (pH = 7.65). H_3PO_4 reference peak is shown in the spectrum of Sub.

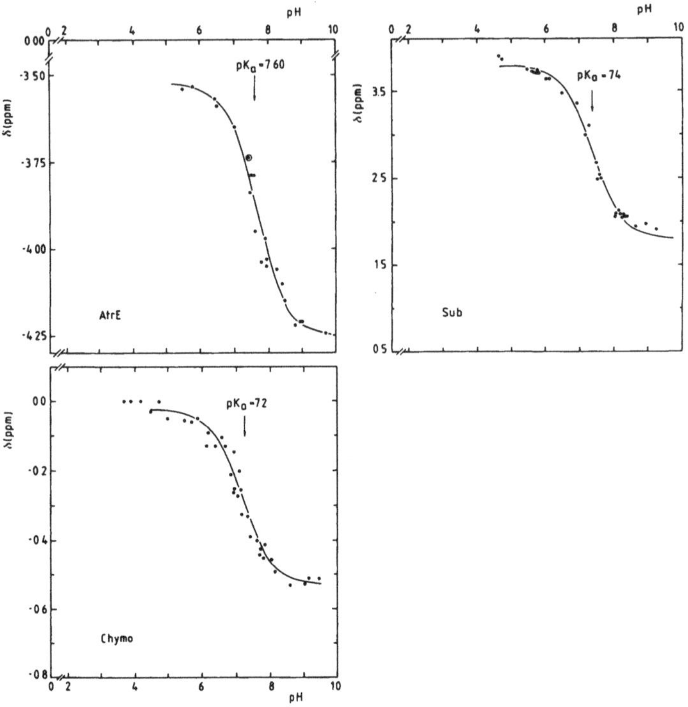

FIGURE 2: pH-Dependence of the chemical shift (δ) in the ^{31}P-NMR spectra of DP-enzymes. Solid lines are theoretical curves calculated according to the Henderson-Hasselbach equation. For abbreviations, see legend to Figure 1.

an increase in δ, i.e. a downfield displacement of the ^{31}P line in the spectrum, according to a simple titration curve of a group with pK_a between 7.2 and 7.6 (Fig. 2).

Prolonged exposure of the DP-enzymes to 36 ^{o}C at pH-values within the titration range led to a second line in the spectrum with a position downfield relative to the first line and different for the three enzymes. Its intensity gradually increased at the expense of the original line, which finally disappeared (Fig. 3). The δ corresponding with this line did not show a pH-dependence, the value being 1.3, 2.1, and 5.2 ppm for AtrE, Chymo and Sub, respectively. On the basis of denaturation studies and NMR data obtained with the model compounds, the second line was identified as belonging to the MP-enzyme, resulting from dealkylation of the

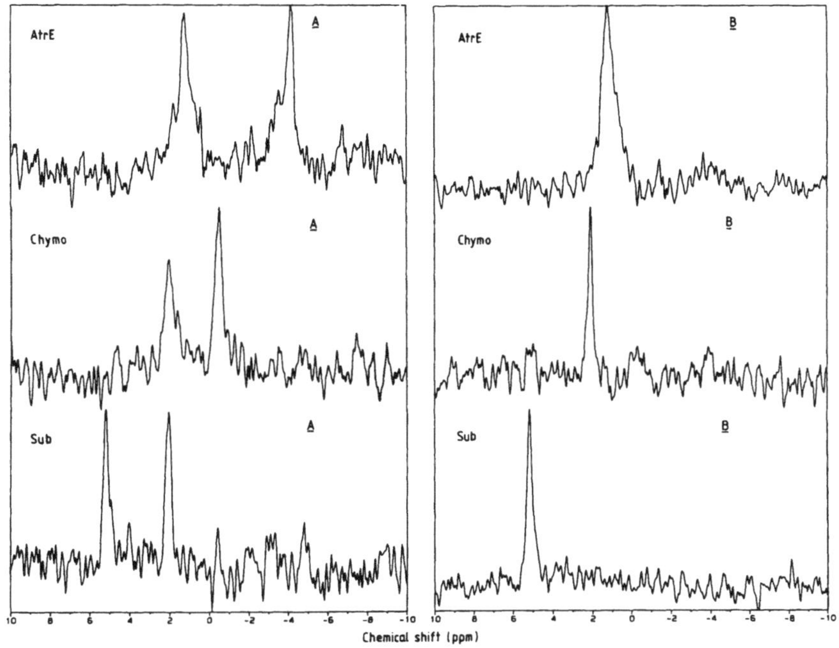

FIGURE 3: ^{31}P-NMR spectra of partially (A) and completely aged (B) DP-enzymes in buffered aqueous solution at 4 °C. AtrE = atropinesterase (pH = 7.95), Chymo = α-chymotrypsin (pH = 7.80), Sub = subtilisin A (pH = 8.20).

DP-enzyme, i.e. from ageing. With denatured DP-enzymes this conversion was not observed.

From NMR spectra of samples of DP-enzymes subjected to ageing for some time, the fraction of unaged enzyme remaining was determined. The decrease in this fraction with time proceeded according to first-order kinetics (Fig. 4). The rate constants, k_a, obtained at different pH's increased with decreasing pH. As plots of δ of unaged DP-enzyme versus k_a for various pH-values in the above-mentioned titration range were linear (Fig. 5), the pH-dependence of k_a followed a similar titration curve as obtained for δ. This allowed the estimation of the maximum value of the rate constant of ageing, k_a^m, obtained when the titratable group is fully

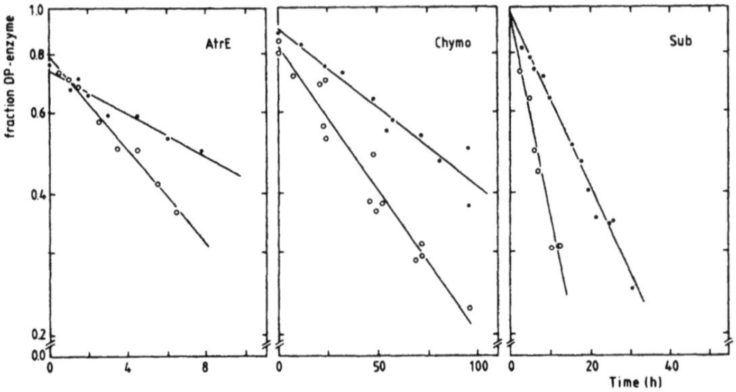

FIGURE 4: Semi-logarithmic plot of the decrease in the fraction of DP-enzyme with time, at 36 ^{o}C and different pH. The fractions of DP-enzymes were obtained from the NMR spectra. AtrE = atropinesterase at pH = 7.6 (●) and pH = 6.7 (O), Chymo = α-chymotrypsin at pH = 7.6 (●) and pH = 3.5 (O), Sub = subtilisin A at pH 7.6 (●) and pH = 5.5 (O). The straight lines were calculated by the method of least squares.

protonated. This value together with the pK_a of the titratable group and the observed difference in chemical shift between MP- and DP-enzyme at high pH are given in Table I for each of the three enzymes. It can be seen that these quantities increase in the order Chymo-Sub-AtrE.

To determine whether loss of the isopropyl group proceeded by P-O or by C-O scission and to investigate the mechanism of this release, the ageing reaction was run to completion at pH $\leq$ 6.7 in the presence of $H_2{}^{18}O$, and the products formed were analysed by mass spectrometry. The results (Table II) show that invariably both isopropanol with an $^{18}O/^{16}O$ ratio equal to that of the solvent and propene were formed; no other products were found. The amount of propene increased and that of isopropanol decreased in the order Chymo-Sub-AtrE. In the same sequence an increase is seen in the difference between the total amount of products actually found and that to be expected upon complete ageing.

TABLE I. Some Characteristic Quantities Related to the Ageing of DP-Enzymes in Aqueous Solution at 36 °C.

Quantity (a)	Chymotrypsin	Subtilisin	Atropinesterase
k_a^m (10^{-6} s^{-1})	6.2	28.0	34.5
pK_a (b)	7.2	7.4	7.6
$\delta_{MP}-\delta_{DP}$ (ppm)	2.7	3.4	5.6

(a) k_a^m = maximum value of the rate constant of ageing obtained from the linear plots in Fig. 5 for a fully protonated histidine at pH = 5.0. $\delta_{MP}-\delta_{DP}$ = difference in chemical shift between MP- and DP-enzyme at high pH.

(b) Based on Fig. 2.

TABLE II: Results of Mass-Spectrometric Analysis of Products Formed upon Ageing of DP-Enzymes in Aqueous Solution at 36.0 °C.

		Chymotrypsin	Subtilisin	Atropinesterase
Isopropanol (%)	(a)	97 ± 13	42 ± 6	18 ± 14
Propene (%)	(a)	1.2 ± 0.2	3.8 ± 0.6	11.3 ± 5.5
$^{18}O/^{16}O$ ratio	(b)	0.99 ± 0.03	1.01 ± 0.01	0.98 ± 0.05

(a) The amount measured after complete ageing at 36.0 °C (± s.e.m.; n = 4-6) expressed as percentage of the amount expected on the basis of the concentration of phosphorylated enzyme. Ageing was completed at pH = 5.7 for chymotrypsin and subtilisin, and at pH = 6.7 for atropinesterase in either 0.1 M citric acid-sodium citrate buffer or 0.05 M sodium sulphate (see Ref. 23). Since the results in these media did not differ significantly, the averaged values of both media are given. In some experiments also a small amount of acetone was found, presumably due to contamination.

(b) The ratio (± s.e.m.; n = 5-6) of ($H_2{}^{18}O/H_2{}^{16}O$) and (^{18}O-isopropanol/^{16}O-isopropanol).

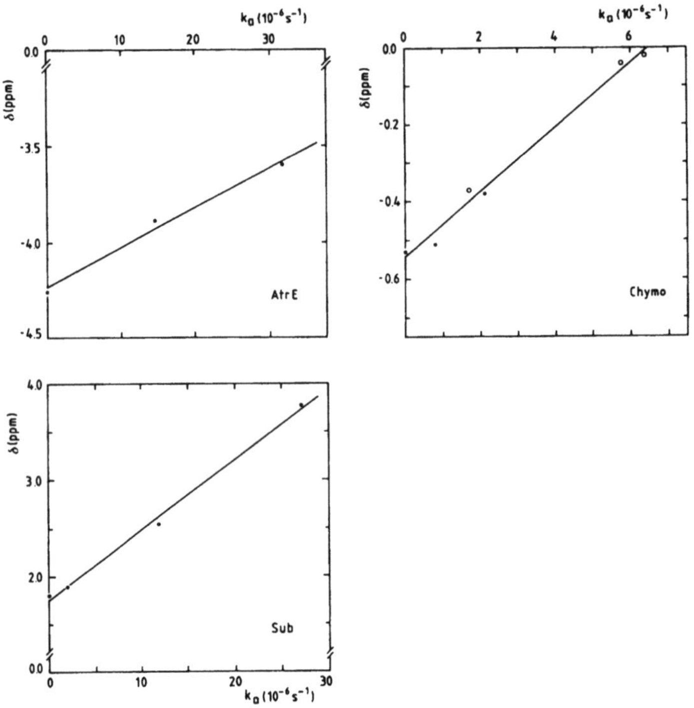

FIGURE 5: Relation between the chemical shift (δ) of unaged inhibited enzyme and the rate constant of ageing (k_a) in their dependence on pH. Values of δ (Fig. 2) were plotted versus the corresponding k_a-values for different pH's (≥ 4.5). ● = k_a determined by ^{31}P NMR. O = k_a obtained from Ref. 4. The straight lines were calculated by the method of least squares, including the limiting value of δ at high pH where k_a = 0. For abbreviations, see legend to Fig. 4.

Discussion

The present data for different but identically phosphorylated enzymes indicate that ageing is a common feature of DFP-inhibited serine-hydrolases even if these are not evolutionary related. The high values of k_a compared to the (unimolecular) rate constants for hydrolysis of non-cyclic trialkyl phosphates to secondary phosphate esters under similar conditions (10,24,25) and the absence of any appreciable ageing after denaturation, support the notion that ageing of DP-enzymes is a catalytic process which depends on the native enzyme structure (4,10). Conversion of the DP-enzyme into an MP-species proceeds exclusively by C-O scission, since the $^{18}O/^{16}O$ ratio of the isopropanol produced is equal to that of the aqueous solvent (Table II). Thus it may be expected that interactions between the scissile C-O bond and structural elements of the active centre which are common in the serine hydrolases investigated, play an important role in ageing.

The increase in k_a with decreasing pH according to a simple titration curve points to an intramolecular general acid catalysis in which the presence of a proton on a titratable group with pK_a around 7.4 is essential. There are substantial indications that this group is the Nε2 of the histidine residue of the charge-relay system (23) which in the active enzymes has a pK_a < 7.0. On the basis of data obtained for intramolecular general acid-catalysed hydrolysis of phosphate esters (26-28), it is expected that ageing of DP-enzymes requires hydrogen bonding between the oxygen of the scissile C-O bond and an adjacent proton donor. The increase in δ upon protonation of Nε2, which increase is restricted to the DP-enzymes, is consistent with a direct interaction between the DP-group and the histidine residue in the active centre under the formation of such a hydrogen bond (23). From these results it is concluded that hydrogen bonding between the ester oxygen of the hydrolysable P-O-C structure and the Nε2 of the histidine residue in the charge-relay system is a prerequisite for ageing of DP-enzymes.

Ageing of DP-enzymes according to an S_N1 type reaction will lead to formation of an isopropyl carbenium ion which may react with water to yield isopropanol, may be captured by other nucleophilic reactants present in the protein or solution, or may loose a proton to give propene (18,19). In case of an S_N2 type reaction, attack of a solvent molecule displaces the phosphorylated enzyme from the secondary carbon atom and only isopropanol is formed (19). Thus, the formation of propene (Table II) unambiguously proves that ageing in all three enzymes has at least some S_N1 character. Thermodynamic data (23) together with the data in Table II suggest that in the case of DP-Chymo the isopropanol results from a reaction of predominantly S_N2 character. Therefore, the decrease in the amounts of isopropanol and total free low-molecular weight products and the increase in the amount of propene in the order Chymo-Sub-AtrE is interpreted as a shift in the ageing from a mainly S_N2 type reaction in Chymo to one with a considerable contribution of S_N1 in Sub and AtrE. Since hydrolysis of phosphate esters by acid-catalysed unimolecular C-O scission will proceed through the ionic conjugate acid of the ester (29), this indicates that proton transfer from the histidine to the ester oxygen is more likely to occur in Sub and AtrE than in Chymo.

The apparent increase in S_N1 character is attended with an increase in k_a^m (Table I). The concomitant increase in pK_a of the active-site histidine in the DP-enzymes (Table I) can be attributed to stronger hydrogen bonding with the ester oxygen (10,30), whereas the increase in chemical shift difference between the MP- and DP-enzymes (Table I) suggests conformational changes upon ageing which lead to enhanced binding of the phosphoryl substituent by non-bonding interactions between the non-scissile isopropoxy group and the substrate binding pocket (23). Apparently, in all cases direct interaction with the active-site histidine is lost upon ageing, since the chemical shift of the MP-enzymes does not appreciably depend on pH between 4 and 10. These results suggest that the larger S_N1 contribution to ageing of DFP-inhibited Sub and AtrE may result from a larger stabilization of the anion in the transition state by hydrogen bonding with the active-site histidine and non-bonding interactions between the remaining isopropyl group and the substrate binding pocket. Upon ageing the first interaction will be weakened and will finally disappear whereas the non-bonding interaction becomes stronger, at least in the case of Sub and AtrE. To what extent conformational changes underlying this interplay of interactions contribute to the refractoriness of aged DP-enzyme to reactivation is not known yet.

Phenomenologically, ageing of DFP-inhibited AChE closely resembles that of the three serine hydrolases studied, in particular with respect to the pH-dependency (cf. 4). This suggests that the active site of AChE has some structural features in common with these enzymes, viz. a so-called charge-relay system, and that the outline of the ageing process presented above is also applicable to DP-AChE.

References

1. Hobbiger, F.. 1963. In: Handbuch der Experimentellen Pharmakologie, Bd. 15. Cholinesterase and Anticholinesterase Agents (G.B. Koelle, sub-ed.). Springer-Verlag, Berlin - Göttingen - Heidelberg. p. 921.

2. Usdin, E.. 1970. In: International Encyclopedia of Pharmacology and Therapeutics, Section 13, Vol. 1. Anticholinesterase Agents (A.G. Karczmar, ed.). Pergamon Press, Oxford - New York. p. 249.

3. Aldridge, W.N. and E. Reiner. 1972. Enzyme Inhibitors as Substrates. Frontiers of Biology, Vol. 26. North-Holland Publishing Company, Amsterdam. p. 79.

4. Berends, F.. 1964. Reactivering en "Veroudering" van Esterasen geremd met Organische Fosforverbindingen. Ph.D. Thesis, University of Leiden. (summary in English)

5. Michel, H.O., B.E. Hackley Jr., L. Berkowitz, G. List, E.B. Hackley, W. Gillilan and M. Pankau. 1967. Arch. Biochem. Biophys. 121, 29.

6. Davies, D.R. and A.L. Green. 1956. Biochem. J. 63, 529.

7. Coult, D.B., D.J. Marsh and G. Read. 1966. Biochem. J. 98, 869.

8. Berry, W.K. and D.R. Davies. 1966. Biochem. J. 100, 572.

9. Schoene, K., J. Steinhanses and A. Wertmann. 1980. Biochim. Biophys. Acta 616, 384.

10. Keyer, J.H.. 1971. Aging of Phosphonylated Cholinesterases, Ph.D. Thesis, University of Leiden.

11. Keyer, J.H., G.Z. Wolring and L.P.A. de Jong. 1974. Biochim. Biophys. Acta 334, 146.

12. Hobbiger, F.. 1956. Brit. J. Pharmacol. 11, 295.

13. Dawson, R.M., H.D. Crone, M.P. Bladen and M. Poretski. 1981. Neurochem. Int. 3, 335.

14. Sterri, S.H.. 1977. Biochem. Pharmacol. 26, 656.

15. Crone, H.D.. 1974. Biochem. Pharmac. 23, 460.

16. Schoene, K.. 1980. Monogr. neural. Sci. 7, 85.

17. Aldridge, W.H.. 1975. Croat. Chim. Acta 47 (3), 215.

18. Breslow, R.. 1969. Organic Reaction Mechanisms. 2nd Edition. W.A. Benjamin, Inc., New York.

19. Badea, F.. 1977. Reaction Mechanisms in Organic Chemistry. Abacus Press, Tunbridge Wells.

20. Beauregard, G., J. Lum and B.D. Roufogalis. 1981. Biochem. Pharmac. 30, 2915.

21. Kaiser, E.T., T.W.S. Lee and F.P. Boer. 1975. J. Amer. Chem. Soc. 93, 2351.

22. Amitai, G., Y. Ashani, A. Gafni and I. Silman. 1982. Biochemistry 21, 2060.

23. Van der Drift, A.C.M.. 1983. Physico-Chemical Characterization of Atropinesterase from Pseudomonas Putida. A comparison with other serine hydrolases. Ph.D. Thesis, University of Utrecht.

24. Barnard, P.W.C., C.A. Bunton, D.R. Llewellyn, C.A. Vernon and V.A. Welch. 1961. J. Chem. Soc., 2670.

25. Cox Jr., J.R. and O.B. Ramsay. 1964. Chem. Revs. 64, 317.

26. Hudson, R.F.. 1965. Structure and Mechanisms in Organo-Phosphorus Chemistry. Academic Press, New York.

27. Kirby, A.J. and S.G. Warren. 1967. The Organic Chemistry of Phosphorus. Reaction Mechanisms in Organic Chemistry, Vol. 5. Elsevier, Amsterdam.

28. Singleton Jr., R.. 1973. J. Chem. Ed. 50, 538.

29. Ingold, C.K.. 1969. Structure and Mechanism in Organic Chemistry. 2nd Edition. Cornell University Press, Ithaca.

30. Jaffé, H.H., L.D. Freedman and G.O. Doak. 1954. J. Amer. Chem. Soc. 76, 1548.

LIPID MECHANISM AND ACETYLCHOLINESTERASE FUNCTION

Konrad Kaufmann

Department of Neurobiology, The Weizmann Institute of Science,
Rehovot, Israel

Abstract

The mechanism of acetylcholine-induced ion channels is described in detail. On the basis of thermodynamic phase diagrams, the free energy coupling, membrane excitation, and permeability fluctuations appear as a consequence of first principles of chemomechanical coupling, hydrodynamic excitations, and thermal fluctuations. The theory is outlined and compared with the observations of proton-induced conformational changes, adiabatic membrane current pulses, and proton-induced ion channels in pure phospholipid bilayers. Local control of permeability and electromotive force across asymmetrically protonated phospholipid membranes is discussed. Enzymatic hydrolysis of acetylcholine thus induces ion channels in agreement with the requirement of a specific receptor mechanism. The mechanism is chemiosmotic and offers a solution to the controversy about localised versus delocalised energy coupling at ATPases.

"Es ist wohl unnötig zu betonen, dass ich mit vielen Ideen nicht einverstanden bin. Aber ich bin in den Haber Colloquien erzogen worden, in denen Controversen als ein wesentliches Element wissenschaftlichen Fortschritts betrachtet wurden."

David Nachmansohn 14.11.1981

Protonic mechanism of membrane transport

Although the chemical basis of nerve activity has been established and molecularly characterised for the cholinergic system by Nachmansohn, Silman, Changeux, and others, the physical mechanism of the transport processes across biological membranes has not yet been identified despite considerable effort.

Molecular Basis of Nerve Activity

Such situation could arise from erroneous a priori hypotheses. One such hypothesis, implicit in most membrane models, is that of an inert lipid bilayer, and the consequent assignment of all elements of the transport system to autonomous proteins. The latters are not measurable autonomously, since any membrane transport is observed in the presence of the associated lipid bilayer. Although it is evident that specific proteins induce specific transport across biological membranes, it is impossible to exclude that the protein activities control specifically a lipid mechanism of transport across the bilayers.

All membranes are constituted by a lipid bilayer. Therefore, any consequence of any protein activity on the physical properties of the lipid monolayers by necessity takes part in the function of the proteins. The lipid properties imply free energy coupling, propagating excitations, and conductivity fluctuations, as derived from first principles below and demonstrated directly experimentally. Local enzymatic proton sources such as acetylcholinesterase (AChase) and ATPases inevitably imply, with certain probability values, coupling of chemical to osmotic free energy as well as induction of specific ion channels, on the basis of the same lipid phase diagrams.

On the other side, the hypothesis of inert lipid membranes requires to assume very different protein mechanisms for free energy coupling, receptor function, voltage-induced excitation, distinguished by postulating, respectively, intrinsically chemiosmotic enzymes, non-catalytic receptors, voltage-dependent channel proteins. Each of such postulated protein mechanisms has, in physical detail, remained completely mysterious.

The assumption of a non-catalytical receptor of ACh violates causality because catalytic hydrolysis has a faster relaxation time than the physiological response; instead, the slow response has to follow, and therefore to be caused by, enzymatic hydrolysis. I have earlier predicted that AChase activity is sufficient to increase the permeability of the lipid component of nerve membranes.

In vitro, the catalytic mechanism has been established in collaboration with Silman. It makes use of the local decrease in pH at AChase as described by Silman and Karlin; it is also compatible with the observation by Podleski and Changeux that physiological responses to ACh are subject to pH buffers. The mechanism of the ion channel induction was identified: ion channels of the same kinetic relaxation times appear at corresponding pH in the absence of proteins across the pure phospholipid bilayers. The lipid component was shown to be capable to produce the ion channel conductivities reported for various lipid-protein membranes.

Although the induction by specific proteins is obvious, the assumption of autonomous protein channels for ion transport must therefore be considered inobservable.

The falsification of inert lipid bilayer hypotheses demands for a membrane theory on the basis of the physical properties of lipids, and of the specific control functions of proteins. I shall outline protonic control functions; the obvious generalisation to control by voltage, calcium, or surface pressure is not attempted.

Active ion transport

Catalytic protons exert a two-fold function on the membrane:

On the one side, protonation of the phospholipid surface influences the thermal fluctuations in the fluid-crystalline order of the polymeric hydrocarbon chains. This proton receptor function suffices to induce ion channels in monolayer states of large compressibility and large consequent lipid area and bilayer thickness fluctuations of the membrane. In particular, such fluctuations diverge at a critical point and become extremal at the phospholipid pK.

On the other side, asymmetric protonation of one surface of the bilayer only creates an electromotive force which drives ion transport across such channels.

By consequence, asymmetric protonation fluctuations cause directed proton transport across proton-induced ion channels. Phospholipid bilayers thus relate chemiosmotic and chemoreceptor phenomena. This relation is demonstrated experimentally, since both AChase and ATPase activity induce ion channels. The protonic receptor mechanism is "active" after Ussing, since the local electrochemical diffusion occurs even in the absence of global bulk asymmetry due to local asymmetric hydrolysis.

The establishment of a stationary active gradient across the membrane does require stationary enzyme activity, i.e. localisation of enzyme and substrate at the same side of the membrane. Therefore, at a cholinergic membrane, only ATPase activity is capable to establish such membrane potentials by the lipid mechanism, in contrast to AChase.

The opposite asymmetry of AChase localisation by consequence induces instationary and depolarising ion transport upon hydrolysis of applied ACh. The agreement with the experimental observations, and in particular transient overshoot of depolarising membrane excitation with reversal of its sign demonstrates the intrinsic relation of chemiosmosis and receptor functions.

Free energy coupling

Non-catalytic models of ACh receptor function admit dissipation of free energy useful for transport. Williams argued that free energy coupling, to avoid dissipation, has to make use of local, "energised" protons at the membrane surface. Eigen and deMaeyer discovered very high proton mobilities in water and ice according to which another mechanism is required in addition to store free energy. The discussion of surface proton mechanisms by Nagle and Morowitz, Haines, and Nicholls and Deamer is in this respect insufficient. The situation is further complicated by the fact that Mitchell's theory was clearly applicable although formulated in terms of aqueous protonic electrochemical potentials, and verified by the observed coupling of bulk gradient to chemical free energy.

How can protonic free energy be stored at the membrane surface?

This question is related to the present scientific controversy on local versus global chemiosmosis. It is also related to the problem how the protonic mechanism of ion transport by ACh could apply in the presence of neutral pH buffers in the global bulk medium.

Chemomechanical coupling to the phospholipids resolves this problem principally, applying for two-dimensional monolayers the one-dimensional polymeric muscle models developed by Katchalsky, Oplatka, and Kedem.

The reversible chemomechanical coupling between local enzymatic proton sources and associated protonable phospholipid monolayers stores the free energy which would, otherwise, drive protons into the bulk solution. Storage at the surface is due to the lipid thermodynamic transitions of long relaxation times, e.g. several msec according to Tsong, and Träuble and Eibl, in order to release protons from the phosphate groups. According to monolayer phase diagrams by Träuble and Teubner, the model lipid methyl phosphatidic acid (dimyristoyl) at the apparent pK stores above 0.1 kcal/mole protonation even for constant surface pressure. For adiabatic coupling, of course, efficiency is 1. Chemomechanical coupling above 1 kcal/mole phospholipid protonation is easily achieved not too far from the apparent critical point of onset of first order phase transitions.

The assumption of efficient free energy coupling thus leads to the hypothesis of specific boundary lipids near the enzymes which are driven by the catalytic activity toward the critical pK range. Examples for specific boundary lipids are phosphatidyl serine at Na^+ - K^+- ATPase, or phosphatidyl inositol at hydrophobic AChase, reported by Papahadjopoulos and Kimelberg and, respectively, by Futerman and Silman.

With respect to these properties of phospholipid bilayers, the controversy between Mitchell and Williams on the role of localised versus global protons resolves as follows:

enhanced proton activity and surface diffusion is a consequence of the negatively charged lipid phosphate groups. The semi-conductor like mobility mechanism of protons in the structured surface water leads to a proton specificity of the equilibrium Gouy-Chapman layer and of non-equilibrium propagating excitations. In thermal equilibrium between surface and bulk, the local proton activity is controlled by the globally applied electrochemical potentials. Enzymatically produced "energised" local protons delocalise rapidly along the surface only, and couple free energy chemiosmotically, by "protomotive" force, along the surface and eventually, due to membrane thickness fluctuations, across the bilayer to the other surface.

Critical osmosis

The protomotive force can only drive ion transport when permeable structures are induced by critical perturbation of the impermeable lipid order. The protons do so in the range of pK of the phospholipids, as observed. In case of asymmetric proton-sources, a proton-induced ion channel appears only in the presence of a protomotive force even if symmetry is observed of the global, neutral bulk pH.

The phospholipid bilayer fluctuations define a very interesting thermodynamic system with respect to the third dimension of transport. It is in most states closed or impermeable, but conditionally open in a critical range only. Since conductivity fluctuations can be induced by asymmetric proton sources, as observed in collaboration with Silman, Nernst-Planck diffusion suffices to establish a non-equilibrium membrane gradient. When considering only the transport one-dimensionally, the lipid membrane appears as a Maxwellian demon, although it directly follows from the thermodynamic principles of fluctuations and diffusion.

The non-enzymatic side of the membrane becomes protonated, driven by catalysis which also induces the open state. The closed state in absence of catalytic protons stores the established osmotic gradient for, in principle, infinite times.

A testable prediction of critical osmosis is a lipid-dependent upper limit for the "static head" electrochemical potential difference which can be built up. It is achieved when the local protomotive force is zero even though the enzyme is actively inducing the critical pH range, say pK, of the phospholipids. Therefore, the reverse electrochemical potential built up must balance the local protomotive force. This value can be estimated from the pK observed in the absence of applied electrostatic potential differences in a monolayer system. If the bulk pH at the enzymatic side is 7, a pK of 6 gives ca. 60 mV, pK 5 ca. 120 mV, pK 2 ca. 300 mV which should never be exceeded for typical phospholipids crucially.

Desensitisation

With respect to the mechanism of ACh-induced ion transport, the prediction from phospholipid monolayer phase diagrams of reduction of fluctuation strength upon full protonation at very acid pH below the pK is of particular interest. Further production of protons has no effect any more on the low strength of thickness fluctuations in such states. A similar desensitisation of proton receptors has been observed in vivo by Gruol and by Krishtal et alii. Since the work of Katz and Thesleff, desensitisation of the physiological ACh receptor mechanism is well-known in cholinergic systems.

Again, the slow lipid relaxation times allow to observe a local membrane state such as full protonation even at rather neutral bulk pH reversibly.

The effect of Ca^{2+} on phospholipid bilayers is more complex, but basically mimicks the effect of protons. Ca^{2+} also produces the phenomenon of desensitisation at too high concentrations, but there is no experimental proof that this is achieved by the phospholipid properties.

Clearly, the observations in collaboration with Silman demonstrate that phospholipid bilayers represent proton receptors, capable of desensitisation. The proton-induced ion channel activity near the pK was observed to disappear or to be reduced dramatically upon further application of acid.

While desensitisation remains an obstacle requiring adjustable parameter models in the hypothesis of autonomous protein receptors, desensitisation naturally follows from the direct observation of a critical range of increased fluctuations of area and thickness, surrounded by states of lower fluctuation strength at both more neutral and more acid pH. This observation is made on monolayer surface pressure-area diagrams as well as, by measurement of membrane current, on bilayers made of synthetic lipids such as methyl phosphatidic acid.

Physiological receptor of acetylcholine

The lipid mechanism described above fulfills the requirements of the physiological receptor of ACh in vivo, i.e. in the presence of AChase activity and the absence of pharmacological modifications.

As described by Nachmansohn, Changeux, Neumann, and others, specific ligand binding does induce a conformational change, via a permeable state, into the desensitised state. Receptor coupling is allosteric, however, the channel subunit appears to be the lipid bilayer associated to the ACh-specific binding site.

Adiabatic excitation

Despite the rather slow release of protons from phospholipid surfaces, the local protons propagate rapidly along the two-dimensional surface according to hydrodynamic principles. The propagation velocity equals the adiabatic density derivative of surface pressure. From isothermal phase diagrams of Träuble and Teubner, velocities up to 20 m/sec can be roughly estimated. Clearly, slowing down is predicted at the apparent pK, even down to 0 m/sec at a critical point, since any local excitation will be of an effective mass corresponding to the diverging coherence length.

It is very interesting to note that sub-threshold excitations do not propagate, as already described half a century ago by Auger and Fessard. Above a critical threshold value, the voltage-induced depolarising pulse gradually increases in amplitude until it appears as an all-or-nothing action potential of defined propagation velocity. Propagation velocities of action potentials as well as of monolayer shock waves, described recently by Moebius et al., achieve maximal velocities in the order of several m/sec.

Propagation velocities may achieve optimal values in myelinated axons due to the insulation from viscous damping. Viscosity is essentially due to coupling to transversal excitation of the associated bulk aqueous medium in contact with the two-dimensional density pulse in the monolayers. As pointed out to me by H.W. Strube, in the worst case of complete coupling to the transversal modes, propagation of the excitation must cease due to dissipation after a few μm, allowing for almost adiabatic propagation of free energy associated to electrical signal in unmyelinated membrane systems for typical cellular dimensions only.

Msec relaxation times have been observed in pure lecithins near the phase transition temperature with high activation energies (60 kcal/mole) by Tsong, using Eigen and deMaeyer's temperature jump technique. 100 msec relaxation time of vanishing activation energy had also been observed. Hill reported similar time scales for rapid "local potentials" and, respectively, slow "threshold relaxations";most remarkably, Hill also discovered small microdegree temperature changes associated to, and of the same shape as, the action potential. Iwasa and Tasaki confirmed these observations in axons and detected reversible changes in membrane pressure and area, too.

Msec membrane current pulses are also routinely observed in pure phospholipid bilayers. As demonstrated below, these pulses as well as the ion channels only require the presence of phospholipids and protons. Propagating density pulses in lipid monolayers therefore represent a physically defined mechanism for propagating electrostatic potential changes associated to the other thermodynamic variables temperature, surface pressure, and so forth. These other variables have, however, no place in a priori electrostatic equivalent circuits. Moreover, these surface variables have no meaning in terms of single protein molecules.

Ion specificity

How could ion-specific transport such as of Na^+ as opposed to K^+ arise across lipid ion channels by likewise inspecific electrochemical diffusion?

Thermodynamically, ion specificity can only be achieved at the expense of free energy. It is therefore intrinsically related to the problem of active ion transport. Passive channel models cannot, as shown by Hille, achieve ion specificity merely by specific binding sites because such a mechanism uses free energy from the applied potential and thus hinders diffusion.

There is no need for a priori ion-specific channels since transport requires both a permeable structure and a thermodynamic driving force. Since the ion channels can kinetically be directly attributed to the rather inspecific lipid bilayer, the specificity of biological membrane transport arises by control of the electromotive forces created by chemiosmotic free energy coupling.

In particular, specific ions which activate catalysis induce ion channels in the presence of specific electromotive forces only. Generalising critical osmosis, discussed above, to Na^+- activated ATPase, it is found theoretically that the electrostatic component of the protomotive force created only in the local presence of Na^+ inavoidably drives specific H^+- Na^+- cotransport toward the non-enzymatic side of the membrane.

Other ions which do not directly activate the enzyme, such as K^+ which is also present in large concentration generally, are not specifically driven by the local chemiosmotic mechanism. Since, however, the ion channels per se are passive and inspecifically uncoupling in principle, the reverse global electromotive force built up does drive ions in the corresponding direction. For example, Na^+- ATPase, which establishes as described a reverse global potential difference positive at the non-enzymatic side, indirectly thus drives reverse K^+ transport toward the enzymatic membrane side. In particular, K^+ approaches by this mechanism an equilibrium potential equal, but opposite in sign, to the electrostatic component of the membrane potential built up by Na^+- ATPase. Though, K^+ appears to be actively driven by the hydrolysis of ATP.

This ion-specific mechanism is clearly testable by the predicted specificities and directions which are due to the enzyme activation and localisation. It represents a crucial test of the theory that these specificities as well as these directions agree with those observed for H^+- ATPases following Mitchell, Schatz and Ernster, for Na^+- K^+- ATPase following Skou, Glynn and Karlish, and also for Ca^{2+}- activated ATPase, which transports Ca^{2+} toward the non-enzymatic side according to Fleischer. AChase, which is not activated by either of those specific ions, should only induce uncoupling of the specific gradients established by Na^+- K^+- ATPase, i.e. immediate Na^+- influx and delayed K^+- efflux after depolarisation of the electrostatic component of the membrane potential. Again, the specificities and directions crucially test the proposed mechanism.

A particularly interesting prediction following from the properties of critical osmosis is active proton transport toward the non-catalytic side by any hydrolase of sufficient activity to induce ion channels at acid phospholipid pK. It were falsified if Na^+ - K^+ - ATPase, as generally assumed, would not transport H^+. Ken-Dror, Schnaiderman and Avi-Dor found such H^+ transport in the predicted direction. That this active H^+ - Na^+ - cotransport was not observed previously may derive from the fact that proton activities are very low usually, as compared to Na^+ and K^+; therefore, large potential differences, i.e. logarithmic changes in proton activity, can be achieved by comparatively small and eventually inobserved proton fluxes. Still, these forces drive by their electrostatic component Na^+ and K^+ transport until the static head or equilibrium potentials described above.

It was not attempted to derive strict stoichiometries. Our mechanism implies stoichiometric catalysis, but diffusive transport across eventually uncoupling ion channels. Pietrobon, Azzone and Walz have described "slips" in active redox pumps which clearly falsify strict stoichiometries. So far, no contradiction arises, however, from these observations with respect to the lipid mechanism described here, although a more detailled analysis is required to settle this question.

Direct observation of the conformational transitions

In order to directly reconstitute the bilayer transport properties predicted from phospholipid monolayer surface pressure- area phase diagrams, two crucial elements of the theory appear particularly accessible: the conformational changes and the transport of protons.

In order to quantitatively control eventual correlations between bilayer conformational changes and area/thickness fluctuations to the induction of discrete ion channel conductivities, I measured the electrical capacitance of pure phospholipid bilayers. Marty and Neher have already reported in vivo membrane capacitance steps in states of ion channel activity. While these observations were made in the presence of vesicles and attributed to vesicle fusion, capacitance steps of similar magnitude do also appear in planar lipid bilayers made from two monolayers. Boheim, Hanke and Eibl reported a ca. 20% change in membrane capacitance at the thermal phase transition of myristoyl-stearoyl phosphatidyl choline to be associated by membrane current fluctuations.

Electrical capacitance clearly represents an order parameter of the lipid component of membrane and cannot be attributed to autonomous proteins. Therefore, the observation of capacitance steps in the presence of ion channel activity demonstrates the correlation of lipid order fluctuations to the induction of ion channels. The membrane conductivity per se does in contrast not represent an order parameter, but rather a phenomenon associated to its fluctuations.

I here report the induction by protons of stepwise changes in membrane capacitance, correlated to the induction of ion channels, of synthetic diphytanoyl phosphatidyl choline bilayers. Planar bilayers were made following Hanke, as described earlier, in 1 M KCl unbuffered at room temperature ca. 22 °C. In addition to an applied +77 mV voltage, ±3.7 mV were superimposed by triangular AC voltage of 10 Hz.

Fig. a) demonstrates the membrane current under these conditions. The step in membrane current every 50 msec is proportional to the electrical capacitance and represents 60 pF, corresponding to 0.5 $\mu F/cm^2$. The capacitance was not crucially altered during successive titration of HCl asymmetrically to the + side of the bilayer, and no significant effects on membrane current were observed above pH 2.5 on this side. On the other side, pH 6.5 was approximately maintained throughout the experiment, in the non-conducting states. Conductivity was only 4 pS.

However, when pH 2.4 was achieved in this membrane, clearly, discrete changes in membrane capacitance were observed. Three such changes are shown in Figs. b) and c), amounting to ca. 50 % and 25 %. Other capacitance steps ranged from few % to above 100 %. They appeared spontaneously, but only near the apparent pK of the phospholipid used.

Fig. c) also demonstrates superimposed small membrane current steps. In order to unambiguously assign these current steps to the appearance of ion channels, controls with disconnected AC voltage generator were routinely undertaken. Fig. d) at +77 mV DC pH 2.4 of the same membrane as in c) does prove the appearance of ion channels which we already reported earlier.

Similar results were obtained with bilayers made of
- synthetic dimyristoyl methyl phosphatidic acid, generously provided by H.Eibl;
- heterogeneous soybean lecithin from Sigma;
- crude Torpedo electric organ membrane fractions which contained native AChase and ATPase.

Induction of ion channels and capacitance steps was always achieved by direct asymmetric acidification with HCl or, in the latter system, by enzymatic hydrolysis of ACh and, likewise, of ATP. ATP-induced ion channels could already be observed, in 0.1 M NaCl unbuffered, at asymmetric pH 5 produced by hydrolysis.

We have reported earlier that ACh-induced ion channels are induced by AChase even at buffered bulk pH 7.4. Kinetically, the capacitance steps, ion channel opening and closing, as well as the msec membrane current pulses appeared very similar, no matter whether they were induced enzymatically or by direct acidification of pure synthetic lipid bilayers.

Protonic control

Eigen and deMaeyer earlier predicted that biological membranes represent almost ideal conditions for protonic control systems. Träuble demonstrated the specific role of protons in controlling the thermodynamic phospholipid states. Control experiments demonstrated, again, that protons control both permeability and ion transport kinetically.

Capacitance and associated ion channels appeared in the absence of salt, and in the virtual absence of OH^- at acid pH 2. Controls in the absence of added HCl demonstrated current pulses of similar kinetic relaxation times in the virtual absence of Cl^-. Protons and protonable phospholipids were the only ions required.

The drop in pH from 6 to 5-4 at the non-titrated side during the conductive state clearly demonstrates proton transport across the membrane, since the electrostatic potential was kept constant. While other ions such as Na^+ and K^+ are obviously conducted across many ion channels, too, the similar relaxation times of current pulses and ion channel opening and closing in most lipid-protein membranes, and the appearance of phenomenologically very similar properties due to proton-phospholipid interactions clearly demonstrates the prediction to be valid for both receptor and transport function of the bilayers.

Interestingly, msec membrane pulses can, as expected for adiabatic excitations, in certain instances clearly be asymmetrically directed against the bulk-applied DC voltage. This is evident from the fact that, after reversal of the electrode positions, and the associated membrane potential, the current direction persists with respect to the membrane for at least several seconds. In the pure lipid membrane, the required free energy may derive from the initial condition, in particular from membrane curvature. The slow kinetics of such active protonic pulses has not yet been consistently determined.

Conclusion

Phospholipid bilayer membranes represent protonic control systems. From Langmuir thermodynamic phase diagrams, free energy coupling, propagation of adiabatic excitations, and area/thickness fluctuations in the monolayers are evident. Slow lipid relaxation times store local protons at the surface, which couple catalytic to osmotic free energy. ACh thus induces ion channel fluctuations by local surface hydrolysis. The responsible proton-induced phospholipid conformational transitions are directly observed as stepwise changes in the membrane capacitance. The lipid mechanism described appears evident and sufficient for specific protein-induced transport across biological membranes.

a) b)

c) d)

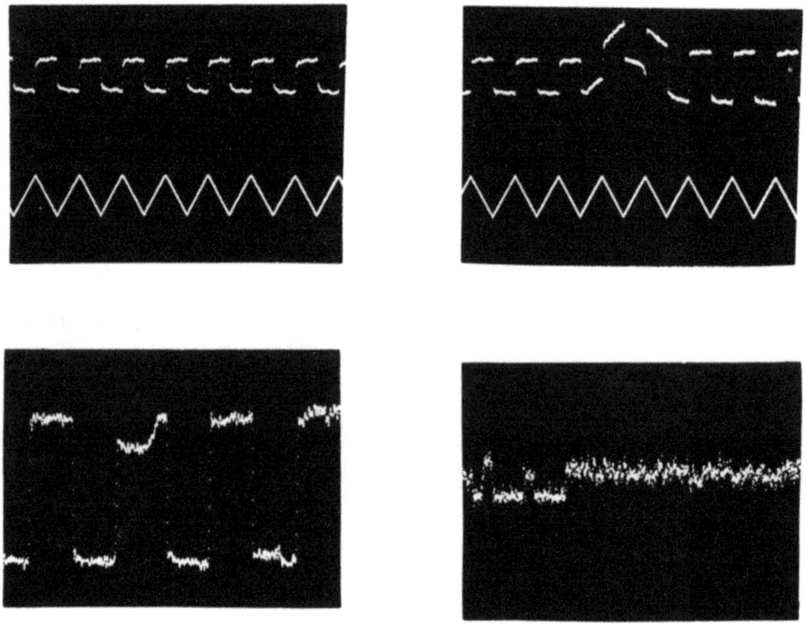

a) Diphytanoyl lecithin (Avanti) planar bilayer in 1 M KCl at +77 ± 3.7 mV. Acidification of the + side to pH 2.5 did not significantly alter the membrane properties. pH 6.5 non-titrated side. Top trace shows the capacitive membrane current response corresponding to 60 pF. Lower trace shows the applied voltage of 10 Hz frequency.
b) Same membrane as in a) but at pH 2.4. First discernible capacitance step from 60 to 90 pF.
c) Same as in b) showing two subsequent capacitance steps between 100 and 75 pF.
d) Same state as in c), but at DC voltage +77 mV. The ion channels on this trace correspond to 100 and 70 pS calculated conductivity. At a similar pH close to the pK of this lipid, unit channels as well as manifold channel sizes had already been observed in collaboration with Silman.

Acknowledgement: The encouragement of several wise men, including David Nachmansohn, has been crucial. The theory developped in particular by discussions with Robert Graham from the theoretical physics group of the University of Essen. The Minerva foundation supported my work in Israel.

References

Abbott, B.C., Hill, A.V., Howarth, J.V.: Proc.Roy.Soc.B 148, 149 (1958)
Antonov, V.F., et al.: Nature 283, 585 (1980)
Auger, D., Fessard, A.: C.R.Soc.Bio.Paris 118, 1059 (1935)
Boheim, G., Hanke, W., Eibl, H.: Proc.Nat.Acad.Sci.USA 77, 3403 (1980)
Changeux, J.-P., Kasai, M., Lee. C.-Y.: Proc.Nat.Acad.Sci.USA 67, 1241 (1970)
Changeux, J.-P., et al.: Symp.Quant.Biol. XL, 211 (1976)
Deamer, D.W., Nichols, J.W.: Proc.Nat.Acad.Sci.USA 80, 165 (1983)
Eibl, H.: in "Polyunsaturated Fatty Acids", Eds. Kunau et al., American Oil Chemist's Society, Champaign, IL (1977), pp. 229-244.
Eigen, M., deMaeyer, L.: Proc.Roy.Soc.London A 247, 505 (1958)
Eigen, M., deMaeyer, L.: in "Technique Organic Chem. VIII, part 2, Interscience Publ, New York 1963, pp. 895-1054
Ernster, L., Schatz, G.: J.Cell Biol. 91, S227-S255 (1981)
Fleischer, S., et al.: in "Function and Molecular Aspects of Biomembrane Transport", Eds. Quagliariello et al., Elsevier/North Holland, Biomedical Press, Amsterdam 1979, p. 465
Futerman, A.H., et al.: in "Molecular Basis of Nerve Activity", Eds. Changeux et al., Walter de Gruyter Verlag, Berlin, to appear.
Gruol, D.L., et al.: Brain Res. 183, 247 (1980)
Haines, T.H.: Proc.Nat.Acad.Sci.USA 80, 160 (1983)
Hanke, W.: Dissertation, Universität Bochum (1981)
Hanke, W., Miller, C.: J.Gen.Physiol. 82, 25 (1983)
Hill, A.V.: Proc.Roy.Soc.B 111, 106 (1932); see also Abbott et al.
Hille, B., Campbell, D.T.: J.Gen.Physiol. 67, 265 (1976)
Hille, B., Schwarz, W.: J.Gen.Physiol. 72, 409 (1978)
Iwasa, K., Tasaki, I.: Biochem.Biophys.Res.Commun. 94, 716; 95, 1328 (1981)
Karlish, S.J.D., Glynn, I.M.: Ann.N.Y.Acad.Sci. 242, 461 (1974)
Katchalsky, A., Oplatka, A.: in "Handbook of Sensory Physiology", Ed. Loewenstein, Springer Berlin 1971, pp. 1-17
Katz, B., Thesleff, S.: J.Physiol.London 138, 63 (1957)
Kaufmann, K.: Int.J.Quant.Chem. 12 Suppl. 2, 169 (1977); "Protonenkontrolle des Ionentransports durch biologische Membranen", Universität Göttingen 1982; in: "Biol.Struct. and Coupled Flows", Eds. Oplatka et al., Acad.Press, p.105
Kaufmann, K., Silman, I.: Naturwissenschaften 67, 608 (1980); Biophys.Chem. 18, 89 (1983)
Kedem, O.: in "Biol.Struct. and Coupled Flows, Eds. Oplatka et al., Academic Press, New York 1983, pp. 27-32
Kedem, O., Katchalsky, A.: J.Gen.Physiol. 45, 143 (1961)
Kell, D.: Biochim.Biophys.Acta 549, 55 (1979)
Ken-Dror, S., Schnaiderman, R., Avi-Dor, Y.: Arch.Biochem.Biophys.229,640(1984)
Kimelberg, H.K., Papahadjopoulos, D.: Biochim.Biophys.Acta 282, 277 (1972)
Krishtal, O.A., Pidoplichko, V.I.: Brain Res. 214, 150 (1981)
Langmuir, I.: J.Am.Chem.Soc. 40, 1361 (1918)
Mitchell, P.: Nature 191, 144 (1961); "Chemiosmotic Coupling in Oxidative and Photosynthetic Phosphorylation", Glynn Research Ltd., Bodmin U.K. 1966
Möbius, D., Grüniger, H.: in "Charge and Field Effects in Biosystems", Eds. Allen et al., Abacus Press, Tunbridge Wells, England 1984, p. 265
Nachmansohn, D., Neumann, E.: "Chemical and Molecular Basis of Nerve Activity", Academic Press, New York 1975

Nachmansohn, D.: "Chemical and Molecular Basis of Nerve Activity", Academic Press, New York 1975
Nagle, J.F., Morowitz, H.J.: Proc.Nat.Acad.Sci.USA 75, 298 (1978)
Neher, E., Marty, A.: Proc.Nat.Acad.Sci.USA 79, 6712 (1982)
Nichols, J.W., Deamer, D.W.: Proc.Nat.Acad.Sci. 77, 2038 (1980)
Pietrobon, D., Azzone, G.F., Walz, D.: Eur.J.Biochem. 117, 389 (1981)
Podleski, T.D., Changeux. J.-P.: Science 157, 1579 (1967)
Silman, I.: Trends in Biochem. Sci. 1, 225 (1976)
Silman, I., Karlin, A.: Proc.Nat.Acad.Sci.USA 58, 1664 (1967)
Skou, J.: Biochim.Biophys.Acta 23, 394 (1957)
Tasaki, I., Iwasa, K.: Biochem.Biophys.Res.Commun. 101, 172 (1981)
Träuble, H.: in "Structure of Biological Membranes", Eds. Abrahamsson et al., Plenum Press, New York 1977, pp. 509-550
Träuble, H., Eibl, H.: Proc.Nat.Acad.Sci.USA 71, 214 (1974)
Tsong, T.Y.: Proc.Nat.Acad.Sci.USA 71, 2684 (1974)
Ussing, H.H.: Adv.Enzymol. 13, 21 (1952)
Williams, R.J.P.: J.Theor.Biol. 3, 209 (1962); see also Kell, D.

"Das bedeutet aber nicht, dass ich mit Ihrer Interpretation übereinstimme. Ich habe vor meinen Mitarbeitern oft Hippokrates zitiert: "ή δὲ πετρα σφαλερή, ή δὲ κρίτος χαλεπή" (der Versuch ist trügerisch, die Interpretation schwierig). Ich habe viele Jahre gebraucht, um zu meiner Theorie zu kommen, ein wirkliches Ende gibt es nicht, wie es Einstein und Planck oft betont haben. Es ist durchaus möglich, Modelle herzustellen ohne Rezeptoreiweiss, in der Ionenkanäle isoliert werden können, nur mit Hilfe von AcCh und AcChE und mit Lipiden!"

David Nachmansohn 12.9.1981

AUTHOR INDEX

SUBJECT INDEX

Bei Fragen zur Produktsicherheit wenden Sie sich bitte an:
If you have any questions regarding product safety,
please contact:

Walter de Gruyter GmbH
Genthiner Straße 13
10785 Berlin
productsafety@degruyterbrill.com